Springer Series in Materials Science

Volume 332

The Springer Series in Materials Science covers the complete spectrum of materials research and technology, including fundamental principles, physical properties, materials theory and design. Recognizing the increasing importance of materials science in future device technologies, the book titles in this series reflect the state-of-the-art in understanding and controlling the structure and properties of all important classes of materials.

Neeraj Kumar · Rashi Gusain · Suprakas Sinha Ray
Editors

Two-Dimensional Materials for Environmental Applications

Springer

Editors
Neeraj Kumar
Centre for Nanostructures and Advanced Materials, DSI-CSIR Nanotechnology Innovation Centre
Council for Scientific and Industrial Research
Pretoria, South Africa

Department of Chemical Sciences
University of Johannesburg
Johannesburg, South Africa

Rashi Gusain
Centre for Nanostructures and Advanced Materials, DSI-CSIR Nanotechnology Innovation Centre
Council for Scientific and Industrial Research
Pretoria, South Africa

Department of Chemical Sciences
University of Johannesburg
Johannesburg, South Africa

Suprakas Sinha Ray
Centre for Nanostructures and Advanced Materials, DSI-CSIR Nanotechnology Innovation Centre
Council for Scientific and Industrial Research
Pretoria, South Africa

Department of Chemical Sciences
University of Johannesburg
Johannesburg, South Africa

ISSN 0933-033X ISSN 2196-2812 (electronic)
Springer Series in Materials Science
ISBN 978-3-031-28755-8 ISBN 978-3-031-28756-5 (eBook)
https://doi.org/10.1007/978-3-031-28756-5

This Springer imprint is published by the registered company Springer Nature Switzerland AG
The registered company address is: Gewerbestrasse 11, 6330 Cham, Switzerland

Preface

With increasing industrialization, escalating population, growing fuel demand, spiraling pollution, and global climate change, environmental challenges have garnered significant global attention, from waste management to sustainable development. The development of effective and affordable materials for environmental remediation technologies is required to tackle complicated and interlinked environmental issues. Due to high surface area, tunable surface chemistry, and distinctive properties, 2D nanomaterials are highly important for environmental remediation applications.

This book highlights how 2D nanomaterials play a significant role in addressing existing environmental problems. It reflects the present state-of-art of 2D nanomaterials employed in environmental applications such as water and wastewater treatment, adsorption, photocatalysis, membrane separation, desalination, deionization, environmental pollutants sensing/detection, the capture of CO_2 and other gaseous molecules, catalytic conversion of CO_2, microbial treatment, and electrochemical remediation. Each chapter provides an essential and comprehensive overview of the recent advances in material development and application. Special attention is given to preparation methods, tuning of physiochemical properties, surface and interface chemistry, structural porosity, assemblies integration for fabrication of devices, and their relationship with overall efficiency. It offers a valuable reference guide for environmental and materials scientists, engineers, and policymakers, working towards environmental sustainability.

This book contains fourteen chapters. Chapter 1 sheds light on the recent advances and accomplishments of MXenes in the environmental field, focusing on improving their properties through various strategies. Moreover, the fundamental aspects, structural features, and different synthesis methodologies of MXenes are summarized. The challenges to overcome and future research directions to realize the full potential of MXene materials for environmental remediation applications are also emphasized. Chapter 2 focuses on the applications of MXenes as well as their composites/hybrids for water treatment and CO_2 capture and conversion. This chapter also discusses the challenges with regard to the utilization of Mxene and its materials in wastewater remediation. Chapter 3 reports the structural, surface, textural, and chemical properties of 2D nanomaterials and their roles in the adsorptive removal and photocatalytic degradation of organic pollutants. The chapter also covers futuristic opportunities, environmental sustainability, and technological challenges in preparing and applying inorganic analogues of graphene for wastewater treatment. Chapter 4 covers graphitic carbon nitride (GCN)-based photocatalysts for environmental applications. In addition, the basic principle of photocatalytic pollutant degradation over a semiconductor surface is discussed. Recent developments in modification strategies to enhance the photoactivity of GCN have been summarized systematically. Photocatalytic applications of GCN-based photocatalysts with respect to environmental remediation are presented in this chapter. The challenges and future perspectives in designing GCN-based photocatalysts for efficient performance towards environmental applications are addressed along with the conclusion. In Chap. 5, the authors discuss various types of 2D nanomaterials used for antibacterial activities. For better understanding, the 2D nanomaterials are classified into carbon and non-carbon based for antibacterial activities. The present challenges and possible future directions for the development of 2D nanomaterials with advanced antimicrobial properties are also reviewed. Chapter 6 covers the recent advances in developing graphene-based photocatalytic materials. The principle of photocatalysis, the basic properties of graphene, and the mechanism of how the photocatalytic efficiency against the removal of the liquid pollutant can be enhanced when coupled with graphene have been discussed in this book chapter. Furthermore, current challenges and future recommendations for developing graphene-based photocatalysts are also discussed. Chapter 7 discusses various graphene-based materials, their adsorption phenomena, and their application to water purification. On the other hand, Chap. 8 reports an overview of the concept of membrane technology, membrane fabrication techniques and the importance of 2D nanomaterials in the desalination and wastewater treatment membranes. Moreover, this chapter emphasizes the wide range of membrane applications as well as their potential and challenges for use in the development of nanotechnology-based environmental remediation. The determination of the most important toxic environmental pollutants in various environmental samples with 2D nanomaterials-based optical and electrochemical sensors is reported in Chap. 9. In Chap. 10, the authors discussed the potential catalytic properties of transition metal dichalcogenides, transition metal oxides, MXenes, graphene, nitrides, carbides, and their hybrids for environmental applications. Further, material performance in terms of the band gap, appropriately positioned valence and conduction bands, stability, economics, etc., are also compared for the wastewater systems. Finally, the outlook of the field is also presented with respect to challenges and research directions to tap this important unused energy source, i.e., wastewater. Chapter 11 focuses on 2D MOFs materials, sensor design strategies, and environmental applications. Moreover, this chapter summarizes and evaluates recent advances in the development and application of 2D MOFs-based sensors for determining toxic environmental pollutants. Chapter 12 provides a platform for the researchers, who are focused on exploring the application of MoS_2-based materials in water purification. The research demands future water applications of 2D MoS_2 nanomaterials are also identified. In Chap. 13, significant advancements in the design, synthesis, structure properties, and photoactivity performance of 2D metal oxides in CO_2 reduction reactions are discussed. Specifically, a comprehensive insight into understanding the interplay between the strategic design of the 2D metal oxides and their multidimensional heterostructures (0D/2D, 1D/2D, 2D/2D, and 2D/3D) with controlled nanoscale interface structure properties correlated to photocatalytic activity performance in CO_2 reduction reaction are highlighted. The construction strategies of 2D metal oxide heterojunctions with other materials such as non-metals, metal dopants, and metal oxides providing high interfacial intimate contacts for improved charge carriers' separations and transfer for improved high-performance CO_2 photocatalytic conversion are critically discussed. Finally, Chap. 14 aims to provide a brief overview of the recent advantages in the nano-engineering of the various 2D materials for CO_2 capture. In particular, the recent development of emerging strategies such as doping, defects engineering, hetero-structural designing, and architectural functionalization of 2D nanomaterials for enhanced CO_2 capture are discussed thoroughly. The challenges and future outcomes have also been highlighted, which will open the directions for future research.

In summary, following are the key insights of this book.

- Provides a state-of-the-art overview of fundamental aspects of 2D nanomaterials for environmental remediation;
- Comparison of the different 2D materials to realize best-fit properties for specific environmental applications and development of practical commercial products and devices;
- Demonstrates challenges and prospects for advancement, experimental strategy, fabrication, and improvement in overall performance;
- This book is ideal for water scientists, material scientists, researchers, engineers (chemical and civil), including under- and post-graduate students who are interested in this exciting field of research; and
- Finally, this book will also help industrial researchers and R&D managers to bring advanced 2D nanomaterials-based environmental remediation solutions into the market.

We express our sincerest appreciation to all authors for their valuable contributions as well as reviewers for their critical evaluation of the proposal and manuscripts. Our special thanks go to Zachary Evenson and Viradasarani Natarajan at Springer Nature for their suggestions, cooperation, and advice during the various stages of manuscript preparation, organization, and production of this book. The financial support from the Council for Scientific and Industrial Research, the Department of Science and Innovation, and the University of Johannesburg is highly appreciated.

Johannesburg, South Africa — Neeraj Kumar
Pretoria, South Africa — Rashi Gusain
Johannesburg, South Africa — Suprakas Sinha Ray

The University of Johannesburg
and the Council for Scientific
and Industrial Research

Contents

Contributors

Lopamudra Acharya Centre for Nano Science and Nano Technology, ITER, Siksha 'O' Anusandhan Deemed to be University, Bhubaneswar, Odisha, India

Rashmi Acharya Department of Chemistry, I.T.E.R., Siksha 'O' Anusandhan Deemed to be University, Bhubaneswar, Odisha, India

Zaeem Bin Babar Institute of Environmental Sciences and Engineering (IESE), School of Civil and Environmental Engineering (SCEE), National University of Sciences and Technology (NUST), Islamabad, Pakistan

Fatma Budak Faculty of Pharmacy, Department of Analytical Chemistry, Ankara University, Ankara, Türkiye;
Graduate School of Health Sciences, Ankara University, Ankara, Türkiye

Ahmet Cetinkaya Faculty of Pharmacy, Department of Analytical Chemistry Ankara, Ankara University, Ankara, Türkiye;
Graduate School of Health Sciences, Ankara University, Ankara, Türkiye

Priyanka Chaudhary School of Basic Sciences, Advanced Materials Research Center, Indian Institute of Technology Mandi, Kamand, Mandi, Himachal Pradesh, India

Pooja Devi Academy of Scientific and Innovative Research, New Delhi, India;
Materials Science and Sensor Application, Central Scientific Instruments Organisation, Chandigarh, India

Neerugatti KrishnaRao Eswar School of Engineering, Department of Nanobiomaterials, Saveetha Institute of Medical and Technical Sciences, Chennai, India

Rashi Gusain Department of Chemical Sciences, University of Johannesburg, Doornfontein, South Africa;
Centre for Nanostructures and Advanced Materials, DSI-CSIR Nanotechnology Innovation Centre, Council for Scientific and Industrial Research, Pretoria, South Africa

Sabir Hussain Department of Space Science, Institute of Space Technology, Islamabad, Pakistan

Pratiksha Joshi CSIR – Indian Institute of Petroleum, Mohkampur, Dehradun, India;
Academy of Scientific and Innovative Research, Ghaziabad, India

S. Irem Kaya Gulhane Faculty of Pharmacy, Department of Analytical Chemistry, University of Health Sciences, Ankara, Türkiye

Om. P. Khatri CSIR – Indian Institute of Petroleum, Mohkampur, Dehradun, India; Academy of Scientific and Innovative Research, Ghaziabad, India

Ajay Kumar School of Engineering and Innovation, The Open University, Milton Keynes, UK

Kuldeep Kumar Department of Physics, School of Basic and Applied Sciences, Maharaja Agrasen University, Solan, Himachal Pradesh, India

Neeraj Kumar Department of Chemical Sciences, University of Johannesburg, Doornfontein, Johannesburg, South Africa;
Centre for Nanostructures and Advanced Materials, DSI-CSIR Nanotechnology Innovation Centre, Council for Scientific and Industrial Research, Pretoria, South Africa

Usman Liaqat School of Chemical and Materials Engineering, National University of Sciences and Technology, Islamabad, Pakistan

Peter Ramashadi Makgwane Centre for Nanostructures and Advanced Materials (CeNAM), Council for Scientific and Industrial Research, Pretoria, South Africa;
Department of Chemistry, University of the Western Cape, Bellville, South Africa

Sweta Mehta CSIR – Indian Institute of Petroleum, Mohkampur, Dehradun, India; Academy of Scientific and Innovative Research, Ghaziabad, India

Waheed Miran School of Chemical and Materials Engineering, National University of Sciences and Technology, Islamabad, Pakistan

Subhasish Mishra Department of Chemistry, I.T.E.R., Siksha 'O' Anusandhan Deemed to be University, Bhubaneswar, Odisha, India

Elishba Noor School of Chemical and Materials Engineering, National University of Sciences and Technology, Islamabad, Pakistan

Vincent Ojijo Centre for Nanostructures and Advanced Materials, DSI-CSIR Nanotechnology Innovation Centre, Council for Scientific and Industrial Research, Pretoria, South Africa

Jonathan Tersur Orasugh Department of Chemical Sciences, University of Johannesburg, Doornfontein, Johannesburg, South Africa;
Centre for Nanostructures and Advanced Materials, DSI-CSIR Nanotechnology Innovation Centre, Council for Scientific and Industrial Research, Pretoria, South Africa

Goksu Ozcelikay Faculty of Pharmacy, Department of Analytical Chemistry Ankara, Ankara University, Ankara, Türkiye

Sibel A. Ozkan Faculty of Pharmacy, Department of Analytical Chemistry Ankara, Ankara University, Ankara, Türkiye

Anchal Pandey CSIR – Indian Institute of Petroleum, Mohkampur, Dehradun, India;
Academy of Scientific and Innovative Research, Ghaziabad, India

Kulamani Parida Centre for Nano Science and Nano Technology, ITER, Siksha 'O' Anusandhan Deemed to be University, Bhubaneswar, Odisha, India

Irwing Ramirez School of Engineering and Innovation, The Open University, Milton Keynes, UK

Rohit Rana Natural Product Chemistry and Process Development Division, CSIR-Institute of Himalayan Bioresource Technology, Palampur, Himachal Pradesh, India

Kashif Rasool Qatar Environment and Energy Research Institute, Hamad Bin Khalifa University (HBKU), Doha, Qatar

Suprakas Sinha Ray Department of Chemical Sciences, University of Johannesburg, Doornfontein, Johannesburg, South Africa;
Centre for Nanostructures and Advanced Materials, DSI-CSIR Nanotechnology Innovation Centre, Council for Scientific and Industrial Research, Pretoria, South Africa

Rotimi Emmanuel Sadiku Division of Polymer Technology, Department of Chemical, Metallurgical and Materials Engineering and Institute of NanoEngineering Research, Tshwane University of Technology, Pretoria, South Africa

Ankeet Premraj Sahare Department of Chemistry, National Institute of Technology Jamshedpur, Jamshedpur, Jharkhand, India

Koena Selatile Centre for Nanostructures and Advanced Materials, DSI-CSIR Nanotechnology Innovation Centre, Council for Scientific and Industrial Research, Pretoria, South Africa;
Division of Polymer Technology, Department of Chemical, Metallurgical and Materials Engineering and Institute of NanoEngineering Research, Tshwane University of Technology, Pretoria, South Africa

Asif Shahzad Department of Materials Science and Engineering, Uppsala University, Uppsala, Sweden

Munish Sharma Department of Physics, School of Basic and Applied Sciences, Maharaja Agrasen University, Solan, Himachal Pradesh, India

Sweta Sharma Academy of Scientific and Innovative Research, New Delhi, India

Pooja Singh Academy of Scientific and Innovative Research, New Delhi, India

Lesego Tabea Temane Centre for Nanostructures and Advanced Materials, DSI-CSIR Nanotechnology Innovation Centre, Council for Scientific and Industrial Research, Pretoria, South Africa;
DST-CSIR National Centre for Nanostructured Materials, DSI-CSIR Nanotechnology Innovation Centre, Council for Scientific and Industrial Research, Pretoria, South Africa

Naveen Kumar Veldurthi Department of Chemistry, National Institute of Technology Jamshedpur, Jamshedpur, Jharkhand, India

Merve Yence Faculty of Pharmacy, Department of Analytical Chemistry Ankara, Ankara University, Ankara, Türkiye;
Graduate School of Health Sciences, Ankara University, Ankara, Türkiye

Waqas Qamar Zaman Institute of Environmental Sciences and Engineering (IESE), School of Civil and Environmental Engineering (SCEE), National University of Sciences and Technology (NUST), Islamabad, Pakistan

Chapter 1
MXenes: An Emerging Class of Materials for Environmental Remediation

Naveen Kumar Veldurthi, Ankeet Premraj Sahare, and Neerugatti KrishnaRao Eswar

Abstract MXenes are an intriguing class of two-dimensional early transition metal (M) carbides, nitrides, or blends with surface terminal groups like –OH, –F, –O, etc. The synthesis of the first MXene Ti_3C_2 was reported in 2011 from its parent Ti_3AlC_2 (MAX phase). Since then, MXenes have gained enormous attention for an extensive range of applications due to their peculiar properties, such as high specific surface area, excellent conductivity, tailorable interlayer spacing, tunable surface chemistry with easy functionalization, hydrophilic nature, and good mechanical strength. Considering the rich pool of lucrative properties of MXenes, they have been widely employed in environmental remediation applications such as photocatalysts, membranes, electrochemical separation techniques like capacitive deionization (CDI), and adsorbents for heavy metals, organic dyes, radioactive ions, etc. MXenes have also been studied for utilization in electrocatalytic sensors for pollutant detection, solar desalination, anti-biofouling and antibacterial agents. This book chapter mainly sheds light on the recent advances and accomplishments of MXenes in the aforementioned fields, focusing on improving their properties through various strategies. Moreover, the fundamental aspects, structural features, and different synthesis methodologies of MXenes are summarized. The challenges to overcome and future research directions to realize the full potential of MXene materials for environmental remediation applications are also emphasized.

N. K. Veldurthi (✉) · A. P. Sahare
Department of Chemistry, National Institute of Technology Jamshedpur, Jamshedpur, Jharkhand, India
e-mail: naveen.chem@nitjsr.ac.in

N. K. Eswar (✉)
School of Engineering, Department of Nanobiomaterials, Saveetha Institute of Medical and Technical Sciences, Chennai, India
e-mail: krishnaraoeswarneerugatti.sse@saveetha.com

N. Kumar et al. (eds.), *Two-Dimensional Materials for Environmental Applications*, Springer Series in Materials Science 332, https://doi.org/10.1007/978-3-031-28756-5_1

Abbreviations

CDI	Capacitive Deionization
AChE	Acetylcholinesterase
OPs	Organophosphate pesticides
E. coli	Escherichia coli
B. subtilis	Bacillus subtilis
PVDF	Polyvinylidene fluoride
MB	Methylene Blue
UF/NF Membrane	Ultrafiltration/Nanofiltration Membrane
MG	Methyl green
BSA	Bovine serum albumin
MR	Methyl red
MO	Methyl orange
SERS substrate	Surface-enhanced Raman scattering substrate

1.1 Introduction

Historically, 2D materials such as graphene and graphene oxide derivatives have been widely explored to remove a variety of pollutants in the environment. In 2011, a new fascinating class of 2D transition metal carbides/nitrides/carbonitrides was discovered, known as MXene. The representative structural formula of MXene is $M_{n+1}X_nT_x$, where M is the early transition metal (such as Ti, V, Zr, and Nb), X is C and N elements, and T represents the surface functional groups (e.g., –O, −OH and –F) [16]. The value of 'n' varies in the range of 1–3 depending on the number of atomic layers in the unit cell, leading to the typical formulas of M_2XT_x, $M_3X_2T_x$, and $M_4X_3T_x$, respectively. The first MXene Ti_3C_2 was reported in the seminal paper published by Naguib et al. [27] using wet-chemical etching in HF and exfoliation of parent MAX phase Ti_3AlC_2. Since then, this MXene two-dimensional material has motivated an upsurge of research, and nearly 30 MXenes have been successfully synthesized to date, including TiNbC [17, 18], Ti_2C [23], V_2C [26], Nb_4C_3 [15], and Mo_2C [36], etc. However, Titanium-based MXenes, particularly $Ti_3C_2T_x$, are more promising for environmental remediation applications.

MXenes are known for their salient properties, such as hydrophilic nature, excellent electrical and metallic conductivities (6,000–8,000 Scm^{-1}), higher surface area, high chemical stability, and easily tuneable structure. Given these fascinating properties possessed by MXene materials, they have been employed in a plethora of applications. For instance, MXene is used for supercapacitors, electrochemical energy storage in batteries, and electromagnetic interference shielding owing to its excellent electrical conductivity [39]. Moreover, the hydrophilic nature of MXenes offers ample active functional sites on their surface, resulting in MXenes as efficient adsorbents for various environmental pollutants [45, 52]. In addition, the choice of surface functional groups makes it possible to improve the desired characteristics in MXenes, thus enhancing activity in removing pollutants from contaminated water. These surface functionalities are significant to exhibit the physicochemical properties of MXene materials, which decide their performance in applications such as adsorption or membrane separation. Moreover, MXenes possess robust redox ability, which is advantageous in developing efficient photocatalysts and electrocatalytic sensors for the abatement and the detection of environmental contaminants, respectively [19, 37]. Therefore, MXenes have received immense attention for employing them in applications for environmental remediation, and the range of their applicability has been extended because of their tuneable interlayer spacing and tailorable surface chemistry [5]. This book chapter mainly focuses on the fundamental aspects, structural features, and various synthesis methodologies of MXenes, including the challenges to overcome and future research directions to realize the full potential of MXene materials for environmental remediation applications.

1.2 Structural Features of MXenes

MXene is a two-dimensional material containing transitional metal, carbide, nitride, and surface functional groups. Whether single or multi-layered, the MXene layers are held together by Van der Waals interaction and hydrogen bond [49], as shown in Fig. 1.1. In MXenes, metal atom 'M' is arranged in a hexagonal structure similar to its parent MAX phase and octahedral sites are occupied by the 'X' atoms. Even so, the pattern of atoms can change with the stoichiometric parameter 'n' [35]. In the MAX phase, the M-X bond has a mixed covalent/metallic/ionic character, whereas the M-A bond is metallic [49].

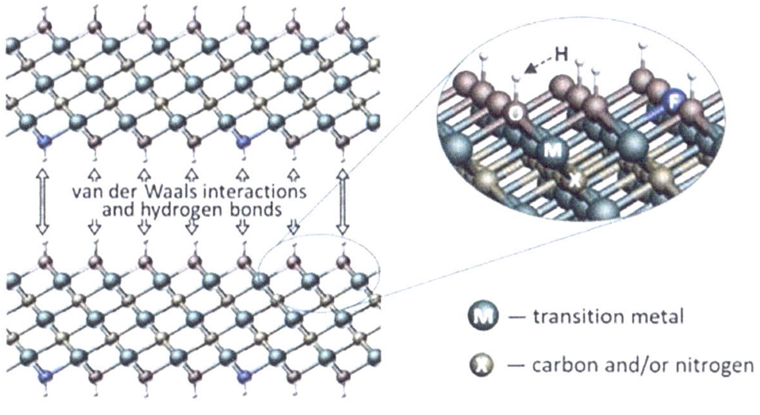

Fig. 1.1 Structure of a typical MXene. Reproduced with permission from [49]. Copyrights © 2020, John Wiley and Sons

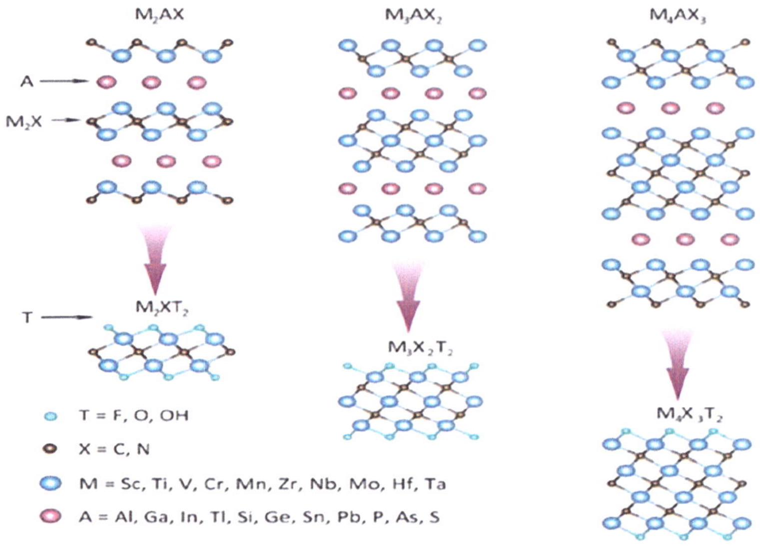

Fig. 1.2 Structure Schematic of M_2X, M_3X_2, M_4X_3. Adapted with permission from [30]. Copyrights © 2019, Elsevier

MXenes mainly exist in three types M_2X, M_3X_2, and M_4X_3. The M atoms in M_2X type MXenes have ABABAB stacking (hexagonal close-packed), whereas M_3X_2 and M_4X_3 have ABCABC packing types with face center cubic structure [30] (Fig. 1.2). The surface of MXene is grafted with a functional group like OH, F and O after the etching process, which is responsible for the various properties and functions of MXene. The surface sensitivity of MXene is highly dependent on the nature of transition metal on the surface, etching, delaminating, and storage conditions.

1.3 Overview of Synthesis Methodology

The most commonly used route to synthesize MXenes is the HF-mediated etching method from its parent MAX phase. So far, 30 MXenes have been derived from different MAX phases, and more than 70 MAX phases have been identified [3]. MAX phase is the hexagonal layered transition metal carbides and nitrides. It has the general formula $M_{n+1}AX_n$ where M represents an early transition metal element, A means a part from group 13 or 14, and X stands for C, N, or their blends. In the MAX phase, the M-X bond is more stable than the A layer, which is relatively weak in bonding to the M-X layer. Therefore, A layer is selectively etched from the MAX phase leading to the formation of the M-X layer [28]. Synthesis of MXene in this method follows a two-step protocol: etching and exfoliation.

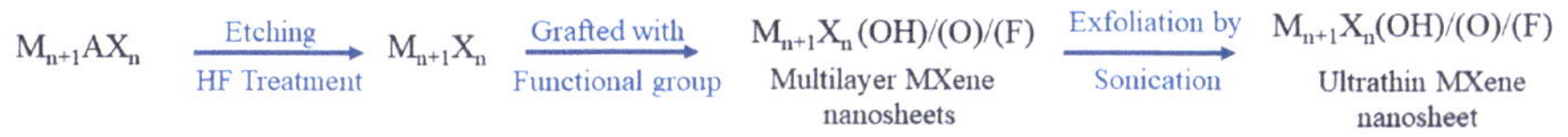

Scheme 1.1 Sequential steps involved in the synthesis of MXenes

In MXene synthesis, etching from the MAX phase is carried out in acidic solutions containing fluoride ions (HF, a mixture of LiF and HCl, or NH_4HF_2) and salts containing fluorine ions (NH_3F, KF, LiF, or NaF) [51]. After etching, the surface is grafted with a functional group like −OH, −F, −O. Then it is exfoliated using a sizeable organic intercalation agent such as urea, hydrazine monohydrate, dimethyl sulfoxide, isopropylamine, and tetrabutylammonium hydroxide [24, 25] with sonication to prepare single MXene nanosheets which are a few atomic layers thick. The steps are summarized in Scheme 1.1.

In early works, MAX phases are so inert that they cannot be etched in the presence of common acids, bases and even molten salts. Interestingly, using the dilute HF solution worked out in the selective removal of the A layers of MAX phases leading to the formation of MXenes. Subsequently, HF has become the etchant of choice for preparing a variety of MXenes. By changing the concentration of the HF solution, the quality and yield of MXenes can be achieved [41] (Fig. 1.3). For example, more sophisticatedly etched $Ti_3C_2T_x$ can be obtained by using higher concentrations of HF solutions [1]. However, due to the health hazards associated with using HF, in situ HF method has been adopted, which does not use HF solution directly but utilizes HF-containing or HF-forming etchants like ammonium bifluoride [11], ammonium fluoride [9], and LiF/HCl [20]. Another advantage of using LiF/HCl instead of HF is that the interlayer spacing between MXene nanosheets rises due to the insertion of water molecules and cations like Li^+ weakening the interlayer attractions [1]. More interestingly, few reports on the synthesis of MXenes originating from non-MAX phase-based precursors. The first is a synthesis of Mo_2C reported by Halim et al. [10] by etching the Al layers from Mo_2Ga_2C. In addition, Zr_3C_2 and Hf_3C_2 are derived from $Zr_3Al_3C_5$ and $Hf_3[Al(-Si)]_4C_6$, respectively [55, 56]. The other important aspect during the synthesis of MXenes is the introduction and optimizing the surface terminal groups. The choice of surface terminal groups determines the properties of as prepared MXenes like bandgap energy, work function properties etc. and thus makes MXenes versatile in employing for various applications [5].

1.4 Environmental Remediation Applications of MXenes

After the discovery of MXenes in 2011, it was first used in batteries and electrochemical capacitors) due to its excellent electrical conductivity [3, 19, 28]. Since then, MXenes have received a great deal of attention due to their appealing properties such as high electrical conductivity, hydrophilic nature, large surface-to-volume ratio, high chemical stability, abundant surface terminations, robust redox ability,

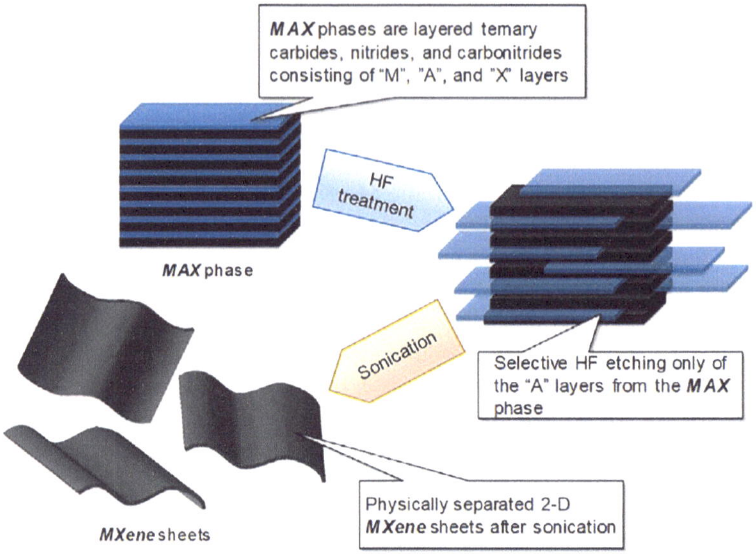

Fig. 1.3 Exfoliation of MAX phases and formation of MXenes. Adapted with permission from [40]. Copyright © 2012, American Chemical Society

and biocompatibility, which make them ideal candidates for employment in environmental applications such as photocatalysis, adsorbents for various pollutants in water, membrane materials, electrocatalytic stoichiometry, and biocompatibility (Fig. 1.4).

Fig. 1.4 Various environmental remediation applications of MXenes. Reproduced with permission from [46] using Creative Commons CC BY License. Copyrights © 2021 Mawada Mohammed Tunesi et al., Springer Nature

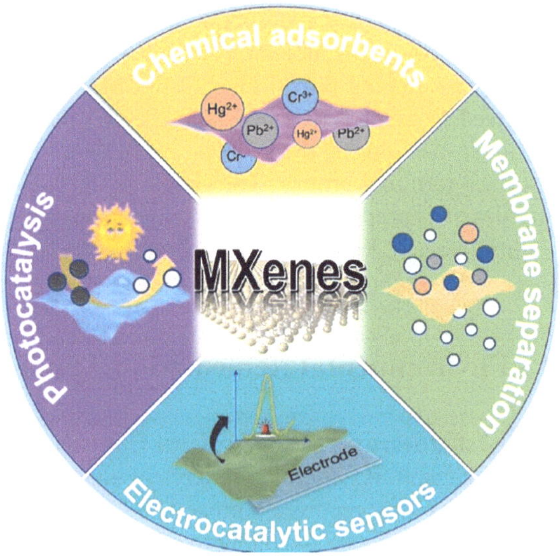

1.4.1 MXenes for Photocatalytic Pollutant Degradation

MXenes have been extensively considered for the photocatalytic degradation of various pollutants in water bodies. Several research articles have been published in this field since 2014 [16]. MXenes are perceived as successful candidates for photocatalysis which can be primarily accredited to the valuable part of MXenes as co-catalysts for the construction of composites between MXene and other semiconductors and tuneable bandgap of MXenes by changing the surface functionalities of MXenes [44]. The main drawback of any photocatalytic process is the inevitable charge carrier recombination after the excitation of photocatalysts by the light source. In this view, MXenes play the photogenerated electron acceptor promoting photogenerated charge carrier separation because of its lower Fermi level than most reported semiconductor photocatalysts. In addition, MXene can also act as robust support for the growth and even dispersion of photocatalysts, like TiO_2 [29, 50], CdS [21, 53] and MoS_2 [50], by avoiding agglomeration [7]. Moreover, the intimate contact occurred by growing the photocatalyst on MXene leads to further separation and migration of photogenerated charge carriers, thereby increasing the efficiency of the photocatalytic system. MXenes also help in the crucial step of reactant adsorption in the photocatalytic degradation of pollutants by tuning the surface functionalities of MXenes depending on the type of concerned pollutant. The possible roles of MXenes in improving photocatalytic performance are illustrated in Fig. 1.5.

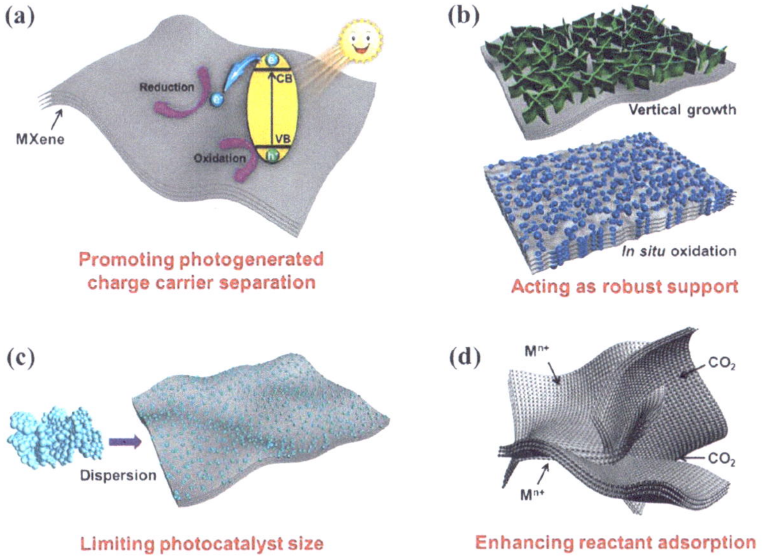

Fig. 1.5 Different roles of MXenes in improving photocatalytic performance. Reproduced with permission from [16]. Copyrights © 2020, Elsevier

1.4.2 MXenes as Adsorbents

Adsorption is one of the effective water purification techniques because of its simple operation, relatively low cost, high removal efficiency, and less harmful secondary sludge [8]. Due to the large specific surface area, hydrophilicity, and abundant active sites on their surface, MXenes can adsorb many kinds of pollutants, including heavy metal ions, organic dyes, and radioactive ions, through electrostatic and chemical interactions [5]. MXenes have shown promising adsorption capability toward dye contaminants in water, and the negatively charged surface of MXenes preferentially adsorbs cationic dye over anionic dye due to the electrostatic interactions as shown in Table 1.1. However, the MXenes surface can be modified either negatively or positively to adsorb target species by adjusting the pH of the solution [7, 8, 22, 26, 38]. MXenes are also considered efficient adsorbents for heavy metal ions, including copper (Cu(I/II)), lead (Pb(II)), mercury (Hg(II)) and chromium (Cr(VI)) in aquatic environments. In contrast, it is complicated to remove them using biological and chemical processes [12].

Moreover, MXenes have exhibited excellent adsorption characteristics for radionuclides because of their resistance to intense radiation and chemical stability with even harsh molten salt solutions [47]. The MXenes have been demonstrated to remove radionuclides with high mobility, and long half-lives disposed into the water environment uranium U(VI), cesium (Cs(I)), palladium (Pd(II)), Thorium (Th(IV)), and Europium III (Eu(III)) [14, 46, 53, 54]. It has been achieved by tuning MXenes by intercalation and surface functionalization to encapsulate the radionuclides whose hydrated radii are more significant than the layer d-spacing for MXenes [47, 48]. The adsorptive remediation capabilities of MXenes are improved than that of traditional pollutant adsorbents like carbon nanotubes and graphene.

1.4.3 MXenes as Separation Membranes

Membrane separation technology is a highly reliable and efficient route for wastewater treatment. MXenes, with hydrophilicity, tuneable interlayer spacing, and surface chemistry, have been recognized to be reliable candidates for water purification membranes (see Table 1.1). In this method, pollutants can be separated by size exclusion through the constrained interlayer spacing of MXenes. The first MXene-based membrane $Ti_3C_2T_x$ was reported by Ren et al. in [33], which is nonpermeable to cations with hydration radii larger than the MXene interlayer spacing (~6 Å) that include Li^+, Na^+, K^+, Mg^{2+}, Ca^{2+}, Ni^{2+}, and Al^{3+} and promoted ultrafast water flux [33]. MXene-based membranes have shown excellent performance in removing various inorganic salts and organic pollutants compared to the pristine MXene sheets owing to combining with other nanomaterials, such as graphene oxide and TiO_2 nanoparticles, which offer more water transport pathways than pristine membranes.

Table 1.1 Summary of MXenes' recent utilization in environmental remediation applications. Reproduced with permission from [32]. Copyrights © 2019, Elsevier

MXene as/for	Material	Applications
Adsorbents	$Ti_3C_2T_x$ Nanosheets	Removal of Cu(II) from water
	Alk-MXene ($Ti_3C_2(OH/ONa)_xF_{2x}$)	Removal of Pb(II) from water
	$Ti_3C_2T_x$ Nanosheets	Removal of Cr(VI), ($K_3[Fe(CN)_6]$, $KMnO_4$ and NaAuCl4) from water
	Urchin-like rutile Titania carbon (u-RTC) from (Ti_3C_2(OH)0.8F1.2)	Removal of Cr(VI) from water
	Ti_3C_2(OH)0.8F1.2)-iron oxide nanocomposite	Removal of phosphate from wastewater
	$Ti_3C_2T_x$	Adsorption and subsequent catalytic decomposition of MB
	Magnetic $Ti_3C_2T_x$ nanocomposite	Removal of toxic Hg(II) from water
	Multi-layered $Ti_3C_2T_x$	Removal of uranium and nuclear waste
Water purification membranes	$Ti_3C_2T_x$/PVDF	Selective sieving of (Li^+, Na^+, K^+, Mg^{2+}, Ca^{2+}), (Ni^{2+} and Al^{3+}), and MB^+ dye
	AgNPs/$Ti_3C_2T_x$	UF/NF membrane with high water flux and favorable rejection of BSA, MG, and multivalent ions
	$Ti_3C_2T_x$/PES	UF/NF membrane with high water flux and high rejection to Congo red and Gentian violet dyes
	Ultrathin 2D $Ti_3C_2T_x$ film	UF/NF membrane with high water flux and salt rejection
	$Ti_3C_2T_x$-GO	Membrane rejection for MR, MB, Rose Bengal, and brilliant blue dye
	TiO_2/Ti_3C_2	Photodegradation of MO under ultraviolet irradiation
Electrochemical separation	Ar plasma-modified 2D $Ti_3C_2T_x$ nanosheets	CDI desalination with a high salt removal rate
	Aerogel-like $Ti_3C_2T_x$	Porous electrode with high electro-adsorption capacity

(continued)

Table 1.1 (continued)

MXene as/for	Material	Applications
Photocatalytic agents	TiO_2/Ti_3C_2	Hydrogen production under Visible-Light Irradiation
	2D Ti_3C_2/TiO_2	Photocatalytic degradation of methyl orange dye
	CdS/Ti_3C_2	Visible-light photocatalytic activity and apparent quantum efficiency
	Ti3+doped rutile TiO_2/Ti_3C_2	Enhanced photocatalytic activity
Sensors	$Ti_3C_2T_x$	Detection of ethanol, methanol, acetone, and ammonia gas
	$Ti_3C_2T_x$ film	Detection of volatile organic compounds at room temperature
	Ti_3C_2	Mediator-free detection of nitrite
	Ti_3C_2	Ultrasensitive detection of phenols
	alk-Ti_3C_2	Detection of Cd(II), Pb(II), Cu(II) and Hg(II)
	Nafion/$Ti_3C_2T_x$	Electrocatalytic detection of bromate in drinking water
	Ag, Au and Pd@$Ti_3C_2T_x$	SERS substrate
	$Ti_3C_2T_x$	SERS substrate
	Ti2N	SERS substrate
Antibacterial agents	$Ti_3C_2T_x$	Antibacterial activity against E. coli and B. subtilis in aqueous media
	$Ti_3C_2T_x$/PVDF	Efficient antibacterial membrane against E. coli and B. subtilis
	Ag@ $Ti_3C_2T_x$	Anti-biofouling composite membrane against E. coli
	$Ti_3C_2T_x$/chitosan electrospun nanofibers	95 and 62% growth inhibition of E. coli and S. aureus, respectively

Moreover, intercalated nanomaterials separate the pores on MXene nanosheets, thereby increasing interlayer separation and creating enormous nanochannels for water flux [2].

1.4.4 MXenes for Capacitive Deionization (CDI)

A new desalination method called capacitive deionization (CDI) uses saltwater to move across the area between two electrodes while using a precise voltage to attract ions to the electrodes' opposing charges. When the voltage is removed or reversed, adsorbed ions are released, and electrodes are regenerated [51]. Due to their high volumetric capacitance, high electrical conductivity, hydrophilicity, tuneable thicknesses and interlayer distances, MXenes have been used in capacitive deionization (CDI) to capture cations and anions concurrently [4, 42, 43]. The first Ti_3C_2-MXene CDI electrode was fabricated by Srimuk et al. in [43] and was used directly without any binder onto the porous separator of the CDI cell and showed steady water desalination performance for 30 CDI cycles with an average salt adsorption capacity of 13 ± 2 mg g^{-1} [43]. Due to their high reversibility of ion intercalation and de-intercalation in an aqueous solution and their high capacitance, MXenes have been studied extensively as an electrode material for water desalination by CDI.

1.4.5 MXenes as Electrocatalytic Sensors for Pollutant Detection

The detection of pollutants is the first and critical stage of the environmental remediation process. In this view, MXene materials have also been extended for sensing trace amounts of pesticides and heavy metal ions (see Table 1.1) [34]. As a result of their electrical conductivity, ion transport capabilities, hydrophilicity, and active functionalization, MXenes have attracted a lot of interest in the field of sensors [6]. Moreover, MXene could be conveniently combined with other materials like metal nanoparticles and carbon-based components to improve selectivity and sensitivity. Alkaline intercalation of Ti_3C_2 MXene exhibited a sensitive and simultaneous electrochemical detection of Cd(II), Pb(II), Cu(II) and Hg(II), which is superior to most of the reported values [57]. In another report, the drawback of poor electron transfer of acetylcholinesterase (AChE)-based electrocatalytic sensors for organophosphate pesticides (OPs) detection has been overcome by utilizing $Ti_3C_2T_x$ MXene nanosheets combined with Ag nanoparticles. The catalytic sensor made of this hybrid system could detect malathion in a linear range from 10^{-14} to 10^{-8} M, with acceptable selectivity, satisfactory reproducibility, and exceptional stability [13]. MXenes have expanded progress in electrocatalytic sensors for various target species by tuning their surface terminations. MXenes deliver superior conductivity and high dispersibility, unlike graphene, which provides good conductivity at the cost of low hydrophilicity [46].

1.4.6 MXenes as Antibacterial Agents

MXenes have also demonstrated antibacterial properties toward various bacteria like Bacillus subtilis (B. subtilis) and Escherichia coli (E. coli) [31]. MXenes are also applied as efficient antimicrobial coatings. $Ti_3C_2T_x$-supported PVDF membranes inhibited bacterial growth against B. subtilis and E. coli [8]. The antibacterial properties of MXenes can be ascribed to the negatively charged surface, the straight destruction of bacteria due to contact with the hydrophilic surface, and hydrogen bonding between the cell membrane and oxygenate groups of MXene that prevent the nutrient consumption in bacteria [26]. MXenes are likely resistant to biofouling and provide bactericidal properties, which make MXenes the best fit for membrane applications.

1.5 Conclusion and Future Perspectives

MXenes have shown fascinating properties and advantages for employment in various environmental remediation applications, including photocatalysis, accelerating the expansion of the photocatalysts, adsorbents for heavy metals, organic dyes, radioactive ions, etc., separation membranes, electrode materials in capacitive deionization (CDI), electrocatalytic sensors for pollutant detection and anti-biofouling or antibacterial agents. MXenes have become potential substitutes for noble metals as co-catalysts in photocatalytic applications. However, Ti_3C_2-based MXenes have been much reported in photocatalytic degradation applications, whereas another novel MXenes like V_2C, Zr_2C and Ti_2C could be further utilized for photocatalytic environmental applications. Although the most studied MXene Ti_3C_2 has its prominent advantages in ecological applications, it still suffers from the stability issue that could be easily oxidized and transformed into TiO_2 after storing it in water for a more extended period of time. Therefore, the stability of MXene materials should be monitored in the long run for being used as adsorbents and membrane materials in practical applications. MXenes flake size is related to the water flux control for membrane applications. Therefore, combining theoretical and experimental studies could achieve an optimal flux rate with a high salt rejection rate.

Moreover, MXene membranes often get damaged with swelling, resulting in low selectivity and rejection of ions. Hence, tuning the interlayer d-spacing of MXene sheets can be a feasible way to avoid swelling or structural disintegration of MXene membranes. Few of the MXenes have been reported to exhibit antibacterial activity against some bacteria, which paves the way for developing anti-biofouling membranes. However, the mechanism under the bactericidal properties of MXenes is needed to be fully understood. Many MAX phase materials have been theoretically reported so far. They need to be verified through experimental methods to realize the expansion of the MXene family with improved properties. Many attempts have been made in recent years to avoid using hazardous HF during the synthesis of MXenes. However, more attention is needed in this direction to make the synthesis of MXenes an environment-friendly technique. Most of the reported MXenes have been prepared in a top-down approach, whereas scanty literature is available on synthesizing the bottom-up approach to synthesize MXenes. Therefore, novel and environment-friendly synthesis methods of the bottom-up approach have to be explored with more control over the MXene characteristics. Rapid progress has been made in spreading the MXenes into various environmental applications. However, studies on the life cycle analysis and potential toxic effects of MXenes in the environment are very scarce. More comprehensive investigations have to be carried out on the possible hazards associated with MXenes before realizing the applications of MXenes on a commercial scale.

References

1. M. Alhabeb et al., Guidelines for synthesis and processing of two-dimensional titanium carbide ($Ti_3C_2T_x$ MXene). Chem. Mater. **29**(18), 7633–7644 (2017)
2. Y.J. Al-Hamadani, B.-M. Jun, M. Yoon, N. Taheri-Qazvini, S.A. Snyder, M. Jang, J. Heo, Y. Yoon, Applications of MXene-based membranes in water purification: a review. Chemosphere **254**(126821), 126821 (2020)
3. B. Anasori et al., 2D metal carbides and nitrides (MXenes) for energy storage. Nat. Rev. Mater. **2**(2), 16098 (2017)
4. W. Bao, X. Tang, X. Guo, S. Choi, C. Wang, Y. Gogotsi, G. Wang, Porous cryo-dried MXene for efficient capacitive deionization. Joule **2**(4), 778–787 (2018)
5. L. Damptey, et al., Surface functionalized MXenes for wastewater treatment—a comprehensive review **6**(6), 2100120 (2022)
6. K. Deshmukh, T. Kovářík, S.K. Khadheer Pasha, State of the art recent progress in two dimensional MXenes based gas sensors and biosensors: a comprehensive review. Coord. Chem. Rev.**424**(213514), 213514 (2020)
7. P. Feng, et al., Synergistic lubrication of few-layer $Ti_3C_2T_x/MoS_2$ heterojunction as a lubricant additive Friction (2022)
8. Y. Ge, Z. Li, Application of lignin and its derivatives in adsorption of heavy metal ions in water: a review. ACS Sustain. Chem. Eng. **6**(5), 7181–7192 (2018)
9. M. Ghidiu et al., Conductive two-dimensional titanium carbide 'clay' with high volumetric capacitance. Nature **516**(7529), 78–81 (2014)
10. J. Halim, et al., Synthesis and characterization of 2D molybdenum carbide (MXene)**26**(18), 3118–3127 (2016)
11. J. Halim et al., Transparent conductive two-dimensional titanium carbide epitaxial thin films. Chem. Mater. **26**(7), 2374–2381 (2014)
12. I. Ihsanullah, MXenes (two-dimensional metal carbides) as emerging nanomaterials for water purification: progress, challenges and prospects. Chem. Eng. J. **388**, 124340 (2020)
13. Y. Jiang, X. Zhang, L. Pei, S. Yue, L. Ma, L. Zhou, Z. Huang, Y. He, J. Gao, Silver nanoparticles modified two-dimensional transition metal carbides as nanocarriers to fabricate acetycholinesterase-based electrochemical biosensor. Chem. Eng. J. (Lausanne, Switzerland: 1996) 339, 547–56 (May 2018)
14. A.R. Khan, S.M. Husnain, F. Shahzad, S. Mujtaba-Ul-Hassan, M. Mehmood, J. Ahmad, M.T. Mehran, S. Rahman, Two-dimensional transition metal carbide ($Ti_3C_2T_x$) as an efficient adsorbent to remove cesium (Cs+). Dalton Trans. (Cambridge, England: 2003) **48**(31), 11803–12 (2019)

15. N. Khossossi, R. Ahuja, MXene-based 2D anode materials for next-generation batteries, in *Next-Generation Materials for Batteries* (2021), pp. 1–7
16. P. Kuang et al., MXene-based photocatalysts. J. Mater. Sci. Technol. **56**, 18–44 (2020)
17. Y.-M. Li et al., Theoretical investigations of TiNbC MXenes as anode materials for Li-ion batteries. J. Alloy. Compd. **778**, 53–60 (2019)
18. Y. Li et al., 1T-MoS_2 nanopatch/Ti_3C_2 MXene/TiO_2 nanosheet hybrids for efficient photocatalytic hydrogen evolution. Mater. Chem. Front. **3**(12), 2673–2680 (2019)
19. Z. Li, Y. Wu, 2D early transition metal carbides (MXenes) for catalysis**15**(29), 1804736 (2019)
20. A. Lipatov, et al.,Effect of synthesis on quality, electronic properties and environmental stability of individual monolayer Ti3C2 MXene flakes **2**(12), 1600255 (2016)
21. Q. Liu et al., MXene as a non-metal charge mediator in 2D layered CdS@Ti_3C_2@TiO_2 composites with superior Z-scheme visible light-driven photocatalytic activity. Environ. Sci. Nano **6**(10), 3158–3169 (2019)
22. T. Liu et al., Two-dimensional MXene incorporated graphene oxide composite membrane with enhanced water purification performance. J. Membr. Sci. **593**, 117431 (2020)
23. X. Liu et al., Anchoring effects of S-terminated Ti_2C MXene for lithium-sulfur batteries: a first-principles study. Appl. Surf. Sci. **455**, 522–526 (2018)
24. O. Mashtalir, et al., Amine-assisted delamination of Nb_2C MXene for li-ion energy storage devices **27**(23), 3501–3506 (2015)
25. O. Mashtalir et al., Intercalation and delamination of layered carbides and carbonitrides. Nat. Commun. **4**(1), 1716 (2013)
26. W. Mu et al., Improving barium ion adsorption on two-dimensional titanium carbide by surface modification. Dalton Trans. **47**(25), 8375–8381 (2018)
27. M. Naguib, et al., Two-dimensional nanocrystals produced by exfoliation of Ti_3AlC_2 **23**(37), 4248–4253 (2011)
28. J. Pang et al., Applications of 2D MXenes in energy conversion and storage systems. Chem. Soc. Rev. **48**(1), 72–133 (2019)
29. C. Peng et al., (111) TiO_{2-x}/Ti_3C_2: synergy of active facets, interfacial charge transfer and Ti^{3+} doping for enhance photocatalytic activity. Mater. Res. Bull. **89**, 16–25 (2017)
30. J. Peng et al., Surface and heterointerface engineering of 2D MXenes and their nanocomposites: insights into electro- and photocatalysis. Chem **5**(1), 18–50 (2019)
31. K. Rasool, M. Helal, A. Ali, C.E. Ren, Y. Gogotsi, K.A. Mahmoud, Antibacterial activity of $Ti_3C_2T_x$ MXene. ACS Nano **10**(3), 3674–3684 (2016)
32. K. Rasool, R.P. Pandey, P. Abdul Rasheed, S. Buczek, Y. Gogotsi, K.A. Mahmoud, Water treatment and environmental remediation applications of two-dimensional metal carbides (MXenes). Mater. Today (Kidlington, England) 30, 80–102 (November 2019)
33. C.E. Ren, K.B. Hatzell, M. Alhabeb, Z. Ling, K.A. Mahmoud, Y. Gogotsi, Charge- and size-selective ion sieving through $Ti_3C_2T_x$ MXene membranes. J. Phys. Chem. Lett. **6**(20), 4026–31 (2015)
34. A. Rhouati, M. Berkani, Y. Vasseghian, N. Golzadeh, MXene-based electrochemical sensors for detection of environmental pollutants: a comprehensive review. Chemosphere **291**(Pt 1), 132921 (2022)
35. R.M. Ronchi et al., Synthesis, structure, properties and applications of MXenes: current status and perspectives. Ceram. Int. **45**(15), 18167–18188 (2019)
36. D. Ruan et al., Exfoliated Mo_2C nanosheets hybridized on CdS with fast electron transfer for efficient photocatalytic H_2 production under visible light irradiation. Appl. Catal. B **264**, 118541 (2020)
37. T. Schultz et al., Surface termination dependent work function and electronic properties of $Ti_3C_2T_x$ MXene. Chem. Mater. **31**(17), 6590–6597 (2019)
38. A. Shahzad et al., Two-dimensional $Ti_3C_2T_x$ MXene nanosheets for efficient copper removal from water. ACS Sustain. Chem. Eng. **5**(12), 11481–11488 (2017)
39. F. Shahzad, et al., Electromagnetic interference shielding with 2D transition metal carbides (MXenes)**353**(6304), 1137–1140 (2016)

40. Q. Shan et al., Two-dimensional vanadium carbide (V2C) MXene as electrode for supercapacitors with aqueous electrolytes. Electrochem. Commun. **96**, 103–107 (2018)
41. A. Sinha et al., MXene: an emerging material for sensing and biosensing. TrAC Trends Anal. Chem. **105**, 424–435 (2018)
42. P. Srimuk, J. Halim, J. Lee, Q. Tao, J. Rosen, V. Presser, Two-dimensional molybdenum carbide (MXene) with divacancy ordering for brackish and seawater desalination via cation and anion intercalation. ACS Sustain. Chem. Eng. **6**(3), 3739–3747 (2018)
43. P. Srimuk, F. Kaasik, B. Krüner, A. Tolosa, S. Fleischmann, N. Jäckel, M.C. Tekeli, M. Aslan, M.E. Suss, V. Presser, MXene as a novel intercalation-type pseudocapacitive cathode and anode for capacitive deionization. J. Mater. Chem. Mater Energy Sustain. **4**(47), 18265–71 (2016)
44. M.M. Tunesi et al., Application of MXenes in environmental remediation technologies. Nano Converg. **8**(1), 5 (2021)
45. L. Wang, H. Song, L. Yuan, Z. Li, Y. Zhang, J.K. Gibson, L. Zheng, Z. Chai, W. Shi, Efficient U(VI) reduction and sequestration by Ti_2CT_x MXene. Environ. Sci. Technol. **52**(18), 10748–10756 (2018)
46. L. Wang, L. Yuan, K. Chen, Y. Zhang, Q. Deng, D. Shiyu, Q. Huang et al., Loading actinides in multi-layered structures for nuclear waste treatment: the first case study of uranium capture with vanadium carbide MXene. ACS Appl. Mater. Interfaces **8**(25), 16396–16403 (2016)
47. X. Xie, N. Zhang, Positioning MXenes in the photocatalysis landscape: competitiveness, challenges, and future perspectives**30**(36), 2002528 (2020)
48. Y. Yang et al., Plasma-modified Ti_3C2T_x/CdS hybrids with oxygen-containing groups for high-efficiency photocatalytic hydrogen production. Nanoscale **11**(40), 18797–18805 (2019)
49. H. Younes, D. Lou, M.M. Rahman, D. Choi, H. Hong, L. Zou, Review on 2D MXene and graphene electrodes in capacitive deionization. Environ. Technol. Innov. 28(102858), 102858 (2022)
50. W. Yuan, et al., Laminated hybrid junction of sulfur-doped TiO_2 and a carbon substrate derived from Ti_3C_2 MXenes: toward highly visible light-driven photocatalytic hydrogen evolution, **5**(6), 1700870 (2018)
51. X. Zhan et al., MXene and MXene-based composites: synthesis, properties and environment-related applications. Nanoscale Horiz. **5**(2), 235–258 (2020)
52. Y. Zhang, et al., Adsorptive environmental applications of MXene nanomaterials: a review. RSC Adv. **8**(36), 19895–19905 (2018)
53. P. Zhang, L. Wang, L.-Y. Yuan, J.-H. Lan, Z.-F. Chai, W.-Q. Shi, Sorption of Eu(III) on MXene-derived titanate structures: the effect of nano-confined space. Chem. Eng. J. (Lausanne, Switzerland: 1996) **370**, 1200–1209 (August 2019)
54. Y.-J. Zhang, J.-H. Lan, L. Wang, Q.-Y. Wu, C.-Z. Wang, T. Bo, Z.-F. Chai, W.-Q. Shi, Adsorption of uranyl species on hydroxylated titanium carbide nanosheet: a first-principles study. J. Hazard. Mater. **308**, 402–10 (May 2016)
55. J. Zhou, et al., A two-dimensional zirconium carbide by selective etching of Al_3C_3 from nanolaminated $Zr_3Al_3C_5$ **55**(16), 5008–5013 (2016)
56. J. Zhou et al., Synthesis and electrochemical properties of two-dimensional hafnium carbide. ACS Nano **11**(4), 3841–3850 (2017)
57. X. Zhu, B. Liu, H. Hou, Z. Huang, K.M. Zeinu, L. Huang, X. Yuan, D. Guo, H. Jingping, J. Yang, Alkaline intercalation of Ti_3C_2 MXene for simultaneous electrochemical detection of Cd(II), Pb(II), Cu(II) and Hg(II). Electrochim. Acta **248**, 46–57 (2017)

Chapter 2
Application of MXenes in Water Purification, CO_2 Capture and Conversion

Jonathan Tersur Orasugh, Lesego Tabea Temane, and Suprakas Sinha Ray

Abstract The number of nanomaterials that are suitable for many applications has increased with the 2011 discovery of two-dimensional (2D) transition metal carbides as well as nitrides (MXenes). MXenes are a new class of 2D materials that are quickly gaining popularity for various uses in the fields of medicine, chemistry, and the environment. MXenes but also MXene-composites or hybrids have several desirable properties, including a large surface area, outstanding chemical stability, hydrophilicity, excellent thermal conductivity, and environmental compatibility. MXenes have therefore been utilized in the creation of lithium-ion batteries, semiconductors, and hydrogen storage. The remediation of contaminated groundwater, surface waters, industrial and municipal wastewaters, as well as the capture and conversion of hydrogen, are just a few of the environmental applications where MXenes have recently been used. These applications frequently outperform those for traditional materials. MXene-composites can deionize via Faradaic capacitive deionization (CDI) as well as adsorb a range of organic and inorganic contaminants when employed for electrochemical applications. The applications of MXenes as well as its composites/hybrids for water treatment and CO_2 capture and conversion, are conversed in this chapter as per the literature. We have also discussed the challenges with regard to the utilization of Mxene and its materials in wastewater remediation, along with drawn conclusions.

J. T. Orasugh (✉) · S. S. Ray
Department of Chemical Sciences, University of Johannesburg, Doornfontein, Johannesburg 2028, South Africa
e-mail: calunivjto@gmail.com

J. T. Orasugh · L. T. Temane · S. S. Ray
Centre for Nanostructures and Advanced Materials, DSI-CSIR Nanotechnology Innovation Centre, Council for Scientific and Industrial Research, Pretoria 0001, South Africa

N. Kumar et al. (eds.), *Two-Dimensional Materials for Environmental Applications*, Springer Series in Materials Science 332, https://doi.org/10.1007/978-3-031-28756-5_2

2.1 Introduction

MXenes, which were first identified in 2011, are new additions to the 2D materials family tree, which is rapidly expanding [1, 2]. These artificial substances, which have the general formula $M_{n+1}X_nT_x$, belong to the transition metal nitrides, carbides, but also carbonitrides family. Here, M is early transition metal (like Mo, Ti, Zr, or W, for example), and X is carbon or nitrogen. T stands for surface terminal groups, including chlorine (Cl), oxygen (O), fluorine (F), as well as hydroxyl (OH) or chlorine (Cl). The quantity of surface functionalities is denoted by the symbol x, while n is an integral integer between 1 and 3. With −F or −OH terminal groups, $Ti_3C_2T_X$ is one of the most prevalent MXenes. MXene materials have high surface areas, exceptional chemical stability, thermal conductivity, and hydrophilicity. As a result, MXenes have been used as innovative materials for several industrial utilization, including lithium ion batteries, semiconductor production, and hydrogen storage [2].

MXenes are reportedly used in environmental remediation, such as desalination, wastewater treatment, and water quality sensing [3, 4]. For instance, 2D Ti_3C_2 MXenes have been extensively studied for the detection of biomolecules, drinking water pathogens, and heavy metals (such as Pb, Hg, and Cd). Similar to MXene-based adsorbents, these substances have a high surface area and many highly active functional sites and are hydrophilic. $Ti_2C_2T_x$ nanosheets can use UV-based photocatalytic reactions to break down micropollutants like diclofenac [5]. As free-standing electrode materials for capacitive de-ionization and desalination applications, MXenes also have incredible promise. Additionally, MXene-based membranes for desalination and drinking water applications offer lower biofouling potential and greater flow performance than traditional membranes [6]. The realm of water treatment and environmental remediation therefore places MXenes in a prominent position.

The current chapter largely concentrates on MXenes' water remediation elements while also providing an up-to-date synthesis of Mxenes and its composites/hybrids towards current water treatment techniques. We also discuss the significance of MXenes for CO_2 sequestration purposes and examine ways to reduce their energy requirements, which will be a major problem for the water sector in the upcoming years.

2.1.1 MAX Phases as Well as MXene Synthesis

According to Ronchi et al. [3] and Hwang et al. [4], MAX phases, which are ternary nitrides or carbides with the general formula $M_{n+1}AX_n$, give rise to MXenes.. Here, X denotes either carbon (C) or else nitrogen (N), while M denotes early d-block transition metals (like V, Ti, Mo, Sc, but also W). MAX phases are a class of layered materials where A-layers connect 2D $M_{n+1}X_n$ (A elements are Al or Si). The special combination of metallic and ceramic characteristics that characterizes these phases. Due to the M-A bond's greater reactivity and relative weakness when compared to the M-X bond. As an outcome, the "A" group atoms from the MAX phases are eliminated, resulting in MXenes. The ensuing metallic bond formed by the $M_{n+1}X_n$ layers is significantly more powerful than the van der Waals interface in layered materials like graphite and transition metal dichalcogenides (TMDs) [4]. MXenes' improved strength and stiffness improve their durability in industrial applications like water treatment.

These phases have 16 A elements as well as 14M elements in them. The recent identification of quaternary in- plus out-of-plane ordered MAX phases pave the way for the identification of numerous more phases. The MAX phases' chemical diversity contains the secret to maximising potential application characteristics. There is still considerable work to be done because many of the more recent quaternary (higher) phases have not yet been defined.

2.1.2 Traditional Synthesis Methods

From the precursor MAX phase, MXenes are made by wet etching bulk crystals into single- or multiple-layer MXene sheets with an etchant, usually hydrofluoric acid (HF) [7]. According to the following equations, the synthesis of MXenes can be explained:

$$M_{n+1} + AX_n(s) + 3HF(aq) \Rightarrow M_{n+1}X_n(s) + AF_3(s) + 3/2H_2(g) \tag{2.1}$$

$$M_{n+1} + X_n(s) + 2H_2O(l) \Rightarrow M_{n+1}X_n(OH)_2(s) + H_2(g) \tag{2.2}$$

$$M_{n+1} + X_n(s) + 2HF(aq) \Rightarrow M_{n+1}X_nF_2(s) + H_2(g) \tag{2.3}$$

Phase 1 MXene layers are created when the "A" layers of the MAX phases interact with HF (or fluoride ions).

Phase 2 The hydrophilic $-OH$, $-O$, and/or $-F$ groups are used to replace the "A" atoms of the MAX phases.

Phase 3 MXenes are loosely packed layers that resemble graphite due to weak interactions between the $M_{n+1}X_n$ layers.

The MXene surfaces become negatively charged due to the development of surface functional groups ($-OH$, $-O$, and/or $-F$), which leads to the formation of stable dispersions [7]. The most popular MAX precursor used in the production of the most extensively researched $Ti_3C_2T_X$ MXene is Ti_3AlC_2 [7]. Many novel MXenes, including Ti_2C, Mo_2C, and Ti_3SiC_2, are reportedly synthesized with the aid of the HF etching approach [8]. But utilizing and handling risks are associated with the MXenes production employing HF etchants. The production of stable nitride-based MXenes

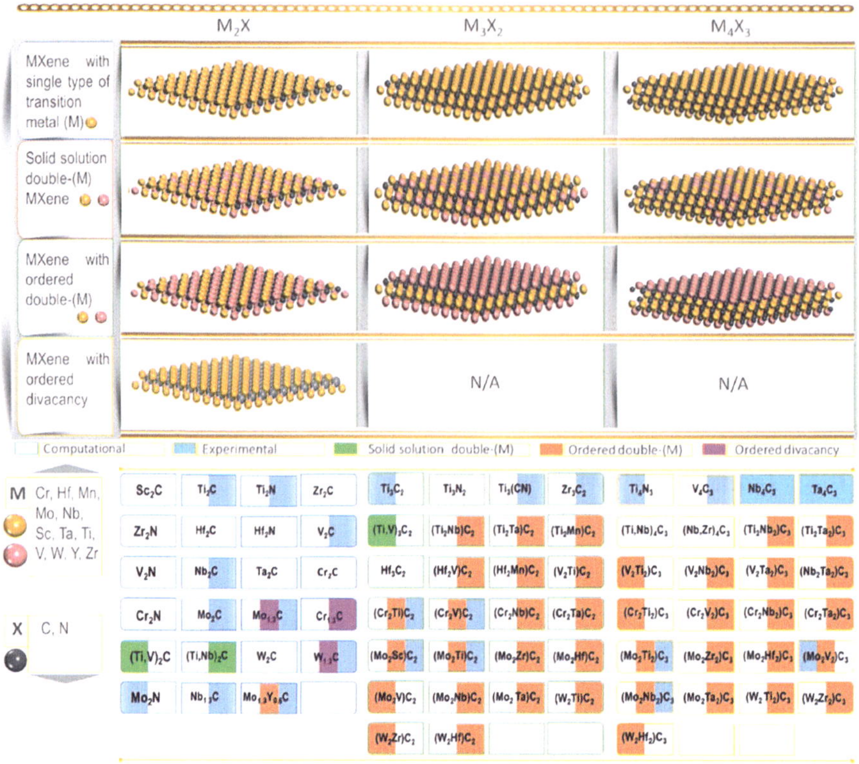

Fig. 2.1 Different types of MXenes synthesized theoretically and experimentally. Reproduced with permission from Adekoya et al. [7], Copyright 2022, ACS Publications

($Ti_{n+1}N_n$) cannot be done using this method [8]. Therefore, creating more adaptable and environmentally friendly protocols for MXene production is necessary.

Diverse kinds of Mxenes produced practically and theoretically are represented in Fig. 2.1.

For the purpose simplicity, some of the major Mxene synthesis approaches as well as their merits and demerits are presented in Table 2.1.

2.1.2.1 Novel Methods for Synthesizing MXene

For the traditional and developing MAX phases' exfoliation, recent research has produced gentler and less harmful etchants than HF [3]. Hydrothermal procedures, fluoride-free MXene synthesis, and MXene synthesis with gentler fluorine salts are three major groups into which these methods can be divided. Hydrothermal procedures, fluoride-free MXene synthesis, and MXene synthesis with gentler fluorine salts are three major groups into which these methods can be divided.

Table 2.1 Comparison of three types of different synthetic methods and evaluation of their merits and demerits. Reproduced with permission from Peng et al. [9]. Copyright 2022, John Wiley and Sons

Synthesis strategies	Regents	Morphology	Merits	Demerits
HF etching	HF	Accordion-like structure with abundant −F terminations and defects	1. Effective for most MAX High yield	1. Dangerous operation Cannot be peeled in situ
Fluoride-based acid etching	LiF/NaF/KF + HCl NH_4HF_2	Clay-like MXene with large interlayer spacing and few −F terminations	1. Relatively safe Direct ultrasonic peeling	1. Long etching time Introducing fluoride salt impurities
Alkaline solution etching	NaOH TMAOH	Accordion-like structure only with −O, −OH terminations (For TMAOH terminated with $Al(OH)_4^-$)	1. No risk of acid corrosion Fluorine-free functional group	1. For NaOH: severe etching conditions For TMAOH: need HF pretreatment
Molten fluoride salt and Lewis acidic molten melts etching	LiF + NaF + KF Lewis Acid	Accordion-like structure (For Lewis Acid terminated with various halogens)	1. For molten fluoride salt: can obtain nitride MXene For Lewis acid: achieve precise control of surface functional groups	1. Introduce salt impurities Severe etching conditions
Electrochemical etching	NH_4Cl + TMAOH	Single or few-lawyer structure without −F functional group	1. Safe etching environment 2. Fluorine-free functional group Get stripped MXene directly	1. Strict etching conditions Introduce salt impurities
Water-free etching	NH_4HF_2 in organic solvents	Accordion-like structure with extremely high −F terminations	1. Conducive to the use of organic systems due to the absence of water Can be delaminated directly by ultrasonication	1. Long etching time Tedious washing steps

(continued)

Table 2.1 (continued)

Synthesis strategies	Regents	Morphology	Merits	Demerits
HCl-based hydrothermal etching	HCl-hydrother-mal	Layered structure with −Cl and −O terminations	1. Simple experiment operation Fluorine-free functional group	1. Severe etching conditions Rely on the prediction of precise reaction conditions by DFT (density functional theory)
Halogen etching	Br_2, I_2, ICl, IBr	Accordion-like structure terminated with various halogens	1. Mild etching environment 2. Precise control of surface functional groups	1. Tedious etching and purification steps 2. Strict etching condition
Chemical vapour deposition (CVD) process	Methane and bimetal foil (Cu/Mo)	Ultra-thin and large-size flake	1. High purity 2. Accurate control of thickness	1. Low productivity 2. High synthesis temperature
Lithiation-expansion-microexplosion mechanism	Lithium-ion	Single-layer or few-layer structure without −F functional group	1. Simple and safe synthesis environment 2. Fluorine-free functional group	1. Low productivity 2. Consume resources
In Situ electrochemical synthesIS	LiTFSI + $Zn(OTF)_2$	In situ etching and stripping in the battery	1. Green synthetic environment friendly 2. Extremely convenient operation	1. Rely on expensive metal ion salts 2. Restricted to be used inside the battery

Abbreviation: DFT, Density functional theory; TMAOH, TetraMethyl Ammonium Hydroxide

2.1.2.2 Milder Fluoride-Based Salts Approach

In order to create an etchant solution, fluoride milder salts like CaF, KF, NaF, and LiF, but also CsF can be combined with HCl or sulfuric acid (H_2SO_4). These procedures can produce multilayered carbides of an early transition metal having various compositions but also altered physico-chemical characteristics. For example, compared to MXenes generated through standard HF etching, monolayers of $Ti_3C_2T_X$ MXene produced using an amalgamation of lithium fluoride (LiF) with HCl provide superior-quality monolayers with defect-free surfaces [10]. Similar to how stronger etchants like ammonium bifluoride (NH_4HF_2) to help create $Ti_3C_2T_X$ MXene with $-NH_4^+$ interlayers, which may be utilized to exclude negatively charged impurities from water, gentler etchant solutions also aid in the creation of these interlayers [11].

2.1.2.3 Hydrothermal-Enabled Approaches

By using this method, harmful HF fumes are avoided. Hydrothermal etching of Ti_3AlC_2 sheets using etchants like NH_4F can produce $Ti_3C_2T_X$ MXene [12]. According to studies, Ti_3C_2 and Nb_2C MXenes can be made to neglect the utilization of HF by hydrothermal etching $NaBF_4$ and HCl [12]. "A" group removal efficiency, interlayer spacing, ease of exfoliation, and c-lattice parameter are all significantly higher in hydrothermally produced MXenes than in HF-etched MXenes [12].

2.1.2.4 Fluoride-Free Procedures

Recent developments in MXene synthesis that do not employ HF include salt-templated (like molten salt) as well as chemical vapor deposition (CVD) [13–15]. Al-layers from MAX phases could as well be leached by immersing Ti_3AlC_2 in a sodium hydroxide (1M) solution at 80 °C for 100 h, trailed by hydrothermal reaction using a 1M H_2SO_4 at 80 °C for 2 h [16]. But because it takes so long, optimization must come before scaling. Alternative methods that use Ti_3AlC_2 anodic corrosion are possible, but they haven't been successfully shown in large-scale usage. Therefore, more study is necessary to develop MXenes more safely and achieve regulated surface functionalization, a considerable surface area, and adequate physio-chemical stability.

Following the dissolution of aluminum and in situ intercalation of NH_4OH, carbide flakes ($Ti_3C_2T_x$, T = O, OH) with diameters up to 18.6 μm and a high yield (over 90%) of mono- and bilayers are extracted in another study grounded upon the anodic corrosion of Ti_3AlC_2 in a binary aqueous electrolyte by Yang and colleagues [15].

Again, a group of researchers have proposed a fluoride-free etching approach to synthesise Mxene "MoS_2/Mo_2CT_X as represented in Fig. 2.2 [17]. Mo_2CT_X, where T is OH, O, or S, was produced by etching the Mo_3AlC_2 MAX phase in a solution of NaOH (15M) and Na_2S (4M) in order to oxidize the Al layers and incorporate

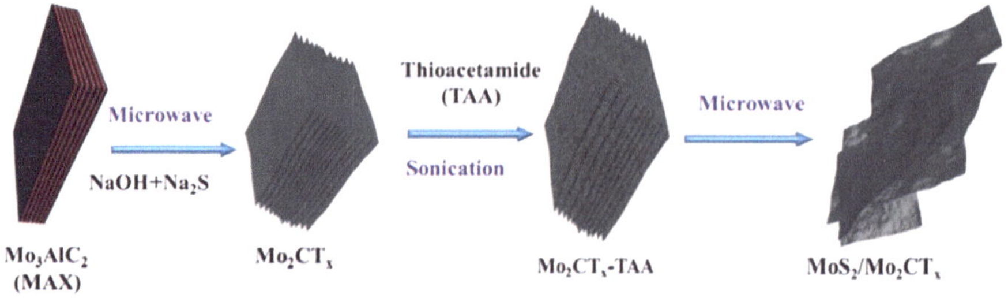

Fig. 2.2 Schematic representation of the microwave-assisted two-step vulcanization method for synthesizing MoS_2/Mo_2CT_X from Mo_3AlC_2. Reproduced with permission from Unnikrishnan et al. [17]. Copyright 2022, Elsevier Science Ltd.

sulfur (S) in situ during a hydrothermal process. "A" appropriate etchant can be used to selectively erode the A layers because, in MAX phases, the metallic MA bonds are weaker as well as more reactive than MX bonds. NaOH's potent reactivity with the amphoteric element Al to produce complexes like $Al(OH)_4^-$ in hydrothermal circumstances suggests that it might be used as an etching agent for Mo_3AlC_2. Furthermore, Mo_2CT_X, where T might be OH, O, or S, is formed when Na_2S promotes in situ vulcanization at the margins and surface.

Additionally, S^{2-} somewhat corroded the surfaces of Mo_2CT_X nanosheets, resulting in the creation of a limited number of Mo-S bonds and a significant amount of O and −OH groups due to the stable Mo–OH groups. To improve sulfur functionalization, TAA was utilized as a sulfur source and intercalating agent. TAA interacts with Mo to create complexes through Mo-S bonds on the MXene's surfaces and edges. Finally, to create MoS_2/Mo_2CT_X, Mo_2CT_X-TAA was microwave-heated for 2 h at a predetermined temperature (180 °C). At the time of heating, TAA generates S_2^- ions that interact with Mo atoms to create MoS_2 [17].

2.1.2.5 Microexplosion-Based Lithiation Expansion Approach

Despite numerous studies on the production of MXene free of fluorine, it is still technically challenging to directly produce high-purity single-layer or few-layer MXene using straightforward and manageable solutions.

Sun and associates recently have created a new technique for making single-layer or few-layer MXene (Fig. 2.3) [18]. The anode at the first stage was made of Ti_3AlC_2, and the cathode and reference electrode were made of lithium foil. Lithium-ion intercalation was used to expand Al-Li alloy following the 0.10 mA discharge process. Li_xAl_y will microexplode after being subjected to ultrasonic treatment in distilled water, evaporating ultra-thin $Ti_3C_2T_x$ MXene in the process. To further demonstrate the method's applicability, MAX Ti_3SiC_2 based on silicon is also employed to create MXene.

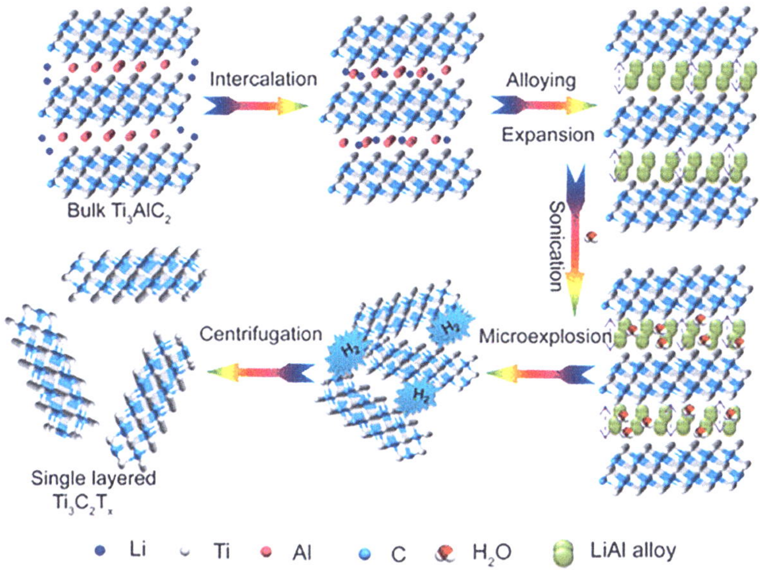

Fig. 2.3 Schematic of the lithiation intercalation–expansion–microexplosion process. Reproduced with permission from Sun et al. [18], copyright 2019, ACS Publications

2.1.2.6 In Situ-Based Electrochemical Approach

Etching is a necessary step in the conventional way of manufacturing MXene, and device assembly follows. This multistep approach is not ecologically friendly. Zhi's group [19] anticipated a recently integrated approach that consists of the in-situ MAX etching along with ions storage in MXene. V_2CT_X MXene is produced inside the battery through an electrochemical reaction employing LiTFSI but also $Zn(OTF)_2$ amalgamated ionic electrolyte adopted as the etchant. It's important to remember that the battery's lifetime has resulted in exfoliation, oxidation of the electrode material, as well as redox reaction of the oxidation product V_2O_5. The resulting ZIB offers an energy density of up to 310.3 Wh kg^{-1} and a 409.7 mAh g^{-1} capacity at 0.5 Ag^{-1}. This MAX-devices-derived MXene-devices approach sets a standard for practical and environmentally friendly MXene device manufacturing, and it is highly motivating.

2.1.2.7 CVD Approach

To create MXene, all etching techniques rely on the precursor to isolate the "A" atomic stratum with precision. The lateral dimension of MXene obtained chemically ranges from several hundred nanometers to ten micrometers. In 2015, Xu et al. [20] developed a bottom-up synthesis method in which CVD was used to directly synthesize large-size (more than 100 microns) and ultra-thin (a few nanometers thick)

$-Mo_2C$ crystals. Methane was the carbon source, while the copper foil underneath contained a layer of molybdenum foil. When the temperature rises over 1085 °C, Cu melts, Mo atoms diffuse to the surface, but also when the surface reacts with the carbon atoms released by the breakdown of methane, Mo_2C crystal was directly created. Later, two years after Mo_2C was created via the CVD technique, its lateral dimensions approached the centimeter range [21].

Thirumal et al. [22] have recently demonstrated a CVD-based synthesis of MXene@CNTs aimed at the degradative removal of RhB dye from water: though the Mxene was not entirely synthesized by the CVD approach. Still, the CNTs in the final composite form (Fig. 2.4) revealed from their study that the composite degraded RhB faster (3.5 times) thansynthesized Mxene.

MXene may be synthesized under control using the CVD process, and the resulting material possess large tangential dimensions, homogeneous depth, but also low defect concentration. The duration, reaction temperature, and gas are a few sophisticated control process parameters. Goknur Cambaz Buke's group conducted a thorough investigation in 2020 into the effects of system impurities, the depth of the Cu

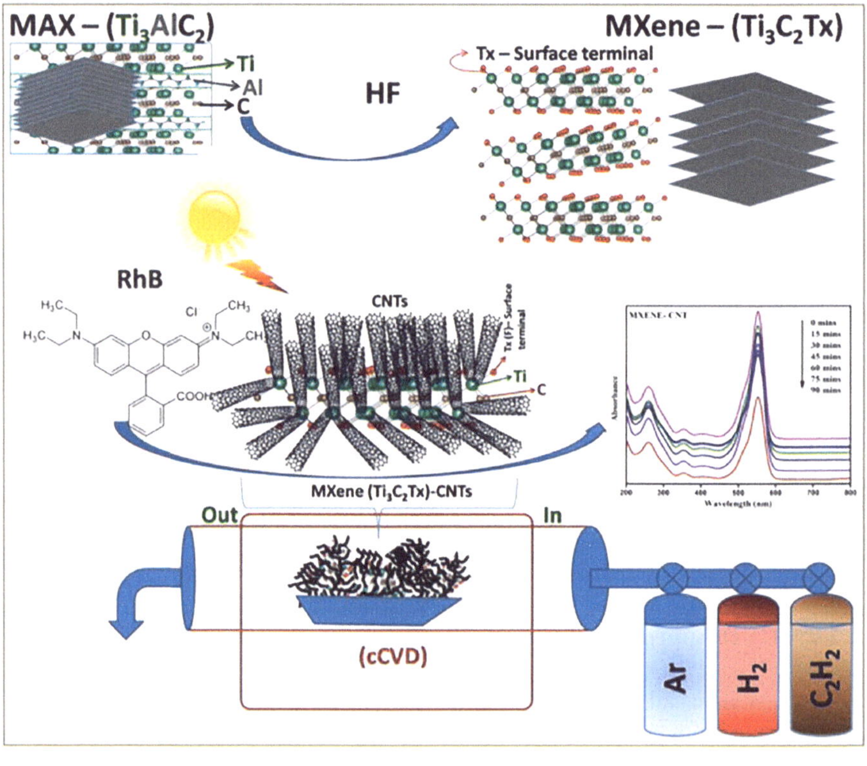

Fig. 2.4 The synthesis of $Ti_3C_2T_x$-CNTs hybrids by CVD technique with the UV absorbance spectra of MXene –CNTs-RhB. Reprocuced with permission from Thirumal et al. [22]. Copyright 2022, Elsevier Science Ltd.

substrate, along with the graphene formation upon the Cu surface prior to the formation of Mo_2C on the morphology and thickness of Mo_2C crystals, being extremely important for its synthesis management [23].

2.2 Application of Mxene-Based Materials for Wastewater Pollutant Removal

Adsorption is a frequently used water filtration method because it is simple to use, relatively inexpensive, very effective at removing contaminants, readily available in a wide range of adsorbent materials, and produces fewer toxic byproducts [24, 25]. To ensure that the procedure is effective and practical for purifying water, it is essential to have a high specific surface area, a wide range of pore sizes, chemical stability, as well as the capability to create the adsorbent material in large quantities [24, 25]. To remove different pollutants from wastewater, several nanomaterials have been used extensively, including graphene, carbon nanotubes (CNTs), zero-valent iron (ZVI), iron oxide, but also chitosan [24, 25]. Adsorption is a frequently used water purification method because it is simple to use, relatively inexpensive, very effective at removing contaminants, readily available in a wide range of adsorbent materials, and produces less toxic byproducts [24, 25]. To ensure that the procedure is effective and practical for purifying water, it is essential to have a large specific surface area, a wide range of pore sizes, stability (chemical), and the ability to create the adsorbents in large amounts.

Adsorptive, antibacterial, reductive, as well as metallic benefits are paired with high electronic conductivity, hydrophilicity, along with processability in MXenes. The use of derivatized MXene in wastewater remediation applications has gained interest in the present decade. In particular, Ti_3C_2Tx and its composites have shown effective removal but also sequestration of dyes, toxic metals, as well as radionuclides, along with functioning as antibacterial agents for water sterilization. As of today, the use of MXenes in the treatment of wastewater can be divided into three primary categories: adsorbents, membranes, and photocatalysts. Table 2.2 presents literature findings on the application of Mxene and its materials in water treatment.

2.2.1 Metals Removal from Wastewater

MXenes large specific surface area, plentiful functionalities of −OH as well as −O groups, and variable surface chemistry allow for successful competition with or superiority over other nanomaterials with respect to adsorption performance. Additionally, surface functionalities on MXene create locations for direct ion exchange but also reduce certain cations and organic compounds [7, 17]. MXenes now have a new potential use for unique reductive-adsorptive pollution remediation thanks to their in situ

Table 2.2 Adsorptive and photocatalytic application of MXenes-based materials

Mxene-based adsorbent	Application	Adsorption capacity	Photocatalytic degradation %	Recyclability		Refs.
				(%)	Cycles	
Remediation of metals						
$Ti_3C_2T_x$ Nanosheets	Cu(II) removal	78. 45 mg g^{-1}	–	30	3	[26]
$Ti_3C_2T_x$ Nanosheets	Cr(VI) removal, as well as ($K_3[Fe(CN)_6]$, $KMnO_4$ and $NaAuCl_4$) removal	250 mg/g	–	–	–	[27]
Urchin-like rutile Titania carbon (u-RTC) from ($Ti_3C_2(OH)_{0.8}F_{1.2}$)	Removal of Cr(VI) from water	225 mg/g	–	~90	5	[28]
$Ti_3C_2T_x$ MXene (MoS_2/MX-II)	Hg^{2+}	7. 21 mmol/g	–	37. 3%	5	[29]
Delaminated-$Ti_3C_2T_x$	Cu^{2+}	78. 45 mg g	–	–	–	[30]
Removal of phosphates/nitrate						
$Ti_3C_2T_x$	Phosphate ions removal	83. 564 mg/g	–	80	6	[31]
	Nitrate ions removal	66. 346 mg/g	–	80	6	[31]
MXene-Ca^{2+}	Phosphate ions removal	823. 6 mg/g	–	–	–	[32]
Removal of dyes						
$Ti_3C_2T_x$	Adsorption of MB plus it catalytic decomposition	19. 963	–	–	–	[33]
Magnetic $Ti_3C_2T_x$ nanocomposite	Hg(II) removal from water	1128. 41 mg/g	–	81. 04%	5	[34]

(continued)

Table 2.2 (continued)

Mxene-based adsorbent	Application	Adsorption capacity	Photocatalytic degradation %	Recyclability		Refs.
				(%)	Cycles	
Multilayered $Ti_3C_2T_x$	Uranium as well as nuclear waste removal	214 mg/g	–	–	–	[35]
Mxene	methylene blue (MB) and	140	–	<80	4	[36]
	acid blue 80	200	–	<80	4	[36]
2D MXenes	methyl orange (MO)	94. 8 mg/g	–	<80	5	[37]
Removal of pharmaceutical residue						
Ti_3C_2TX MXene	Amitriptyline (AMT)	58. 7 mg/g	100	~90	5	[38]
Sodium intercalated $Ti_3C_2T_x$ (SI-$Ti_3C_2T_x$)	Ciprofloxacin (CPX)	208. 2 mg/g	–	99. 7	5	[39]
$Ti_3C_2T_X$ MXene	Ddiclofenac					[40]
$Ti_3C_2T_X$ MXene	Verapamil (VRP)	–	85. 1	80. 9	4	[40]
Mxene-based Photocatalyst						
Degradation of dyes						
2D Ti_3C_2/TiO_2	MO dye photocatalytic degradation	–	97. 4	92. 50	4	[41]
$Ti_3C_2T_x$	Adsorption of MB plus it catalytic decomposition	19. 963	–	–	–	[33]

(continued)

Table 2.2 (continued)

Mxene-based adsorbent	Application	Adsorption capacity	Photocatalytic degradation %	Recyclability		Refs.
				(%)	Cycles	
$Ti_3C_2T_X$	MB	–	81. 2	–	–	[42]
$Ti_3C_2T_X$	MB	–	100	>80	5	[43]
Pharmaceuticals Degradation						
$Ti_3C_2T_X$ MXene	Diclofenac	–	100	91. 3	7	[5]
$Ti_3C_2T_X$ MXene	AMT	–	100	–	–	[44]
$Ti_3C_2T_X$ MXene	Ibuprofen (IBP)	–	100	–	–	[44]
$Ti_3C_2T_X$ MXene	CBZ	–	~60	–	–	[45]
$TiO_2/Ti_3C_2T_X$ (001-T/MX)	CBZ	–	98. 67	–	–	[45]
	Generation of hydrogen	*H_2 selectivity*	*Hydrogen generation rate*	*Catalytic stability*		
TiO_2/Ti_3C_2	Photocatalytic generation of hydrogen	Extremely high	17. 8 μmol h^{-1} $g_{catalyst}{}^{-1}$	25 h	3	[46]
Mxene	Photocatalytic generation of hydrogen	100%	401 mmol g^{-1} h^{-1}	45 h	–	[47]
CdS-nanosheet (NS)@titanium carbide (Ti_3C_2) MXene	Photocatalytic generation of hydrogen	–	1. 73 mmol h^{-1} g^{-1}	–	–	[48]

(continued)

Table 2.2 (continued)

Mxene-based adsorbent	Application	Adsorption capacity	Photocatalytic degradation %	Recyclability		Refs.
				(%)	Cycles	
"Ti_3C_2@CdS"	Photocatalytic generation of hydrogen	High	88. 162 μmol g^{-1} h^{-1}	18 h	3	[49]
CdS@Ti_3C_2	Photocatalytic generation of hydrogen	87. 61%	63. 5276 μmol h^{-1}	36 h	3	[50]
CdS/Ti_3C_2	Improved visible-light photocatalytic H_2 production	–	14,342 μmol h^{-1} g^{-1}	28 h	7	[51]

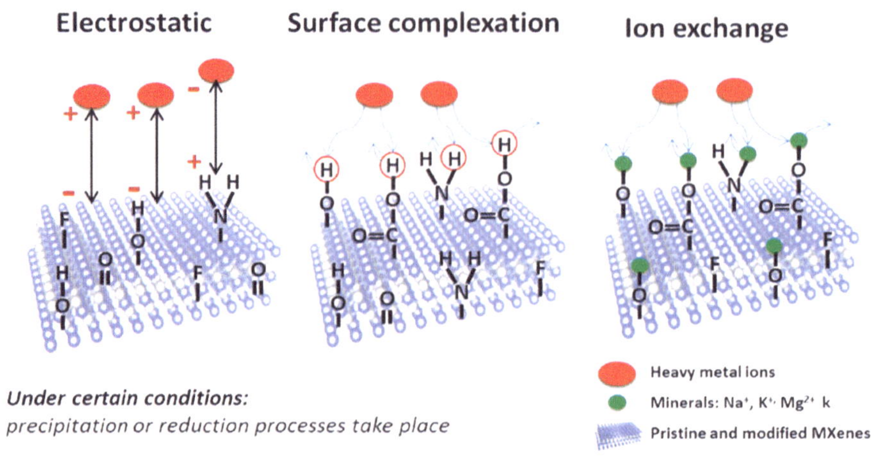

Fig. 2.5 Schematic illustrations of main adsorption mechanisms of heavy metal ions on MXene-based adsorbents. Reproduced with permission from Othman, Mackey, & Mahmoud [54]. Copyright 2022, Elsevier Science Ltd.

reduction capability and adsorption, which is seen as advantageous over many other nanomaterials.

Adsorbents made of MXene have distinct characteristics and architectures that have a high ability to bind a variety of heavy metals. Therefore, in order to further enhance their performance, it is crucial to understand their adsorption behavior and mechanisms. Adsorption mechanisms are used to describe how heavy metals interact with the molecules of the adsorbents. It is crucial to comprehend the physical or chemical interaction mechanism because it is related to the adsorption capacity, isotherms, kinetics, thermodynamics, and selectivity. The three primary forms that these processes can take are electrostatic interactions, ion exchange, and surface complexation. As shown in Fig. 2.5, they can vary depending on the kind of adsorbents used and the heavy metals involved.

Because of their very negative surface charges, MXenes have demonstrated outstanding ability to spontaneously intercalate polar organic molecules like hydrazine, urea, and DMSO as well as Na^+, K^+, NH^{4+}, Mg^{2+}, and Al^{3+} cations between their layers (zeta-potential between 30 and 80 mV). This suggests the possibility of using them as adsorbents.

Depending on the wet-chemical etching technique, the resulting MXene sheets often have surfaces rich in functional groups like –OH, –F, O, but also –Cl. The sheets become negatively charged as a result, which aids in the heavy metal ions' electrostatic attraction to them. Jun et al. evaluated the $Ti_3C_2T_x$ MXene's selectivity for the removal of Ba(II) and Sr(II) metals both individually and in mixtures. MXene has the ability to remove Ba(II) and Sr(II) metals from single electrolyte solutions by over 85%. It was noted that this elimination efficiency was high and broadly comparable. In mixed electrolyte solutions, MXene removed Sr(II) more efficiently than Ba(II) in mixed-ion solutions [52]. Due to Sr(IIstronger)'s electronegativity

(1.0) compared to Ba(II(0.9),)'s, it was able to connect electrostatically with MXene's negative surfaces more effectively. Similar to this, in another work, MXene preferred to adsorbed Pb(II) and Cu(II) ions over Zn(II) and Cd(II) ions from a mixed electrolyte solution due to the metallic electronegativity of heavy metals [53].

Lead (Pb) [55], copper (Cu), chromium (Cr) [27], Cu(II) [26], and mercury [56], among other heavy metals, have all been shown to be capable of being absorbed by MXenes. MXenes' surface groups interact with the specific pollutant to absorb it [56]. In contrast to other 2D and porous materials, the MXene nanosheets' possess a small interlayer spacing <2 Å resulting in significant cation trapping with radii of 4.5 Å [57]. The intercalation of cations and/or tiny organic molecules may result in an even greater expansion of the interlayer space by enhancing the c-LP for faster adsorption. A reversible rise in the c-LP with monolayer to triple layers of water inserted between MXene nanosheets can be caused by the presence of hydrophilic groups and weak bonding forces in hydrated MXene, allowing more ions to collect between the sheets [56].

Functionalization, which affects a material's physical characteristics and capacity to adsorb specific species, allows the surface of MXenes to be customized. By soaking $Ti_3C_2T_X$ in a solution of 5% sodium hydroxide (NaOH) for two hours, Peng et al. [55] created alk-MXene ($Ti_3C_2(OH/ONa)_XF_{2-X}$) with activated hydroxyl groups on the surface. Pb^{2+} was removed selectively using the alk-MXene, which demonstrated good average adsorption capacity as well as selectivity towards Pb^{2+} adsorption in the presence of high concentrations of competing cations, including Ca^{2+} and Mg^{2+} [55]. Alk-layered MXene's structure and the abundance of activated Ti–OH sites, where Pb^{2+} has been trapped in a hexagonal hydroxyl trap by hydroxyl and oxygen groups, were credited with this high removal capacity. According to DFT predictions, alk-MXene can be a great option for the adsorption of harmful heavy metals due to its negative energies formation towards the adsorption of Zn^{2+}, Pb^{2+}, Pd^{2+}, Cu^{2+}, and Cd^{2+} [58].

DFT was used to investigate the removal of Pb^{2+} by M_2X-type MXenes and the function of the M and X sites (where M = Zr, Sc, V, Ti, Nb, Cr, Hf, Mo, Ta but also X is carbon/nitrogen) [28]. $Ti_2C(OH)_2$ was anticipated to be the most effective MXene for heavy metals removal out of all $M_2X(OH)_2$ structures. With a predicted theoretical adsorption capacity of 2560 mg/g, the adsorption behavior of Pb^{2+} and Cu^{2+} on V_2C and Ti_2C MXenes has been further explored by molecular dynamics (MD) simulation [59].

MXenes' external surface groups not only offer direct ion exchange sites, but may also decrease some organic compounds and cations like Cu^{2+}, Ag^{+}, and Cr^{6+} [27], etc. This in-situ reduction capability exposes a brand-new, highly-promising MXenes function for pollutant removal. For instance, DL-$Ti_3C_2T_X$ nanosheets showed effective reductive removal of Cr^{6+}, with a maximal removal capacity of 250 mg/g and a residual concentration of Cr^{6+} after treatment of fewer than 5 parts per billion (ppb) [27]. This method was able to extract Cr^{3+} from an aqueous solution by adsorption while also reducing Cr^{6+} to Cr^{3+} in a single step. Additionally, $K_3[Fe(CN)_6]$, $KMnO_4$, and $NaAuCl_4$ can be reductively removed from a variety of water/wastewater streams using $Ti_3C_2T_X$ nanosheets [27]. The selective reduction of bromate (BrO_3^-) to bromide (Br^-) in aqueous media was also accomplished using $Ti_3C_2T_X$, with a reduced capacity of 321.8 mg BrO_3^- per g $Ti_3C_2T_X$ at pH 7 and 25 °C. The Ti-C outer layer of $Ti_3C_2T_X$ reacted with BrO_3^- in the reduction mechanism, causing a partial but irreversible oxidation of $Ti_3C_2T_X$ to TiO_2 and a subsequent reduction of BrO_3^- to Br^- [60]. According to Shehzad et al., DL-$Ti_3C_2T_x$ efficiently and quickly removed Cu^{2+} from the aqueous medium. The reduction of Cu^{2+} to Cu_2O and CuO, followed by the adsorption of reduced products, was made possible by the surface functional groups on MXene nanosheets. With 2.7-fold better reuseability than commercially activated carbon in this procedure, DL-$Ti_3C_2T_X$ had a experimental adsorption capacity of 78.45 mg/g.

The synthesis of nanocomposites of MXene with other appropriate materials has further increased the stability and recyclability of MXene. One example is the layered rutile TiO_2-C/TiC nanocomposite (u-RTC) that Zou et al. created by in situ solvothermal alcoholysis of $Ti_3C_2T_X$. The u-RTC demonstrated selective adsorption of Cr^{6+} with a maximum removal capacity of 225 mg/g, which was significantly higher than the pristine $Ti_3C_2T_X$'s removal capacity of 62 mg/g in the presence of anions like chloride (Cl^-), sulfate (SO_4^{2-}), and nitrate (NO_3^-). By bridging oxy groups and creating a potent inner-sphere complexation with $Cr_2O_7^{2-}$ ions, u-RTC removes Cr^{6+} by this method. Additionally, the high adsorption capacity of the u-RTC nanocomposite of 1120 kg kg^{-1} demonstrated outstanding efficiency in the treatment of industrial wastewater [28]. For the adsorption of mercury ions (Hg^{2+}) from the model and actual industrial effluent, a hydrothermally created $Fe_2O_3/Ti_3C_2T_x$ nanocomposite was used [34]. With a maximum adsorption of 1128.41 mg/g, the nanocomposite displayed remarkable Hg^{2+} removal efficiency. For the removal of phosphate from aqueous solutions, Zhang et al. used a comparable magnetic $Fe_2O_3/Ti_3C_2T_x$ nanocomposite and demonstrated outstanding sorption selectivity, quick kinetics, and highly selective removal capability. With regards to the criterion (0.5 mg/L) of the wastewater discharge standards, the composite was able to cleanse 2400 kg of water with phosphate per kg of $Fe_2O_3/Ti_3C_2T_x$. The synergy between $Ti_3C_2T_x$ and Fe_2O_3 nanoparticles (NPs) with Fe–OH sites was thought to be the cause of the high adsorption capacity [61].

Due to their substantial specific surface areas, hydrophilicity, and unique surface functionalities, delaminated-$Ti_3C_2T_x$ have shown promising potentials for the removal of copper (Cu) [30]. The reductive adsorption of Cu^{2+} to create Cu_2O and CuO species may be facilitated by oxygenated moieties in the layered structure of MXene. The greatest experimental adsorption capacity of these produced MXenes was 78.45 mg/g, and 80% of the total quantity of metal ions was adsorbed within one minute. They also demonstrated remarkably quick copper absorption. The Freundlich adsorption isotherm and a pseudo-second-order kinetic model, respectively precisely determined the maximum copper uptake onto the adsorbent material. Delaminated-$Ti_3C_2T_X$ had an adsorption capacity of 2.7 times more than commercially available activated carbon, further demonstrating through thermodynamic analysis that the adsorption process was endothermic [30].

Due to limitations in the adsorption capabilities of MXenes and MXenes-based composites, it is crucial to develop simple methods for making MXenes-based composites with greater adsorption capacity [62]. To do this, levodopa (DOPA) and MXenes ($Ti_3C_2T_X$) were mixed under relatively mild reaction conditions to create MXenes-based polymeric composites. In this reaction system, DOPA could self-polymerize to cling to MXene surfaces, and different carboxyl groups could be introduced during the polymerization process to produce poly (DOPA). In comparison to raw $Ti_3C_2T_X$, the produced composites showed greater adsorption capability in the removal of heavy metal ions. Because multiple reactive functional groups were added to MXenes, it is noteworthy that these composites might be further functionalized with other functional materials via a subsequent reaction [62].

To remove harmful mercuric ions (Hg^{2+}), a heterogeneous nanoadsorbent made of 2D $Ti_3C_2T_X$ MXene nanosheets functionalized with nanolayered molybdenum disulfide was created [29]. Mercury was adsorbed by the composite's oxygenated terminal groups of $Ti_3C_2T_X$ and the sulfur (disulfide) present in the mixture. The composite's ability to be removed was enhanced by ultrasonication thanks to an increase in the MXene nanosheets' surface area and interlayer distance. The reduction of 50 mol/L of Hg^{2+} to 0.01 mol/L in around 120 s was found to be an unheard-of kinetic behavior for mercury adsorption. Additionally, the proposed composite demonstrated significant recyclability because even after five cycles, 0.08 mM of Hg^{2+} was completely removed [29]. In order to enhance the adsorption capabilities of MXene nanosheets employed to remove Pb(II) ions, Wang et al. reported the usage of biosurfactants [63]. The Ti_2CT_X nanosheets were functionalized using chitosan, a cationic surfactant, lignosulfonate, an anionic surfactant, and enzymatic hydrolysis lignin, a nonionic surfactant. The enzymatic hydrolysis lignin-functionalized Ti_2CT_X displayed a maximum adsorption capacity of 232.9 mg g1 for Pb(II) ions; it was discovered that nonionic surfactant could prevent the nanosheets from restacking and introduced active functional groups for increasing their adsorption activity [63]. In order to enhance the adsorption capabilities of MXene nanosheets employed to remove Pb(II) ions, Wang et al. reported the usage of biosurfactants [63]. The Ti_2CT_X nanosheets were functionalized using chitosan, a cationic surfactant, lignosulfonate, an anionic surfactant, and enzymatic hydrolysis lignin, a nonionic surfactant. The enzymatic hydrolysis lignin-functionalized Ti_2CT_X displayed a maximum adsorption capacity of 232.9 mg/g for Pb(II) ions; it was discovered that nonionic surfactant could prevent the nanosheets from restacking and introduced active functional groups for increasing their adsorption activity [63].

2.2.2 Organic Contaminants Removal from Wastewater

Since the unearthing of 2D transition metal carbides and nitrides (MXenes) in 2011, there are now more nanomaterials that can be used for a wide range of purposes. High conductivity, hydrophilicity, and catalytic activity characteristic features of MXenes have sparked an increase in interest in the use of these compounds in water treatment as well as environmental remediation.

2.2.2.1 Adsorption of Dyes

The utilization of Mxene and their composites or hybrids for adsorptive removal of contaminants from wastewater has proven to be of great importance within the recent decade. Dyes or coloured contaminants removal using these systems has thus been of utmost interest to researchers globally [36]. MXenes have been reported to absorb Cr^{6+} and MO through chemical adsorption, including complexation, electrostatic adsorption, surface contacts, as well as ion exchange processes (Fig. 2.6) in a fashion similar to their adsorption mechanism for metals [37].

MXenes have reportedly displayed a remarkable capacity for the adsorption of MB and acid blue 80 due to their electrostatic attractions; these materials displayed quick kinetics and substantial selectivity [36]. As per this report, MXene and MOF had surface areas of 9 and 630 m^2/g, respectively, and their isoelectric points were roughly pH 3 and 9. As a result, at concentrations of the adsorbents and adsorbates of 25 and 10 mg/L, MXene and MOF demonstrated a high capacity for MB (140 mg/g) and AB (200 mg/g) adsorption, respectively, due to their electrostatic attractions. Furthermore, the authors revealed that hydrophobic interactions were primarily responsible for the MOF's ability to capture the MB [36].

In another report, a group of authors aimed to remove Cr^{6+} and MO from fluids; 2D MXenes have also been used as adsorbents [37]. The adsorption isotherm closely followed the Langmuir isotherm and also the highest exclusion efficiencies were 104 and 94.8 mg/g for Cr^{6+} and MO, respectively. It was also discovered that the rate of Cr^{6+} along with MO removal was very fast and that the kinetic system was driven by pseudo-second-order kinetics. The MXenes could be frequently recycled and were successfully regenerated by 0.1M NaOH aqueous solution.

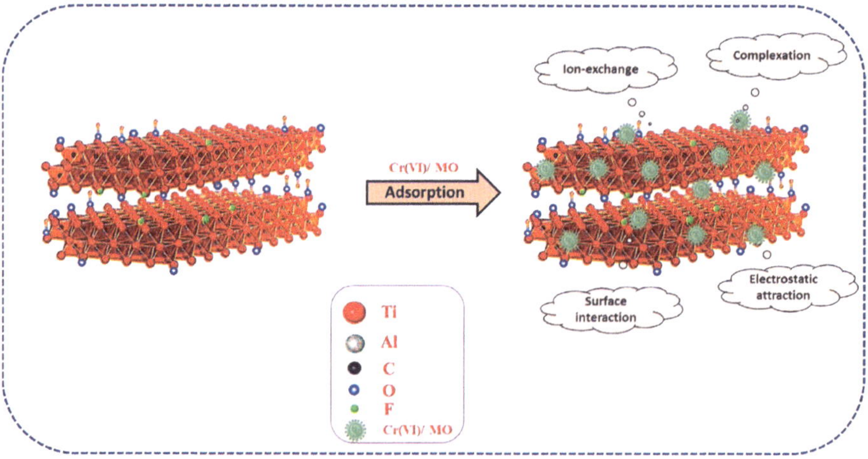

Fig. 2.6 Schematic representation of Cr(VI) adsorption and MO on MXenes. Reproduced with from Karthikeyan et al. [37], Copyright 2021, Elsevier Science Ltd.

Expanding the interlayer gap of MXene and tuning its surface functionalities using hot alkaline solution chemical treatment were two simple techniques used to increase the adsorption activity of $Ti_2C_2T_X$ MXene [64]. As a result, the treatment increased the MXene interlayer spacing by up to 29% using LiOH, and the surface functional groups were tweaked by changing −F to −OH. NaOH-$Ti_3C_2T_x$ has the maximum adsorption capacity (189 mg/g) for MB via surface and intercalation adsorption among the MXenes, followed by LiOH-$Ti_2C_2T_X$ and NaOH-$Ti_2C_2T_X$. According to analyses, a Langmuir isotherm could account for the improved adsorption toward cationic dyes [64]. In a different work, 2D MXenes adorned with Fe_3O_4 (2D-MX@Fe_3O_4) were created utilizing an in-situ growing technique in order to assess the impact of temperature on the efficacy of MXenes at removing dye [65]. The synthesized complex showed a typical 2D lamellar structure with superparamagnetic property (20.3 emu.g^{-1}), and zeta potential measurement showed a negatively charged surface of 2D-MXene@Fe_3O_4 at a neutral medium, which was advantageous for cationic dye removal. Furthermore, the removal process demonstrated an outstanding decolorization at high temperatures with 91.93% (55 °C) compared to other lower temperatures. Adsorption isotherms showed that the removal of MB at high temperatures (40 and 55 °C) and the removal at low temperatures (25 °C) both fit the Freundlich isotherm model well. Thermodynamic analyses showed exothermic and chemisorption processes for dye removal, indicating that a rise in temperature might considerably boost the produced complex's ability to remove the dye. According to the results, surface adsorption through electrostatic interaction contributed to the dye adsorptive exclusion process at 25 °C, while the MB decolorization was improved by hydrogen bonding (Ti−OHN) and electrostatic attraction in the high-temperature system, whereas the many Ti–OH groups on the complex surface played a major role in this [65]. Additionally, by thermally processing Ti_3C_2 nanosheets to add Fe_3O_4 nanoparticles (diameter: 5 nm) to them, magnetic MXene composites were created [66]. Thermodynamic analyses showed exothermic and chemisorption processes for dye removal, indicating that a rise in temperature might considerably boost the produced complex's ability to remove the dye. According to the results, surface adsorption via electrostatic interaction contributed to the dye removal process in the case of 25 °C. In contrast, the abundant Ti–OH groups on the complex surface played a dominant role in improving the MB decolourization via hydrogen bonding (Ti−OHN) and electrostatic attraction at high-temperature system [66]. Additionally, by thermally processing Ti_3C_2 nanosheets to add Fe_3O_4 nanoparticles (diameter: 5 nm) to them, magnetic MXene composites were created [66]. These composites were used in sophisticated oxidation techniques for the catalytic breakdown of organic pigments. The catalytic degradation findings showed that the MXene composite had a very high degradation efficiency toward several organic dyes under ideal circumstances. The degradation process was therefore found to involve both hydroperoxyl radicals (•OH) as well as superoxide radicals (•O_2) radicals, according to the electron spin resonance (ESR) spectra. Furthermore, for all batch trials, the ideal experimental conditions were as follows: Temperature = 40 °C, pH = 3, catalyst concentration was 0.5 g/L and H_2O_2 concentration was 10 mmol/L, respectively [66].

For the treatment of synthetic colors in model wastewater, ultrasonication-assisted $Ti_3C_2T_X$ MXenes were used [67]. As a result, the viability of using MXenes with ultrasonication assistance for wastewater treatment was examined using two different types of ultrasonication frequencies, 28 and 580 kHz. As a result, when compared to virgin and ultrasonication-assisted MXene at 580 kHz, 28 kHz MXene had more oxygenated functional groups and dispersion. These MXenes were used to remove positively charged MB but also negatively charged MO; they have strong MB adsorption activity. The notable benefits of ultrasonication-assisted MXenes for water treatment include their outstanding selectivity for positively charged target molecules and rapid kinetics at low doses. Although electrostatic interaction (attraction) between MXene and MB was the primary adsorption mechanism, physical properties (such as aggregation kinetics and hydrodynamic diameter) were measured through dynamic light scattering, found to be crucial factors in regulating the adsorption activity of the system. On the basis of the fourth adsorption–desorption cycle, MXenes that were helped by ultrasonication also showed considerable regeneration properties [67]. The extraordinary removal efficacy of MB by functionalized Ti_3C_2 with sulfonic groups (referred to as Ti_3C_2-SO_3H) was discovered (highest adsorption capacity = around 4 times). The results of the kinetics and isotherms analyses demonstrated that the experimental data were best explained by the pseudo-first-order model and the Langmuir isotherm adsorption model, respectively. These findings indicated that the process of MB adsorbing onto the surface of the adsorbents was endothermic and spontaneous. The capacity of the dye to bind to the aqueous solution was better when the pH of the solution was alkaline, according to experimental pH studies.

Another work used low toxicity etching chemicals to create Ti_3C_2 MXenes using a hydrothermal technique [78]. These MXenes have high c lattice parameters, greater interlayer distances, and higher Brunauer–Emmett–Teller (BET) specific surface areas because of the slow-release mechanism used during the hydrothermal process. The used etching method avoided using highly concentrated hydrofluoric acid and can be used for Ti_3C_2 flakes as well as other MXene materials (such as Nb_2C). MB and MO dyes were well-adsorbed by the produced MXenes [68].

The Gogotsi group were the first to report that cationic dyes, such as MB, may be efficiently adsorbated by ML-Ti_3C_2Tx and then decomposed catalytically [33]. According to the Freundlich model, MB was adsorbed with constant values for adsorption capacity (KF) and intensity (v) of 19.96 and 48.15, respectively. The production of TiO_2 following partial oxidation of $Ti_3C_2T_X$ and wedging of the layered structure may be the causes of the increase in stacking disorder brought on by the adsorption of MB molecules onto $Ti_3C_2T_X$ surfaces. In addition, the dye can be eliminated through a process called photocatalytic degradation, which is triggered when $Ti(OH)_4$ and/or TiO_2 develop on the surface of MXene when it is exposed to UV light. The quick electron–hole pair recombination, limited stability in aqueous conditions, and restricted light-absorption range of MXene have, however, all served to constrain its photocatalytic degradation activities. Due to this, the TiO_2/$Ti_3C_2T_X$ nanocomposite underwent a greater photocatalytic degradation of MO than TiO_2 and $Ti_3C_2T_X$ [69]. Compared to pure TiO_2 NPs and $Ti_3C_2T_X$, which were degraded by the same experimental conditions only 77 and 42%, respectively, MO was roughly 98%

degraded in 30 min. The $TiO_2/Ti_3C_2T_X$ nanocomposite's improved photocatalytic performance can be related to more efficient electron–hole separation and the creation of a $Ti_3C_2T_X$ and TiO_2 heterojunction in the composite when exposed to UV light, as opposed to TiO_2 or $Ti_3C_2T_X$ alone.

2.2.2.2 Adsorption of Pharmaceuticals

The application of Mxene and its composites/hybrids in adsorptive water treatment have been largely investigated by researchers globally within the recent decade. Researchers have postulated the adsorption mechanism of Mxene and its materials towards pharmaceuticals based on hydrogen bonding, hydrophobic interaction, and electrostatic interaction as represented in Fig. 2.7 [38].

The viability of several medicinal compounds being adsorbed on $Ti_3C_2T_X$ MXene was examined by Yoon et al. in their study [38]. The adsorption capacity for amitriptyline, therefore, showed the maximum value (58.7 mg/g) at pH 7 as a result of the electrostatic interaction between negatively charged MXene and positively charged amitriptyline. Additionally, MXene sonicated at different frequencies (0, 28, and 580 kHz) was used to increase adsorption activity. Amitriptyline's maximal adsorption capacity was discovered in the order of 28 kHz (214 mg/g) preceding 580 kHz

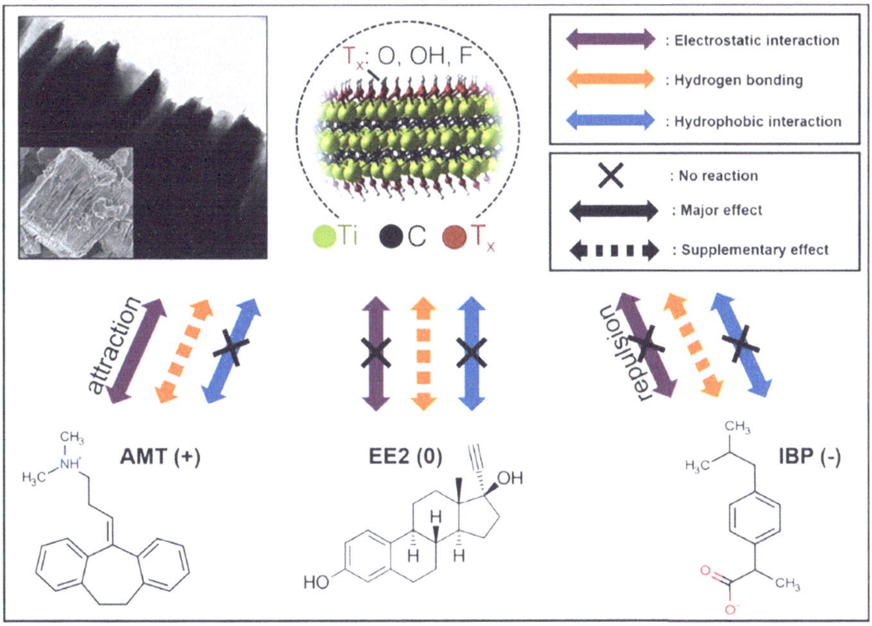

Fig. 2.7 Schematic diagram for plausible adsorption mechanisms of selected pharmaceutical compounds by MXene. Reproduced with permission from Kim et al. [38], Copyright 2021, Elsevier Science Ltd.

(172 mg/g) and 0 kHz (138 mg/g). This is noteworthy because sonication cavitation bubbles can produce well-dispersed MXene and more by generating oxygenated functional groups on MXene. The strongest activity, in particular, was shown at lower frequencies by forming larger cavitation bubbles. Although background inorganics had a negative impact on the adsorption activity, background organics such as natural organic matter had the potential to improve it. Furthermore, by demonstrating a competitive effect between cetylpyridinium chloride and pharmaceutical chemicals, cetylpyridinium chloride, a cationic surfactant used in the pharmaceutical industry, demonstrated a negative effect on adsorption activity [38].

Using a one-step chemical co-precipitation approach and ammonium bifluoride as a mild etchant, magnetic titanium carbide ($Ti_3C_2T_X$) MXene materials were created [5]. Through the use of ultraviolet (UV) and chlorine, these materials were used to photocatalytically degrade diclofenac. Comparing the diclofenac degradation caused by UV and chlorination treatment alone, as well as UV/H_2O_2 operations at pH 7, the diclofenac degradation was improved by the formation of active radicals, including hydroxyl radical and reactive chlorine species. As a result, the UV/chlorine process' first-order rate constant was 0.1025 min^{-1}, which was 12.7 and 6.8 times more than its corresponding values for the UV/only and UV/H_2O_2 processes. Using a one-step chemical co-precipitation approach and ammonium bifluoride as a mild etchant, magnetic titanium carbide ($Ti_3C_2T_X$) MXene materials were created [5]. Through the use of ultraviolet (UV) and chlorine, these materials were used to photocatalytically degrade diclofenac. Comparing the diclofenac degradation caused by UV and chlorination treatment alone, as well as UV/H_2O_2 operations at pH 7, By producing active radicals like the hydroxyl radical but also reactive chlorine species, the breakdown of diclofenac was improved. As a result, the UV/chlorine process' first-order rate constant was 0.1025 min^{-1}, which was 12.7 and 6.8 times more than its corresponding values for the UV/only and UV/H_2O_2 processes. Notably, adding magnetic nanoparticles to the surfaces of MXene nanosheets could increase the developed composites' capacity for adsorption and speed up the flow of electrons between molecules in solution. Throughout seven successive rounds of the degradation procedure, the proposed MXenes showed astounding stability and photodegradation efficiency [5]. Since it is still in the early stages and is being tested in lab settings, MXenes should be further researched for the removal of medicinal substances. Various MXenes materials and composites should be taken into consideration for this application. It was discovered that sonicated MXenes outperformed unprocessed MXenes in terms of reuse and removal efficiency. Notably, adding magnetic nanoparticles to the surfaces of MXene nanosheets could increase the developed composites' capacity for adsorption and speed up the flow of electrons between molecules in solution. Throughout seven successive rounds of the degradation procedure, the proposed MXenes showed astounding stability and photodegradation efficiency [5]. Since it is still in the early stages and is being tested in lab settings, MXenes should be further researched for the removal of medicinal substances. Various MXenes materials and composites should be taken into consideration for this application. It was discovered that sonicated MXenes outperformed unprocessed MXenes in terms of reuse and removal efficiency [38].

2.2.3 Oil/Water Separation

The environment and human health may be endangered by oily wastewater from industrial water discharge and oil spills. MXenes' distinct layered structure and adaptable interface chemistry made them useful as nano-fillers for enhancing the properties of polymers [70]. One study used freeze-drying of polyamide acid/MXene solutions and thermal imidization to create sturdy, light, and hydrophobic 3D structures out of polyimide/MXene. Ti_3AlC_2 was etched and then exfoliated using ultrasound to create the MXene ($Ti_3C_2T_X$) dispersion [70]. As a result of the strong connections between polyimide chains as well as MXene nanosheets, interconnected, extremely porous, as well as hydrophobic aerogels having low density were created, resulting in improved compressive activity, a high capacity for absorbing oil, and effective separation of oil and water. After 50 compression-release cycles, the designed aerogel miraculously recovered to its original height, demonstrating extremely high elasticity and exceptional fatigue resistance. In addition, high absorption capabilities for various organic liquids were discovered, ranging from roughly 18 to 58 times their weight. These recently created structures have good thermal stability and reusability in hostile settings and can quickly extract liquid paraffin, soybean oil, and chloroform from the water–oil system [70].

For oil-in-water emulsion separation, ultra-thin 2D titanium carbide MXene membranes (30 nm) supported on porous polyethersulfone were created. These membranes have excellent activity, stability, and recyclability, but also separation activity [100]. Particularly, the MXene membranes' exceptional antifouling resistance, alluring oil/water separation, and high reusability were derived from their inherent hydrophilicity, low oil droplet adhesion, and regular stacking of the 2D lamellar architecture; the MXene membranes also demonstrated exceptional oil rejection with low compromised water permeance, because of the reduced d-spacing due to the ion intercalation [71]. These MXene membranes can be used to separate oil-in-water emulsions because they have a large hydrophilic surface area, little oil adhesive force, and nm-sized interlayer spacing for water transportation [71]. A water layer can stick to the MXene layer surface, and oil droplets can be repelled thanks to its relatively rough surface structure, strong hydrophilicity, and under-water oleophobicity. The layered MXene membranes have nm-sized slit channels with compact and coordinated structures that restrict and block the entry of oil droplets and also promote rapid water transport. The ion intercalation capabilities of MXene membranes can enhance the separation activity further in the case of the challenging separation of oil-in-salt water emulsions by shrinking the interlayer spacing, which results in higher oil removal efficiency with only marginally lessened water permeance [71].

$Ti_3C_2T_X$ MXene was used as the functional layer, and regular print paper as the substrate in Saththasivam et al.'s design of inexpensive, flexible membranes [72]. A membrane that is considerably hydrophilic and oleophobic and has an underwater oil contact angle of 137° was created utilizing a straightforward coating technique using MXene ink. The design of MXene membranes was also reported by coating $Ti_3C_2T_X$ MXene 2D nanosheets carbides onto porous PVDF films through vacuum filtration. These membranes showed a remarkable separation efficiency for oil/water emulsions of over 99% and a high water permeation flux of over 450 L per m^2 per h per bar [72]. With oil contact angles close to 158° and oil slide angles under 7°, the suggested membranes showed outstanding underwater oleophobicity. Additionally, various stable emulsions, including emulsified crude oil-in-water mixtures, were separated using multilayer MXene membranes, demonstrating an impressive penetration flow of 887 L/m^2/h bar1 and considerable separation efficiency of 99.4%. These membranes were extremely resistant to corrosive liquids like acidic, alkaline, and salty ones and could also extract oil droplets from water in corrosive environments [73].

Facet construction was used to create a multitasking hierarchical MXene@UIO-66-$(COOH)_2$ composite membrane supported on a nylon 66 microporous substrate with good chemical stability and effective multi-component pollutant-oil–water emulsion separation activity [74]. This composite membrane's inherent hydrophilicity, low oil adhesion property, and hierarchical intercalation structure allowed it to successfully separate a variety of oil-in-water emulsions and dyes while also exhibiting good antifouling resistance and high recyclability. Additionally, under strong acid (3M HCl), alkaline (1M NaOH), and salty (Saturated NaCl) conditions, this chemically stable composite membrane displayed significant corrosion resistance [74]. A multifunctional hierarchical MXene@UIO-66-$(COOH)_2$ composite membrane with good chemical stability as well as efficient multi-component pollutant-oil–water emulsion separation activity was made through facet construction. It was supported on a nylon 66 microporous substrate [74]. This composite membrane's inherent hydrophilicity, low oil adhesion property, and hierarchical intercalation structure allowed it to successfully separate a variety of oil-in-water emulsions and dyes while also exhibiting good antifouling resistance and high recyclability. Additionally, under strong acid (3M HCl), alkaline (1M NaOH), and salty (Saturated NaCl) conditions, this chemically stable composite membrane displayed significant corrosion resistance [74]. In a further investigation, multilayered and single-layered MXenes ($Ti_3C_2T_X$) were used to separate water/vegetable oil emulsions. It was found that underwater, the membranes were significantly oleophobic (>124°), in contrast to the air, where they showed substantial oleophobicity (5°). The isolation efficiency of water–oil emulsions for both kinds of coated membranes was achieved above 98%, with a surface coverage of 3.2 mg/cm^2 for multilayered MXene and 2.9 mg/cm^2 for single-layered-MXene. For separation membranes coated with 2.65 mg/cm^2 of multilayered-MXene, the separation efficiency was greater than 98%, while it was less than 90% for membranes with 1.89 and 0.77 mg/cm^2 at all tested emulsion concentrations [75].

Innovative hybrid systems, such as polyimide aerogels having MXenes, can separate water and oily compounds while maintaining acceptable safety properties, thermal stability, as well as high reusability [70]. MXenes membranes, such as the $Ti_3C_2T_X$ membrane, also showed high durability to corrosive liquids, alongwith alkaline, acidic, as well as salty, but also could effectively eliminate oil droplets from water-based corrosive environments.

2.2.4 Photocatalysis

Photocatalytic degradation of colourants/dyes using Mxene-based materials has been of great interest to researchers in recent decades. We herewith discuss a handful of Mxene-based photocatalysts for colorant/dye degradation as per available literature.

A group of researchers in their work have shown the degradation of MB and Rhodamine B (RB) along with their removal using a straightforward solvothermal process, resulting in the creation of a cube-shaped Co_3O_4 particle self-assembled modified Ti_3C_2 MXene nanocomposite [76]; the Ti_3C_2 sheets' surface and interior layers had evenly distributed Co_3O_4 particles, which prevented the sheets from being stacked again and created a well-organized composite structure. The ability of the created nanocomposites to be recycled for roughly eight consecutive cycles shows potential uses in composite materials and water and wastewater treatment [76]. Notably, MXene-based nanocomposites' chemical alterations and well-organized self-assembly revealed their crucial significance for various applications [77].

For the removal of carbamazepine, a hybridized photocatalyst made of $Ti_3C_2T_X$ MXene nanosheets was developed [45]. The produced photocatalyst's heterostructure was obtained by carefully controlling the oxidation action through the Schottky junction established between the TiO2-MXene surfaces. The dye degradation capacity was greatly reduced under acidic pH 3.0–5.0, and the assessed carbamazepine Kapp value of under ultraviolet light and found to be 0.0304 min^{-1}, higher than that under natural sunlight. The carbamazepine molecule was damaged by radicals OH and O_2 during the photocatalytic degradation; the degradation process was proposed [45].

A typical photocatalyst should have good carrier mobilities, sufficient bandgaps, visible-light absorption, the high redox potential of photoinduced holes, and suitable bandgaps. MXenes display metallic characteristics because they have a high Fermi-level electron density [7]. Additionally, MXene's large and anisotropic carrier mobility makes it easier for photogenerated electron–hole pairs to move and separate, which makes it promising for photocatalytic uses [5]. MXenes' bandgaps can be adjusted to match the UV light spectrum by changing their surface terminations, with a 50% photothermal conversion efficiency [5, 51]. Besides changing the surface groups, it can be helpful to arrange the surface groups in relation to the M atoms to meet the requirements for photocatalysis application [51]. MXenes, which are very metallically conductive, have also been thought of as a potential substrate for surface-enhanced Raman spectroscopy (SERS) because they exhibit a strong localized surface plasmonic resonance (LSPR) effect [78]. MXenes were projected to exhibit light absorption at wavelengths between 300 and 500 nm due to their bandgap, which ranges from 0.92 to 1.75 eV [6]. Theoretically examined for photocatalysis were the structural, electrical, as well as optical properties of MXenes (M_2CT_2 where $M = Ti$, Hf, or Zr; $T = -O-$, $-F$, or $-OH$) with various geometries [6]. The M_2CT_2 that has the functional groups on the opposite sides of the metal atoms is best for photocatalytic use. This finding implies that Ti_2CO_2, Zr_2CO_2, and Hf_2CO_2 may be useful in visible-light-driven photocatalysis [6]. As a result of their outstanding anisotropic carrier mobility, strong optical absorption at both visible as well as ultraviolet wavelengths, but also stability in water, 2D Zr_2CO_2 and Hf_2CO_2 are said to be equally employable as prospective photocatalytic materials for water splitting [79]. Through the use of computational techniques, the influence of functional groups on the photo-electrochemical characteristics of Sc_2CT_2 MXene and the potential photocatalytic activities of the variously terminated compounds was evaluated [80]. Sc_2CF_2 has a good band structure and a stronger redox capability of photo-activated carriers, resulting in better photocatalytic performance, according to comparisons of the photocatalytic properties of Sc_2CF_2, Sc_2CO_2, and $Sc_2C(OH)_2$ structures with similar (energetically) advantageous geometries. $TiO_2/Ti_3C_2T_X$ nanocomposites have also been utilized to make hydrogen through photocatalytic water splitting [46]. The $Ti_3C_2T_X$ in this nanocomposite served as an electron sink to encourage the separation of photo-generated charge carriers and to offer a 2D platform for close interactions with TiO_2 NPs. The $TiO_2/Ti_3C_2T_X$ (5 wt.%) composite outperformed pure rutile TiO_2 in the photocatalytic hydrogen evolution reaction by a factor of 400 when exposed to visible light.

A powerful method to boost photocatalytic performance is generally thought to be exposing the extremely active planes to the photo-excitation source. The photocatalytic activity of a composite made of layered $Ti_3C_2T_X$ and (001) faceted TiO_2 nanosheets was assessed [41]. High-efficacy photogeneration of electron–hole pairs is possible thanks to the composite's highly active (001) TiO_2 facets. Additionally, the interfacial Schottky junction, which serves as a hole reservoir and has the effect of trapping holes, considerably enhances carrier separation. According to DFT calculations, the overlap between the Ti d-orbital and the O p-orbital caused the interface to form, as illustrated. Compared to p-$TiO_2/Ti_3C_2T_X$, the higher photocatalytic activity of (001) $TiO_2/Ti_3C_2T_X$ resulted in a 2.3-fold increase in the photocatalytic degradation of MO. When compared to TiO_2 (4.924 eV for (001) surface and 6.578 eV for (101) surface, which may also produce a Schottky barrier through their interfaces, the work function of −OH-terminated $Ti_3C_2T_X$ sheets (1.8 eV) is lower. Instead of injecting electrons, the photogenerated holes may come from TiO_2 and go to Ti_3C_2, which had a −OH termination.

$Ti_3C_2T_X$ has also been used as metal sulfide photo-absorbers for the formation of visible-light photocatalytic hydrogen [51]. $Ti_3C_2T_X$ had its −F terminations changed to −O/−OH terminations using a hydrothermal process, and the resulting structure was then linked with cadmium sulfide (CdS) to create a $CdS/Ti_3C_2T_X$ nanocomposite. With an apparent quantum efficiency of 40.1% at 420 nm, this composite demonstrated a high photocatalytic hydrogen generation activity of 14,342 mol h g I in visible light. In order to achieve superior photoredox efficiency in photocatalysis and energy conversion applications, it is essential to design a well-contacted heterojunction interface between MXene and the other NPs. This understanding is provided by the work being done to develop MXene-based nano-architectures. The fact that MXenes, such as $Ti_3C_2T_X$, absorb light in both the UV and visible ranges and exhibit plasmon resonance peaks in either the visible or IR range is also significant to note [81]. They may convert light to heat effectively as a result. Consequently, some of the effects attributed to photo-catalysis may be caused by heating and thermally accelerating chemical reactions in accordance with Arrhenius law.

Photocatalytic degradation of pharmaceutical residues such as diclofenac have been proposed by Jang et al. [5], as depicted in Fig. 2.8, where the authors postulated that the main mechanism of degradation was through hydroxylation, C-N bond cleavage, ring opening, dichlorination, as well as decarboxylation. They projected $Ti_3C_2T_X$ as an excellent recyclable but also an efficient photocatalyst for diclofenac degradation.

In another work, Shahzad and colleagues [45] revealed that with the aid of UV irradiated light, Schottky junction formed $TiO_2/Ti_3C_2T_X$ (001-T/MX) composite could degrade 98.67% of carbamazepine (CBZ) where the photocatalyst degradation

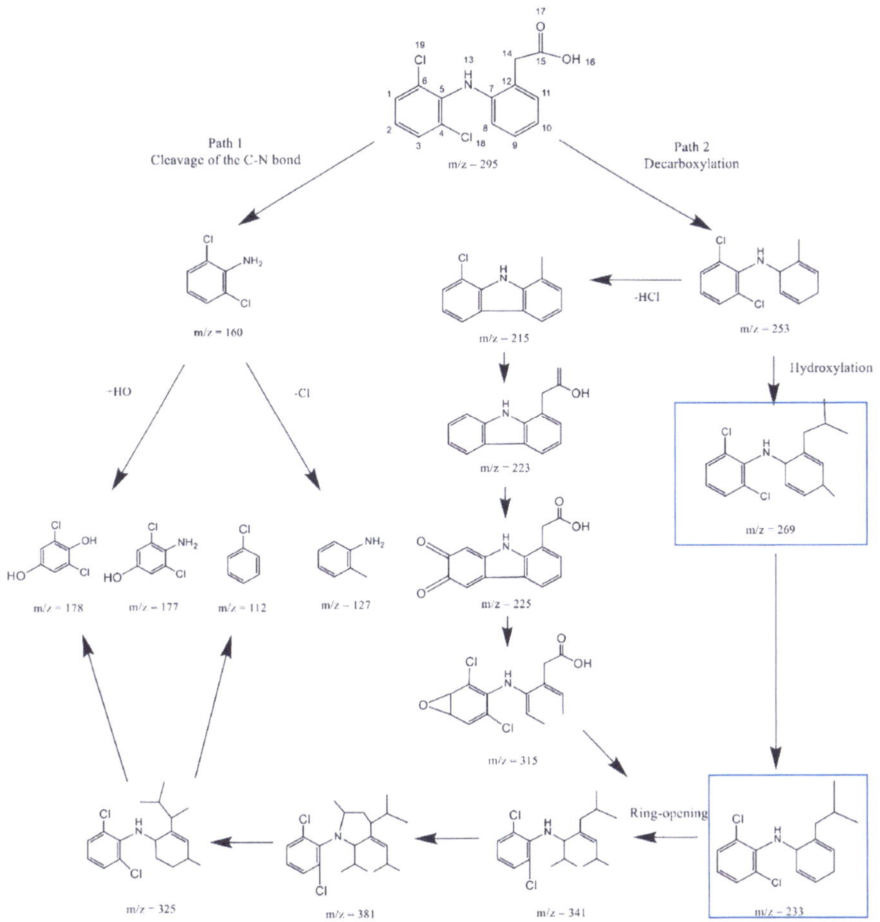

Fig. 2.8 Proposed pathway of DCF degradation by UV/chlorine process. Reproduced with permission from Jang et al. [5], Copyright 2020, Elsevier Science Ltd.

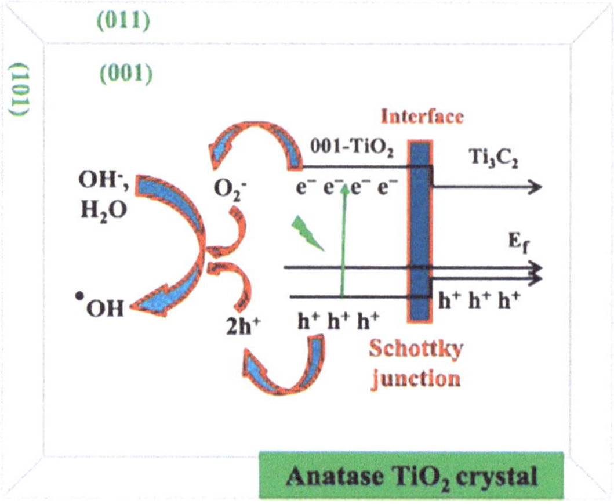

Fig. 2.9 Schematic illustration of the degradation mechanism of CBZ by the 001-T/MX photocatalyst. Reproduced with permission from Shahzad et al. [45], Copyright 2018, Elsevier Science Ltd.

capacity governed pH 3.0–5.0 enabling radical. OH and radical. O_2 attacked the CBZ molecule, as depicted in Fig. 2.9. These authors observed that while the virgin Mxene degraded only ~60% of CBZ, its composite 001-T/MX outperformed it by degrading 98.67% of the contaminant as earlier conversed [45].

As per literature, $Ti_3C_2T_X$ has been the main Mxene material largely adopted by researchers globally as either adsorbent or photocatalyst: this calls for more digging towards developing other forms of Mxenes having comparable or better performance towards the same application.

2.2.5 Removal of Radionuclides

For the removal of radioactive pollutants and impurities, researchers have developed novel architectures and composites [82–84]. MXenes have demonstrated strong resistance to radiation-induced damage and acceptable thermal stability; as a result, it is anticipated that they will function well as adsorbents for the removal of radionuclides from nuclear waste. $Ti_3C_2(OH)_2$ and V_2CT_X MXenes have been theoretically predicted to remove uranyl ions $[UO_2(H_2O)n]^{2+}$ effectively using DFT [85]. Additionally, theoretical calculations predicted a maximum capacity for uranium adsorption of 536 mg/g on $V_2C(OH)_2$ nanosheets, which is a significant improvement over the investigational isolative performance of 174 mg/g [86]. The authors postulated that uranyl ions, as well as the OH functionalities of V_2CT_X nanosheets have a strong connection, and high elimination of all hydroxylated and carbonated uranyl species. However, because the U–F bonding on the adsorption site is weaker than the U–O interaction, fluoride-terminated MXenes were predicted to be less advantageous for uranyl adsorption applications. Another study looked at the actinide U^{6+} adsorption on DMSO intercalated $Ti_3C_2T_x$, and the results showed that the synthesized adsorbent had a Q_{Max} of 214 mg/g, which was five times greater than that of pure $Ti_3C_2T_x$ [35]. MXene nanosheets enhanced interlamellar spacing following intercalation along with hydration activation was attributable to the greater actinide U^{6+} removal. The theoretical value of the experimental removal capacity was around 54.5% greater, indicating that hydrated $Ti_3C_2T_x$ may be able to attain even higher adsorption capacities [35].

The effective exclusion of metallic ions, dyes, salts, as well as radionucleotides via adsorption was demonstrated elsewhere with both virgin and functionalized MXenes, to sum up, this section [87]. MXenes' adjustable abundance functionalities, surface chemistry, along with the ability to adapt to local electrical as well as chemical perturbations are all factors in the effective utilization of MXenes as efficient adsorbents. MXenes are also capable of in-situ oxidation, making them useful for the reductive exclusion of various ions. MXenes and their nanocomposites can also be effective photodegradation and adsorption agents various pollutants, including dyes, thanks to their great optical absorption. More varieties of MXenes with added physicochemical properties should be synthesized and functionalized in order to advance the research. As an illustration, the synthesis of $Mo_2TiC_2T_X$, $Mo_2Ti_2C_3T_X$, and $Cr_2T_iC_2T_X$ has theoretically predicted and proved the existence of two novel varieties of ordered double transition metal carbide MXenes, designated M′2M″C2 but also M′2M″2C3, respectively [87].

Al-based MOFs, CAU-1, created via the solvothermal method of complexation between 2-aminoterephthalic acid but also Al^{3+}, has been thought of as a useful material for the sequestration of radioactive contaminants from aqueous environments [88]. With respect to this, MXenes offer auspicious capabilities as adsorbents for radionuclides exclusion, such as cesium (Cs^+), U^{6+}, thorium Th^{4+}, palladium Pd^{2+}, as well as europium III (Eu^{3+}). For example, a number of MXene materials with sizable surface areas have been created and used to exclude Pd^{2+} from HNO_3 aqueous solution; investigations showed that the adsorption performance more closely matched the Langmuir model [89]. In simulated nuclear wastes, the produced MXenes were able to maintain outstanding Pd^{2+} selectivity and thermodynamic analysis revealed that Pd^{2+} low temperatures may enhance adsorption. Notably, after 5 cycles of adsorption, the Mxenes generated at 45 °C showed no discernible loss and excellent regeneration activity [89]. Furthermore, radioactive Cs^+ from model nuclear effluent was removed using the produced Mxenes ($Ti_3C_2T_X$) [90]. Because equilibrium was reached in less than an hour, this meant that Mxenes adsorption had exceptionally quick kinetics. At neutral pH conditions, they had a substantial Q_{max} (148 mg/g) at adsorbent as well as adsorbate concentrations of 5 as well 2 mg/L. NaCl, KCl, $CaCl_2$, and $MgCl_2$ were the four distinct ions that were studied in the Mxenes Cs^+ adsorption. It was discovered that ion exchange was involved in the Cs^+ adsorption by Mxene as a result of the Cs^+ removal rate changing in the presence of these components [90]. In a different study, a MXene composite was created using dried ceramic powder made from hydroxyapatite, and it demonstrated higher thermal stability than pure

MXene that adsorbs Cs. The normalized leaching rate of simulated radioactive Cs was therefore estimated to be 2.02 (±0.09) × 10^{-4} g/m^2/d. When combined with hydroxyapatite along with cold sintering at 200 °C, MXene that had been adsorbed by Cs was thermally stable [91].

Environmental contamination by $99Tc^{7+}$ from radioactive wastewater sources is a severe problem because of the extensively prolonged half-life of 99Tc and the great mobility of pertechnetate [92]. A 3D MXene-polyelectrolyte nanocomposite was developed in an intriguing work to remove perrhenate, a pertechnetate simulant [92]. Re(VII) exclusion capacity of ~363 mg/g along with quick sorption kinetics were produced by the addition of poly(diallyldimethylammonium chloride, or PDDA), which controlled the surface charge as well as increasing the stability of Ti_2CT_X nanoplatelets. Whenever competing anions, like Cl^- or SO_4^{2-}, exist together at a concentration of 1800 times, the produced nanocomposite showed appropriate selectivity for ReO_4^-. Batch sorption assessments and X-ray photoelectron spectroscopy were used to establish the immobilization mechanism as a sorption-reduction process. When the pH rose, the local environment gradually changed from octahedral-coordinated Re^{4+} to tetrahedral-coordinated Re, demonstrating the pH-dependent lowering capability of Ti_2CT_X/PDDA nanocomposite toward Re^{6+} and RE^{7+} [92]. The carboxyl functionalized $Ti_3C_2T_X$ exhibited high exclusion possibilities for U^{6+} but also Eu^{3+}, defined by very fast adsorption kinetics of 3 min, high Q_{Max} of ~344.8 mg/g for U and 97.1 mg/g for Eu), along with a significant exclusion percentage of the radionuclides from a test sample. The pseudo-second-order kinetic model but also the Langmuir adsorption isotherm model both accurately predicted U^{6+} and Eu^{3+} adsorption on MXenes. It was demonstrated that the adsorption of U^{6+} on TCCH followed an inner-sphere configuration, while the adsorption of Eu^{3+} was evaluated by both inner-sphere complexation and electrostatic interaction. A carboxyl-ended diazonium salt grafted naturally on MXene's surface. MXene supplied the diazonium salt with electrons, which led the diazonium to cleave along with nitrogen release, creating an aromatic radical (I) [93]. The H-atom on the surface of MXene then joined this free radical to form Ti–OH, which then turned into Ti–O radical (II). Ultimately, the Ti–O radical joined another recently produced aromatic radical to create a covalent mode of Ti–O–C. (III). Different carboxyl-modified MXenes (known as TCCH-1, TCCH-5, TCCH-10, TCCH-15, but also TCCH-20) were created by varying the reaction ratio of diazonium salt to MXene to 1:1, 5:1, 10:1, 15:1, but also 20:1 [93]. A 2D transition metal carbide, Ti_2CT_X, was also used to effectively remove uranium via a sorption-reduction approach; Ti_2CT_X showed exceptional U(VI) removal throughout a wide pH range, with an absorption capacity of 470 mg/g at pH 3.0. Bidendate binding to the MXene substrate allowed the decreased U(IV) species to be identified as mononuclear at low pH. Nanoparticles of the UO_{2+X} phases that have undergone some Ti_2CT_X conversion to amorphous titanium oxide adsorbs to the substrate at close to neutral pH. When treating wastewater from uranium mining, these materials can be thought of as prospective choices for porous reactive barriers [94].

For concurrent adsorption along with reduction of U(VI) from the aqueous mixture in anoxic circumstances, a novel composite was built employing zero-valent iron nanoparticles anchored upon alkalized Ti_3C_2Tx nanoflakes. The result was that the produced composite showed quick removal kinetics, notable selectivity, 100% removal efficacy, but also up to 1315 mg/g uptake capacity for U(VI) apprehension [95]. In comparison to other traditional pollution adsorbents, MXenes showed notable radionuclide removal potentials [96]. However, small interlayer spacing along with restacking of pure MXenes, might prevent radioactive nuclides whose hydrated ionic radii are greater than the d-space between the MXene sheets from adhering to them. By adding the right intercalants along with cross-linking particles, the interlayer spacing amongst the stacked MXenes sheets may be effectively changed for encasing the radionuclides with greater hydrated ionic radii, as well as the adsorption potential of MXenes could be noticeably improved [4]. It was discovered that certain functionalities, such as amidoxime chelating functionalities, could significantly increase the MXene nanosheets' selectivity towards uranyl ions and their aqueous solutions stability. This allowed for the well-planned, quick, as well as the recyclable exclusion of uranium from aqueous mixtures containing competing metal ions [97].

There could be at least 20 more double MXenes of transition metals. For eliminating radionuclides as well as other contaminants with enormous hydrated ionic radii, pure MXenes' restacking and narrow interlayer spacing are the key obstacles which restrict radionuclide adsorption. One strategy to upsurge the surface accessibility along with considerably enhancing the adsorbates capacity of adsorption is the intercalation as well as delamination of MXenes into smaller nano-sheets. MXene adsorbents are projected to be produced at a cost that is comparable to other nanomaterials of high-performance adsorbent for tackling the problems associated with water as well as wastewater remediation due to their extremely efficient and selective adsorption capabilities.

2.3 Mxene-Based Materials Issues and Prospects in Water Remediation

MXenes have demonstrated enormous potential in adsorption, deionization (CDI), ion-sieving membranes, capacitive photocatalytic destruction of contaminants, catalysis, as well as sensors. This viewpoint addresses the main obstacles that stand in the way of MXene applications in environmental remediation. Additionally, the paths for future research that will concentrate on resolving these fundamental issues with MXene applications have conversed.

Despite multiple studies documenting MXenes' beneficial use in environmental applications, several technological obstacles still need to be overcome in order to fully utilize these wonderful materials. Figure 2.10 and the following summary show the main obstacles in MXene applications in environmental remediation.

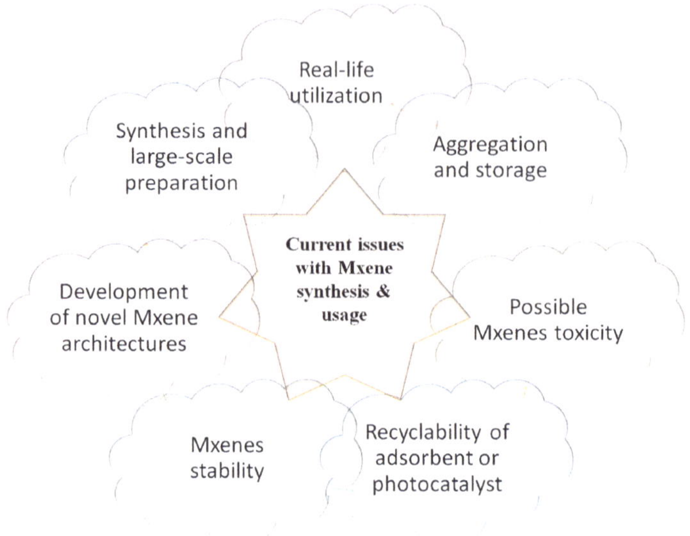

Fig. 2.10 Major issues regarding MXenes utilization in wastewater environmental remediation

2.3.1 Synthesis and Large-Scale Preparation

Significant environmental and health risks are associated with the method frequently employed to synthesize MXenes using hydrofluoric acid (HF). A potential innovation in this area would be to synthesize MXenes using green ingredients. In the past few years, progress has been made in examining the potential uses of less harmful compounds as HF substitutes in the synthesis of MXenes [7, 10]. However, this topic needs to receive logical research attention to find simpler solutions. In advancing green production methods for MXenes, applying green chemistry can play a significant role.

There hasn't been much investigation on the bottom-up strategy for MXene synthesis. Focusing research on this approach could lead to significant advancements in producing high-quality MXenes with the requisite characteristics. The choice of appropriate conditions is essential to produce MXene structures with the necessary properties since the synthesis conditions of MXenes have a significant impact their surface chemistry [7].

Additionally, the difficulty is MXenes synthesis with the necessary characteristics. According to studies, MXenes typically exhibit multilayered structures and an accordion-like morphology rather than 2D nanosheets. This prevents the full application of MXene's capabilities in various industries. The researchers also need to address another potential issue with MXenes' declining conductivity as the number of layers increases in electrical applications.

Creating a successful, inexpensive system for the industrial-scale synthesis of MXenes presents another significant obstacle. This would significantly advance the use of MXenes in industry and help to address pressing environmental problems. It will also broaden the applications of MXenes in several disciplines.

2.3.2 Recyclability of Adsorbent or Photocatalyst

One of the main challenges facing the utilization of Mxene-based materials in an environmental application is their recyclability. As per available works of literature, the efficiency of Mxene-based adsorbents or photocatalyst reduces to a great extent and, in some instances, beyond reconsideration. Kim et al. appear to partly proffer a solution to this challenge in their work, where they observed that with further regeneration cycles for P-MXene, the adsorption capacity steadily fell from 58.7 to 40.6 $mg.g^{-1}$. The sonicated MXene adsorption capacity, however, held steady over the course of 5 cycles, as depicted in Fig. 2.11 [38]. Since $Ti_3C_2T_X$ MXene is a relatively new material, an accurate cost comparison is rather difficult, however, from an economic standpoint, sonicated MXene would be preferable. Weaker binding energies between adsorbed AMT but also sonicated MXene may account for the enhanced performance on the reclaimability of MXene. It can be deduced here that sonication would be a great tool in enhancing the regeneration of Mxene-based adsorbent materials for water remediation. In a recent study by Ghani et al. [39], for the purpose of removing ciprofloxacin from wastewater, the group suggested a unique electrochemically enabled complete regeneration method for intercalated $Ti_3C_2T_X$ MXene. However, they proposed that sodium intercalated $Ti_3C_2T_x$ (SI-$Ti_3C_2T_x$) proved better stability even over five cycles of regeneration [39].

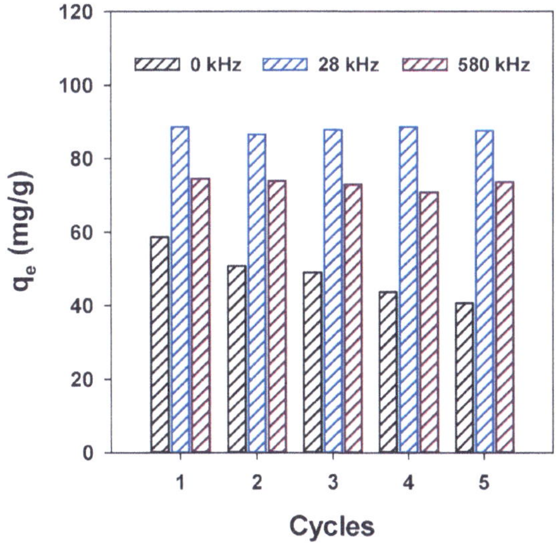

Fig. 2.11 Reusability of pristine and sonicated Ti_3C_2TX MXene for AMT adsorption at pH 7. With permission from Kim et al. [38], Copyright 2021, Elsevier Science Ltd.

2.3.3 Storage and Aggregation

The storage and aggregation of MXenes are additional problems that require research consideration; overcoming them would not only enhance their functionality but also broaden the range of uses for these 2D materials [98]: another problem and chance for researchers to study and address is the prerequisite for sub-zero temperature for MXene dispersions storage.

2.3.4 Possible Mxenes Toxicity

The analysis of MXenes' life cycle within the biota as well as the assessment of their potentially harmful consequences, present another significant problem. MXene exposure and its derivatives may affect people and other living things. The main routes of exposure to nanomaterials are the gastrointestinal tract, skin, and respiratory system. Despite the paucity of information regarding the potential toxicity of MXenes, some recent research have raised questions about the potential negative effects of MXenes on humans and the aquatic environment.

The environmental toxicity of MXenes must be thoroughly evaluated, and mitigation strategies must be put forth. To ensure the security of scientists working in the lab and foresee the potential negative effects of their practical applications, much research is needed in this direction.

It has been said that because of their better colloidal stability, decreased toxicity, and longer circulation inside the body, modifications to MXene surfaces may increase their in vivo efficiency [99]. MXenes developed for biomedical applications may have structural as well as dose-dependent antibacterial activity when utilized in targeted drug delivery, photothermal therapy, photoacoustic as well as optical imaging, as well as for implant formation.

2.3.5 Mxenes Stability

The stability of MXenes under thermal and oxidative conditions in aqueous media is another important issue that requires the researchers' full focus [99]. MXenes are thermodynamically metastable in a ventilated aqueous setting, even at ambient settings, and generally have considerable resistance to oxidation [99]. Because it has a low oxygen resistance, $Ti_3C_2T_X$, for instance, oxidizes easily in an aqueous solution.

The practical uses of MXenes on a wide scale are thought to be hindered by their metastability. When oxygen is present, and the temperature is above 200 °C, MXene particles ($Ti_3C_2T_X$) are thermally unstable and can degrade to carbon, TiO_2, and CO_2 (at high temperatures). The stability of MXenes may be significantly impacted by surface modification, which can be investigated further to gain more knowledge. Developing effective methods for long-term, deterioration-free storage of MXene solution is also crucial.

2.3.6 Development of Novel Mxene Architectures

Additionally, there is a necessity to create more MXenes structural architecture, particularly for environmental cleanup, that may have improved characteristics. There are still study areas where MXene applications may be possible, even though they have been discovered various environmental domains.

The most used MXene structure for environmental applications is Ti_3C_2 [6, 7, 99]. Researchers for use in environmental applications must also investigate the intriguing properties of other MXenes that are now in use. Additionally, procedures must be created to acquire and enhance the desirable properties of MXenes for applications in environmental remediation. To possibly create novel MXene structures, it is also necessary to investigate the other MAX phases.

2.3.7 Real-Life Utilization

Majority of the MXenes' existing environmental usage are founded upon theoretical research or experiments conducted in the lab. Even if it is still early for practical applications of MXene materials, the element of genuine environmentally friendly usage needs to be investigated to see whether they are feasible in the real world. For instance, it is necessary to research how MXene-based materials perform in actual wastewater treatment applications where coexisting ions and natural organic matter (NOM) are present.

Before MXenes are used commercially, a few important concerns must be handled in addition to those already mentioned. To get the needed qualities for certain environmental applications, MXene-based materials' properties must be modified. For instance, the performance of MXene-based membranes for applications in water treatment and desalination might be negatively impacted by the swelling of these membranes in an aqueous medium. To solve this issue, it is essential to develop effective methods for adjusting the membrane's properties to provide the appropriate gap between the layers. The manufacture of non-swelling MXene membranes has recently received some attention [6]. As an anti-biofouling agent in water treatment, MXenes have also shown to have enormous promise [6, 64, 99, 100]. To completely comprehend the interactions of MXenes but also biofoulants, it is necessary to investigatively study the anti-biofouling mechanism of MXenes.

2.4 Application of Mxene as Well as Its Materials for Carbon Dioxide (CO_2) Capture but also Conversion

Owning to the carbon to oxygen atoms covalent double bond, CO_2 has the inert and stable characteristics of a greenhouse gas. Given that CO_2 has a double bond, it is thermodynamically stable and therefore challenging to activate for the production of products with added value. In addition, CO_2 is a key carbon source for the production of value-added goods and fuels for industrial purposes because it is the primary carbon building block for life on Earth [101]. The volume proportion of CO_2 in the atmosphere is around 400 ppm, and its concentration continues to rise [102]. The melting and even extinction of the iceberg show that the environmental risks posed by greenhouse gases (such as CO_2) are becoming more obvious recently. Both the greenhouse effect and the production of renewable energy can be reduced through CO_2 capture as well as conversion into important chemicals but also fuels [103]. Due to the great thermodynamic stability of CO_2, however, CO_2 conversion is adversely affected by conditions such as high temperature, high-energy substrates, high pressure, or prolonged reactions [103]. For CO_2 high-efficiency conversion under favorable conditions, the development of materials with a high CO_2 capacity and catalytic impact is crucial. Numerous materials have been proposed as effective CO_2 sorbents but also catalysts, comprising of solid, liquid, as well as gel-like substances [103–106]. The liquid approach for CO_2 capture and conversion is straightforward and effective, but the separation benefits significantly from the use of solid absorbents and CO_2 utilization catalysts. For CO_2 capture and conversion, it is still important to create unique, high-performance materials [107]. Amongst solid materials, MXene has shown the potential to be an effective absorbent for capturing CO_2.

2.4.1 MXene-Based Materials for CO_2 Capture

2.4.1.1 Single Component MXene

As earlier mentioned, Mxene and its materials possess the potential as CO_2 amongst CO_2 collection materials, especially in its form as transition metal carbides. However, they have a very small, exposed surface area [108]. CO_2 storage on M_2C MXene (M = Ti, Zr, Hf, V, Nb, Ta, Mo, W) (0001) surfaces has been reported in a study by Morales-Garca and colleagues using the most recent first-principles density functional theory (DFT) calculations within the generalized gradient approximation (GGA) by the Perdew-Burke-Ernzerhof (PBE) exchange–correlation functional with D3 Grim [109]. High adsorption energies of up to 3.69 eV were observed with CO_2 activation. This resulted in anionic $CO2^{\delta-}$ species with bending structures, elongated (CO) bonds, and an MXene–CO_2 charge transfer, unprecedented above 2*e* for Hf_2C. These high adsorption energies indicated that M_2C MXenes could absorb, store, and activate CO_2 more efficiently than their 3D bulk counterparts. Theoretically, these materials can adsorb CO_2 up to high temperatures and low partial pressures, according to the estimated adsorption but also desorption rates from transition state theory (TST) models. Additionally, MXenes were found to be extremely competitive with other currently available material solutions, producing CO_2 uptakes of 2.34–8.25 mol CO_2 kg^{-1} of a substrate. Ti_2CT_x, V_2CT_x, Zr_2CT_x, Nb_2CT_x, Mo_2CT_x, Hr_2CT_x, Ta_2CT_x, and W_2CT_x are possible orders for MXene's CO_2 capture capacity. Even at low CO_2 partial pressure as well as relatively high temperature, the adsorption capacity was still considerable. The increased CO_2 capture capability of MXene was attributed to its larger surface area, extra sites for adsorption, MXene/CO_2 charge transfer, along with higher energy for adsorption [109].

Wang and colleagues prepared MXene powders ($Ti_3C_2T_x$ and V_2CT_x) by exfoliating MAX phases (Ti_3AlC_2 and V_2AlC, respectively) powders in a solution of NaF and hydrochloric acid (HCl). MXenes have been postulated as a potential sorbent for CO_2 collection [110]. Compared to MXenes that were synthesized as-is, intercalated MXenes had a larger amount of CO_2 adsorbed. The capacity of Intercalated $Ti_3C_2T_x$, which had an SSA of 66 m^2/g, was 5.79 mmol/g, comparable to the capacity of many common sorbents. $Ti_3C_2T_x$ has a theoretical capacity of 44.2 mmol/g with an SSA of 496 m^2/g, which is comparable to metal organic frameworks (MOF's) theoretical capacity. MXenes also have a very high volume-uptake capability because of their high pack density. This study measured 502 $V.v^{-1}$ as the capacity of intercalated $Ti_3C_2T_x$. This value exceeded the volume capacity of the majority of currently understood sorbents [110]. The CO_2 adsorption characteristics of Mo_2C MXene were first examined in a recent study by Jin et al. Despite having a lower specific surface area (SSA), the experimental findings demonstrated that Mo_2C MXene had a greater CO_2 adsorbed capacity than Ti_3C_2 and V_2C MXenes. Overall, in the gas capture process of Mo_2C MXene, physical absorption caused by capillary condensation in mixed pores (micropores as well as mesopores) and slit-like pores between two Mo_2C MXene flakes coexists with CO_2 adsorption generated by functional groups with some chemisorption properties. Mo_2C produced in this study exhibited stronger CO_2 adsorption properties than Ti_3C_2 and V_2C because of the high O/F ratio. Mo_2C-Li had the largest CO_2 residual capacity among the as-prepared samples and the best CO_2 adsorption performance. According to the theoretical calculation, Li^+ interacted more strongly with Mo_2C MXene, and CO_2 interacted more strongly with Li^+ than Na^+ and K^+. Therefore, the introduction of diffident cations via fluoride salts had a considerable impact on Mo_2C MXene's ability to adsorb CO_2 [111].

It is obvious from Fig. 2.12a–d that there was no visible hysteresis loop at reasonably high pressures, and Mo_2C MXene N_2 adsorption/desorption isotherm was of type II (an invertible adsorption–desorption curve) [112], demonstrating that the N_2 adsorption and desorption processes for these coincide. It is assumed that the functional groups or cations on the surface of Mo_2C MXene can function as a surfactant in the N_2 adsorption process, comparable to prior studies of silicas with type II isotherms but also surfactant-containing materials [113, 114]. The associated insets of Fig. 2.12 display the computed pore size distributions. Most pores were less than 10 nm in size, which suggested mesopores. Additionally, the pore size distributions saw a distinct peak corresponding to a homogeneous pore size of 1.8 nm. As a result,

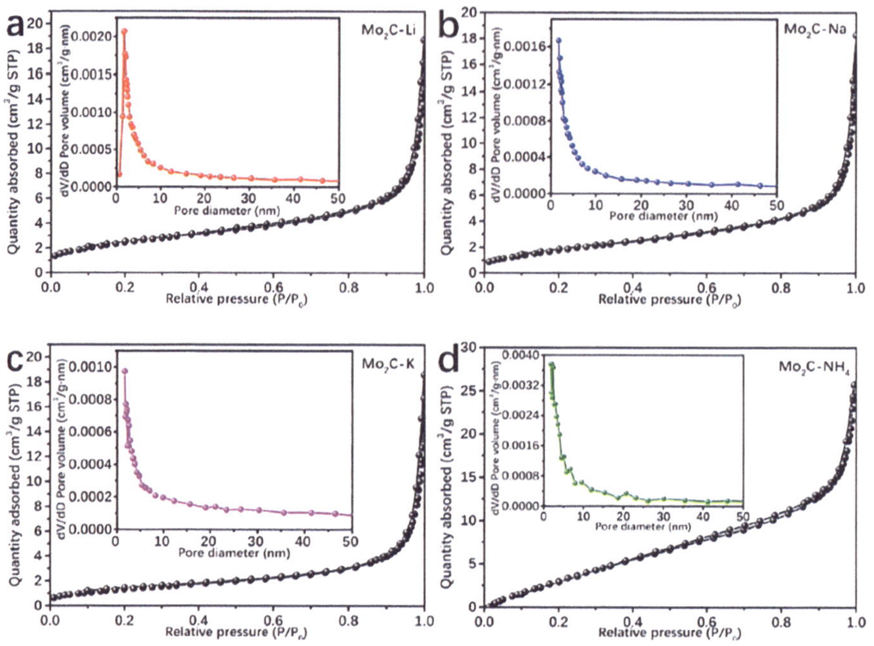

Fig. 2.12 N_2 sorption isotherms and pore size distributions (the insets) of samples: **a** Mo_2C-Li, **b** Mo_2C-Na, **c** Mo_2C-K, **d** Mo_2C-NH_4. Reproduced with permission from Jin et al. [111], Copyright 2022, Elsevier Science Ltd.

in Mo_2C MXene as it is, micropores and mesopores coexist. The samples' SSA was estimated using the N_2 adsorption/desorption test to range from 4.9 to 8.9 m^2g^{-1}, as well as the pore volume (PV) were reportedly 0.02 cm^3 g^{-1} at the relative pressure $P/P_0 = 0.99$ [111].

Figure 2.13 displays the samples' CO_2 adsorption and desorption isotherms. In contrast to V_2C, which had no CO_2 adsorption ability below 2 MPa, Mo_2C MXene had been reported to present an obvious capacity to adsorb CO_2 in the low-pressure region [115]. In the low and intermediate range, the amount of CO_2 that was adsorbed generally rose linearly as pressure increased. A hysteresis loop was present during the desorption process. Almost parallel to the adsorption curve, the desorption curve gradually declined. It is evident from the fact that the hysteresis loop could not shut even at ambient pressure that the CO_2 bound to Mo_2C MXene was not released entirely after desorption. This shows that Mo_2C MXene was a significant sorbent in collecting CO_2. The Mo_2C MXene adsorption–desorption curves in Fig. 2.13 are type V isotherms. The amount of CO_2 adsorbed then quickly increased as a result of capillary condensation in some mesopores under relatively high pressure ($P/P_0 > 0.8$). Hysteresis loops could also occur because capillary condensation and evaporation typically do not occur at the same pressure. The H_3 hysteresis loop was thought to be suitable for CO_2 adsorption of Mo_2C MXene based on the geometry of the loop in Fig. 2.13 [111]. The type H_3 hysteresis loop is known to be seen in aggregates of plate-like particles (as observed in Fig. 2.14b) that give rise to slit-shaped pores and do not show any limiting adsorption at high P/P_0 [112]. Pressure caused the desorption curve to decline gradually (rather than quickly). It is possible to think of the interlayer space between two Mo_2C MXene sheets as slit-like pores, which can also be treated as a space to adsorb and store CO_2 molecules and cause the development of an H_3 hysteresis loop [115].

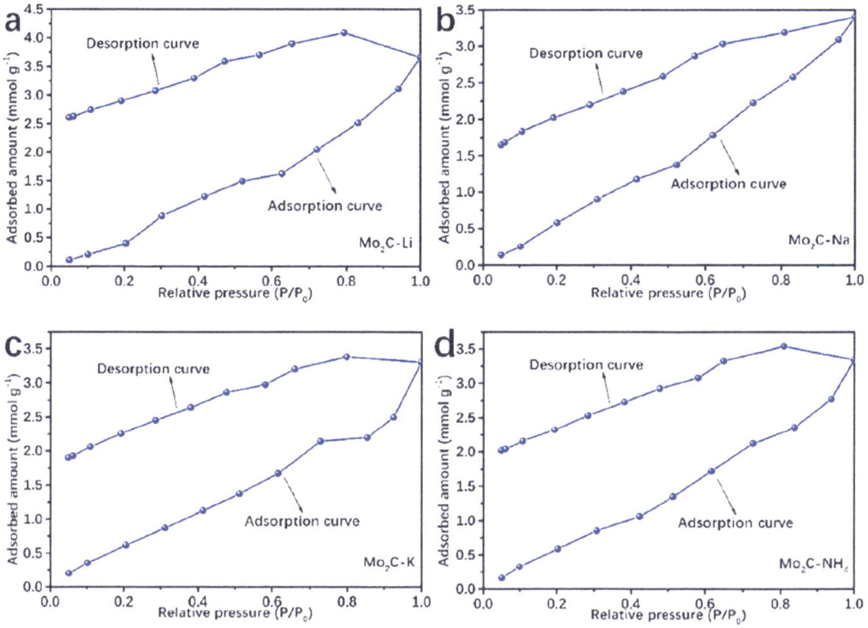

Fig. 2.13 CO_2 adsorption and desorption curves of Mo_2C MXene at room temperature under 0–4 MPa: **a** Mo_2C-Li, **b** Mo_2C-Na, **c** Mo_2C-K, **d** Mo_2C-NH_4. Reproduced with permission from Jin et al. [111], Copyright 2022, Elsevier Science Ltd.

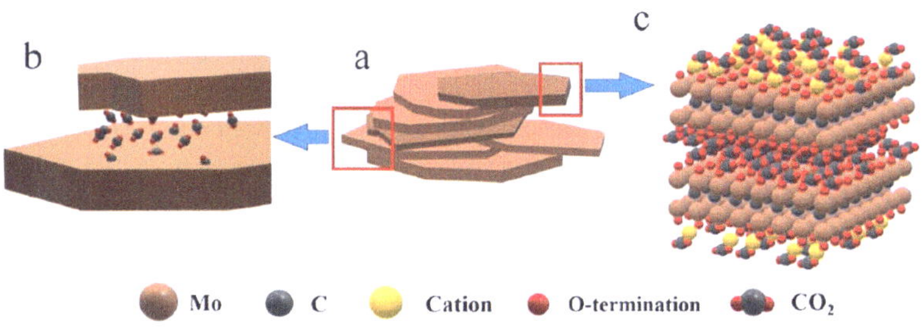

Fig. 2.14 Schematic of CO_2 adsorption of Mo_2C MXene. **a** Laminated structure of Mo_2C MXene shown in SEM image **b** Slitlike pores with adsorbed CO_2 molecules between the flakes of Mo_2C MXene **c** Interlayer space inside a Mo_2C MXene flake with the adsorbed CO_2 molecules. Reproduced with permission from Jin et al. [111], Copyright 2022, Elsevier Science Ltd.

2.4.1.2 Brief on MXene-Based Composites

Nevertheless inorganic films have shown outstanding gas separation efficacy with great thermal as well as chemical stability [116–118], their main limitations include brittleness, high cost, non-scalability, and difficulties customizing their pore sizes. The idea of mixed matrix membranes (MMMs) is an intriguing method for enhancing polymer membrane isolation capabilities and overcoming inorganic membrane difficulties.

Strong, high-performance MMMs-based on $Ti_3C_2T_x$ MXene as well as a tripple copolymers Pebax-1657 (Pebax), Pebax-2533, along with polyurethane (PU) were described by Shamsabadi and colleagues. These co-polymers were chosen to study the impact of embedding $Ti_3C_2T_x$ into co-polymers with hard and soft segments without functionalization on the resultant MMMs [117]. The free-standing membranes' CO_2/N_2 and CO_2/H_2 separation capabilities exceeded Robeson's upper bounds, while the thin-membrane composite (TFC) performances were within the desired range for CO_2 collection that is both effective and affordable. Characterization and molecular dynamics simulation results point to the creation of H-bonds between Pebax chains and $Ti_3C_2T_x$ as the cause of the superior performances of the Pebax-$Ti_3C_2T_x$ membranes. These bonds then resulted in the formation of well-molded galleries of $Ti_3C_2T_x$ nanosheets in the hard parts of the Pebax. Quick and specialized CO_2 transport was made possible by the interfacial contacts and $Ti_3C_2T_x$ nanochannels.

The transport of CO_2 along with N_2 gas molecules across membranes is shown in Fig. 2.15 and is most likely controlled by the process. Given that CO_2 molecules have a higher quadrupole moment than N_2 molecules, they may be more favorably adsorbed on the $Ti_3C_2T_x$ surface [119]. This has been reported for MXene sensors that detect nitrogen atmosphere diluted CO_2 and it has been confirmed by the first-principles density functional theory [109, 119]. The selectivity of CO_2 solubility was also further improved by the availability of the nanosheet interlayer. These ultramicroporous interlayers additionally offer molecular sieving, increasing CO_2 permeability and diffusivity selectivity. About 0.35 nm was the interlayer gallery between the $Ti_3C_2T_x$ layers, which is larger than the kinetic diameter of CO_2 and smaller than that of N_2 [120]. N_2 molecules are effectively blocked from passing through the molecular sieving framework. The crystallinity of the hard segments is also reduced when $Ti_3C_2T_x$ and Pebax are mixed, but the polymer was reported to experience an increase in phase separation, which enhanced CO_2 permeability. Last but not least, the barrier effect of $Ti_3C_2T_x$ nanosheets was revealed to lengthen the gas molecules transport paths (due to the tortuosity effect), improving CO_2/N_2 diffusivity selectivity [117].

Most of the Mxene-based composite materials applied in CO_2 capture and conversion have been those of an inorganic nature, such as 2D/2D $Ti_3C_2T_{A/R}$ MXene coupled g–C_3N_4 [121], Co–$Ti_3C_2T_x$ [122], Ti_3C_2/Boron–Doped g–C_3N_4 [123], and $CsPbBr_3$/MXene [124]: for instance, Que et al. reportedly produced a highly stable $FAPbBr_3/Ti_3C_2$ nanocomposite aimed at efficient CO_2 capture and conversion anchoring $FAPbBr_3$ quantum dots on the surface of Ti_3C_2 [125]. The formed

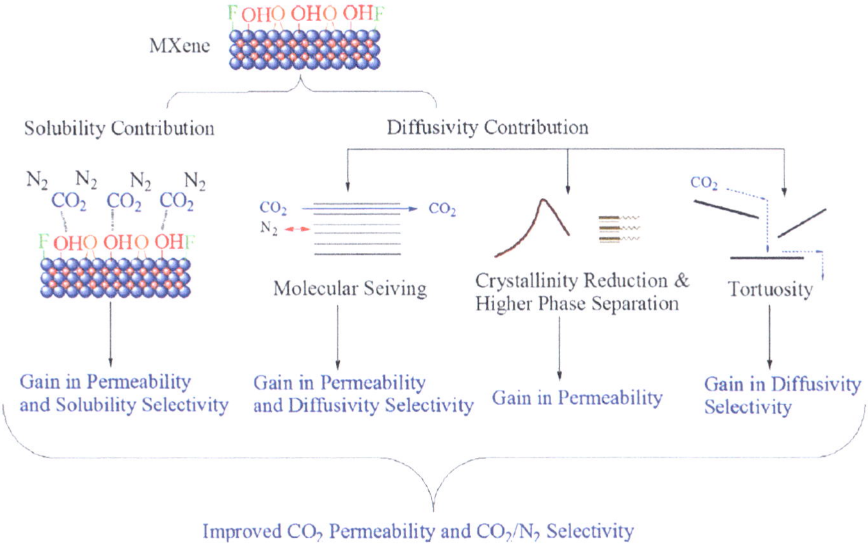

Fig. 2.15 Proposed mechanism for transport of CO_2 and N_2 gas molecules across $Ti_3C_2T_x$ – Pebax membranes. Reproduced with permission from Shamsabadi et al. [117], Copyright 2019, ACS publications

nanocomposite was said to be within a Schottky heterojunction for CO_2 photocatalytic reduction. The composite was said to present a better performance in comparison to the quantum dots alone. When exposed to visible light and deionised water, the $FAPbBr_3/Ti_3C_2$ composite photocatalyst has an interesting photocatalytic piece. As an electron acceptor, the Ti_3C_2 nanosheet facilitates the quick separation of excitons and supplies particular catalytic sites. The $FAPbBr_3/0.2\text{-}Ti_3C_2$ composite achieves an ideal electron consumption rate of 717.18 $\mu mol.g^{-1}\ h^{-1}$, a 2.08-fold increase over the rate of the purest FAPbBr3 QDs (343.90 $\mu mol.g^{-1}\ h^{-1}$) [125]. The approach used by these authors for the fabrication of their nanocomposite simple solution intercalation-casting is presented in Fig. 2.16. The CO selectivity was reported to be 92.74%.

Ali Khan and Tahir in their work on CoAlLa–LDH Anchored on Titania–Sandwiched Ti_3C_2 nanocomposite as presented in Fig. 2.17, prepared their nanocomposite system using electrostatic enabled self-assembly thus: The already-prepared $Ti_3C_2T_{A/R}$ was added to a 20 mL amount of methanol solution in an usual technique, and the mixture was agitated for 30 min. The 20 mL of methanol solution was added simultaneously as the previously created $Co_2Al_{0.95}LA_{0.05}$-LDH. For proper mixing, each solution was agitated for 30 min. The $Ti_3C_2T_{A/R}$ solution was combined with the $Co_2Al_{0.95}LA_{0.05}$-LDH solution, which was then agitated for an additional hour to ensure optimal mixing using the self-assembly method. To achieve full dispersion, $Co_2Al_{0.95}$La0.05-LDH was loaded with $Ti_3C_2T_{A/R}$ at different percentages (5; 10; 15; 20; and 25%). To create the $Co_2Al_{0.95}LA_{0.05}$-LDH/TiO_2/Ti_3C_2 composite, the

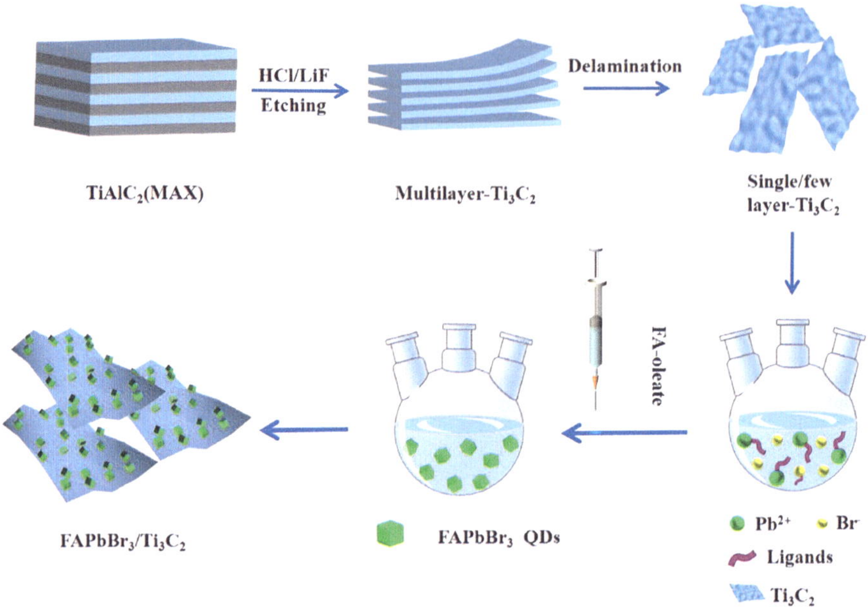

Fig. 2.16 Schematic illustration of the preparation process of FAPbBr3/Ti3C2 composite photocatalysts. Reproduced with permission from Que et al. [125], Copyright 2021, ACS Publications

obtained solution was then dried for 24 h at 100 °C in an oven. According to these authors, the presence of electron-rich La^{3+} allowed trimetallic $Co_2Al_{0.95}LA_{0.05}$-LDH to attain greater photoactivity than bimetallic Co2Al1-LDH. The yield rates of CO and CH_4 were significantly greater than pristine samples over the $Co_2Al_{0.95}LA_{0.05}$-LDH/TiO_2/Ti_3C_2 MXene (nano)composite following the CO_2 photocatalytic reduction with H_2O. These yield rates were 46.32 and 31.02 mol g^{-1} h^{-1}, respectively. This significantly higher performance was made possible by a more effective sorption method, which also resulted in superior charge carrier separation due to oxygen-deficient locations, strong interfacial contact, and the presence of dual-phase titania as a bridge for separating charges. Bireforming of methane and photocatalytic dry reforming of methane (DRM) were used to further evaluate the composite performance (BRM). BRM produced more CO and H_2 due to the efficient attachment of reactants to electron-rich faulty sites. In addition, the quantum efficacy as well as stability analysis validated the Ti_3C_2T/CoAlLa-LDH composite catalyst's great durability over several cycles. This high durability is due to the stable Ti_3C_2TA/R structures with CoAlLa-fundamental LDH's properties [121].

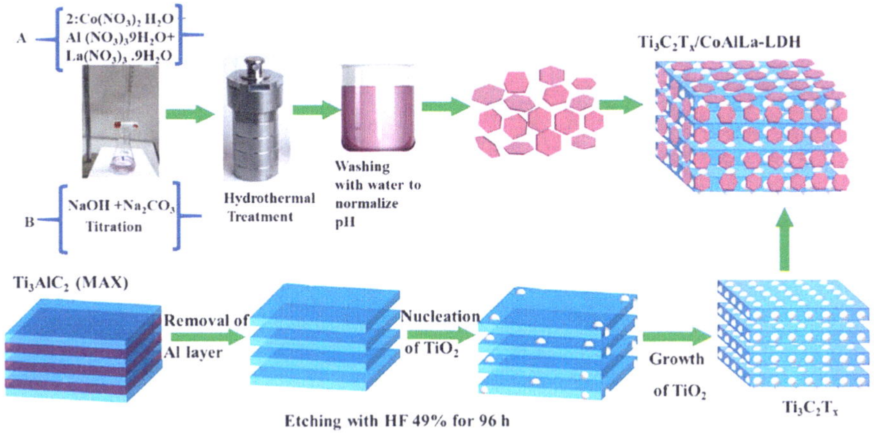

Fig. 2.17 Schematic illustration of the $Ti_3C_2T_{A/R}$/CoAlLa-LDH nanocomposite. Reproduced with permission from Ali Khan and Tahir [121], Copyright 2021, ACS publications

2.4.1.3 Atomic Defects in MXene

In addition to helping to provide a more efficient active site for CO_2 adsorption and activation, defects are well-known to play a important part in increasing a material's photocatalytic activity. In comparison to 2D materials, the impact of inherent defects on MXenes hasn't been thoroughly studied [126, 127].

High-throughput screening was used in the study by Parey and colleagues to analyze the role of atomic defects in M_2X-type MXenes (M = Sc, Ti, V, Cr, Mn, Y, Zr, Nb, and Mo; X = C and N) for possible CO_2 capture, activation, and dissociation [128]. Through DFT-D3 dispersion correction, well-resolved first-principles simulations were used to analyze the two-dimensional MXenes. As illustrated in Fig. 2.18a, several M_2X (Sc_2C/N, Ti_2C/N, V_2C/N, Cr_2C/N, Mn_2C/N, Y_2C/N, Zr_2C/N, Nb_2C/N, and Mo_2C/N) structures were initially built in a hexagonal atomic lattice with $P6_3/mmc$ symmetry. The X layers are alternately sandwiched between the M layers. The MX bonding interactions used had an average bond length of 2.12 Å and connected the layers. The anti-ferromagnetic (FM) ground state, which was necessary throughout the wet chemical reaction, was typically unstable for MXenes. Three different types of vacancies were taken into account (see Fig. 2.18b–d) to examine the function of defects: (i) metal vacancies (VM), (ii) carbon/nitrogen vacancies (VC/N), and (iii) metal and one neighboring carbon/nitrogen vacancy (VMC/N). A total of 72 MXenes were created, with 18 structures in each of the four categories (pure, VM, VC/N, and VMC/N). Figure 2.18e, f showed the ideal bond lengths for defects of all kinds and pure bonds. They considered the typical M-C/N bond lengths in relation to the vacancy that was significantly impacted by the flaws. These bond lengths are primarily influenced by the metal atoms' growing nuclear charge and consistent shielding, which increases their atomic radius. For instance, the MX bond length clearly decreased in the following order: $Sc_2X > Ti_2X > V_2X > Cr_2X > Mn_2X$.

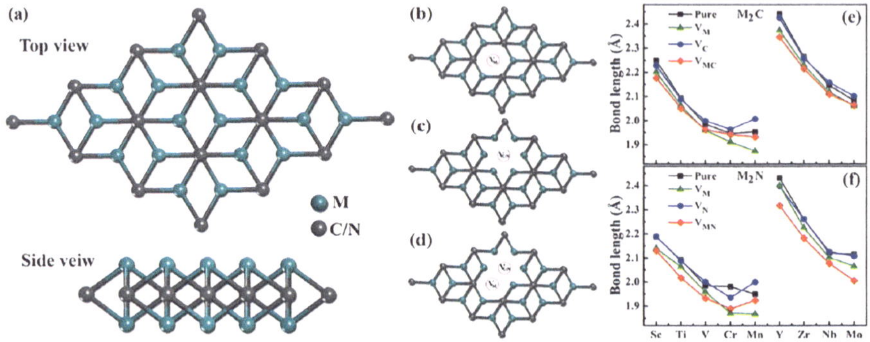

Fig. 2.18 **a** Top and side views of pure M_2X-type (M = Sc, Ti, V, Cr, Mn, Y, Zr, Nb, and Mo; X = C and N) MXenes along with **b** metal vacancy (VM), **c** carbon/nitrogen vacancy (VC/N), and **d** metal and nearby carbon/nitrogen vacancy (VMC/N). Optimized bond lengths of pure and defective **e** M_2C and **f** M_2N-type MXenes. Reproduced with permission from Parey, V., Abraham, Mir, & Singh [128]. Copyright 2021, ACS publications

Strong carbide/nitride formation elements 4d (Y, Zr, Nb, and Mo) had a similar pattern of activity. The MX bond length decreases in defect structures compared to pure systems. For example, for VM, VC/N, and VMC/N, the MC/N bond length for pure Sc_2C/N reduces from 2.25/2.18 to 2.22/2.17, 2.20/2.14, and 2.17/2.13 Å, respectively. Adding more vacancies allows the MX bond lengths of defect structures to be shortened.

In a work conducted by Khaledialidusti and colleagues, they discovered that the intrinsic $MO_2TiC_2T_X$ MXene flaws promoted CO_2 adsorption due to the unique surface ending with F, O, or OH [129]. The results demonstrated that, as suggested by the theoretical formation energy of the defect, the synthesized structural flaws were governed by the terminators types. The molybdenum layer was the easier of the two layers to generate defects in than the inner layer (i.e., the titanium layer). Furthermore, compared to the development of F- and OH-typed terminators, the formation of an O-typed terminator required more energy. Through an exothermic procedure, CO_2 was tightly linked to the MXene defects. On the other hand, MXene's ideal sites engaged in a mild, non-spontaneous endothermic reaction with CO_2. The capacity to capture CO_2 was projected to increase as the ratio of atomic defects on the MXene increased.

Finally, it should be noted that external variables like temperature, pressure, and contaminants can generally impact CO_2 collection capability. Adsorption conformation, adsorption strength, the amount of interactive sites, as well as the ratio or size of the pores in MXene-based hybrids were the structural parameters that improved CO_2 capture. Undoubtedly, functionalizing MXene and adding CO_2-philic elements to the MXene-based composite might increase the CO_2 collection capability. The collection of CO_2 with high capacity as well as good selectivity at moderate temperatures and low CO_2 percentage, specifically in gases mixed with CO_2 deserves more study. One of the practical as well as challenging types of materials for of CO_2 collection as

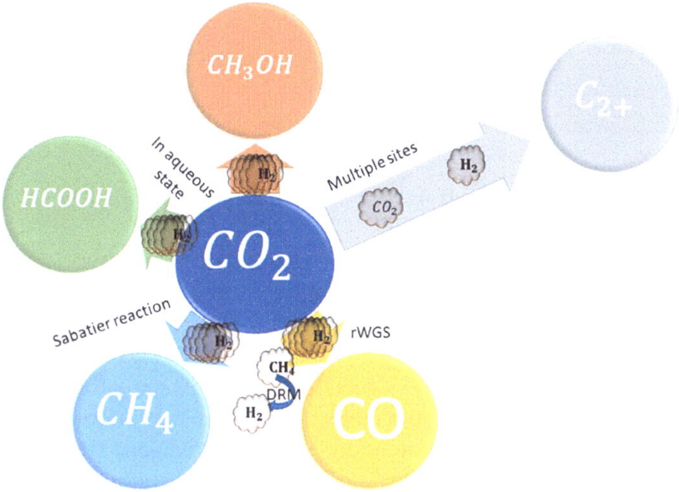

Fig. 2.19 Plausible CO_2 conversion reactions

well as conversion application industrially towards value-added goods sustainably was other gas' potential to dominate the MXene-based absorbents/catalysts sites of interaction designed for the capture as well as conversion of CO_2 [109].

2.4.2 CO_2 Conversion by MXene-Based Materials

2.4.2.1 CO_2 Thermo-Conversion

Single-Atom@MXene

The frequent CO_2 reduction processes list is shown in Fig. 2.19 and equations i-vi. Thermodynamics data and a significant amount of study demonstrate that CO and CH_4 are the most straightforward and easy products in the reduction of CO_2. The reactions that produce CO and methane directly from hydrogen are referred to as the reverse water-to-gas shift reaction (rWGS) as well as Sabatier reaction, respectively [130].

$$2CO_2 + 6H_2 \rightarrow C_2H_5OH + 3H_2O \quad (2.4)$$

$$CO_2 + H_2 \rightarrow HCOOH \quad (2.5)$$

$$CO_2 + 3H_2 \rightarrow CH_3OH + H_2O \quad (2.6)$$

$$CO_2 + CH_4 \rightarrow 2CO + 2H_2 \tag{2.7}$$

$$CO_2 + 4H_2 \rightarrow CH_4 + 2H_2O \tag{2.8}$$

$$CO_2 + H_2 \rightarrow CO + H_2O \tag{2.9}$$

The secret to successfully converting CO_2 using heat is catalysts. Single-atom catalysts were very effective and frequently used to convert CO_2 thermally into compounds with additional value. They are unstable, though, and have a propensity to agglomerate, losing their catalytic function. Single-atom catalysts require support in the form of metal–organic frameworks, metal oxides, and heteroatom-doped carbon, all of which often require high temperatures or hazardous H_2 atmospheres [131].

In order to stabilize a single Pt atom on Ti-deficient MXene ($Pt_1/Ti_{3\text{-}X}C_2T_y$) at ambient temperature without the use of a hazardous H_2 atmosphere [132]. More specifically, a Pt-C bond was formed to stabilize a single Pt atom at the shortage locations in MXene (Fig. 2.20). $Pt_1/Ti_{3\text{-}X}C_2T_y$ may catalyze the reaction of amines and silane with 1 atm CO_2 to make formamides with a nearly 100% conversion rate along with 100% selectivity (Fig. 2.21a). Nearly little formamide was generated when $Ti_{3\text{-}X}C_2T_y$ or the Pt_1 precursor were the only catalysts. Additionally, Pt nanoparticles supported on MXene demonstrated a conversion rate of approximately 18%, which is significantly lower compared to ~100% $Pt_1/Ti_{3\text{-}X}C_2T_y$ catalyzed conversion rate (Fig. 2.21b). In addition, the amount of CO_2 converted by MXene or the precursor of single atom Pt was insignificant (Fig. 2.21b). By functionalizing anilines with electron-donating groups rather than electron-withdrawing groups, the conversion rate would rise rather than fall. Additionally, aliphatic and secondary amines increased product yield. It is important to note that the CO_2 conversion process required a rather high temperature (140 °C) and a lengthy reaction period (10 h). More critically, the CO_2 conversion process involved the use of environmentally unfriendly volatile organic chemicals (like DMF, for example). Other single-atom MXene hybrids are projected to be designed in order to efficiently convert CO_2 at ambient temperature as well as high rate atmospheric pressure without the need for organic solvents. Although the selectivity remained extremely high, it appeared that $Pt_1/Ti_{3\text{-}X}C_2T_y$'s recyclability needed to be enhanced because there was a 20% reduction in CO_2 conversion rate after 5 times of recyclability (Fig. 2.21c) [132].

2.4.2.2 Electroconversion of CO_2

Single Component MXene

As far as we are aware, MXene was the only component mentioned in nearly all papers on the electroreduction of CO_2. Among all the $M_3C_2T_x$ MXene studied, Sun et al.'s DFT theoretical calculation findings demonstrated that $Cr_3C_2T_x$ and $Mo_3C_2T_x$

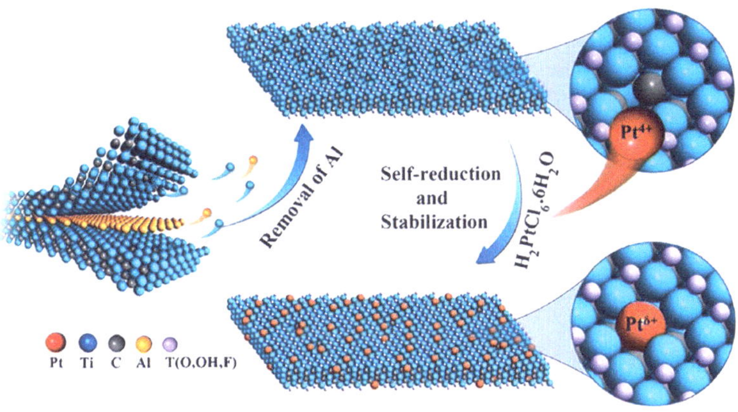

Fig. 2.20 Illustration of the Simultaneous Self-Reduction-Stabilization Process for the Preparation of $Pt_1/Ti_{3-x}C_2T_y$. Reproduced with permission from Zhao et al. [132], copyright 2019, ACS Publications

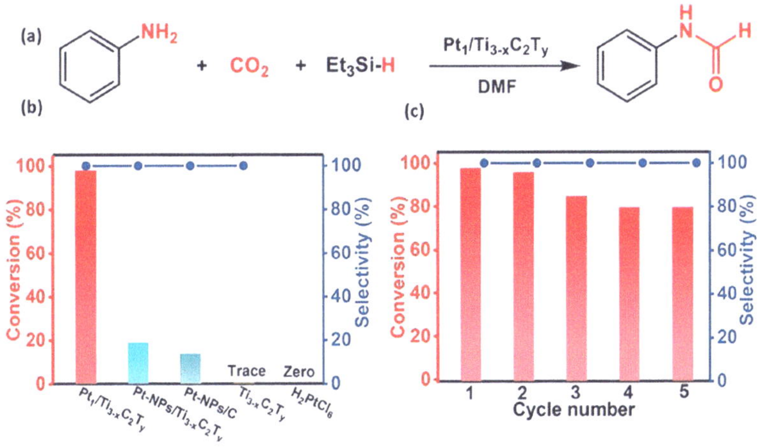

Fig. 2.21 **a** Chemical equation for the N-formylation of aniline with CO_2 and Et_3SiH. **b** Catalytic performance of the N-formylation of aniline using different catalysts. **c** Recycling test of $Pt_1/Ti_{3-x}C_2T_y$ for the catalytic N-formylation of aniline. Reproduced with permission from Zhao et al. [132], copyright 2019 ACS Publications

MXene provided the most effective proficiency for the electrochemical conversion of CO_2 to CH_4. The energy barrier for CO_2 electroconversion to CH_4 would be lower if the MXene surface would have been functionalized with OH or O. The authors revealed that MXene tended to be dehydroxylated because less energy was required to liberate H_2O from MXene covered with OH compared to MXene covered with O [133].

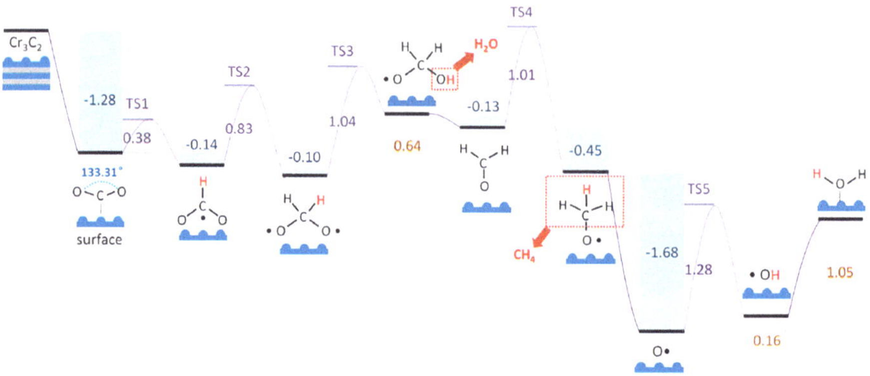

Fig. 2.22 Minimum energy path for the CO_2 conversion into $*CH_4$ and $**H_2O$ catalyzed by Cr_3C_2 calculated with PBE/DFT-D3. The intermediates and transition states (TSs) are indicated. Gibbs free energies for reaction (black line) and activation (purple line) are given, in eV. Different shadings indicate spontaneous (blue) versus nonspontaneous (brown) reactions and the barrier (purple) of activation for each reaction. Reproduced with permission from Li et al. [133], Copyright 2017, ACS Publications

For the electroreduction of CO_2 to CH_4 (Fig. 2.22), the sixth H^+/e pair gain on the $CH_3O^•$ radical occurred on the CH_3 moiety in this pathway, first releasing CH_4 and inserting an O atom into the material, then continuing with two additional hydrogenations to form an $OH^•$-doped solid and ultimately a captured H_2O molecule. Even if the H_2O product contaminated the material, water desorption would only require a modest amount of reaction energy (in comparison to the limiting $**OH^•/**H_2O$ step), and on top of that, prospective CO_2 molecules will thermodynamically displace the caught water [133].

Theoretical investigation of 2D transition metal carbides as well as nitrides (MXenes) as potential electrocatalysts for the conversion of CO_2 to CH_4 was carried out in a different work by Handoko and colleagues. Due to typically weaker *CO binding, a novel CO_2 reduction route via –H coordinated *HCOOH was found on O-terminated MXene catalysts [134]. The electro-reduction of CO_2 to CH_4 could be described via the expressions in the next line, which calls for 8 proton/electron transfer steps: $CO_2 + 8H^+ + 8e^- \rightarrow CH_4 + 2H_2O$. The computed free energy diagram for CO_2 reduction on Ti_2CO_2 MXene, with the most advantageous pathway, denoted in dark brown (Fig. 2.22). At zero potential, it was discovered that the initial reaction step of CO_2 gas adsorption as well as protonation on this surface was exergonic. Following CO_2 adsorption, incoming H would bind to CO_2's terminal oxygen to create the intermediate *COOH, which is then linked to Ti_2CO_2 via the C atom. The spontaneous initial step demonstrated on Ti_2CO_2 has been recognized as being the exception rather than the rule, as 15 out of the 19 M_2XO_2 MXene families investigated in this work required energy input for the synthesis of *COOH. Two options were taken into consideration for the second hydrogenation process. This is similar to transition metal catalysts recognized in the first scenario, in which entering hydrogen binds to the –OH side of the *COOH to produce *CO while simultaneously releasing water[34]. This route was unfavorable on the Ti_2CO_2 surface and was less likely to succeed due to the high reaction energy (PG = 1.73 eV). The likelihood of a second alternative, in which incoming hydrogen attaches to the C atom of *COOH to generate *HCOOH, was shown to be much more favorable. Following the *HCOOH pathway, we discover that the least energy-consuming process for producing CH_4 on Ti_2CO_2 is as follows:

$$CO_2(g) \rightarrow^* COOH \rightarrow^* HCOOH \rightarrow^* CHO \rightarrow^* H_2CO \rightarrow^* CH_2OH \rightarrow^* HOCH_3 \rightarrow^* CH_3 \rightarrow CH_4(g).$$

It was also noted that the intermediates' binding to the O-terminated MXene surface alternates between −C and −H coordination: $^*COOH$, $^*CHO$, $^*CH_2OH$, and $^*CH_3$ bind to the surface of the MXene through the C atom, whereas * HCOOH, $^*H_2CO$, and $^*HOCH_3$ bind through the H atom. As a result, the route through $^*HCOOH$ is easier to reach and has more exergonic reaction energy than $^*CO$ [134].

Despite the ease and low cost of theoretical prediction, it is crucial to design the intended MXene to produce the targeted goods. This is because MXene structure in nature would differ from how it is created synthetically. The electrolyte was also a crucial component in the electro-reduction of CO_2, but it was challenging to model the electrolyte's impact on the electroconversion of CO_2 in a theoretical calculation [135–137]; however, it was challenging to model the impact of electrolyte on the CO_2 electro-conversion in the theoretical computation. To acquire the high effectiveness of CO_2 electroreduction but also mechanism interpretation, it was still intriguing to integrate theoretical calculations with experimental data. Handoko et al. employed Ti_2CT_x and MO_2CT_x as the electrocatalysts for CO_2 reduction and discovered that HCOOH was the main product by combining tests and theoretical calculations [38]. Ti_2CT_x electrocatalysts could achieve over 56% Faradaic efficiency, while MO_2CT_x electrocatalysts could achieve partial current densities of 2.5 mA/cm^2. Additionally, Seh demonstrated that functionalizing F on MXene led to a reduced overpotential of CO_2 electroreduction [138].

MXene-Based (Nano)composite

Another promising electrocatalyst for CO_2 electroreduction was a nanocomposite made of MXene. In comparison to ZnO, MXene, ZnO-Fe, and ZnO-MXene, Kannan et al. discovered that ZnO-Fe/MXene nano-composite had a greater current density "18.7 mA/cm^2" of CO_2 electroreduction [139]. Instead, there was no current peak for ZnO-Fe/MXene under the N_2 gas environment during the cyclic voltammetry (CV) studies. Due to its large surface area, high conductivity, as well as numerous surface functionalities, ZnO-Fe/MXene also showed a more favorable current density when compared to other electrocatalysts (such as Cu/Sn and B-doped graphene). The high retention rate (88%) of oxidation current for ZnO-Fe/MXene after 1000 cycles also served as evidence that it was more stable than ZnO-MXene, MXene, ZnO, and ZnO-Fe. In ZnO-Fe@MXene for CO2 electroreduction, the ternary components served distinct purposes. ZnO was beneficial for the adsorption of CO_2. Fe encouraged charge transfer while inhibiting H_2 development. As mentioned previously, MXene performed activities such as electron transport. The three components (Fe, ZnO, and MXene) worked together synergistically to produce extremely effective CO_2 electroreduction [139].

In a different study, Qu and colleagues demonstrated how a simple NH3-etching pyrolysis method might be used to introduce titanium vacancies (V_{Ti}) and N-doping into Ti_3C_2 MXene nanosheets. With a noteworthy 92% faradaic proficiency and a -16.2 mA/cm^2 partial current density for CO generation, these nanosheets exhibited exceptional CO_2 reduction reaction (CO_2RR) characteristics in seawater that are comparable to those of electrodes made of noble metal. Mechanistic investigations showed that the presence of N dopants as well as V_{Ti} synergistically modify the Ti active site electronic structure, where the free energy blockades for the important processes of *COOH production as well as *CO desorption were significantly lowered, resulting in a large increase in CO_2RR [140].

2.5 Conclusion

MXenes are excellent candidates for applications in diverse niches within future decades, especially in environmental remediation. To fully capitalize on the exceptional qualities of these fascinating materials, "Mxenes", there are a number of difficulties as well as unexplored research areas that demand rapid attention. It is anticipated that existing obstacles will be overcome, and MXenes will become one of the stand-out materials for future water-based environmental remediation applications as a result of the expanding research in this field along with technological improvement, as presented in this chapter.

Acknowledgements The authors wish to acknowledge the Department of Chemical Sciences, University of Johannesburg, Doornfontein, Johannesburg 2028, South Africa, and DSI-CSIR Nanotechnology Innovation Centre, Council for Scientific and Industrial Research, Pretoria 0001, South Africa.

Declaration The authors have no conflict of interest.

References

1. M. Naguib et al., Two-dimensional nanocrystals produced by exfoliation of Ti3AlC2. Adv. Mater. **23**(37), 4248–4253 (2011)
2. M. Boota et al., Interaction of polar and nonpolar polyfluorenes with layers of two-dimensional titanium carbide (MXene): intercalation and pseudocapacitance. Chem. Mater. **29**(7), 2731–2738 (2017)

3. R.M. Ronchi, J.T. Arantes, S.F. Santos, Synthesis, structure, properties and applications of MXenes: current status and perspectives. Ceram. Int. **45**(15), 18167–18188 (2019)
4. S.K. Hwang et al., MXene: an emerging two-dimensional layered material for removal of radioactive pollutants. Chem. Eng. J. **397**, 125428 (2020)
5. J. Jang et al., Magnetic $Ti_3C_2T_x$ (Mxene) for diclofenac degradation via the ultraviolet/chlorine advanced oxidation process. Environ. Res. **182**, 108990 (2020)
6. Y.A. Al-Hamadani, et al., Applications of MXene-based membranes in water purification: a review. Chemosphere **254**, 126821 (2020)
7. G.J. Adekoya, et al., *Applications of MXene-Containing Polypyrrole Nanocomposites in Electrochemical Energy Storage and Conversion* (ACS Omega, 2022)
8. F. Dixit et al., Application of MXenes for water treatment and energy-efficient desalination: a review. J. Hazard. Mater. **423**, 127050 (2022)
9. J. Chen et al., Synthesis of MXene and its application for zinc-ion storage. SusMat **2**(3), 293–318 (2022)
10. A. Lipatov et al., Effect of synthesis on quality, electronic properties and environmental stability of individual monolayer Ti3C2 MXene flakes. Adv. Electron. Mater. **2**(12), 1600255 (2016)
11. J. Halim et al., Transparent conductive two-dimensional titanium carbide epitaxial thin films. Chem. Mater. **26**(7), 2374–2381 (2014)
12. Y. Guo et al., Synthesis of two-dimensional carbide Mo_2CT_x MXene by hydrothermal etching with fluorides and its thermal stability. Ceram. Int. **46**(11), 19550–19556 (2020)
13. Y. Gogotsi, Transition metal carbides go 2D. Nat. Mater. **14**(11), 1079–1080 (2015)
14. P. Urbankowski et al., Synthesis of two-dimensional titanium nitride Ti_4N_3 (MXene). Nanoscale **8**(22), 11385–11391 (2016)
15. S. Yang et al., Fluoride-free synthesis of two-dimensional titanium carbide (MXene) using a binary aqueous system. Angew. Chem. **130**(47), 15717–15721 (2018)
16. V. Natu et al., 2D Ti_3C_2Tz MXene synthesized by water-free etching of Ti_3AlC_2 in polar organic solvents. Chem **6**(3), 616–630 (2020)
17. B. Unnikrishnan et al., Synthesis and in situ sulfidation of molybdenum carbide MXene using fluorine-free etchant for electrocatalytic hydrogen evolution reactions. J. Colloid Interface Sci. **628**, 849–857 (2022)
18. Z. Sun et al., Selective lithiation–expansion–microexplosion synthesis of two-dimensional fluoride-free Mxene. ACS Mater. Lett. **1**(6), 628–632 (2019)
19. Z. Li et al., Impacts of oxygen vacancies on zinc ion intercalation in VO_2. ACS Nano **14**(5), 5581–5589 (2020)
20. C. Xu et al., Large-area high-quality 2D ultrathin Mo_2C superconducting crystals. Nat. Mater. **14**(11), 1135–1141 (2015)
21. D. Geng et al., Direct synthesis of large-area 2D Mo_2C on in situ grown graphene. Adv. Mater. **29**(35), 1700072 (2017)
22. V. Thirumal et al., Facile single-step synthesis of MXene@CNTs hybrid nanocomposite by CVD method to remove hazardous pollutants. Chemosphere **286**, 131733 (2022)
23. F. Turker et al., CVD synthesis and characterization of thin Mo_2C crystals. J. Am. Ceram. Soc. **103**(10), 5586–5593 (2020)
24. A. Zaman et al., Biopolymer-based nanocomposites for removal of hazardous dyes from water bodies, in *Innovations in Environmental Biotechnology*. ed. by S. Arora et al. (Springer Nature Singapore, Singapore, 2022), pp.759–783
25. J.T. Orasugh, S.S. Ray, *Nanocellulose-Graphene Oxide-Based Nanocomposite for Adsorptive Water Treatment*, in *Functional Polymer Nanocomposites for Wastewater Treatment*, ed. by M.J. Hato, S. Sinha Ray (Springer International Publishing, Cham, 2022), pp. 1–53
26. A. Shahzad et al., Two-dimensional $Ti_3C_2T_x$ MXene nanosheets for efficient copper removal from water. ACS Sustain. Chem. Eng. **5**(12), 11481–11488 (2017)
27. Y. Ying et al., Two-dimensional titanium carbide for efficiently reductive removal of highly toxic Chromium(VI) from water. ACS Appl. Mater. Interfaces. **7**(3), 1795–1803 (2015)

28. G. Zou et al., Synthesis of urchin-like rutile titania carbon nanocomposites by iron-facilitated phase transformation of MXene for environmental remediation. J. Mater. Chem. A **4**(2), 489–499 (2016)
29. A. Shahzad et al., Unique selectivity and rapid uptake of molybdenum-disulfide-functionalized MXene nanocomposite for mercury adsorption. Environ. Res. **182**, 109005 (2020)
30. A. Shahzad et al., Two-dimensional $Ti_3C_2T_x$ MXene nanosheets for efficient copper removal from water. ACS Sustain. Chem. Eng. **5**(12), 11481–11488 (2017)
31. P. Karthikeyan et al., Two-dimensional (2D) $Ti_3C_2T_x$ MXene nanosheets with superior adsorption behavior for phosphate and nitrate ions from the aqueous environment. Ceram. Int. **47**(1), 732–739 (2021)
32. Z. He et al., Ca2+ induced 3D porous MXene gel for continuous removal of phosphate and uranium. Appl. Surf. Sci. **570**, 150804 (2021)
33. O. Mashtalir et al., Dye adsorption and decomposition on two-dimensional titanium carbide in aqueous media. J. Mater. Chem. A **2**(35), 14334–14338 (2014)
34. A. Shahzad et al., Mercuric ion capturing by recoverable titanium carbide magnetic nanocomposite. J. Hazard. Mater. **344**, 811–818 (2018)
35. L. Wang et al., Rational control of the interlayer space inside two-dimensional titanium carbides for highly efficient uranium removal and imprisonment. Chem. Commun. **53**(89), 12084–12087 (2017)
36. B.-M. Jun et al., Adsorption of selected dyes on $Ti_3C_2T_x$ MXene and Al-based metal-organic framework. Ceram. Int. **46**(3), 2960–2968 (2020)
37. P. Karthikeyan et al., Effective removal of Cr (VI) and methyl orange from the aqueous environment using two-dimensional (2D) $Ti_3C_2T_x$ MXene nanosheets. Ceram. Int. **47**(3), 3692–3698 (2021)
38. S. Kim et al., Enhanced adsorption performance for selected pharmaceutical compounds by sonicated $Ti_3C_2T_x$ MXene. Chem. Eng. J. **406**, 126789 (2021)
39. A.A. Ghani et al., Adsorption and electrochemical regeneration of intercalated $Ti_3C_2T_x$ MXene for the removal of ciprofloxacin from wastewater. Chem. Eng. J. **421**, 127780 (2021)
40. S. Kim et al., Effect of single and multilayered $Ti_3C_2T_x$ MXene as a catalyst and adsorbent on enhanced sonodegradation of diclofenac and verapamil. J. Hazard. Mater. **426**, 128120 (2022)
41. C. Peng et al., Hybrids of two-dimensional Ti_3C_2 and TiO_2 exposing 001 facets toward enhanced photocatalytic activity. ACS Appl. Mater. Interfaces **8**(9), 6051–6060 (2016)
42. J. Qu et al., Preparation and regulation of two-dimensional $Ti_3C_2T_x$ MXene for enhanced adsorption–photocatalytic degradation of organic dyes in wastewater. Ceram. Int. **48**(10), 14451–14459 (2022)
43. B.-M. Jun et al., Ultrasonic degradation of selected dyes using $Ti_3C_2T_x$ MXene as a sonocatalyst. Ultrason. Sonochem. **64**, 104993 (2020)
44. M. Jeon et al., Sonodegradation of amitriptyline and ibuprofen in the presence of $Ti_3C_2T_x$ MXene. J. Hazard. Mater. Lett. **2**, 100028 (2021)
45. A. Shahzad et al., Heterostructural $TiO_2/Ti_3C_2T_x$ (MXene) for photocatalytic degradation of antiepileptic drug carbamazepine. Chem. Eng. J. **349**, 748–755 (2018)
46. H. Wang et al., Titania composites with 2D transition metal carbides as photocatalysts for hydrogen production under visible-light irradiation. Chemsuschem **9**(12), 1490–1497 (2016)
47. Q. Zhang et al., Defect-engineered MXene monolith enabling interfacial photothermal catalysis for high-yield solar hydrogen generation. Cell Rep. Phys. Sci. **3**(5), 100877 (2022)
48. X. Chen et al., Titanium carbide MXenes coupled with cadmium sulfide nanosheets as two-dimensional/two-dimensional heterostructures for photocatalytic hydrogen production. J. Colloid Interface Sci. **613**, 644–651 (2022)
49. Y. Wang et al., Ti_3C_2 MXene coupled with CdS nanoflowers as 2D/3D heterostructures for enhanced photocatalytic hydrogen production activity. Int. J. Hydrog. Energy **47**(52), 22045–22053 (2022)

50. B. Sun et al., The fabrication of 1D/2D CdS nanorod@Ti_3C_2 MXene composites for good photocatalytic activity of hydrogen generation and ammonia synthesis. Chem. Eng. J. **406**, 127177 (2021)
51. J. Ran et al., Ti_3C_2 MXene co-catalyst on metal sulfide photo-absorbers for enhanced visible-light photocatalytic hydrogen production. Nat. Commun. **8**(1), 1–10 (2017)
52. B.-M. Jun et al., Adsorption of Ba^{2+} and Sr^{2+} on $Ti_3C_2T_x$ MXene in model fracking wastewater. J. Environ. Manag. **256**, 109940 (2020)
53. B.-M. Jun et al., Effective removal of Pb (ii) from synthetic wastewater using $Ti_3C_2T_x$ MXene. Environ. Sci. Water Res. Technol. **6**(1), 173–180 (2020)
54. Z. Othman, H.R. Mackey, K.A. Mahmoud, A critical overview of MXenes adsorption behavior toward heavy metals. Chemosphere **295**, 133849 (2022)
55. Q. Peng et al., Unique lead adsorption behavior of activated hydroxyl group in two-dimensional titanium carbide. J. Am. Chem. Soc. **136**(11), 4113–4116 (2014)
56. A. Shahzad et al., $Ti_3C_2T_x$ MXene core-shell spheres for ultrahigh removal of mercuric ions. Chem. Eng. J. **368**, 400–408 (2019)
57. C.E. Ren et al., Charge- and size-selective ion sieving through $Ti_3C_2T_x$ MXene membranes. J. Phys. Chem. Lett. **6**(20), 4026–4031 (2015)
58. J. Guo et al., Heavy-metal adsorption behavior of two-dimensional alkalization-intercalated MXene by first-principles calculations. J. Phys. Chem. C **119**(36), 20923–20930 (2015)
59. X. Guo et al., High adsorption capacity of heavy metals on two-dimensional MXenes: an ab initio study with molecular dynamics simulation. Phys. Chem. Chem. Phys. **18**(1), 228–233 (2016)
60. R.P. Pandey et al., Reductive sequestration of toxic bromate from drinking water using lamellar two-dimensional $Ti_3C_2T_x$ (MXene). ACS Sustain. Chem. Eng. **6**(6), 7910–7917 (2018)
61. Q. Zhang et al., Efficient phosphate sequestration for water purification by unique sandwich-like MXene/magnetic iron oxide nanocomposites. Nanoscale **8**(13), 7085–7093 (2016)
62. D. Gan et al., Bioinspired functionalization of MXenes ($Ti_3C_2T_x$) with amino acids for efficient removal of heavy metal ions. Appl. Surf. Sci. **504**, 144603 (2020)
63. S. Wang et al., Facile preparation of biosurfactant-functionalized Ti_2CT_X MXene nanosheets with an enhanced adsorption performance for Pb (II) ions. J. Mol. Liq. **297**, 111810 (2020)
64. Z. Wei et al., Alkali treated $Ti_3C_2T_x$ MXenes and their dye adsorption performance. Mater. Chem. Phys. **206**, 270–276 (2018)
65. Z. Zhu et al., Effect of temperature on methylene blue removal with novel 2D-Magnetism titanium carbide. J. Solid State Chem. **280**, 120989 (2019)
66. Y. Cui et al., A novel one-step strategy for preparation of Fe_3O_4-loaded Ti_3C_2 MXenes with high efficiency for removal organic dyes. Ceram. Int. **46**(8), 11593–11601 (2020)
67. B.-M. Jun et al., Ultrasound-assisted $Ti_3C_2T_x$ MXene adsorption of dyes: removal performance and mechanism analyses via dynamic light scattering. Chemosphere **254**, 126827 (2020)
68. Y. Lei et al., Facile preparation of sulfonic groups functionalized Mxenes for efficient removal of methylene blue. Ceram. Int. **45**(14), 17653–17661 (2019)
69. Y. Gao et al., Hydrothermal synthesis of TiO_2/Ti_3C_2 nanocomposites with enhanced photocatalytic activity. Mater. Lett. **150**, 62–64 (2015)
70. N.-N. Wang et al., Robust, lightweight, hydrophobic, and fire-retarded polyimide/MXene aerogels for effective oil/water separation. ACS Appl. Mater. Interfaces **11**(43), 40512–40523 (2019)
71. Z.-K. Li et al., Ultra-thin titanium carbide (MXene) sheet membranes for high-efficient oil/water emulsions separation. J. Membr. Sci. **592**, 117361 (2019)
72. J. Saththasivam et al., A flexible $Ti_3C_2T_x$ (MXene)/paper membrane for efficient oil/water separation. RSC Adv. **9**(29), 16296–16304 (2019)
73. H. Zhang et al., Ultrathin 2D $Ti_3C_2T_x$ MXene membrane for effective separation of oil-in-water emulsions in acidic, alkaline, and salty environment. J. Colloid Interface Sci. **561**, 861–869 (2020)

74. S. He et al., Chemically stable two-dimensional MXene@UIO-66-(COOH)$_2$ composite lamellar membrane for multi-component pollutant-oil-water emulsion separation. Compos. B Eng. **197**, 108188 (2020)
75. A. Moghaddasi et al., Separation of water/oil emulsions by an electrospun copolyamide mat covered with a 2D $Ti_3C_2T_x$ MXene. Materials **13**(14), 3171 (2020)
76. S. Luo et al., Preparation and dye degradation performances of self-assembled MXene-Co_3O_4 nanocomposites synthesized via solvothermal approach. ACS Omega **4**(2), 3946–3953 (2019)
77. K. Li et al., Self-assembled MXene-based nanocomposites via layer-by-layer strategy for elevated adsorption capacities. Colloids Surf. A Phys. Chem. Eng. Asp. **553**, 105–113 (2018)
78. A. Sarycheva et al., Two-dimensional titanium carbide (MXene) as surface-enhanced Raman scattering substrate. J. Phys. Chem. C **121**(36), 19983–19988 (2017)
79. H. Zhang et al., Computational studies on the structural, electronic and optical properties of graphene-like MXenes (M_2CT_2, M = Ti, Zr, Hf; T = O, F, OH) and their potential applications as visible-light driven photocatalysts. J. Mater. Chem. A **4**(33), 12913–12920 (2016)
80. K. Xiong et al., Functional group effects on the photoelectronic properties of MXene (Sc_2CT_2, T = O, F, OH) and their possible photocatalytic activities. Sci. Rep. **7**(1), 1–8 (2017)
81. R. Li et al., MXene Ti_3C_2: an effective 2D light-to-heat conversion material. ACS Nano **11**(4), 3752–3759 (2017)
82. X. Zhong et al., The fabrication of 3D hierarchical flower-like δ-MnO_2@COF nanocomposites for the efficient and ultra-fast removal of UO_2^{2+} ions from aqueous solution. Environ. Sci. Nano **7**(11), 3303–3317 (2020)
83. S. Li et al., Adsorption and mechanistic study of the invasive plant-derived biochar functionalized with CaAl-LDH for Eu (III) in water. J. Environ. Sci. **96**, 127–137 (2020)
84. X. Zhong et al., The magnetic covalent organic framework as a platform for high-performance extraction of Cr (VI) and bisphenol a from aqueous solution. J. Hazard. Mater. **393**, 122353 (2020)
85. Y.-J. Zhang et al., Theoretical insights into the uranyl adsorption behavior on vanadium carbide MXene. Appl. Surf. Sci. **426**, 572–578 (2017)
86. L. Wang et al., Porous carbon-supported gold nanoparticles for oxygen reduction reaction: effects of nanoparticle size. ACS Appl. Mater. Interfaces **8**(32), 20635–20641 (2016)
87. B. Anasori et al., Two-dimensional, ordered, double transition metals carbides (MXenes). ACS Nano **9**(10), 9507–9516 (2015)
88. X. Zhong et al., Aluminum-based metal-organic frameworks (CAU-1) highly efficient UO_2^{2+} and TcO_4^- ions immobilization from aqueous solution. J. Hazard. Mater. **407**, 124729 (2021)
89. W. Mu et al., Removal of radioactive palladium based on novel 2D titanium carbides. Chem. Eng. J. **358**, 283–290 (2019)
90. B.-M. Jun et al., Selective adsorption of Cs+ by MXene ($Ti_3C_2T_x$) from model low-level radioactive wastewater. Nucl. Eng. Technol. **52**(6), 1201–1207 (2020)
91. M. ul Hassan, et al., Post-decontamination treatment of MXene after adsorbing Cs from contaminated water with the enhanced thermal stability to form a stable radioactive waste matrix. J. Nucl. Mater. **543**, 152566 (2021)
92. L. Wang et al., Effective removal of anionic Re (VII) by surface-modified Ti_2CT_x MXene nanocomposites: implications for Tc (VII) sequestration. Environ. Sci. Technol. **53**(7), 3739–3747 (2019)
93. P. Zhang et al., Effective removal of U (VI) and Eu (III) by carboxyl functionalized MXene nanosheets. J. Hazard. Mater. **396**, 122731 (2020)
94. L. Wang et al., Efficient U (VI) reduction and sequestration by Ti_2CT_x MXene. Environ. Sci. Technol. **52**(18), 10748–10756 (2018)
95. S. Wang et al., Highly efficient adsorption and immobilization of U (VI) from aqueous solution by alkalized MXene-supported nanoscale zero-valent iron. J. Hazard. Mater. **408**, 124949 (2021)
96. P. Zhang et al., Aryl diazonium-assisted amidoximation of MXene for boosting water stability and uranyl sequestration via electrochemical sorption. ACS Appl. Mater. Interfaces **12**(13), 15579–15587 (2020)

97. L. Wang et al., Layered structure-based materials: challenges and opportunities for radionuclide sequestration. Environ. Sci. Nano **7**(3), 724–752 (2020)
98. L. Ding et al., Effective ion sieving with $Ti_3C_2T_x$ MXene membranes for production of drinking water from seawater. Nat. Sustain. **3**(4), 296–302 (2020)
99. I.A. Vasyukova, et al., Synthesis, toxicity assessment, environmental and biomedical applications of MXenes: a review. Nanomaterials **12** (2022). https://doi.org/10.3390/nano12111797
100. H. Zhou et al., Water permeability in MXene membranes: process matters. Chin. Chem. Lett. **31**(6), 1665–1669 (2020)
101. S. Jiang et al., Synthesis of polyurea from 1, 6-hexanediamine with CO_2 through a two-step polymerization. Green Energy Environ. **2**(4), 370–376 (2017)
102. Y. Chen, et al., CO_2 capture and conversion to value-added products promoted by MXene-based materials. Green Energy Environ. (2020)
103. M. He, Y. Sun, B. Han, Green carbon science: scientific basis for integrating carbon resource processing, utilization, and recycling. Angew. Chem. Int. Ed. **52**(37), 9620–9633 (2013)
104. D. Voiry et al., Low-dimensional catalysts for hydrogen evolution and CO_2 reduction. Nat. Rev. Chem. **2**(1), 1–17 (2018)
105. L. Zhang et al., Nano-designed semiconductors for electro-and photoelectro-catalytic conversion of carbon dioxide. Chem. Soc. Rev. **47**(14), 5423–5443 (2018)
106. C. Yoo, Y.-E. Kim, Y. Lee, Selective transformation of CO_2 to CO at a single nickel center. Acc. Chem. Res. **51**(5), 1144–1152 (2018)
107. D.M. Weekes et al., Electrolytic CO_2 reduction in a flow cell. Acc. Chem. Res. **51**(4), 910–918 (2018)
108. C. Giordano et al., Metal nitride and metal carbide nanoparticles by a soft urea pathway. Chem. Mater. **21**(21), 5136–5144 (2009)
109. Á. Morales-García et al., CO_2 abatement using two-dimensional MXene carbides. J. Mater. Chem. A **6**(8), 3381–3385 (2018)
110. B. Wang et al., Carbon dioxide adsorption of two-dimensional carbide MXenes. J. Adv. Ceram. **7**(3), 237–245 (2018)
111. S. Jin et al., Carbon dioxide adsorption of two-dimensional Mo_2C MXene. Diam. Relat. Mater. **128**, 109277 (2022)
112. K.S. Sing, Reporting physisorption data for gas/solid systems with special reference to the determination of surface area and porosity (Recommendations 1984). Pure Appl. Chem. **57**(4), 603–619 (1985)
113. M. Kruk et al., Characterization of MCM-48 silicas with tailored pore sizes synthesized via a highly efficient procedure. Chem. Mater. **12**(5), 1414–1421 (2000)
114. M. Park, S. Komarneni, Stepwise functionalization of mesoporous crystalline silica materials. Microporous Mesoporous Mater. **25**(1–3), 75–80 (1998)
115. A. Kurlov et al., Exploiting two-dimensional morphology of molybdenum oxycarbide to enable efficient catalytic dry reforming of methane. Nat. Commun. **11**(1), 1–11 (2020)
116. M. Kruk, M. Jaroniec, Gas adsorption characterization of ordered organic−inorganic nanocomposite materials. Chem. Mater. **13**(10), 3169–3183 (2001)
117. A.A. Shamsabadi et al., Pushing rubbery polymer membranes to be economic for CO_2 separation: embedment with $Ti_3C_2T_x$ MXene Nanosheets. ACS Appl. Mater. Interfaces **12**(3), 3984–3992 (2019)
118. M.S. Boutilier et al., Implications of permeation through intrinsic defects in graphene on the design of defect-tolerant membranes for gas separation. ACS Nano **8**(1), 841–849 (2014)
119. S.J. Kim et al., Metallic Ti3C2T x MXene gas sensors with ultrahigh signal-to-noise ratio. ACS Nano **12**(2), 986–993 (2018)
120. L. Ding et al., MXene molecular sieving membranes for highly efficient gas separation. Nat. Commun. **9**(1), 1–7 (2018)
121. A. Ali Khan, M. Tahir, Construction of an S-Scheme heterojunction with oxygen-vacancy-rich trimetallic CoAlLa-LDH anchored on titania-sandwiched Ti_3C_2 multilayers for boosting photocatalytic CO_2 reduction under visible light. Ind. Eng. Chem. Res. **60**(45), 16201–16223 (2021)

122. Y. Chen, E.A. Silva, Smart transport: a comparative analysis using the most used indicators in the literature juxtaposed with interventions in English metropolitan areas. Transp. Res. Interdiscip. Perspect. **10**, 100371 (2021)
123. H. Wang, Q. Tang, Z. Wu, Construction of few-layer Ti_3C_2 MXene and boron-doped g-C_3N_4 for enhanced photocatalytic CO_2 reduction. ACS Sustain. Chem. Eng. **9**(25), 8425–8434 (2021)
124. A. Pan et al., $CsPbBr_3$ perovskite nanocrystal grown on MXene nanosheets for enhanced photoelectric detection and photocatalytic CO_2 reduction. J. Phys. Chem. Lett. **10**(21), 6590–6597 (2019)
125. M. Que et al., Anchoring of formamidinium lead bromide quantum dots on Ti_3C_2 nanosheets for efficient photocatalytic reduction of CO_2. ACS Appl. Mater. Interfaces **13**(5), 6180–6187 (2021)
126. X. Sang et al., Atomic defects in monolayer titanium carbide ($Ti_3C_2T_x$) MXene. ACS Nano **10**(10), 9193–9200 (2016)
127. W. Cui et al., Atomic defects, functional groups and properties in MXenes. Chin. Chem. Lett. **32**(1), 339–344 (2021)
128. V. Parey et al., High-throughput screening of atomic defects in MXenes for CO_2 capture, activation, and dissociation. ACS Appl. Mater. Interfaces **13**(30), 35585–35594 (2021)
129. R. Khaledialidusti, A.K. Mishra, A. Barnoush, Atomic defects in monolayer ordered double transition metal carbide ($Mo_2TiC_2T_x$) MXene and CO_2 adsorption. J. Mater. Chem. C **8**(14), 4771–4779 (2020)
130. Q. Wang et al., Recent progress in thermal conversion of CO_2 via single-atom site catalysis. Small Struct. **3**(9), 2200059 (2022)
131. J. Jones et al., Thermally stable single-atom platinum-on-ceria catalysts via atom trapping. Science **353**(6295), 150–154 (2016)
132. D. Zhao et al., MXene (Ti_3C_2) vacancy-confined single-atom catalyst for efficient functionalization of CO_2. J. Am. Chem. Soc. **141**(9), 4086–4093 (2019)
133. N. Li et al., Understanding of electrochemical mechanisms for CO_2 capture and conversion into hydrocarbon fuels in transition-metal carbides (MXenes). ACS Nano **11**(11), 10825–10833 (2017)
134. A.D. Handoko et al., Establishing new scaling relations on two-dimensional MXenes for CO_2 electroreduction. J. Mater. Chem. A **6**(44), 21885–21890 (2018)
135. D. Yang et al., Selective electroreduction of carbon dioxide to methanol on copper selenide nanocatalysts. Nat. Commun. **10**(1), 1–9 (2019)
136. Q. Zhu et al., Efficient reduction of CO_2 into formic acid on a lead or tin electrode using an ionic liquid catholyte mixture. Angew. Chem. **128**(31), 9158–9162 (2016)
137. Q. Zhu et al., Hollow metal–organic-framework-mediated in situ architecture of copper dendrites for enhanced CO_2 electroreduction. Angew. Chem. **132**(23), 8981–8986 (2020)
138. A.D. Handoko et al., Two-dimensional titanium and molybdenum carbide MXenes as electrocatalysts for CO_2 reduction. Iscience **23**(6), 101181 (2020)
139. K. Kannan et al., Fabrication of ZnO-Fe-MXene based nanocomposites for efficient CO_2 reduction. Catalysts **10**(5), 549 (2020)
140. D. Qu et al., Nitrogen doping and titanium vacancies synergistically promote CO_2 fixation in seawater. Nanoscale **12**(33), 17191–17195 (2020)

Chapter 3
Inorganic Analogues of Graphene and Their Nanocomposites for Wastewater Treatment

Pratiksha Joshi, Sweta Mehta, Anchal Pandey, and Om. P. Khatri

Abstract Water scarcity has been a grave concern because of increasing urbanization and industrialization activities, unrestrained exploration of natural sources, and depletion of water table. Water pollution causes severe health hazards and jeopardizes biodiversity and the aquatic ecosystem. Therefore, wastewater treatment technologies for providing clean water through sustainable and economic approaches are gaining increasing interest. Adsorption and photocatalytic degradation of organic pollutants by a wide range of activated carbons and nanostructured materials have been addressed for their adsorptive separation and mineralization. The invention of graphene in 2004 has propelled immense interest in different types of two-dimensional (2D) nanostructured materials for a diversified range of energy and environmental applications. The 2D nanostructured analogues of graphene, viz. MoS_2, WS_2, *h*-BN, g-C_3N_4, MXenes, and their composites of high accessible surface area, controlled surface functionalities, and tunable band-gap have shown excellent performance for wastewater treatment. The chapter covers a comprehensive overview of various types of pollutants in water and recent developments on their adsorptive removal and photocatalytic mineralization/conversion by nanostructured MoS_2, WS_2, *h*-BN, g-C_3N_4, MXenes, and their nanocomposites, heterostructures, and hybrids. The structural, surface, textural, and chemical properties of 2D nanomaterials are reviewed to highlight their roles in the adsorptive removal and photocatalytic degradation of organic pollutants. The chapter also covers futuristic opportunities, environmental sustainability, and technological challenges in preparing and applying inorganic analogues of graphene for wastewater treatment.

P. Joshi · S. Mehta · A. Pandey · Om. P. Khatri (✉)
CSIR – Indian Institute of Petroleum, Mohkampur 248005, Dehradun, India
e-mail: opkhatri@iip.res.in

Academy of Scientific and Innovative Research, Ghaziabad 201002, India

N. Kumar et al. (eds.), *Two-Dimensional Materials for Environmental Applications*, Springer Series in Materials Science 332, https://doi.org/10.1007/978-3-031-28756-5_3

3.1 Introduction

The rapid pace of industrialization, urban intensification, unrestricted use of natural resources, population rise, and technological advancement causes several adverse impacts on our surroundings. The water bodies like ponds, waterways, rivers, seas, etc., are often contaminated by industrial effluents, domestic discharges, and agricultural runoffs. More than 30% of the worldwide population lacks access to clean and fresh water [1]. Every day, a million gallons of either poorly treated or non-treated effluents are discharged by textile, refinery, leather, pharmaceutical, fertilizer, pesticide, metal plating, and chemical industries, which contaminate nearby terrestrial lands and water bodies. The industrial effluents carry numerous inorganic, organic, and biological pollutants like dyes, heavy metals, pesticides, oxyanions, pathogens, surfactants, active pharmaceutical ingredients (APIs), polycyclic aromatic hydrocarbons (PAHs), endocrine disruptors, phenolic compounds, personal care products, fire retardants, fluoro compounds, and so on [2]. These contaminants permeate the environment and show intense concerns due to ecological threats, health hazards, and adverse impacts on flora and fauna. Organic dyes, heavy metals, oxyanions, pesticides, and APIs are major pollutants discharged into water.

Organic dyes are vital for various applications in textile, leather, paints, cosmetics, chemicals, pharmaceuticals, and food processing industries. These dyes are water-soluble, and their presence governs the color of effluent streams. Most dyes are heat and detergent-resistant, making them hard to degrade [3]. Dyes in water bodies obstruct sunlight penetration and severally imperil the growth of marine plants. Moreover, these dyes induce toxicity to aquatic life [4]. The dyes-contaminated water poses various health hazards, including diseases associated with critical organs like the kidney, liver, brain, and nervous system. Many organic dyes are mutagenic, teratogenic, and carcinogenic toxins [3].

Pesticides and APIs are other important classes of organic pollutants in water bodies. Agro-food, fruit-packaging, seed-handling, and agricultural runoffs are major sources of fungicides, herbicides, and rodenticides in aquatic bodies [5]. Pesticides in the human body mimic human hormones, leading to poor immunogenicity, hormonal changes, respiratory and reproductive ailments, carcinogenesis, genetic disorders, and intelligence quotient decay, particularly in children [6]. APIs have been recognised as major water pollutants because of their poor biodegradability and bioaccumulation. The APIs include steroidal drugs, anti-bacterials, analgesics, antibiotics, antipyretics, etc. The continual discharge of these contaminants into the environment adversely impacts aquatic systems and human health [7]. Inorganic pollutants, including heavy metals, metalloids, and oxyanions, are another major category of pollutants in aquatic bodies. Their yearly contamination is comparatively higher than the collective contamination of organic and radioactive pollutants [8]. Arsenic, cadmium, mercury, lead, chromium, and fluorides are prioritised pollutants due to their severer health hazards. The permitted limit of these pollutants in water is very stringent. The consumption of these metal ions and their oxyanions beyond the permissible limit interrupts the critical functions of several vital organs, like the kidney, brain, liver, lungs, and nervous system, causing severe diseases, including arsenicosis, Parkinson's, Itai-Itai, and fluorosis [9].

Water contamination has been a major cause of water scarcity besides the uncontrolled exploitation of groundwater and inadequate charge of the water table, particularly in urban localities. Therefore, the efficient removal of different types of water pollutants in economic and scalable approaches is gaining immense interest for a sustainable supply of clean water. Photocatalytic degradation and mineralization, ion exchange, electrocatalytic decomposition, reverse osmosis, adsorption, membrane separation, bioremediation, etc., have been addressed for wastewater remediation applications to remove and decompose various contaminants [9]. The chapter is focused on adsorptive separation and photocatalytic degradation of water pollutants.

Adsorption is the adhesion of contaminates (adsorbate) over the surface of the adsorbent. Different types of adsorbents like metal–organic frameworks, biomass, biochar, activated carbons, zeolites, clays, metal oxides, graphene and its analogues like *h*-BN, MoS_2, WS_2, MXenes, and their nanocomposites, biopolymers like chitosan, cellulose, polyacrylic acid, poly(vinyl alcohol), starch, gelatin, agar, etc., are explored to remove various organic and inorganic pollutants. Photocatalytic degradation is a practical approach for removing non-biodegradable water pollutants. It converts the recalcitrant toxic pollutants into small molecules (including complete mineralization) and non-toxic products [8]. The large surface area having ample surface-active sites with high affinity for targeted pollutants are some of the most desired properties required in photocatalysts besides their photo activities under light exposure [10]. In a photocatalytic process, light exposure causes the excitation of electrons from the valence band to the conduction band. The highly active species ($OH^{\bullet}$, $O_2^{\bullet -}$, and h^+) generated during photocatalysis have a short lifetime. These species are formed in an aqueous medium due to the complex redox reactions and subsequently decompose the toxic pollutants into smaller non-hazardous molecules. Numerous semiconductors like MoS_2, metal oxides, metal–metal complexes, metal–organic framework, MXenes, and their hybrids, heterostructures, and nanocomposites have been demonstrated as photocatalysts [3, 11–13].

Graphene, a 2D honeycomb arrangement of carbon atoms, has shown vast interest because of its unmatchable mechanical properties, excellent electrical and thermal conductivity, remarkable optical transparency, large surface area, and rich surface chemistry [14]. These properties made graphene a material of the decade with its utilization in a diversified range of applications, including electronics, adsorption, sensors, energy conversion and storage devices (supercapacitors, batteries, fuel cells, solar cells), water purification, lubrication, catalysis (electrocatalysis, photocatalysis), hydrogen generation, coatings, etc. [15]. The remarkable features of graphene have fascinated the interest of scientific community with other 2D materials. The inorganic analogues of graphene, a new category of 2D materials showed immense potential for basic and applied research in various disciplines such as electronics, energy, wastewater remediation, photonics, optoelectronic devices, sensors, photocatalysts, lubrication, and so on [16]. Environmental applications, especially the use of inorganic analogues of graphene for the photocatalytic degradation and adsorption of variable pollutants from wastewater and air, have gained the interest of the scientific fraternity. Textural attributes, surface characteristics, a wide range of band gaps, structural defects, and tuneable morphologies of 2D nanostructured materials are the primary features that influence and select their prospects for environmental applications.

Hexagonal boron nitride (*h*-BN), an sp^2 hybridised hexagonal arrangement of B and N at alternate positions, has been identified as a promising material for wastewater remediation [17]. The *h*-BN, also known as "white graphene," has several advantages over graphene, including a wide band gap, electrical insulation, and superior stability [18]. Moreover, water molecules easily permeate through the layers of *h*-BN compared to graphene because of its low surface tension [19]. The nanostructured *h*-BN can also be recycled after burning or heating because of its excellent oxidative stability, which enables its reusability [20].

Molybdenum disulphide (MoS_2), a transition metal dichalcogenide, has been explored for wastewater remediation applications, including the development of membranes for water treatment-related applications [21]. The dangling sulphur atoms of MoS_2 effectively adsorb heavy metals like Hg^{+2} and Ag^+ via Lewis acid–base interactions [22]. The nanostructured MoS_2 effectively adsorbs organic dyes via Van der Waals and electrostatic interactions [23]. The MoS_2 nanosheets are the most potential photocatalysts that can effectively mineralise organic pollutants [23]. The WS_2, a structural analogue of MoS_2, displays excellent potential for adsorptive separation and photocatalytic decomposition of organic dyes. Moreover, the biocompatibility of MoS_2 and WS_2 makes them promising materials for wastewater remediation [24]. The graphitic carbon nitride (g-C_3N_4) is a metal-free photocatalyst with its wide band gap and visible light adsorption ability to decompose water pollutants [25]. The g-C_3N_4 exhibits a layered structure and consists of two basic structural units: s-triazine and tri-s-triazine [26]. MXene has become a prominent player in water purification technology owing to its layered structure, photocatalytic characteristics, and ample hydroxyl sites [27]. The high surface area, chemical stability, and antimicrobial activity make MXenes promising materials for many applications, including wastewater remediation [28].

The chapter presents a thorough insight into various types of organic pollutants in water and recent advances in adsorptive removal and photocatalytic mineralisation of these pollutants using 2D nanostructured materials such as MoS_2, WS_2, *h*-BN, g-C_3N_4, MXenes, and their composites, hybrids, and heterostructures. The roles of structural, surface, textural, and chemical properties of these 2D nanomaterials in the adsorptive removal and photocatalytic degradation of organic pollutants are presented. The chapter also discussed the bottlenecks for technological development in employing inorganic analogues of graphene for wastewater treatment at industrial scales, as well as highlighted their opportunities for the future.

3.2 Adsorptive Removal of Water Pollutants

Industrial effluents, domestic discharges, and agricultural runoffs are major contributors to water pollution. The source of contamination governs the type of pollutants in the wastewater. Adsorption has been the most effective, scalable, and economical approach for separating various pollutants from wastewater. The textural, microstructural, morphological, and chemical features of adsorbents play important roles in the adsorption process. Graphene and its 2D analogues like MoS_2, WS_2, *h*-BN, g-C_3N_4, and MXenes are explored as novel adsorbents for wastewater treatment applications. The textural properties, surface-active sites, structural defects, and tuneable morphologies of inorganic analogous of graphene and their nanocomposites make them suitable materials for adsorptive separation of water pollutants, viz., organic dyes, pesticides, APIs, heavy metals, and oxyanions.

3.2.1 Adsorptive Removal of Organic Dyes

The 2D transition metal dichalcogenides like MoS_2 and WS_2 nanosheets have shown interest as adsorbent materials because of their large surface area, structural defects, dangling sulfur sites, and polarity in the Mo/W-S bond. The polarity of Mo/W-S bond facilitates the organic dye adsorption [29–31]. The microspheres of MoS_2 nanosheets adsorbed both cationic and anionic dyes from an aqueous medium via electrostatic and Van der Waals interactions between MoS_2 and dye molecules [32]. Although MoS_2 is a potential material for dye adsorption; however, the agglomeration of MoS_2 nanosheets in water compromises the adsorption performance. The incorporation and integration of nanostructured MoS_2 with polymers, 2D materials, metals, carbon, and biopolymers improved the adsorption performance and increased the reach of micropollutants towards the adsorption sites efficiently. The MoS_2-rGO nanocomposite prepared through a facile hydrothermal route showed excellent adsorptive separation of Congo red dye with a maximum adsorption capacity of 410 $mg.g^{-1}$ [33]. The integration of MoS_2 with polylactic acid furnished a composite membrane, which reduced the stacking of MoS_2, increased the distribution of edge sites, and offered easily accessible ample active sites for adsorptive separation [34]. The saponin surfactant-assisted exfoliated MoS_2 showed good dispersibility in water. The MoS_2 aerogel prepared by lyophilization of aqueous dispersion of saponin surfactant-assisted exfoliated MoS_2 adsorbed 94% methylene blue dye within five minutes. The MoS_2 aerogel decomposed 80% of adsorbed dye via photocatalytic process [35]. A nanohybrid of polyaniline (PANI) with nanostructured MoS_2, i.e., PANI-MoS_2 showed good adsorptive separation of anionic Congo red dye. The electrostatic interactions between cationic amino functionalities and anionic dye in an acidic medium, π-π interactions, and hydrogen linkages facilitated the dye adsorption (Fig. 3.1). Protonation and deprotonation of PANI-MoS_2 at lower and higher pH expedited the adsorption and recyclability of the prepared material [35].

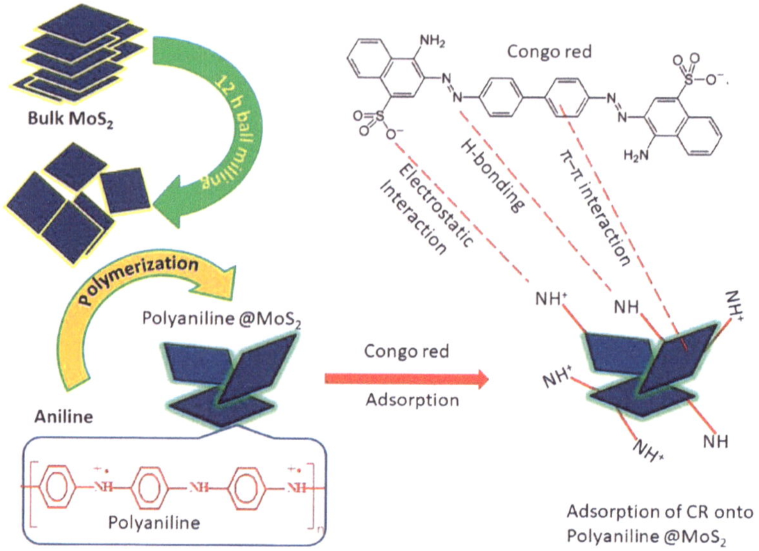

Fig. 3.1 Adsorption of Congo red dye by PANI-MoS_2 nanocomposite and demonstration of plausible adsorptive interaction pathways between PANI-MoS_2 and dye molecules. Reproduced with permission from ref. [35]. Copyright 2018, Royal Society of Chemistry

The nanostructured WS_2, a structural analogue of MoS_2, exhibits similar morphology and polar W-S bond. The WS_2 is not explored extensively for the adsorptive removal of organic dyes regardless of having properties comparable to MoS_2. Flower-like nanostructured WS_2 effectively adsorbed methylene blue dye with an adsorption capacity of 210 $mg.g^{-1}$, and the adsorption equilibrium was established within 2 min [36]. The WS_2/WO_3 heterostructure showed significantly higher adsorption of Rhodamine B dye than WS_2. The incorporation of WO_3 increased the adsorption capacity from 16.4 to 237.1 $mg.g^{-1}$ for Rhodamine B dye. The WO_3 in WS_2/WO_3 heterostructure significantly improved the wettability and negatively charged surface-active sites, which facilitated the adsorption of cationic dye molecules over the surface-active sites of hybrid materials via electrostatic interactions [30].

The *h*-BN, also known as white graphene, exhibits excellent wastewater remediation performance. The hollow spheres of *h*-BN adsorbed the Basic Yellow 1 (191.7 $mg.g^{-1}$) and methylene blue (116.5 $mg.g^{-1}$) dyes. The adsorption of organic dyes on the surface of *h*-BN hollow spheres is attributed to the electrostatic attraction between polar B-N and positively charged dye beside the π-π stacking [37]. The *h*-BN micro rods displayed excellent adsorptive separation of methylene blue dye with an adsorption capacity of 925 $mg.g^{-1}$ [38]. Carbon doping in *the h*-BN network (BCN) enhanced the electron delocalisation and produced ample surface-active zigzag edges. Consequently, nitrogen and sulphur atoms (base sites) of dye molecules showed an excellent affinity with boron atoms of BCN (acid sites) to expedite the adsorptive removal of organic dyes [23]. The porous BCN structure with ample amine sites showed remarkable enhancement in organic dye adsorption. The carbon doping and enrichment of amino sites in nanostructured BCN enhanced the adsorption events, including the dye adsorption rate [39].

MXenes, a new class of 2D inorganic compounds, have garnered considerable attention because of their distinctive 2D surface chemistry, high wettability, and surface-functionalisation potential [40]. The surface functionalities like -OH, -F, and -O^- in MXenes furnish hydrophilicity and good ion-exchange activity to facilitate adsorption. The multi-layered $Ti_3C_2T_x$ showed the adsorption of methylene blue and Acid blue 80 dyes [41]. The alkali (LiOH/NaOH) treatment increased interlayer spacing in $Ti_3C_2T_x$ MXene with the substitution of fluoro by hydroxyl functionalities. The synergy of surface functionalisation and increased inter-layer spacing of MXene improved the adsorption capacity to 189 $mg.g^{-1}$ for methylene blue [42]. Alkali-treated MXene with acrylic acid modification showed good adsorption of Congo red and methylene blue dyes driven by electrostatic interaction, inter-layer adsorption, and hydrogen bonding [43]. Although applying MXenes for the adsorptive removal of organic dyes is gaining considerable interest, the preparation of MXenes with controlled thickness, intercalation features, and surface chemistry is still in the infancy phase compared to other 2D analogues. The advancement in surface chemistry and engineering of MXenes can make them promising candidates for effective and fast adsorptive separation and selective extraction of organic dyes. The hydrophilicity and negative surface charge of MXenes make them highly efficient compared to other 2D nanomaterials. However, agglomeration and stacking of nanostructured MXenes inhibit the mass transfer pathways for effective adsorption of organic pollutants [44].

The low toxicity, ample structural defects, and surface-active sites make g-C_3N_4 a promising candidate for wastewater remediation [45]. The GO/g-C_3N_4 nanocomposite functionalised with Fe_3O_4 nanoparticles via hydrothermal method adsorbed methylene blue dye with a maximum adsorption capacity of 220 $mg.g^{-1}$. The introduction of Fe_3O_4 nanoparticles increased the magnetic force-driven separability of the nanocomposite after adsorption from an aqueous medium. The surface functionalities and π-conjugated planar layer structure of g-C_3N_4 provide abundant interaction sites for methylene blue and tetracycline adsorption. Besides hydrogen interactions, the electrostatic interaction between positively charged methylene blue and negatively charged adsorbent primarily contributed to the adsorptive separation of organic pollutants, as depicted in Fig. 3.2 [46]. Table 3.1 displays the representative examples of nanostructured MoS_2, WS_2, *h*-BN, MXenes, and g-C_3N_4 and their nanocomposites for adsorptive separation of organic dyes.

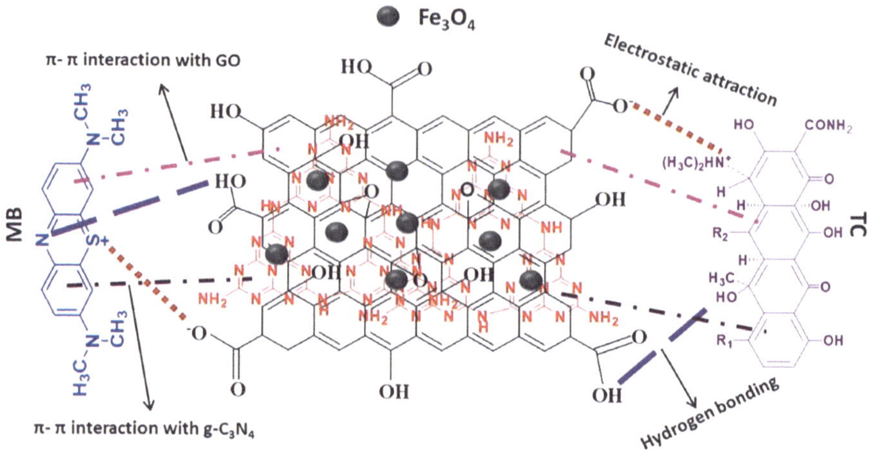

Fig. 3.2 Adsorption of methylene blue and tetracycline on the surface of $Fe_3O_4@GO/g\text{-}C_3N_4$ nanocomposites by different adsorptive interactions. Reproduced with permission from ref. [46] Copyright 2020, Elsevier Publishers

3.2.2 *Adsorptive Removal of Heavy Metal Ions and Oxyanions*

2D nanomaterials and their composites are demonstrated as potential adsorbents for heavy metal cations and oxyanions. The MoS_2, with the abundance of S atoms (Lewis base), possesses a high affinity with heavy metal ions (soft Lewis acids) [23]. The MoS_2 exhibits negative Zeta potential and is advantageous for the adsorptive separation of heavy metal ions [54]. The larger size of hydrated metal ions than the interlayer distance of MoS_2 curbs their reach to sulfur atoms of inner layers, limiting the metal ion adsorption [55]. Different strategies have been developed to design and prepare MoS_2-derived nanocomposites of highly exfoliated sheets with enhanced exposure of sulphur sites for metal ion adsorption. Wang et al. synthesised a chemically exfoliated MoS_2 enriched membrane consisting of 1 T and 2H phases to remove silver ions from wastewater and natural water. The distribution of MoS_2 in the membrane provides abundant surface-active sites for metal ion adsorption [56]. The 1 T and 2H phases of MoS_2 synthesised by hydrothermal route were also demonstrated for the adsorptive separation of Pb (II) and Cu (II). The 1 T phase showed excellent and ultrafast adsorption compared to the 2H phase, supported by a high affinity of metal ions towards surface active sulfur sites of 1 T-MoS_2 [57]. The MoS_2-derived adsorbents are highly specific towards heavy metal ions because of specific soft–soft interactions [58]. The Li-intercalated MoS_2 showed adsorption selectivity in order of Hg (II) > Pb (II) > Cd (II) > Zn (II), in line with their softness trend [59].

The nanostructured MoS_2 grown on thiol-functionalized multi-walled carbon nanotubes (MoS_2/MWCNT) showed excellent adsorption of heavy metal ions [Pb (II)

Table 3.1 Adsorption of organic dyes by nanostructured MoS_2, WS_2, *h*-BN, MXenes, and g-C_3N_4 and their nanocomposites for wastewater remediation

Adsorbent	Organic dyes	Morphology	Adsorption capacity, mg.g^{-1}	References
MoS_2-rGO	Congo Red	Flower-like porous microstructure	410	[33]
PLA-Zein/MoS_2	Methylene Blue	Nanoflakes	111	[34]
MoS_2@PDA@IL	Congo Red	Ultrathin nanosheets	112	[47]
MoS_2-PDOPA	Methylene Blue	Ultrathin nanosheets	244	[48]
1T-WS_2	Methylene Blue	Flower-like nanosheets	210	[36]
WS_2/WO_3	Rhodamine-B	Flower-like morphology	237	[30]
3D Carbon-*h*-BN	Methylene Blue Congo Red	Cheese-like 3D structure	408 307	[49]
Net-like Fibrous-*h*-BN	Methylene Blue	Net-like structure	328	[50]
Boron Nitride	Methylene Blue Rhodamine-B	Cotton flower-like porous structure	471 313	[51]
V_2CTx MXene	Methylene Blue	Accordion-like structure	111	[52]
Fe_3O_4@GO/g-C_3N_4 Nanocomposite	Methylene Blue	Compact sheet-like structure	220	[46]
CeO_2/Fe_3O_4/g-C_3N_4 nanocomposite	Rose Bengal	Nanocomposite	83	[53]

and Cd (II)] from industrial mining water. The intercalation of sodium and sulphate ions during the synthesis of MoS_2/MWCNT nanocomposite exfoliated the MoS_2. The metal-sulphur complex formation through ion exchange and electrostatic interactions, as shown in Fig. 3.3, are revealed as major interactive pathways for the adsorptive separation of Pb (II) and Cd (II) from wastewater [60]. The flower-like WS_2 microcrystals prepared by hydrothermal route were used for wastewater remediation by removal of heavy metal ions [Ag (I), Pd (II), and Au (III)]. The intercalation of Ag (I) ions between the molecular layers of WS_2, followed by interaction with sulphur atoms, formed the Ag_2S. The Au (III) and Pd (II) adsorption followed the

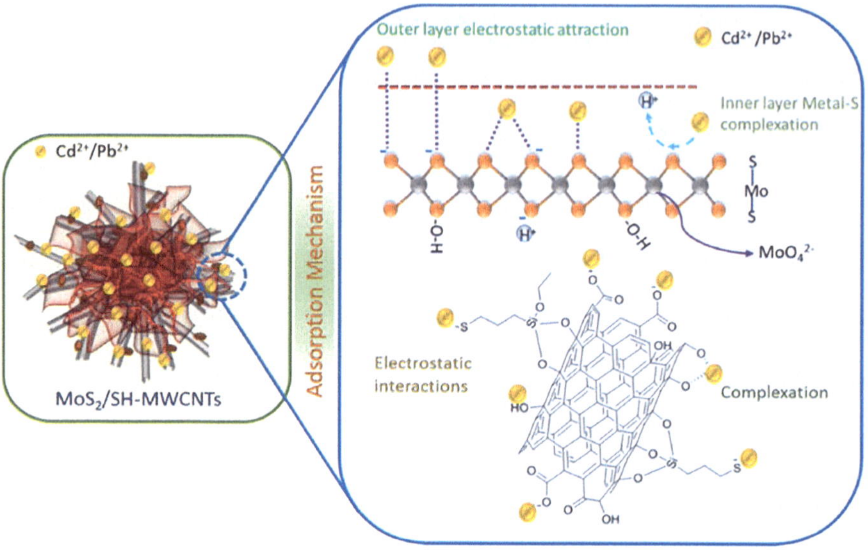

Fig. 3.3 Plausible interaction pathways for adsorptive separation of Pb (II) and Cd (II) from wastewater using MoS_2/MWCNT nanocomposite as an adsorbent [60]

reduction coupled adsorption mechanism as major interaction pathways. The selectivity of these ions follows the adsorptive separation order of Au (III) > Ag (I) > Pd (II) [61].

The nanostructured *h*-BN has been a promising candidate for removing heavy metal ions. The polarity of the B-N bond in *h*-BN induces dipole moment, which increases the adsorption efficiency for metal ions. The excellent chemical stability aids the potential of *h*-BN for the adsorptive removal of various pollutants [62]. The nanostructured boron nitride spheres (NSBNSs) effectively adsorbed Cu (II), Cd (II), and Pd (II) metal ions with removal capacity of 678.7, 536.7, and 107.0 $mg.g^{-1}$, respectively. The intercalation of metal ions between *h*-BN lamellae and the polarity of the B-N bond facilitated the adsorption of metal ions [29]. The nanostructured *h*-BN fibers prepared by using a luffa sponge as a template showed excellent adsorption of Cd (II), Zn (II), Cr (III), and Pb (II) ions with their remarkably high removal capacity of 2989, 1885, 723, and 453 $mg.g^{-1}$, respectively. The outstanding adsorption performance was primarily attributed to their specific interconnected porous attributes, which include closely packed parallel macro channels with mesoporosity on the surface of these channels [63]. The thiol functionalization on the *h*-BN surface efficiently adsorbed the Hg (II) ions. The higher affinity of Hg (II) towards the thiol group improved the adsorption efficiency and selectivity of porous *h*-BN for Hg (II) ions [62].

MXenes and their nanocomposites are promising adsorbents for heavy metal cations due to their rich surface chemistry and high surface area. The surface modification of MXenes by alkali introduces ample hydroxyl groups, which act as good

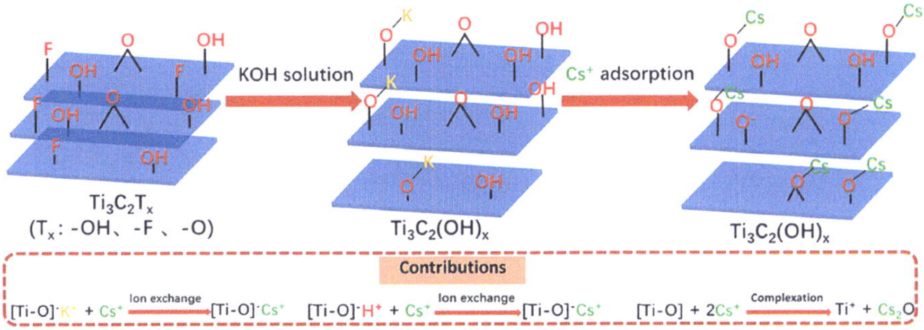

Fig. 3.4 Plausible adsorption mechanism for the removal of Cs (II) by hydroxyl functionalised-$Ti_3C_2T_x$ MXene. Reprinted with permission from [64]. Copyright 2022, American Chemical Society

adsorption sites for heavy metals. The hydroxyl-functionalized $Ti_3C_2T_x$ showed excellent adsorption of Cs (I), supported by ion exchange, complexation, and electrostatic interactions. The major adsorptive pathway of Cs (I) adsorption is the complexation between the oxygen functionalities of MXene and the Cs (I), as displayed in Fig. 3.4 [64]. The $Ti_3C_2T_x$ prepared by etching and exfoliation processes showed excellent removal of carcinogenic chromium oxyanions. The residual concentration of Cr (VI) in the treated water was estimated to be <5 ppb compared to the standard of 0.05 ppm, demonstrating the remarkable effectiveness of the prepared MXene adsorbent [65]. The alkali-treated MXene ($Ti_3C_2T_x$) decorated with layered double hydroxide (LDH) adsorbed the Ni (II) ions with a removal capacity of 222.7 $mg.g^{-1}$. The pH of the aqueous medium governed the adsorption of Ni (II) over the surface of LDH-grafted MXene [66]. The graphitic carbon nitride decorated with polyaniline nanofiber (Ox-g-C_3N_4/PANI-NF) selectively adsorbed hexavalent chromium with a maximum removal capacity of 178.6 $mg.g^{-1}$. The introduction of fibrous PANI improved the solubility of Ox-g-C_3N_4 and created a positive surface charge on the adsorbent. The positively charged surface of Ox-g-C_3N_4/PANI-NF electrostatically adsorbed the Cr (VI) oxyanions. Moreover, nitrogen and oxygen functionalities on the surface of Ox-g-C_3N_4/PANI-NF nanocomposite also contributed to the adsorptive removal of Cr (VI) oxyanions [45]. Table 3.2 shows nanostructured MoS_2, WS_2, *h*-BN, MXenes, g-C_3N_4, and their nanocomposites for the adsorptive removal of heavy metal ions and oxyanions.

3.2.3 Adsorptive Removal of APIs and Pesticides

2D materials and their composites are found to be potential adsorbents for the removal of different APIs and pesticides. The MoS_2 nanocomposite with montmorillonite (MoS_2-Mt) exhibited adsorption of atenolol (ATE) and acebutolol (ACE) with removal capacities of 132.1 and 113.8 $mg.g^{-1}$, respectively. The weak van der

Table 3.2 Application of nanostructured MoS_2, WS_2, *h*-BN, MXenes, and g-C_3N_4 and their nanocomposites for adsorptive removal of heavy metal ions and oxyanions

Adsorbent	Morphology	Heavy metal ions	Adsorption capacity, $mg.g^{-1}$	References
MoS_2/SH-MWCNT	Cross-linked 3D network	Pb (II) Cd (II)	90.0 66.6	[60]
MoS_2 Nanosheet	Honeycomb structure	Ag (I)	4000	[56]
MoS_2@Fe_3O_4NPs	Cloud-like structure	Cr (IV) Cr (III)	218.0 119.0	[67]
Flower-like WS_2	Micro flower-like structure	Ag (I) Pd (II) Au (III)	186.2 67.3 1340.6	[61]
h-BN-SH	Honeycomb-like porous structure	Hg (II)	412.9	[62]
h-BN	Nanospheres	Cu (II) Pb (II) Cd (II)	678.7 536.7 107.0	[29]
$Ti_3C_2(OH)_x$ MXene	Ultrathin layered nanoflakes	Cs (I)	37.3	[64]
Ti_3C_2TX-PDOPA	Packed and stacked lamellar structures	Cu (II)	46.6	[68]
Ox-g-C_3N_4/PANI-nanofiber	Nanofiber interconnected with layered nanosheet	Cr (VI)	178.6	[45]
g-C_3N_4/ATP	Nanorod-wrapped layered seats	Cd (II)	61.1	[69]

Waals interactions with montmorillonite and the hydrogen bonding between amine groups of pollutants and oxygen functionalities of montmorillonite clay facilitated the adsorption of ATE and ACE. The higher adsorption of ATE than ACE was attributed to larger degrees of hydrogen bonding and van der Waals interactions with sulphur atoms of MoS_2 [70]. The polyethylene glycol templated MoS_2 is explored for wastewater remediation. However, poor dispersibility of PEGlayted-MoS_2 compromised the accessibility of micropollutants towards the surface-active sites [32]. The decoration of MoS_2 over the surface of activated carbon furnishes a large surface area and surface-active sites for pollutant adsorption. The PEGylated-MoS_2 nanosheet-decorated biochar (MoS_2-BC) displayed excellent adsorption of tetracycline hydrochloride (TC) with an optimum removal capacity of 249.5 $mg.g^{-1}$. The adsorption mechanism includes electrostatic interactions between TC and the polar groups of MoS_2-BC, conjugated π-π interactions between the aromatic framework of TC and π-electrons-rich domains of MoS_2-BC, and adequate hydrogen bonding, as displayed in Fig. 3.5. Furthermore, the mesoporous structure of MoS_2-BC enhanced the removal efficiency by reducing the hindrance effect [71].

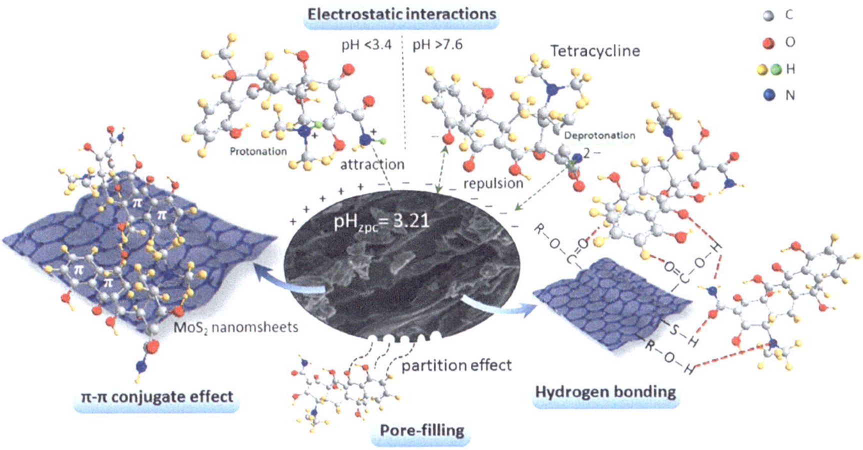

Fig. 3.5 Adsorption of tetracycline hydrochloride on the surface of MoS_2-BC nanocomposites via different interaction pathways, viz. electrostatic interactions, hydrogen bonding, π-π conjugate effect, and pore-filling. Reproduced with permission from ref. [71]. Copyright 2019, Elsevier Science ltd

The enrichment of N-defects in *h*-BN nanosheets (BNNSs) via a single-step alkali-assisted strategy enhanced the adsorption of TC with a maximum removal capacity of 1101 $mg.g^{-1}$. The adsorption isotherm revealed multilayer adsorption over the heterogeneous surface of BNNSs and suggested the dominance of π-π interaction between aromatic rings of TC and N-defects-rich BNNSs [72]. Likewise, porous boron nitride having boron vacancies (B_V-BNNSs) adsorbed TC with a maximum adsorption capacity of 438 $mg.g^{-1}$. The TC is adsorbed over the surface and pores of B_V-BNNSs by electrostatic and π-π interactions. Boron vacancy in B_V-BNNSs improved the adsorption efficiency by 38% compared to pristine BNNSs. The higher adsorption capacity is ascribed to the increased surface area of B_V-BNNSs (1104 $m^2.g^{-1}$) than pristine BNNSs (646 $m^2.g^{-1}$), besides the higher degree of electrostatic interaction pathways [73]. Ghani et al. intercalated 2D-$Ti_3C_2T_x$ MXene nanosheets with sodium ions (SI-$Ti_3C_2T_x$) and used them to remove Ciprofloxacin (CPX). The intercalation of sodium doubled the adsorption capacity of $Ti_3C_2T_x$ MXene for CPX [74]. The surface area and surface-active sites of graphitic carbon nitride can be enhanced by preparing its nanocomposites and hybrid materials. The g-C_3N_4-WO_3-ZnO nanocomposite efficiently adsorbed the imidacloprid pesticide with 99.9% removal efficiency. The conjugated π electron-rich g-C_3N_4 adsorbed imidacloprid via π-π interaction with aromatic rings of imidacloprid and hydrophobic effect. At the same time, ZnO and WO_3 showed coordinated interaction with oxygen and nitrogen-based functionalities of imidacloprid [75]. Table 3.3 reports nanostructured MoS_2, *h*-BN, MXenes, g-C_3N_4, and their nanocomposites for the adsorption of APIs and pesticides.

Table 3.3 Application of nanostructured MoS_2, *h*-BN, MXenes, g-C_3N_4, and their nanocomposites for adsorptive removal of APIs and pesticides

Adsorbent	Morphology	APIs and pesticides	Adsorption capacity, $mg.g^{-1}$	References
PEG-MoS_2 nanosheet-decorated biochar	Hierarchical crumpled structure with curved edges	Tetracycline	249.5	[71]
Nanostructured *h*-BN with boron vacancies	Fluffy and cloud-like with hollow micro-tubes	Tetracycline	992.0	[73]
N-defects in BNNSs	Wrinkled porous nanosheets	Tetracycline	1101.0	[72]
MoS_2/montmorillonite composite	Distorted microspheres	Atenolol Acebutolol	132.1 113.8	[70]
Sodium lignosulfonate functionalised MXene	–	Doxorubicin hydrochloride	190.8	[76]
Sodium Intercalated $Ti_3C_2T_x$ MXene	Ultrathin layered stacking	Ciprofloxacin	208.2	[74]
Ti_3C_2-MoS_2 composite	Layered nanocomposite	Paraquat pesticides	105.5	[77]
Fe_3O_4@GO/g-C_3N_4 nanocomposite	Compact sheet-like structure	Tetracycline	120.0	[46]
rGO/MoS_2/Fe_3O_4	Uniform nanoparticles	Fenitrothion pesticides	33.4	[78]

3.3 Photocatalytic Degradation of Pollutants

Photocatalytic decomposition, conversion, and mineralization are the most promising approaches for converting hazardous pollutants into non-hazardous smaller-size molecules utilizing photocatalysts under light exposure. The 2D nanomaterials and their composites, hybrids, and heterostructures showed excellent photocatalytic activity. The graphene, MoS_2, MXenes, *h*-BN, g-C_3N_4, WS_2, and their hybrid/composite materials are explored for the photocatalytic degradation and conversion of dyes, APIs, and pesticides into their non-hazardous and simplified forms [79].

3.3.1 Photocatalytic Degradation of Organic Dyes

Nanostructured 2D materials and their composites have emerged as the most promising photocatalysts for the mineralisation and decomposition of organic dyes. Among recently explored photocatalysts, MoS_2 has shown much interest because of its large surface area, cost-effectiveness, ease of surface functionalization and doping, tuneable morphological features, interlayer spacing, and band-gap [11, 80]. The narrower band-gap of MoS_2 leads to fast recombination of charge carriers (e^- and h^+ pairs), resulting in low quantum efficiency. The photocatalytic activity of nanostructured MoS_2 can be improved by modifying its structure, controlling the defect sites, metallic and non-metallic doping, and crystal face engineering [11]. Kumbhakar et al. [81] revealed the effect of morphology on the photocatalytic activity of MoS_2 nanostructures (nanorods, nanoplatelets, and nanosheets). Among them, MoS_2 nanosheets exhibited the highest photocatalytic degradation capacity for methylene blue dye [81].

Transition metal doping can introduce structural defects and widen the light adsorption range. The photocatalytic activity of Co-doped MoS_2 is found to be 2.71 times higher than pristine MoS_2 [82]. The N-doped MoS_2 showed excellent photocatalytic activity under visible light for the mineralisation of Rhodamine B dye. The N-doped heterostructure increased the photocatalytic activity rate constant by 27 times to pristine MoS_2 [18]. The pH of the synthesis medium significantly affects the charge distribution and size of MoS_2, thus influencing photocatalytic performance. The size of MoS_2 micro flowers decreased with the lowering of pH, and at a pH of 1, it displayed the highest photocatalytic performance for Rhodamine B dye degradation [83]. The size of the MoS_2 influences the band gap based on the size quantisation effect. The reduction in MoS_2 size increases the band gap and shifts the redox potential for reactive oxygen generation for photocatalytic degradation [84]. Arthi et al. [85] explored the photodegradation of methylene blue and Rhodamine B dyes mixture using MoS_2/r-GO/Cu_2O nanocomposite. The direct contact between MoS_2 and Cu_2O formed a p-n hetero junction where electrons from the conduction band of MoS_2 are transferred to the conduction band of Cu_2O and finally to the surface of the photocatalyst via conductive r-GO structure. Simultaneously, the holes in the valence band of Cu_2O transferred to the valence band of MoS_2 sheets resulting in efficient charge transfer. The electron–hole pairs react with H_2O to form reactive radical species, which facilitate the degradation of organic dyes. Consequently, MoS_2/r-GO/Cu_2O nanocomposite showed 98% photodegradation in 45 min under natural light exposure [85].

The WS_2 is widely explored due to its large surface-to-volume ratio and superior hole and electron transport capacity [86]. The WS_2-PANI nanocomposite showed excellent photocatalytic performance with 93% degradation of methylene blue dye within 90 min [86]. The integration of CdS with WS_2 was examined for photocatalytic decomposition of crystal violet dye. The CdS nanospheres over the WS_2 sheets promoted dye absorption and improved the photocatalytic activity under the visible range because of the surface plasmon effect of CdS nanospheres [87].

Titanium carbide-based MXenes, a novel class of 2D materials with excellent conductivity and stability exhibit good photocatalytic properties [88]. The MXene-loaded g-C_3N_4 photocatalyst as a heterostructure showed remarkably higher photodegradation of methylene blue dye than bare g-C_3N_4. The improved photocatalytic activity is attributed to a higher number of reactive sites. The g-C_3N_4 absorbed the photon energy and generated electron–hole pair, whereas MXene acts as an electron acceptor, and the excited electrons degrade the methylene blue dye. However, excessive loading of MXene reduced the photocatalytic performance [89]. The SnO_2-decorated MoS_2 nanocomposite photo-catalytically decomposed the 94% methylene blue and 58.5% red dyes within 2 h under visible light irradiation. The heterostructure of SnO_2-MoS_2 improved the electron transfer to dye molecules, which enhanced the photocatalytic degradation of organic dyes [90]. The Ti_3C_4 MXene synthesized using HF etchants and then modified to replace fluoro groups with hydroxyl functionalities showed excellent potential for photocatalytic degradation of organic dyes under visible light irradiation [91]. The MoS_2-ZnO nanocomposite photo-catalytically decomposed 95% methylene blue dye within 1 h under visible light exposure (Fig. 3.6). The separation of the electron–hole pair occurs at the interface of MoS_2-ZnO, leading to the effective depletion of the electrons from VB to CB and subsequent retention of the holes. Consequently, MoS_2-ZnO nanocomposite efficiently mineralized the methylene blue dye. Moreover, MoS_2 in the composite prevents photo corrosion of ZnO and enhances the photostability and catalytic activity of the MoS_2-ZnO nanocomposite [92]. Table 3.4 displays representative nanostructured MoS_2, *h*-BN, WS_2, MXenes, and g-C_3N_4, along with their nanocomposites, hybrids, and heterostructures for photocatalytic decomposition and mineralization of organic dyes.

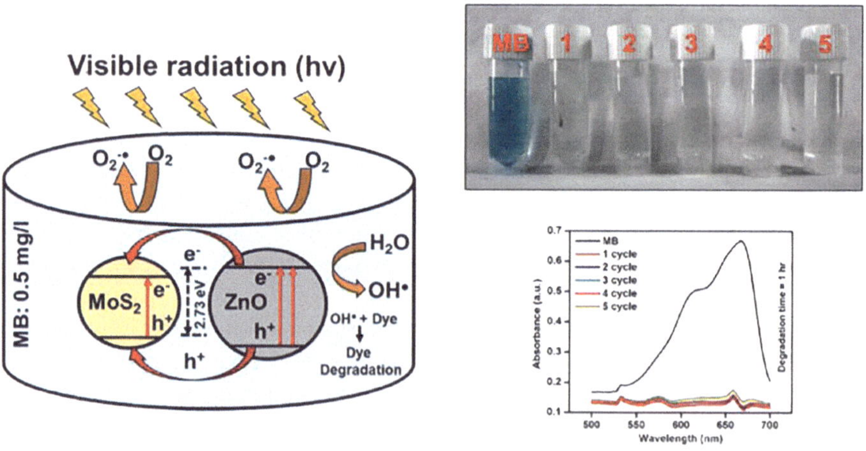

Fig. 3.6 Photocatalytic degradation of methylene blue dye using MoS_2-ZnO-based heterostructure. Reproduced with permission from ref. [92] Copyright 2019, Springer Nature

Table 3.4 Application of nanostructured MoS_2, *h*-BN, WS_2, MXenes, g-C_3N_4, and their nanocomposites, hybrids, and heterostructures for photocatalytic decomposition of organic dyes

Photocatalysts	Morphology	Targeted pollutant	Light source	Degradation efficiency (%)	References
Ti_3C_2-TiO_2	Densely layered structure	Methyl Orange	Solar	99	[93]
g-C_3N_4/MoS_2/Gr	Thin irregular nanosheets	Rhodamine-B	Visible	95	[94]
Mn-MoS_2/rGO	Aggregated nanosheets	Rhodamine-B	Visible	90	[95]
MoS_2/Graphene	Stratified structure made of MoS_2 and curled graphene	Methylene Blue	Natural	97.8	[96]
PANI/*h*-BN	Granular	Methylene Blue Methyl Orange	UV	93 95	[97]
Ag NPs/TiO_2/Ti_3C_2Tx	Layered	Methylene Blue Rhodamine B	Solar	96 84	[98]
PVA/TiO_2/MoS_2/Au	Uniform porous network	Rhodamine B	Visible	86	[99]
La and Mn co-doped Bi Ferrite/Ti_3C_2 MXene	Coagulated nanoparticles on MXene sheets	Congo Red	Visible	100	[100]
Carbon supported TiO_2@ $TiOF_2$	2D Layered structure	Methylene Blue	Visible	90	[101]
WS_2 Nanoparticles	Flake nanoparticle structure	Ternary Mixture (Methyl Orange, Methylene blue, Congo Red	Visible	>95	[102]
MoS_2/r-GO/ Cu_2O	Cu_2O decorated micro-structure over Carbon paper	Methylene Blue + Rhodamine B	Natural	98	[85]

3.3.2 Photocatalytic Degradation of APIs and Pesticides

Over the recent past, research interests have grown immensely in photocatalytic degradation of APIs using 2D nanostructured materials and their composites. The MoS_2-ZnS nanocomposite synthesized by hydrothermal processing showed photocatalytic decomposition of oxytetracycline. The DFT calculation revealed the transfer of electrons from ZnS to MoS_2 under visible light irradiation. Moreover, high charge carrier mobility in MoS_2-ZnS nanocomposite than bare ZnS extended superior photocatalytic activity [103]. The MoS_2-BiOCl nanocomposite photo-catalytically decomposed a hypertensive drug Valsartan (VLS). The presence of BiOCl in MoS_2-BiOCl improved the photocatalytic activity of MoS_2. The rate constant for photodegradation of VLS by MoS_2 (0.0305 min^{-1}) increased to 0.0517 min^{-1} by MoS_2-BiOCl. Moreover, MoS_2-BiOCl showed excellent stability even after 360 min of light irradiation [104].

The C_3N_4-Ti_3C_2 nanocomposite is demonstrated for photocatalytic degradation of CPX antibiotic. The degradation rate is increased by ~10% when Ti_3C_2 was integrated with g-C_3N_4. The mechanistic study revealed the generation of electrons under visible light irradiation, which could rapidly transfer from g-C_3N_4 to Ti_3C_2, and the holes generated over g-C_3N_4 efficiently decomposed the CPX. At the same time, the superoxide ions formed from dissolved oxygen by extracting electrons from the Ti_3C_2 surface facilitated the CPX degradation [105]. Photocatalytic degradation of the pharmaceutical drug carbamazepine was carried out using ZnO-TiO_2-Mxene nanocomposite with a removal efficiency of 99.6%. The TiO_2-MXene enhanced the photocatalytic activity of ZnO by reducing the recombination rate and increasing the electron–hole pair transportation rate. The nanocomposite showed excellent performance even after eight cycles [106]. Catechol, one of the widely used reagents in the pharmaceutical and food industry, is frequently detected in wastewater. The Ti_3C_2-TiO_2 nanocomposite as a photocatalyst decomposed the catechol with a degradation efficiency of 92% under solar light irradiation. The high surface area and metallic nature of nanocomposite enhanced the formation rate of superoxide radical anion ($O_2^{\bullet -}$) and hydroxide radical ($OH^{\bullet}$), accountable for the photocatalytic decomposition of catechol [107].

The WS_2, an inorganic 2D analogue of graphene, has been explored for photocatalytic degradation of APIs. The WS_2-PANI nanocomposite showed highly efficient photocatalytic degradation of antibiotic Nitrofurantoin (NFT). The excellent photocatalytic activity of WS_2-PANI nanocomposite is attributed to its favourable band-gap (2.6 eV) and high charge separation [108]. The WS_2-PANI nanocomposite was also probed for degradation of organic dyes and photocatalytic conversion of heavy metal oxyanions. The Ag@Ag_2S grown on WS_2 multilayers by hydrothermal approach significantly enhanced the photocatalytic performance. The outstanding photodegradation was attributed to the coupling of WS_2 to Ag_2S and the preparation of p-n heterojunctions [109].

The Au nanoparticles/*h*-BN nanocomposite decomposed the Levofloxacin with a removal efficiency of 84.4% under visible light irradiation. The Au nanoparticles enhanced the interfacial electron transfer with *h*-BN in the nanocomposite [110]. Photocatalytic degradation of the Cefoxitin antibiotic was carried out using an *h*-BN/Cd Al_2O_4 nanocomposite. In this degradation process, $OH^\bullet$ radical and the hole plays vital roles in decomposing 100% Cefoxitin within 240 min at neutral pH [111]. Tongue et al. [112] successfully prepared 2D/2D Bi_2WO_6/*h*-BN nanocomposite by solvothermal method and then utilised it for the degradation of tetracycline under visible light irradiation with a degradation efficiency of 94.7%. The incorporation of *h*-BN played an important role in providing more active sites and enhanced the electron/hole pair separation for the improved photocatalytic performance of Bi_2WO_6/*h*-BN nanocomposite. Moreover, the nanocomposite showed excellent stability and recyclability [112].

The $TiO_2/Ti_3C_2T_x$ heterostructure synthesised via hydrothermal method followed by controlled oxidation via Schottky junction formation between the TiO_2-MXene interfaces showed excellent potential for photocatalytic degradation of carbamazepine (CBZ), an antiepileptic drug. The $TiO_2/Ti_3C_2T_x$ heterostructure decomposed 98.7% CBZ under UV-light irradiation. The charge carriers generated by TiO_2 formed the active radicals such as OH• and $O_2{\cdot}^-$ by photocatalysis of water, which eventually decomposed the CBZ (Fig. 3.7) [113]. The MoS_2/ZnS embedded in Sulphur/Nitrogen-doped carbon decomposed the dicofol pesticide under visible light irradiation. The study further revealed that superoxide anion radical ($O_2{\cdot}^-$), electron/hole pairs, and hydroxyl radical prepared during photocatalysis are primarily responsible for decomposing dicofol pesticide [114]. The MoS_2-loaded ZnO-g-C_3N_4 nanocomposite was explored for photodegradation of atrazine, a potent herbicide. The ternary nanocomposite prepared by co-stacked g-C_3N_4/MoS_2 nanosheets with ZnO nanospheres furnished efficient charge carrier transport and separation across the ZnO/g-C_3N_4 interface of heterojunction. Consequently, the ternary nanocomposite exhibited the photocatalytic degradation of atrazine under visible light illumination [115]. Table 3.5 presents examples of inorganic analogues of graphene and their composites, hybrids, and heterojunctions for photocatalytic degradation of APIs and pesticides.

3.4 Conclusion and Future Perspective

Graphene and its 2D analogues like MoS_2, WS_2, *h*-BN, g-C_3N_4, and MXenes are widely explored for the adsorptive separation and photocatalytic degradation of different organic and inorganic pollutants. The chapter furnished an overview of wastewater, type of water pollutants, various adsorbents, and photocatalytic materials by highlighting the adsorption and photocatalytic degradation of both organic and inorganic contaminants, as well as recent developments related to the use of inorganic analogues of graphene as photocatalysts and adsorbents. Various inorganic analogues of graphene and their nanocomposites, hybrids, and heterostructures are emphasized for photocatalytic decomposition and adsorptive removal of organic

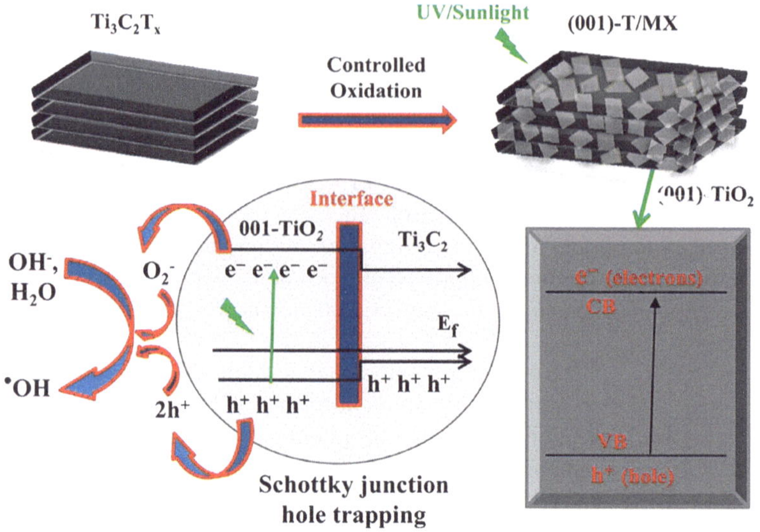

Fig. 3.7 Photocatalytic degradation of carbamazepine drug using TiO_2/MXene nanocomposite. Reproduced with permission from ref. [113] Copyright 2018, Elsevier Publishers

Table 3.5 2D inorganic nanomaterials and their composites, hybrids, and heterojunctions for photocatalytic degradation of APIs and pesticides

Photocatalysts	Pollutants	Light source	Degradation efficiency, %	References
C_3N_4/Ti_3C_2	Ciprofloxacin	Visible	–	[105]
ZnO-TiO_2-Mxene	Carbamazepine	Solar	99.6	[106]
Ti_3C_2-TiO_2	Catechol	Solar	92.0	[107]
Au NPs/*h*-BN	Levofloxacin	Visible	84.4	[110]
BN/Cd Al_2O_4	Cefoxitin Sodium	UV	100	[111]
2D/2D Bi_2WO_6/*h*-BN	Tetracycline	Visible	94.7	[112]
WO_3-TiO_2	Paracetamol	Visible	100	[116]
ZnO-MoS_2	Tetracycline	Visible	84.0	[117]

pollutants such as dyes, APIs, and pesticides, along with inorganic contaminants such as heavy metals and their oxyanions from contaminated water.

Much laboratory-scale work has been carried out on inorganic analogues of graphene; however, the implementation of these materials for water purification on an industrial scale is still in its early phase. The scalable synthesis of many of these materials is in its infancy stage, which makes their applications on an industrial scale expensive; therefore, most studies are restricted to the lab level. Massive efforts are required to switch from conventional materials to novel graphene-like materials and their composites for utilization in industrial water treatment plants. Developing and implementing these 2D nanomaterials and their composites with high water remediation efficiency and sustainability will undoubtedly open the window for wastewater treatment applications.

Acknowledgements The authors are thankful to the Director, CSIR-Indian Institute of Petroleum, Dehradun, for encouraging us to publish this work. PJ, SM, and AP acknowledge the DST-INSPIRE and UGC, Govt. of India, for the fellowship support.

References

1. N. Pichel, M. Vivar, M. Fuentes, The problem of drinking water access: a review of disinfection technologies with an emphasis on solar treatment methods. Chemosphere **218**, 1014–1030 (2019). https://doi.org/10.1016/j.chemosphere.2018.11.205
2. U. Qumar, J.Z. Hassan, R.A. Bhatti, A. Raza, G. Nazir, W. Nabgan, M. Ikram, Photocatalysis vs adsorption by metal oxide nanoparticles. J. Mater. Sci. Technol. **131**, 122–166 (2022). https://doi.org/10.1016/j.jmst.2022.05.020
3. R. Gusain, K. Gupta, P. Joshi, O.P. Khatri, Adsorptive removal and photocatalytic degradation of organic pollutants using metal oxides and their composites: a comprehensive review. Adv. Colloid Interface Sci. **272**, 102009 (2019). https://doi.org/10.1016/j.cis.2019.102009
4. S. Mani, R.N. Bharagava, Textile industry wastewater: environmental and health hazards and treatment approaches, Recent advances in environmental management, CRC Press, pp. 47–69 (2018).
5. C.V. Papazlatani, P.A. Karas, G. Tucat, D.G. Karpouzas, Expanding the use of biobeds: degradation and adsorption of pesticides contained in effluents from seed-coating, bulb disinfestation and fruit-packaging activities. J. Environ. Manage. **248**, 109221 (2019). https://doi.org/10.1016/j.jenvman.2019.06.122
6. A.K. Rana, Y.K. Mishra, V.K. Gupta, V.K. Thakur, Sustainable materials in the removal of pesticides from contaminated water: perspective on macro to nanoscale cellulose. Sci. Total Environ. **797**, 149129 (2021). https://doi.org/10.1016/j.scitotenv.2021.149129
7. M.J. Ahmed, B.H. Hameed, removal of emerging pharmaceutical contaminants by adsorption in a fixed-bed column: a review. Ecotoxicol. Environ. Saf. **149**, 257–266 (2018). https://doi.org/10.1016/j.ecoenv.2017.12.012
8. P. Joshi, K. Gupta, R. Gusain, O.P. Khatri, *Metal Oxide Nanocomposites for Wastewater Treatment, Metal Oxide Nanocomposites*, pp. 361–397 (2020), https://doi.org/10.1002/978 1119364726.ch11
9. K. Gupta, P. Joshi, R. Gusain, O.P. Khatri, Recent advances in adsorptive removal of heavy metal and metalloid ions by metal oxide-based nanomaterials. Coord. Chem. Rev. **445**, 214100 (2021). https://doi.org/10.1016/j.ccr.2021.214100
10. X. Zhao, G. Zhang, Z. Zhang, TiO_2-based catalysts for photocatalytic reduction of aqueous oxyanions: state-of-the-art and future prospects. Environ. Int. **136**, 105453 (2020), https://doi.org/10.1016/j.envint.2019.105453
11. Y. Yuan, R.-T. Guo, L.-F. Hong, X.-Y. Ji, Z.-S. Li, Z.-D. Lin, W.-G. Pan, Recent advances and perspectives of MoS2-based materials for photocatalytic dyes degradation: a review. Colloids Surf. A **611**, 125836 (2021). https://doi.org/10.1016/j.colsurfa.2020.125836
12. H. Zhao, Q. Xia, H. Xing, D. Chen, H. Wang, Construction of pillared-layer MOF as efficient visible-light photocatalysts for aqueous Cr (VI) reduction and dye degradation. ACS Sustain. Chem. Eng. **5**(5), 4449–4456 (2017). https://doi.org/10.1021/acssuschemeng.7b00641

13. V. Thirumal, R. Yuvakkumar, P.S. Kumar, S. Keerthana, G. Ravi, D. Velauthapillai, B. Saravanakumar, Efficient photocatalytic degradation of hazardous pollutants by home-made kitchen blender novel technique via 2D-material of few-layer MXene nanosheets. Chemosphere **281**, 130984 (2021). https://doi.org/10.1016/j.chemosphere.2021.130984
14. A. Chouhan, H.P. Mungse, O.P. Khatri, Surface chemistry of graphene and graphene oxide: a versatile route for their dispersion and tribological applications. Adv. Colloid Interface Sci. **283**, 102215 (2020). https://doi.org/10.1016/j.cis.2020.102215
15. P. Joshi, O.P. Sharma, S.K. Ganguly, M. Srivastava, O.P. Khatri, Fruit waste-derived cellulose and graphene-based aerogels: plausible adsorption pathways for fast and efficient removal of organic dyes. J. Colloid Interface Sci. **608**, 2870–2883 (2022). https://doi.org/10.1016/j.jcis.2021.11.016
16. M.K. Jana, C.N.R. Rao, Two-dimensional inorganic analogues of graphene: transition metal dichalcogenides, Philosophical Transactions of the Royal Society A: Mathematical. Phys. Eng. Sci. **374**(2076), 20150318 (2016). https://doi.org/10.1098/rsta.2015.0318
17. K. Gupta, P. Joshi, O.P. Khatri, h-BN and graphene-based ultralight hybrid aerogels: highly efficient sorbent for recovery of hydrocarbon oils and organic solvents. J. Environ. Chem. Eng. **9**(6), 106788 (2021). https://doi.org/10.1016/j.jece.2021.106788
18. A. Pakdel, C. Zhi, Y. Bando, D. Golberg, Low-dimensional boron nitride nanomaterials. Mater. Today **15**(6), 256–265 (2012). https://doi.org/10.1016/S1369-7021(12)70116-5
19. L. Garnier, A. Szymczyk, P. Malfreyt, A. Ghoufi, Physics behind water transport through nanoporous boron nitride and graphene. J. Phys. Chem. Lett. **7**(17), 3371–3376 (2016). https://doi.org/10.1021/acs.jpclett.6b01365
20. S. Dervin, D.D. Dionysiou, S.C. Pillai, 2D nanostructures for water purification: graphene and beyond. Nanoscale **8**(33), 15115–15131 (2016). https://doi.org/10.1039/C6NR04508A
21. Z. Wang, Q. Tu, S. Zheng, J.J. Urban, S. Li, B. Mi, Understanding the aqueous stability and filtration capability of MoS2 membranes. Nano Lett. **17**(12), 7289–7298 (2017). https://doi.org/10.1021/acs.nanolett.7b02804
22. K. Ai, C. Ruan, M. Shen, L. Lu, MoS2 nanosheets with widened interlayer spacing for high-efficiency removal of mercury in aquatic systems. Adv. Funct. Mater. **26**(30), 5542–5549 (2016). https://doi.org/10.1002/adfm.201601338
23. Z. Wang, B. Mi, Environmental Applications of 2D molybdenum disulfide (MoS2) nanosheets. Environ. Sci. Technol. **51**(15), 8229–8244 (2017). https://doi.org/10.1021/acs.est.7b01466
24. T. Krasian, W. Punyodom, P. Worajittiphon, A hybrid of 2D materials (MoS2 and WS2) as an effective performance enhancer for poly(lactic acid) fibrous mats in oil adsorption and oil/water separation. Chem. Eng. J. **369**, 563–575 (2019). https://doi.org/10.1016/j.cej.2019.03.092
25. X. Yuan, C. Zhou, Y. Jin, Q. Jing, Y. Yang, X. Shen, Q. Tang, Y. Mu, A.-K. Du, Facile synthesis of 3D porous thermally exfoliated g-C3N4 nanosheet with enhanced photocatalytic degradation of organic dye. J. Colloid Interface Sci. **468**, 211–219 (2016). https://doi.org/10.1016/j.jcis.2016.01.048
26. M. Deifallah, P.F. McMillan, F. Corà, Electronic and structural properties of two-dimensional carbon nitride graphenes. J. Phys. Chem. C **112**(14), 5447–5453 (2008), https://doi.org/10.1021/jp711483t
27. Y. Ibrahim, M. Meslam, K. Eid, B. Salah, A.M. Abdullah, K.I. Ozoemena, A. Elzatahry, M.A. Sharaf, M. Sillanpää, A review of MXenes as emergent materials for dye removal from wastewater. Sep. Purif. Technol. **282**, 120083 (2022). https://doi.org/10.1016/j.seppur.2021.120083
28. M. Khatami, S. Iravani, MXenes and MXene-based materials for the removal of water pollutants: challenges and opportunities. Comments Inorg. Chem. **41**(4), 213–248 (2021). https://doi.org/10.1080/02603594.2021.1922396
29. F. Liu, J. Yu, X. Ji, M. Qian, Nanosheet-structured boron nitride spheres with a versatile adsorption capacity for water cleaning. ACS Appl. Mater. Interfaces. **7**(3), 1824–1832 (2015). https://doi.org/10.1021/am507491z

30. G. Li, Y. Wang, J. Bi, X. Huang, Y. Mao, L. Luo, H. Hao, Partial oxidation strategy to synthesise WS (2)/WO (3) heterostructure with enhanced adsorption performance for organic dyes: synthesis. Modell. Mech. Nanomaterials (Basel, Switzerland) **10**(2) (2020), https://doi.org/10.3390/nano10020278
31. D. Panchal, A. Sharma, P. Mondal, O. Prakash, S. Pal, Heterolayered TiO2@layered double hydroxide-MoS2 nanostructure for simultaneous adsorption-photocatalysis of co-existing water contaminants. Appl. Surf. Sci. **553**, 149577 (2021). https://doi.org/10.1016/j.apsusc.2021.149577
32. A.T. Massey, R. Gusain, S. Kumari, O.P. Khatri, Hierarchical microspheres of MoS2 Nanosheets: efficient and regenerative adsorbent for removal of water-soluble dyes. Ind. Eng. Chem. Res. **55**(26), 7124–7131 (2016). https://doi.org/10.1021/acs.iecr.6b01115
33. H. Xie, X. Xiong, A porous molybdenum disulfide and reduced graphene oxide nanocomposite (MoS2-rGO) with high adsorption capacity for fast and preferential adsorption towards Congo red. J. Environ. Chem. Eng. **5**(1), 1150–1158 (2017). https://doi.org/10.1016/j.jece.2017.01.044
34. R. Schneider, M.H.M. Facure, A.D. Alvarenga, P.A.M. Chagas, D.M. dos Santos, D.S. Correa, Dye adsorption capacity of MoS2 nanoflakes immobilized on Poly(lactic acid) fibrous membranes. ACS Applied Nano Materials **4**(5), 4881–4894 (2021). https://doi.org/10.1021/acsanm.1c00442
35. R. Kumar, S.A. Ansari, M.A. Barakat, A. Aljaafari, M.H. Cho, A polyaniline@MoS2-based organic–inorganic nanohybrid for the removal of Congo red: adsorption kinetic, thermodynamic and isotherm studies. New J. Chem. **42**, 18802–18809 (2018). https://doi.org/10.1039/C8NJ02803F
36. H. Dang, L. Chen, L. Chen, M. Yuan, Z. Yan, M. Li, Hydrothermal synthesis of 1T-WS2 nanosheets with excellent adsorption performance for dye removal from wastewater. Mater. Lett. **254**, 42–45 (2019). https://doi.org/10.1016/j.matlet.2019.07.026
37. G. Lian, X. Zhang, S. Zhang, D. Liu, D. Cui, Q. Wang, Controlled fabrication of ultrathin-shell BN hollow spheres with excellent performance in hydrogen storage and wastewater treatment. Energy Environ. Sci. **5**(5), 7072–7080 (2012). https://doi.org/10.1039/C2EE03240F
38. Z. Zhao, C. Bai, L. An, X. Zhang, F. Wang, Y. Huang, M. Qu, Y. Yu, Biocompatible porous boron nitride nano/microrods with ultrafast selective adsorption for dyes. J. Environ. Chem. Eng. **9**(1), 104797 (2021). https://doi.org/10.1016/j.jece.2020.104797
39. L. Tian, F. Liang, L. Dong, J. Li, Q. Jia, H. Zhang, S. Yan, S. Zhang, Preparation and enhanced adsorption properties for CO_2 and dyes of amino-decorated hierarchical porous BCN aerogels. J. Am. Ceram. Soc. **104**(2), 1110–1119 (2021). https://doi.org/10.1111/jace.17501
40. M.M. Tunesi, R.A. Soomro, X. Han, Q. Zhu, Y. Wei, B. Xu, Application of MXenes in environmental remediation technologies. Nano Convergence **8**(1), 5 (2021). https://doi.org/10.1186/s40580-021-00255-w
41. O. Mashtalir, K.M. Cook, V.N. Mochalin, M. Crowe, M.W. Barsoum, Y. Gogotsi, Dye adsorption and decomposition on two-dimensional titanium carbide in aqueous media. J. Mater. Chem. A **2**(35), 14334–14338 (2014). https://doi.org/10.1039/C4TA02638A
42. Z. Wei, Z. Peigen, T. Wubian, Q. Xia, Z. Yamei, S. ZhengMing, Alkali treated Ti3C2Tx MXenes and their dye adsorption performance. Mater. Chem. Phys. **206**, 270–276 (2018). https://doi.org/10.1016/j.matchemphys.2017.12.034
43. C. Hao, G. Li, G. Wang, W. Chen, S. Wang, Preparation of acrylic acid modified alkalised MXene adsorbent and study on its dye adsorption performance. Colloids Surf. A **632**, 127730 (2022). https://doi.org/10.1016/j.colsurfa.2021.127730
44. Y.-Z. Zhang, J.K. El-Demellawi, Q. Jiang, G. Ge, H. Liang, K. Lee, X. Dong, H.N. Alshareef, MXene hydrogels: fundamentals and applications. Chem. Soc. Rev. **49**(20), 7229–7251 (2020). https://doi.org/10.1039/D0CS00022A
45. R. Kumar, M.A. Barakat, F.A. Alseroury, Oxidised g-C3N4/polyaniline nanofiber composite for the selective removal of hexavalent chromium. Sci. Rep. **7**(1), 12850 (2017). https://doi.org/10.1038/s41598-017-12850-1

46. S.K. Sahoo, S. Padhiari, S.K. Biswal, B.B. Panda, G. Hota, Fe3O4 nanoparticles functionalised GO/g-C3N4 nanocomposite: an efficient magnetic nanoadsorbent for adsorptive removal of organic pollutants. Mater. Chem. Phys. **244**, 122710 (2020). https://doi.org/10.1016/j.matchemphys.2020.122710
47. F. Deng, J. Liang, G. Yang, X. Hu, Q. Huang, J. Dou, Y. Wen, M. Liu, X. Zhang, Y. Wei, Gamma-ray initiated polymerisation from polydopamine-modified MoS2 nanosheets with poly (ionic liquid) and their utilisation for adsorptive organic dyes with enhanced efficiency. Chem. Eng. J. Adv. **7**, 100134 (2021). https://doi.org/10.1016/j.ceja.2021.100134
48. Q. Huang, M. Liu, J. Chen, Q. Wan, J. Tian, L. Huang, R. Jiang, Y. Wen, X. Zhang, Y. Wei, Facile preparation of MoS2 based polymer composites via mussel inspired chemistry and their high efficiency for removal of organic dyes. Appl. Surf. Sci. **419**, 35–44 (2017). https://doi.org/10.1016/j.apsusc.2017.05.006
49. Z. Liu, Y. Fang, H. Jia, C. Wang, Q. Song, L. Li, J. Lin, Y. Huang, C. Yu, C. Tang, Novel multi-functional cheese-like 3D carbon-BN as a highly efficient adsorbent for water purification. Sci. Rep. **8**(1), 1104 (2018). https://doi.org/10.1038/s41598-018-19541-5
50. G. Lian, X. Zhang, H. Si, J. Wang, D. Cui, Q. Wang, Boron nitride ultrathin fibrous nanonets: one-step synthesis and applications for ultrafast adsorption for water treatment and selective filtration of nanoparticles. ACS Appl. Mater. Interfaces. **5**(24), 12773–12778 (2013). https://doi.org/10.1021/am403789c
51. K. Maiti, T.D. Thanh, K. Sharma, D. Hui, N.H. Kim, J.H. Lee, Highly efficient adsorbent based on novel cotton flower-like porous boron nitride for organic pollutant removal. Compos. B Eng. **123**, 45–54 (2017). https://doi.org/10.1016/j.compositesb.2017.05.018
52. H. Lei, Z. Hao, K. Chen, Y. Chen, J. Zhang, Z. Hu, Y. Song, P. Rao, Q. Huang, Insight into adsorption performance and mechanism on efficient removal of methylene blue by accordion-like V2CTx MXene. J. Phys. Chem. Lett. **11**(11), 4253–4260 (2020). https://doi.org/10.1021/acs.jpclett.0c00973
53. S.V. Gupta, M. Ahmaruzzaman, CeO2/Fe3O4/g-C3N4 nanohybrid for adsorptive removal of Rose Bengal from aqueous stream. Int. J. Environ. Analyt. Chem., 1–20 (2022), https://doi.org/10.1080/03067319.2022.2039649
54. F. Jia, Q. Wang, J. Wu, Y. Li, S. Song, Two-dimensional molybdenum disulfide as a superb adsorbent for removing Hg2+ from water. ACS Sustain. Chem. Eng. **5**(8), 7410–7419 (2017). https://doi.org/10.1021/acssuschemeng.7b01880
55. K. Ai, C. Ruan, M. Shen, L. Lu, MoS2 nanosheets with widened interlayer spacing for high-efficiency removal of mercury in aquatic systems. Adv. Func. Mater. **26**(30), 5542–5549 (2016). https://doi.org/10.1002/adfm.201601338
56. Z. Wang, A. Sim, J.J. Urban, B. Mi, Removal and recovery of heavy metal ions by two-dimensional MoS2 Nanosheets: performance and mechanisms. Environ. Sci. Technol. **52**(17), 9741–9748 (2018). https://doi.org/10.1021/acs.est.8b01705
57. J. Luo, K. Fu, M. Sun, K. Yin, D. Wang, X. Liu, J.C. Crittenden, Phase-mediated heavy metal adsorption from aqueous solutions using two-dimensional layered MoS2. ACS Appl. Mater. Interfaces. **11**(42), 38789–38797 (2019). https://doi.org/10.1021/acsami.9b14019
58. J. Wang, W. Zhang, X. Yue, Q. Yang, F. Liu, Y. Wang, D. Zhang, Z. Li, J. Wang, One-pot synthesis of multifunctional magnetic ferrite–MoS2–carbon dot nanohybrid adsorbent for efficient Pb(ii) removal. J. Mater. Chem. A **4**(10), 3893–3900 (2016). https://doi.org/10.1039/C6TA00269B
59. A.E. Gash, A.L. Spain, L.M. Dysleski, C.J. Flaschenriem, A. Kalaveshi, P.K. Dorhout, S.H. Strauss, Efficient recovery of elemental mercury from Hg(II)-contaminated aqueous media using a redox-recyclable ion-exchange material. Environ. Sci. Technol. **32**(7), 1007–1012 (1998). https://doi.org/10.1021/es970804n
60. R. Gusain, N. Kumar, E. Fosso-Kankeu, S.S. Ray, Efficient removal of Pb(II) and Cd(II) from industrial mine water by a hierarchical MoS2/SH-MWCNT nanocomposite. ACS Omega **4**(9), 13922–13935 (2019). https://doi.org/10.1021/acsomega.9b01603
61. L. Wang, K. Wang, R. Huang, Z. Qin, Y. Su, S. Tong, Hierarchically flower-like WS2 microcrystals for capture and recovery of Au (III), Ag (I) and Pd (II). Chemosphere **252**, 126578 (2020). https://doi.org/10.1016/j.chemosphere.2020.126578

62. G. Wang, K. Guo, B. Wang, F. Han, Z. Guo, Z. Song, J. Ji, C. Tang, Mercury adsorption on thiol-modified porous boron nitride: a combined experimental and theoretical investigation. Ind. Eng. Chem. Res. **60**(35), 12984–12998 (2021). https://doi.org/10.1021/acs.iecr.1c01530
63. S. Li, F. Liu, Y. Su, N. Shao, D. Yu, Y. Liu, W. Liu, Z. Zhang, Luffa sponge-derived hierarchical meso/macroporous boron nitride fibers as superior sorbents for heavy metal sequestration. J. Hazard. Mater. **378**, 120669 (2019). https://doi.org/10.1016/j.jhazmat.2019.05.062
64. Y. Chi, Y. Xu, C. Xu, J. Tian, Y. Li, B. Gu, H. Song, H. Zhang, Adsorptive removal of radioactive cesium from model nuclear wastewater over hydroxyl-functionalized Mxene Ti3C2Tx. Ind. Eng. Chem. Res. **61**(25), 9054–9066 (2022). https://doi.org/10.1021/acs.iecr.2c01297
65. Y. Ying, Y. Liu, X. Wang, Y. Mao, W. Cao, P. Hu, X. Peng, Two-Dimensional titanium carbide for efficiently reductive removal of highly toxic chromium(VI) from water. ACS Appl. Mater. Interfaces. **7**(3), 1795–1803 (2015). https://doi.org/10.1021/am5074722
66. X. Feng, Z. Yu, R. Long, X. Li, L. Shao, H. Zeng, G. Zeng, Y. Zuo, Self-assembling 2D/2D (MXene/LDH) materials achieve ultra-high adsorption of heavy metals Ni2+ through terminal group modification. Sep. Purif. Technol. **253**, 117525 (2020). https://doi.org/10.1016/j.seppur.2020.117525
67. A.S. Krishna Kumar, S.-J. Jiang, J.K. Warchoł, Synthesis and characterization of two-dimensional transition metal dichalcogenide magnetic MoS2@Fe3O4 nanoparticles for adsorption of Cr(VI)/Cr(III). ACS Omega **2**(9), 6187–6200 (2017). https://doi.org/10.1021/acsomega.7b00757
68. D. Gan, Q. Huang, J. Dou, H. Huang, J. Chen, M. Liu, Y. Wen, Z. Yang, X. Zhang, Y. Wei, Bioinspired functionalisation of MXenes (Ti3C2TX) with amino acids for efficient removal of heavy metal ions. Appl. Surf. Sci. **504**, 144603 (2020). https://doi.org/10.1016/j.apsusc.2019.144603
69. H. Xie, J. Zhang, D. Wang, J. Liu, L. Wang, H. Xiao, Construction of three-dimensional g-C3N4/attapulgite hybrids for Cd(II) adsorption and the reutilization of waste adsorbent. Appl. Surf. Sci. **504**, 144456 (2020). https://doi.org/10.1016/j.apsusc.2019.144456
70. W. Shi, Y. Chu, M. Xia, F. Wang, C. Fu, The adsorption performance and micro-mechanism of MoS2/montmorillonite composite to atenolol and acebutolol: adsorption experiments and a novel visual study of interaction. Ecotoxicol. Environ. Saf. **213**, 111993 (2021). https://doi.org/10.1016/j.ecoenv.2021.111993
71. Z. Zeng, S. Ye, H. Wu, R. Xiao, G. Zeng, J. Liang, C. Zhang, J. Yu, Y. Fang, B. Song, Research on the sustainable efficacy of g-MoS2 decorated biochar nanocomposites for removing tetracycline hydrochloride from antibiotic-polluted aqueous solution. Sci. Total Environ. **648**, 206–217 (2019). https://doi.org/10.1016/j.scitotenv.2018.08.108
72. Y. Chao, J. Zhang, H. Li, P. Wu, X. Li, H. Chang, J. He, H. Wu, H. Li, W. Zhu, Synthesis of boron nitride nanosheets with N-defects for efficient tetracycline antibiotics adsorptive removal. Chem. Eng. J. **387**, 124138 (2020). https://doi.org/10.1016/j.cej.2020.124138
73. Y. Chao, B. Tang, J. Luo, P. Wu, D. Tao, H. Chang, X. Chu, Y. Huang, H. Li, W. Zhu, Hierarchical porous boron nitride with boron vacancies for improved adsorption performance to antibiotics. J. Colloid Interface Sci. **584**, 154–163 (2021). https://doi.org/10.1016/j.jcis.2020.09.075
74. A.A. Ghani, A. Shahzad, M. Moztahida, K. Tahir, H. Jeon, B. Kim, D.S. Lee, Adsorption and electrochemical regeneration of intercalated Ti3C2Tx MXene for the removal of ciprofloxacin from wastewater. Chem. Eng. J. **421**, 127780 (2021). https://doi.org/10.1016/j.cej.2020.127780
75. H. Mirhosseini, A. Mostafavi, T. Shamspur, G. Sargazi, Preparation of novel ternary g-C3N4/WO3/ZnO nanocomposite adsorbent with highly effective imidacloprid removal: optimisation design and a controllable systematic study. J. Mater. Sci.: Mater. Electron. **31**(20), 17903–17920 (2020). https://doi.org/10.1007/s10854-020-04343-1
76. R. Luo, W. Zhang, X. Hu, Y. Liang, J. Fu, M. Liu, F. Deng, Q.-Y. Cao, X. Zhang, Y. Wei, Preparation of sodium ligninsulfonate functionalised MXene using hexachlorocyclotriphosphazene as linkage and its adsorption applications. Appl. Surf. Sci. **602**, 154197 (2022). https://doi.org/10.1016/j.apsusc.2022.154197

77. T. Guo, Y. Lei, X. Hu, G. Yang, J. Liang, Q. Huang, X. Li, M. Liu, X. Zhang, Y. Wei, Hydrothermal synthesis of MXene-MoS2 composites for highly efficient removal of pesticides. Appl. Surf. Sci. **588**, 152597 (2022). https://doi.org/10.1016/j.apsusc.2022.152597
78. A. Mangeli, A. Mostafavi, T. Shamspur, F. Fathirad, F. Mehrabi, Decontamination of fenitrothion from aqueous solutions using rGO/MoS2/Fe3O4 magnetic nanosorbent: synthesis, characterisation and removal application. J. Environ. Health Sci. Eng. **19**(2), 1505–1511 (2021). https://doi.org/10.1007/s40201-021-00706-w
79. J. Low, S. Cao, J. Yu, S. Wageh, Two-dimensional layered composite photocatalysts. Chem. Commun. **50**(74), 10768–10777 (2014). https://doi.org/10.1039/C4CC02553A
80. S. Mohana Roopan, M.A. Khan, MoS2 based ternary composites: review on heterogeneous materials as catalyst for photocatalytic degradation. Catalysis Rev., 1–74 (2021). https://doi.org/10.1080/01614940.2021.1962493
81. S. Karmakar, S. Biswas, P. Kumbhakar, A comparison of temperature dependent photoluminescence and photocatalytic properties of different MoS2 nanostructures. Appl. Surf. Sci. **455**, 379–391 (2018). https://doi.org/10.1016/j.apsusc.2018.05.204
82. W.T. Yein, Q. Wang, Y. Li, X. Wu, Piezoelectric potential induced the improved micropollutant dye degradation of Co-doped MoS2 ultrathin nanosheets in dark. Catal. Commun. **125**, 61–65 (2019). https://doi.org/10.1016/j.catcom.2019.03.023
83. J. Wang, S. Dong, T. Guo, J. Jin, J. Sun, pH-dictated synthesis of novel flower-like MoS2 with augmented natural sunlight photocatalytic activity. Mater. Lett. **191**, 22–25 (2017). https://doi.org/10.1016/j.matlet.2016.12.090
84. T. Thurston, J. Wilcoxon, Photooxidation of organic chemicals catalysed by nanoscale MoS2. J. Phys. Chem. B **103**(1), 11–17 (1999). https://doi.org/10.1021/jp982337h
85. A. Gopalakrishnan, S. Singh, S. Badhulika, Reusable, free-standing MoS2/rGO/Cu2O ternary composite films for fast and highly efficient sunlight driven photocatalytic degradation. ChemistrySelect **5**, 1997–2007 (2020). https://doi.org/10.1002/slct.201904932
86. S. Shahabuddin, S. Mehmood, I. Ahmad, N. Sridewi, Synthesis and characterization of 2D-WS2 incorporated polyaniline nanocomposites as photo catalyst for methylene blue degradation. Nanomaterials **12**(12), 2090 (2022). https://doi.org/10.3390/nano12122090
87. S.V.P. Vattikuti, I.-L. Ngo, C. Byon, Physicochemical characteristic of CdS-anchored porous WS2 hybrid in the photocatalytic degradation of crystal violet under UV and visible light irradiation. Solid State Sci. **61**, 121–130 (2016). https://doi.org/10.1016/j.solidstatesciences.2016.09.015
88. X. Feng, Z. Yu, Y. Sun, R. Long, M. Shan, X. Li, Y. Liu, J. Liu, Review MXenes as a new type of nanomaterial for environmental applications in the photocatalytic degradation of water pollutants. Ceram. Int. **47**(6), 7321–7343 (2021). https://doi.org/10.1016/j.ceramint.2020.11.151
89. MSI. Nasri, M.F.R. Samsudin, A.A. Tahir, S. Sufian, Effect of MXene loaded on g-C3N4 photocatalyst for the photocatalytic degradation of methylene blue. Energies **15**(3), 955 (2022). https://doi.org/10.3390/en15030955
90. A. Rani, K. Singh, A.S. Patel, A. Chakraborti, S. Kumar, K. Ghosh, P. Sharma, Visible light driven photocatalysis of organic dyes using SnO2 decorated MoS2 nanocomposites. Chem. Phys. Lett. **738**, 136874 (2020). https://doi.org/10.1016/j.cplett.2019.136874
91. J. Qu, D. Teng, X. Zhang, Q. Yang, P. Li, Y. Cao, Preparation and regulation of two-dimensional Ti3C2Tx MXene for enhanced adsorption–photocatalytic degradation of organic dyes in wastewater. Ceram. Int. **48**(10), 14451–14459 (2022). https://doi.org/10.1016/j.ceramint.2022.01.338
92. S. Kapatel, C.K. Sumesh, Two-step facile preparation of MoS2·ZnO nanocomposite as efficient photocatalyst for methylene blue (Dye) degradation. Electron. Mater. Lett. **15**(1), 119–132 (2019). https://doi.org/10.1007/s13391-018-00101-y
93. V.Q. Hieu, T.K. Phung, T.-Q. Nguyen, A. Khan, V.D. Doan, V.A. Tran, V.T. Le, Photocatalytic degradation of methyl orange dye by Ti3C2–TiO2 heterojunction under solar light. Chemosphere **276**, 130154 (2021). https://doi.org/10.1016/j.chemosphere.2021.130154

94. H.W. Tian, M. Liu, W. Zheng, Constructing 2D graphitic carbon nitride nanosheets/layered MoS 2 /graphene ternary nanojunction with enhanced photocatalytic activity. Appl. Catal. B **225**, 468–476 (2018). https://doi.org/10.1016/j.apcatb.2017.12.019
95. T.T. Trang Phan, T.T. Truong, H.T. Huu, L.T. Nguyen, V.T. Nguyen, H.L. Nguyen, V. Vo, Visible light-driven Mn-MoS2/rGO composite Photocatalysts for the photocatalytic degradation of rhodamine B. J. Chem., 6285484 (2020), https://doi.org/10.1155/2020/6285484
96. W. Zhang, M. Gao, H. Zhu, Y. Hui, Q. Cao, C. Zhai, Z. Zhang, Microwave-driven construction of MoS 2 /graphene heterostructure for enhanced photodegradation under natural light. Physica Status Solidi (a) 219 (2022), https://doi.org/10.1002/pssa.202100767
97. S. Shahabuddin, R. Khanam, M. Khalid, N.M. Sarih, J.J. Ching, S. Mohamad, R. Saidur, Synthesis of 2D boron nitride doped polyaniline hybrid nanocomposites for photocatalytic degradation of carcinogenic dyes from aqueous solution. Arab. J. Chem. **11**(6), 1000–1016 (2018). https://doi.org/10.1016/j.arabjc.2018.05.004
98. Z. Othman, A. Sinopoli, H.R. Mackey, K.A. Mahmoud, Efficient photocatalytic degradation of organic dyes by AgNPs/TiO2/Ti3C2Tx MXene composites under UV and solar light. ACS Omega **6**(49), 33325–33338 (2021). https://doi.org/10.1021/acsomega.1c03189
99. Y. Zheng, H. Qi, L. Zhang, Y. Zhang, L. Zhong, X. Zhang, Y. Feng, J. Xue, Photocatalytic degradation of dye wastewater by stepwise assembling PVA aerogel/TiO2/MoS2/Au composites in visible light. Water Sci. Technol. **85**(9), 2625–2638 (2022). https://doi.org/10.2166/wst.2022.106
100. M.A. Iqbal, S.I. Ali, F. Amin, A. Tariq, M.Z. Iqbal, S. Rizwan, La- and Mn-codoped bismuth ferrite/Ti3C2 MXene composites for efficient photocatalytic degradation of congo red dye. ACS Omega **4**(5), 8661–8668 (2019). https://doi.org/10.1021/acsomega.9b00493
101. T. Liu, L. Li, X. Geng, C. Yang, X. Zhang, X. Lin, P. Lv, Y. Mu, S. Huang, Heterostructured MXene-derived oxides as superior photocatalysts for MB degradation. J. Alloy. Compd. **919**, 165629 (2022). https://doi.org/10.1016/j.jallcom.2022.165629
102. M. Patel, M. Patel, S. C K, Visible light enhanced photocatalytic performance of WS2 catalyst for the degradation of ternary dye mixture. AIP Conf. Proc., 020047 (2020), https://doi.org/10.1063/5.0001154
103. R. Gusain, N. Kumar, F. Opoku, P.P. Govender, S.S. Ray, MoS2 Nanosheet/ZnS composites for the visible-light-assisted photocatalytic degradation of oxytetracycline. ACS Appl. Nano Mater. **4**(5), 4721–4734 (2021). https://doi.org/10.1021/acsanm.1c00330
104. E. Grilla, M.N. Kagialari, A. Petala, Z. Frontistis, D. Mantzavinos, Photocatalytic degradation of valsartan by MoS2/BiOCl heterojunctions. Catalysts **11**(6), 650 (2021). https://doi.org/10.3390/catal11060650
105. N. Liu, N. Lu, Y. Su, P. Wang, X. Quan, Fabrication of g-C3N4/Ti3C2 composite and its visible-light photocatalytic capability for ciprofloxacin degradation. Sep. Purif. Technol. **211**, 782–789 (2019). https://doi.org/10.1016/j.seppur.2018.10.027
106. E.H. Rdewi, K.K. Abbas, A.M.H. AbdulkadhimAl-Ghaban, Removal pharmaceutical carbamazepine from wastewater using ZnO-TiO2-MXene heterostructural nanophotocatalyst under solar light irradiation. Materials Today: Proc. **60**, 1702–1711 (2022). https://doi.org/10.1016/j.matpr.2021.12.229
107. Z. Wu, W. Fu, H. Xu, R. Zheng, F. Han, Z. Liang, D. Han, D. Han, F. Li, L. Niu, A simple preparation method of in situ oxidised titanium carbide MXene for photocatalytic degradation of catechol. New J. Chem. **46**(19), 9364–9371 (2022). https://doi.org/10.1039/D2NJ01033J
108. T. Fatima, S. Husain, M. Khanuja, Superior photocatalytic and electrochemical activity of novel WS2/PANI nanocomposite for the degradation and detection of pollutants: antibiotic, heavy metal ions, and dyes. Chem. Eng. J. Adv., 100373 (2022), https://doi.org/10.1016/j.ceja.2022.100373
109. E. Torad, M.M. Mohamed, M.M.H. Khalil, E.H. Ismail, Optimal design of silver@silver sulfide-modified WS2 and its application in photocatalytic diclofenac degradation and H2 generation. J. Environ. Chem. Eng. **9**(6), 106446 (2021). https://doi.org/10.1016/j.jece.2021.106446

110. Y. He, N. Xu, L.B. Junior, X. Hao, B. Yao, Q. Yang, D. Liu, Z. Ma, Construction of AuNPs/h-BN nanocomposites by using gold as interfacial electron transfer mediator with highly efficient degradation for levofloxacin hydrochloride and hydrogen generation. Appl. Surf. Sci. **520**, 146336 (2020). https://doi.org/10.1016/j.apsusc.2020.146336
111. R. Kumar, M.A. Barakat, B.A. Al-Mur, F.A. Alseroury, J.O. Eniola, Photocatalytic degradation of cefoxitin sodium antibiotic using novel BN/CdAl2O4 composite. J. Clean. Prod. **246**, 119076 (2020). https://doi.org/10.1016/j.jclepro.2019.119076
112. T. Yan, Z. Du, J. Wang, H. Cai, D. Bi, Z. Guo, Z. Liu, C. Tang, Y. Fang, Construction of 2D/2D Bi2WO6/BN heterojunction for effective improvement on photocatalytic degradation of tetracycline. J. Alloy. Compd. **894**, 162487 (2022). https://doi.org/10.1016/j.jallcom.2021.162487
113. A. Shahzad, K. Rasool, M. Nawaz, W. Miran, J. Jang, M. Moztahida, K.A. Mahmoud, D.S. Lee, Heterostructural TiO2/Ti3C2Tx (MXene) for photocatalytic degradation of antiepileptic drug carbamazepine. Chem. Eng. J. **349**, 748–755 (2018). https://doi.org/10.1016/j.cej.2018.05.148
114. T. Ahamad, M. Naushad, S.I. Al-Saeedi, S. Almotairi, S.M. Alshehri, Fabrication of MoS2/ZnS embedded in N/S doped carbon for the photocatalytic degradation of pesticide. Mater. Lett. **263**, 127271 (2020). https://doi.org/10.1016/j.matlet.2019.127271
115. W.-K. Jo, J.Y. Lee, N.C.S. Selvam, Synthesis of MoS2 nanosheets loaded ZnO–g-C3N4 nanocomposites for enhanced photocatalytic applications. Chem. Eng. J. **289**, 306–318 (2016). https://doi.org/10.1016/j.cej.2015.12.080
116. K. Namshah, R. Mohamed, WO3–TiO2 nanocomposites for paracetamol degradation under visible light. Appl. Nanosci. **8**(2018), https://doi.org/10.1007/s13204-018-0888-4
117. S. Kumar, V. Sharma, K. Bhattacharyya, V. Krishnan, N-doped ZnO–MoS2 binary heterojunctions: the dual role of 2D MoS2 in the enhancement of photostability and photocatalytic activity under visible light irradiation for tetracycline degradation. Mater. Chem. Frontiers **1**(6), 1093–1106 (2017). https://doi.org/10.1039/C6QM00274A

Chapter 4
Graphitic Carbon Nitride ($g\text{-}C_3N_4$)-Based Photocatalysts for Environmental Applications

Rashmi Acharya, Subhasish Mishra, Lopamudra Acharya, and Kulamani Parida

Abstract The alarmingly rising environmental pollution adversely affects the sustainable growth of modern civilization. Scientists have persistently been putting tremendous efforts over the decades to develop environment benevolent technologies to overcome this major challenge. Photocatalysis is one such technology which needs renewable solar energy and abundantly available water resources as driving forces for pollutants' degradation. In addition, the selection of an appropriate semiconductor is highly essential to degrade toxic organic compounds, hazardous heavy metals and noxious gases into harmless products efficiently. Among various semiconductor photocatalysts, $g\text{-}C_3N_4$ (GCN) is considered a robust photocatalyst because of several fascinating properties like metal-free chemical nature, visible-light-responsive activity with moderate band gap of 2.7 eV, tunable electronic structure, facile synthesis, low cost, high thermal and chemical stability. However, low surface area ($\sim$10 m^2 g^{-1}), high rate of charge carriers recombination, incomplete solar spectrum absorbance and inadequate valence band position (1.4 eV vs NHE) are some of the limitations due to which expected photocatalytic performance of GCN is yet to be achieved. Therefore, modification strategies such as exfoliating bulk GCN into nanosheets, incorporating foreign elements into its crystal structure and heterostructure formation have been employed to overcome these limitations to achieve high photocatalytic efficiency. In this chapter discusses the basic principle of photocatalytic pollutant degradation over a semiconductor surface. Recent developments in modification strategies to enhance the photoactivity of GCN have been summarised systematically. Photocatalytic applications of GCN-based photocatalysts with respect to environmental remediation are presented in this chapter. The challenges and future perspectives in designing GCN-based photocatalysts for efficient performance towards environmental applications are addressed along with the conclusion.

R. Acharya (✉) · S. Mishra
Department of Chemistry, I.T.E.R., Siksha 'O' Anusandhan Deemed to be University, Bhubaneswar 751030, Odisha, India
e-mail: drrashmiacharya75@gmail.com; rashmiacharya@soa.ac.in

L. Acharya · K. Parida
Centre for Nano Science and Nano Technology, ITER, Siksha 'O' Anusandhan Deemed to be University, Bhubaneswar 751030, Odisha, India

N. Kumar et al. (eds.), *Two-Dimensional Materials for Environmental Applications*, Springer Series in Materials Science 332, https://doi.org/10.1007/978-3-031-28756-5_4

4.1 Introduction

Recently, the world has been facing serious environmental issues and acute energy shortages due to the rapid advancement in industrialization and massive growth in the population [1]. One of the earth's most important natural resources is water, which is getting polluted day by day through the contamination of effluents containing organic dyes, antibiotics, pesticides, organic compounds and heavy metal ions [2]. For instance, various industries consume more than 100,000 commercial dyes annually. Approximately, 100 tons of dyes are discharged into waste streams per year [3]. These cause serious problems to human health, plants and aquatics owing to their acute toxicity. The need for clean water has put tremendous task for the removal of these noxious pollutants from the natural water bodies. Additionally, increased emissions containing CO_2, NO_X etc. are gradually destroying the air quality of our environment. Moreover, CO_2 level increas in the atmosphere caused by the rapid fossil fuels consumption leads to an energy crunch. This is also responsible for the global warming [4]. Several techniques like adsorption, absorption with or without chemical reaction, biological degradation and condensation have already been used to treat water [5–10]. However, several limitations have restricted these techniques for large scale applications. Therefore, an ideal technique which does not emit toxic by-products or consume much energy and possess potential to remove all kinds of pollutants, needs to be implemented [11, 12]. Advanced oxidation processes (AOPs) are considered as suitable techniques for practical application because the reactive species such as $^{\cdot}OH$, $O_2^{\cdot-}$, $SO_4^{\cdot-}$ etc., resulted from these exhibit strong redox ability to decompose almost all the water contaminants. However, most of the AOPs need large amount of chemicals and energies, which stood as the major disadvantages of the process [13, 14]. On the other hand, photocatalysis has emerged as a promising technology, which converts abundantly available solar energy into chemical energy. This technique has been applied to divergent fields including pollutant degradation, heavy metals removal, hydrogen generation, carbon dioxide conversion and organic transformations [15–18].

The concept of photocatalysis was first developed in 1972, when Fujishima and Honda produced H_2 through water splitting over TiO_2 electrode under UV illumination [19]. After this the photocatalytic technology was gradually applied in the field of environmental remediation. Carey and co-workers carried out the photocatalytic decomposition of organic pollutants by using n-type TiO_2 first in 1976 [20]. However, the expected photocatalytic performance of TiO_2 has not yet been achieved mainly due to fast rate of charge carriers' recombination and poor visible light absorptivity [21, 22]. In order to utilize visible region of solar spectrum, several visible light responsive semiconductor photocatalysts have been used for photocatalytic pollutants degradation. Among these, GCN is contemplated as a robust photocatalyst after the pioneering work of Wang and his co-workers' to produce hydrogen over it by splitting water under visible light irradiation in 2009 [23]. Carbon nitride exists in various allotropic phases such as a-C_3N_4, b-C_3N_4, GCN, cubic-C_3N_4 and pseudo-cubic-C_3N_4. Among these, GCN is the most stable form at ambient conditions [24]. Noble GCN photocatalysts can be prepared conveniently via the facile thermal condensation of carbon and nitrogen rich precursors like ammonium thiocyanate, melamine, thiourea, cyanamide, urea and dicyandiamide [25]. The two dimensional (2D) sheets with honeycomb architecture of GCN are exist as a polymerized form of s-triazine (Fig. 4.1a) or tri-s-triazine (heptazine) units (Fig. 4.1b) as the monomers. These melon units of GCN are well connected through tertiary amine groups to form a nanosheet like structure. The so-formed nanosheets get stacked with each other by Van-der-Waals forces, which is considered as a weak force of attraction [26]. Its metal free nature, facile fabrication routes, inexpensiveness, non-toxicity, moderate band gap energy of 2.7–2.8 eV has enabled it for extensive use in solar driven photocatalysis. Moreover, the positive Valence Band level (1.4 V vs NHE at pH 7.0) the negative Conduction band position (−1.3 V vs NHE at pH 7.0) of GCN are helpful for various photoredox reactions [27–31].

However, the practical photocatalytic applications of GCN are limited mainly due to following reasons: (1) As per the density functional theory (DFT) calculations, valence band maximum (VBM) of GCN is made up of N 2p state and the conduction band minimum (CBM) is the result of hybridization between N 2p and C 2p states.

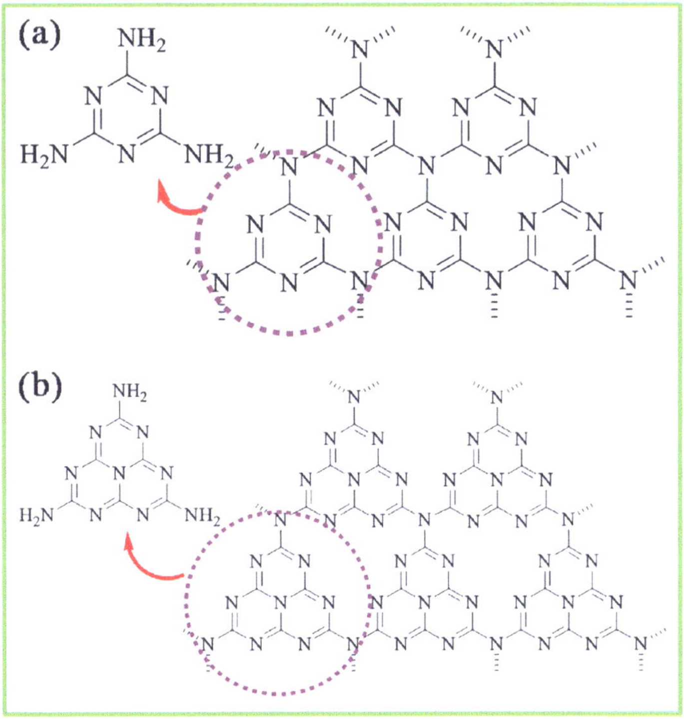

Fig. 4.1 Structure of GCN with **a** s-Triazine and **b** tri-s-triazine as building blocks. Reproduced with permission from ref. [26]. Copyright 2019, Elsevier Science Ltd.

To be precise, the photoproduced electrons distributed between the N 2p and C 2p hybridized orbitals are more vulnerable to charge carrier recombination. This process allegedly reduces the efficiency of the charge carriers and thus affects the performance of GCN as a photocatalyst (2) The band gap energy of GCN is around 2.7 eV, which corresponds to the wave length of 450 nm. Thus complete utilization of broad solar spectrum is limited. (3) It possesses weakly oxidative valence band potential of around +1.4 V, which cannot provide strong driving force to the photoinduced holes to carry out water oxidation reaction in order to generate strong reactive ·OH species required particularly for pollutant degradation. (4) The high temperature solid-phase polycondensation of organic precursors forms agglomerates of bulk GCN with relatively large size and small surface area (10–15 m^2 g^{-1}) resulting in poor photocatalyic performance. (5) The difficulty in separation from treated solution also limits its practical applications [32–34].

Several modification approaches like band gap engineering, exfoliation into nanosheets and heterojunction formation with other semiconductors are implemented to augment photocatalytic activity of GCN based photocatalyts [35–37]. In this chapter, we have addressed these strategies systematically after depicting the basic principle of photocatalytic pollutant degradation over a semiconductor photocatalyst. Thereafter, divergent applications related to environmental remediation over GCN based photocatalyts were discussed. Finally, the present chapter was concluded with advantages, challenges and future perspectives of GCN based photocatalyts.

4.2 Basic Principle of Semiconductor Based Photocatalytic Pollutant Degradation

The photocatalytic pollutant degradation over a semiconductor surface usually proceeds through four important steps as presented in Scheme 4.1. Step 1 involves harvestation of light energy by the semiconductor photocatalyst (SP). The extent of light absorption depends on the textural and microstructural properties of the prepared semiconductor [38]. In step 2, electrons from the valence band (VB) of the semiconductor are excited to the conduction band (CB) by absorbing energy equal to or greater than the band gap energy (E_g) with the formation of equivalent number of holes (h^+) at the VB as shown in Eq. 4.1.

$$sp \underset{\longrightarrow}{h\upsilon} h^+_{VB} + e^-_{CB} \tag{4.1}$$

Step 3 is responsible for effective migration of photoinduced charge carriers in such a way that the recombination rate of charges should be suppressed to a greater extent [39]. In step 4, the degradation of pollutants is carried out through a number of redox reactions by proper utilization of charge carriers [40]. For instance, the photoinduced electrons released from CB with adequate redox potential undergo reduction reaction with surface adsorbed O_2 molecules (ORR) to liberate reactive super oxide anions as per Eq. 4.2.

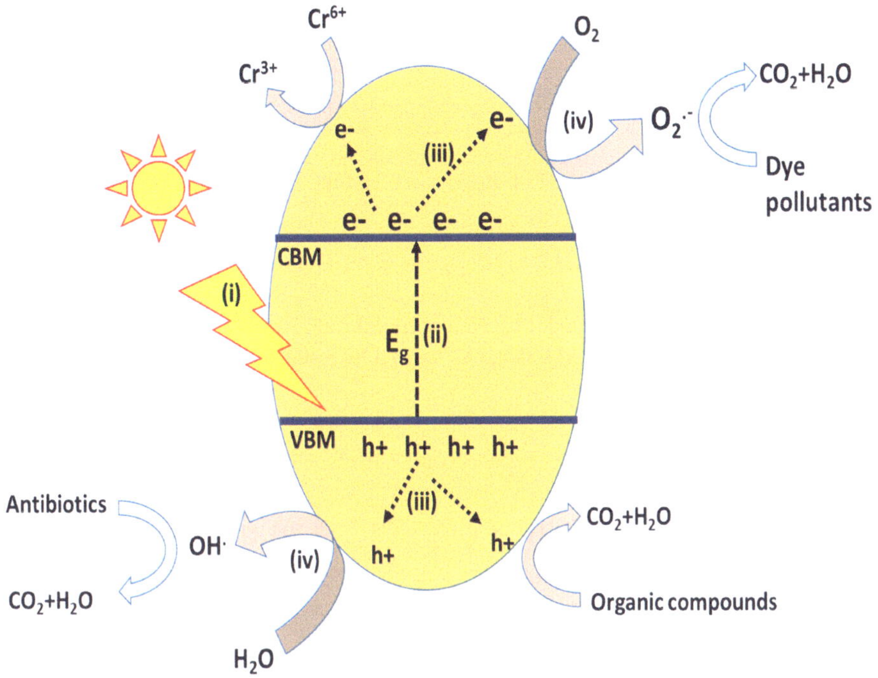

Scheme 4.1 The four step basic mechanism of semiconductor photocatalysis for pollutants removal

$$O_2 + e^-_{CB} \rightarrow O_2^{\cdot -} \tag{4.2}$$

Similarly, the surface adsorbed H_2O molecules are oxidized by the holes released from VB with desired oxidation potential to form strongly reactive ·OH radicals (Eq. 4.3).

$$H_2O + h^+_{VB} \rightarrow OH^{\cdot} + H^+ \tag{4.3}$$

These reactive species decompose the toxic organic compounds, dyes and antibiotics into harmless products (Eq. 4.4) [33]. There also appears a possibility of decomposition of these noxious substances directly by the oxidation reaction with the photoinduced holes as mentioned in Eq. 4.5 [41].

$$\text{Organic compounds/dyes/antibiotics} + O_2^{\cdot -}/OH^{\cdot} \rightarrow H_2O + CO_2 \tag{4.4}$$

$$\text{Organic compounds/dyes/antibiotics} + h^+_{VB} \rightarrow H_2O + CO_2 \tag{4.5}$$

Besides these, the photoinduced electrons are utilized to reduce CO_2 into value added products such as CH_4, CO, CH_3OH, C_2H_5OH etc. (Eq. 4.6–4.9) [42]. In addition, these electrons can be utilized for the reduction of heavy metals ions like Cr (VI), U (VI) etc. in to

$$CO_2 + 8H^+ + 8e^-_{CB} \rightarrow CH_4 + H_2O \tag{4.6}$$

$$CO_2 + 2H^+ + 2e^-_{CB} \rightarrow CO + H_2O \tag{4.7}$$

$$CO_2 + 6H^+ + 6e^-_{CB} \rightarrow CH_3OH + H_2O \tag{4.8}$$

$$2CO_2 + 12H^+ + 12e^-_{CB} \rightarrow C_2H_5OH + 3H_2O \tag{4.9}$$

their nontoxic forms (Eq. 4.10–4.11). Heavy metal ions like As (III) can be oxidized to As (V) by the photogenerated holes, $O_2^{\cdot-}$ and ·OH radicals as presented in Eq. 4.12 [18, 43, 44].

$$Cr_2O_7^{2-} + 14H^+ + 6e^-_{CB} \rightarrow 2Cr^{3+} + 7H_2O \tag{4.10}$$

$$UO_2^{2+} + 4H^+ + 2e^-_{CB} \rightarrow U^{4+} + 2H_2O \tag{4.11}$$

$$As(III) + h^+_{VB}/O_2^{\cdot-}/OH^{\cdot} \rightarrow As(V) \tag{4.12}$$

4.3 Modification Strategies for GCN

Metal free nature, facile synthesis approach, visible light responsive ability are some of the vital features that lead GCN a promising photocatalyst in diversified applications and environmental remediation in particular. However, low quantum efficiency restricts its scale up. Therefore, marvellous efforts from various quarters of researchers are made to promote its photocatalytic efficiency. Among the modification strategies, exfoliation into nanosheets, doping of foreign elements and heterojunction construction with other semiconductors are found beneficial. These strategies are depicted in this section.

4.3.1 Exfoliation of GCN

Single layer nanosheets fabricated from exfoliation of layered materials have been attracting so many eyes as a novel class of nanostructured materials [45, 46]. They possess ultrathin 2D structures in nanometer range. The quantum confinement effect thus generated from size of nanosheets contributes in upscaling the physicochemical properties [47, 48]. For instance, graphene shows extraordinary thermal, mechanical, electrical and optical properties than its bulky counterpart, graphite [49]. In semiconductor photocatalysis, the specially designed monolayer nanosheets have numerous advantage over stacked 3D materials. The exfoliated two dimensional nanosheets possess better surface area thus have abundant number of active sites for the redox reactions to take place [50]. The bulk diffusion length of photoproduced charge carriers are also shorter in case of exfoliated sheets. Its helps in reducing the rate of recombination of charge carriers. More importantly, due to quantum confinement in nanosheets, conduction and valence bands shift in opposite directions making it a higher bandgap semiconductor [51, 52]. This helps in producing charge carriers with higher redox potential for better photocatalytic activity. Thus photocatalytic activity of bulk GCN was anticipated to enhance drastically in an exfoliated 2D nanosheet form. Taking inspiration from graphene, delamination of bulk GCN into monolayer nanosheets will be an effective strategy to improve the surface area and thus better adsorption of pollutants. Various methods including thermal, ultrasonic, chemical oxidation are used for the fabrication of GCN nanosheets. Some of the methods are disused below.

4.3.1.1 Solid State Thermal Exfoliation

The very first attempt to delaminate GCN was reported by Niu and his co-workers in 2012 via a thermal oxidation process. They prepared the nanosheets by heating bulk GCN in an open air atmosphere at 500 °C. After two hours of heating, they obtained nanosheets of ~2 nm thickness due to thermal oxidation and etching of layered material. The basic principle of this method is the hydrogen bond between the polymeric melon units get oxidized away by heating in open air from the bulk material to give exfoliated nanosheets. From the XRD plot, they reported that the peak appeared at 13.1° seemed to be less pronounced in case of GCN nanosheets. This indicated the decrease in size of the layers due to thermal oxidation. The (002) peak appeared at 27.34° shifted to 27.67° in case of nanosheets showing decreased gallery distance between the basic sheets. In comparison to bulk GCN, the nanosheets showed much higher specific surface area (SSA). From the N_2 sorption–desorption graph, they found the SSA of nanosheets was 306m^{2}g^{-1}, whereas the bulk one has only 50 m^{2}g^{-1}. Higher SSA helps in providing more number of reactive sites to perform various reactions. They also examined the photocatalytic activity by detecting the hydroxyl (OH) radicals generated by both the bulk and exfoliated GCN. The nanosheets was observed to show better $^{\cdot}$OH radical production [53]. Dong et al. studied the effect of temperature on the thermal exfoliation process. They prepared GCN nanosheets at different temperatures (i.e. 450, 500 and 550 °C) by heating thiourea in open air for 2 h. The sample synthesized at 550 °C was found to have better characteristics. From the SEM data they studied the textural and microstructural properties of the nanosheets prepared at different exfoliation temperature. The bulk GCN showed disorderly stalked and curved bulks with average particle size in micrometer range. Figure 4.2a shows the SEM images for sample prepared at 550 °C with several cracks over its surface. This depicted the separation between the layers and exfoliation into nanosheets of ~22 nm.

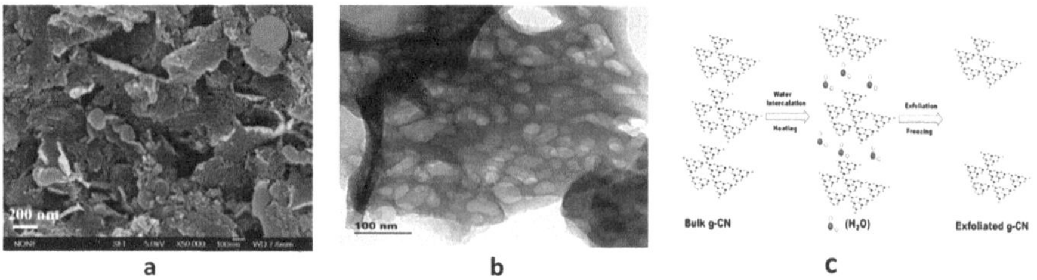

Fig. 4.2 **a** SEM micrograph for exfoliated GCN nanosheets (Reproduced with permission from ref. [54]. Copyright 2015, Elsevier), **b** HRTEM image showing plate like 2D structure of exfoliated GCN (Reproduced with permission from ref. [63]. Copyright 2019, Springer Nature), **c** Schematic diagram for bi-thermal green exfoliation of bulk GCN (Reproduced with permission from ref. [64]. Copyright 2019, Elsevier)

The specific surface area, pore volume and pore size of nanosheets were also higher than other prepared samples. They studied the photocatalytic removal of gaseous NO for each of the samples. The exfoliated GCN nanosheets were observed to perform better than the bulk one. Bulk GCN was able to remove only 18.9% of gaseous NO. Whereas, the sample fabricated at 550 °C shows a 33.5% removal efficiency [54]. They proposed a layer by layer oxidation couple mechanism with a layer splitting process for the thermal oxidation of bulk GCN into thin nanosheets. The CN layers, attached through weak van der Waals force of attraction, are constructed by polymeric melon units having weak hydrogen bonding. These bonds are too weak to stand against heavy thermal oxidation. This helps in the exfoliation of bulk GCN into nanosheets. They concluded that with increase in exfoliation temperature the size and thickness of the bulk sample get reduced.

Although, solid state thermal exfoliation process was the earliest and easiest process to synthesize nanosheets from bulk GCN, the exfoliation efficiency was low. The nanosheets formed via thermal etching constitute around 6 host layers. To synthesize single layered nanosheets, the bulk materials need to be dispersed into a solution phase. That's how the use of ultrasonic and chemical oxidation process are come into play.

4.3.1.2 Liquid Exfoliation

Ultrasound assisted methods have widely been used as a prominent liquid exfoliation method for layered materials [55, 56]. Solvent plays a vital role in ultrasonic exfoliation. When the enthalpy of mixing is minimized and solvent surface energy matches with the GCN nanoflakes, the bulk GCN starts exfoliating slowly under ultrasound [57]. Yang et al. reported an ultrasound method for liquid exfoliation of GCN by sonicating the dispersed bulk sample for 10 h. Five solvents namely methanol, isopropanol (IPA), water, N-methyl-pyrrolidone (NMP) and acetone were used separately for the purpose. The surface area of exfoliated sheets was found to be as high as 384m^2g^{-1} [58]. Still the thickness of exfoliated nanolayers was with ~ 2 nm range which is not a unilamellar structure. Also the undispersed bulk GCN left inside the solution interferes in the exfoliation process. Thus to enhance the exfoliation efficiency and to form single layer nanosheets, chemical oxidation methods are being used. Although $KMnO_4$ was being popularly used as a strong oxidants for chemical exfoliation of layered materials, the GCN layers seems to be weakly resistant to it. Therefore mild oxidants like H_2SO_4 were chosen as a promising candidate for the chemical exfoliation process [59, 60]. Zhu and his co-workers achieved chemically exfoliated GCN nanosheets through H_2SO_4 intercalation between the host layers of GCN. The measured thickness of the GCN nanosheets was about 0.4 nm which lies very much close to the theoretically calculated value of 0.325 nm. Thus, the formation of a single layered GCN nanosheet was confirmed using a chemical exfoliation technique [61]. The H_2SO_4 assisted chemical exfoliation process was able to produce uni-lamellar GCN nanosheets with an exfoliation efficiency of nearly 60%. Yet the efficiency was still not sufficient enough. Rather, the unexfoliated and leftover particles got reassembled to form various unnecessary precipitates. This leads to deposition of sulphate in the processing solution. However, these precipitates can easily be re-dispersed in the solution via continuous water washing and complete exfoliation of GCN can be achieved using chemical methods [62].

Liquid exfoliation methods were proved to be effective for the synthesis of monolayered GCN from its bulk counterpart. Yet the use of toxic and costly chemicals is still a concern. Thus a harsh chemical free green route is necessary for the exfoliation process.

4.3.1.3 Green Exfoliation

Pattanik et al. in 2019 reported a chemical free bi-thermal route to exfoliate bulk GCN into nanosheets. The bulk sample was refluxed in water for 6 h followed by freezing for the same time interval. HRTEM images (Fig. 4.2b) showed plate like structure of exfoliated GCN sheets with d value calculated from the fringe to be 0.326 nm. This value coincided with the XRD pattern of (002) plane in exfoliated nanosheets. The specific surface area for the exfoliated nanosheets was calculated to be 63.8 m^2g^{-1}. To evaluate the photocatalytic performance of as prepared samples, degradation of ciprofloxacin was carried out. As can be expected from the SSA values, the photocatalytic degradation of ciprofloxacin was two times higher in case of exfoliated nanosheets [63]. In another work, they synthesized GCN sheets via a modified bi-thermal heating cooling cycle route. The bulk samples were refluxed in water with continuous heating and stirring for 8 h. After that, the samples were subjected to freezing for another 8 h.This heating and freezing cycles was repeated for 10 times to obtain exfoliated GCN samples. The van der Waals force of attraction between layers of bulk GCN got destructed due to continuous heating and cooling effect. Figure 4.2c shows the schematic diagram of whole synthesis process. The surface area, pore volume and peak pore size of the exfoliated sample was way much higher than the bulk one. Thus, the exfoliated nanosheets perform better photocatalytic reactions under visible light. To elucidate this, photocatalytic degradation of

Congo red (CR) was performed. The exfoliated sheet managed to decolourize 95% of the CR in the aqueous solution whereas bulk GCN was able to degrade maximum of 81.8% of CR [64].

4.3.2 Doping in GCN

The noble GCN is a polymeric nano-structured material with low band gap energy of 2.7 eV. Its appropriate band structure has drawn boundless attention of researchers in the field of photocatalysis. Additionally, the intriguing properties possessed by GCN include cost-effective, non-toxicity, earth abundance and easy preparation methods [65–67]. However, low specific surface area, minimal light absorption property and quick charge carrier recombination of GCN prevent its photocatalytic activity from being suitable for practical application [68, 69]. Hence, for the improvement of photocatalytic activity, modification processes including band gap engineering, morphology adjustment and heterojunction formation have been developed so far [70, 71]. Recently, doping i.e. introduction of impurities in to the GCN lattice has been regarded as the most effective strategy for tuning the band gap energy along with broadening the light absorption range and escalating the separation efficiency of charge carriers [72, 73]. Various metal and non-metal elements have been doped in to GCN framework in order to improve its photocatalytic activity which we have summarized in upcoming parts.

During the preparation process of metal doped GCN, the corresponding metal soluble salt is thoroughly mixed with the precursor for simultaneous doping of metal ion in to the framework of GCN [74]. However, to preserve the metal free character of GCN, non-metal element doping has attracted enormous attention of researchers. Furthermore, the existence of high ionization energy with better electronegativity, non-metal elements can generally form covalent bond by accepting electrons from the host compound [75].

Metal and non-metal doping can alter the optical and electronic behaviour, by the introduction of extra element in to framework of GCN. Besides, doped GCN can increase the separation efficiency of charge carriers, alongside improving the light absorption range and photocatalytic performance as compared to bulk GCN. Characterization techniques like PXRD, UV–Vis DRS and PL analysis have been employed to investigate the remarkable structural and optical changes of doped GCN [76, 77]. The two distinct diffraction peaks obtained in PXRD pattern of GCN are at $2\theta = 13.1$ and $27.4°$ correspond to (100) and (002) plane respectively [78]. The (100) and (002) plane of GCN represent the in-planar structural units of tri-s-triazine and interlayer stacking of the aromatic rings in that order [79]. As can be seen from Fig. 4.3, the increasing content of dopants resulted in the reducing peak intensity of (002) plane along with the gradual disappearance of (001) plane. This fact indicated the host-gust interaction between GCN and dopant ion hinders the crystal growth of GCN. In the PXRD pattern, no peak for dopant species has been observed although a slight shift of 2θ value to lower angle has been witnessed. This observation is

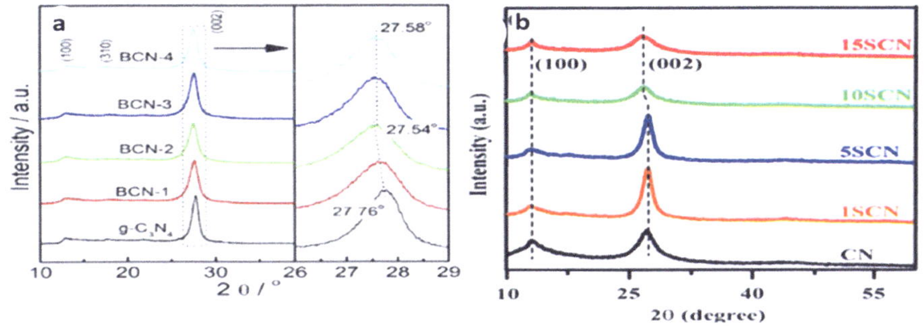

Fig. 4.3 PXRD pattern of **a** B and (Reproduced with permission from ref. [80]. Copyright 2018, Elsevier) **b** S doped GCN (Reproduced with permission from ref. [81]. Copyright 2020, Elsevier) in comparison with that of pristine GCN

ascribed to the doping of metal and non-metal ion in to the interstitial site of GCN resulting in the increase of interlayer spacing [81].

The optical properties of bulk and doped GCN are studied through UV–Vis DRS and PL analyses. Pristine GCN shows light absorption up to 460 nm. This value can be assigned to the $n\pi^*$ transition of the electron pairs present on the non-bonding orbitals of nitrogen atom and anti-bonding orbitals of the aromatic structure. In case of doped GCN, a red shift in the absorption edge to higher wavelength has been observed as shown in Fig. 4.4 [82, 83].

Moreover, the band gap energy of the doped GCN is narrowed as compared to the neat counterpart [20]. The above-mentioned results indicated the strong interaction between dopant and GCN which significantly affect the electronic arrangement and band structure of GCN. The distortion in the electronic structure strongly impacted the recombination rate of e^-/h^+ pairs. As displayed in Fig. 4.5a, doped GCN exhibits diminished PL peak intensity than that of pristine material [79, 85]. The current density of a material is highly dependent on the flow of electrons. More the number of electrons available, greater is the amount of current density [86]. Doped GCN is anticipated to exhibit improved photocurrent density in comparison to bare GCN due to its greater light absorption capacity and superior charge carrier separation efficiency. As presented in Fig. 4.5b, metal and non-metal doped GCN exhibited noticeable increased photocurrent density than that of bulk GCN [84, 87].

4.3.3 GCN/Semiconductor Heterojunction Construction

Despite the significant improvement in GCN based photocatalysis, the recombination rate of electron–hole pairs in the the semiconductor is still a major issue. Maximum amount of photoproduced electrons and holes got recombined within a few nanoseconds of their production. Thus, the photocatalytic activity is significantly reduced [88]. Formation of heterostructures with other materials could be an effective strategy

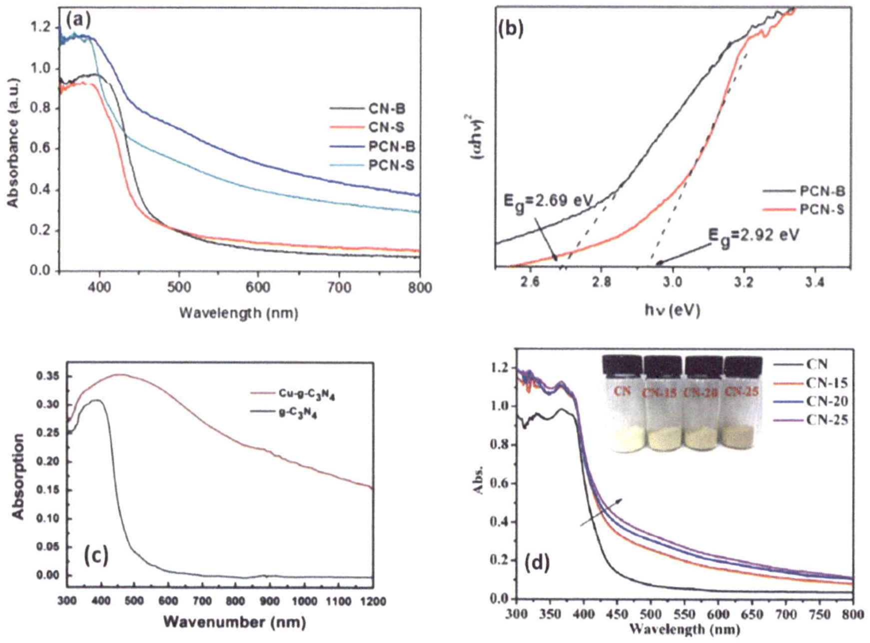

Fig. 4.4 **a** UV–Vis DRS and **b** Tauc plot for bulk (CN-B), exfoliated (CN-S), P- doped GCN (P-CN) and P-doped exfoliated (P-CNS) GCN (Reproduced with permission from ref. [82]. Copyright 2017, Elsevier), **c** solid state UV–vis–NIR diffuse reflectance spectra of GCN and Cu doped GCN (Reproduced with permission from ref. [83]. Copyright 2015, Elsevier) **d** UV- vis DRS of GCN samples prepared with different N ratios. Reproduced with permission from ref. [84]. Copyright 2016, Elsevier Science Ltd.

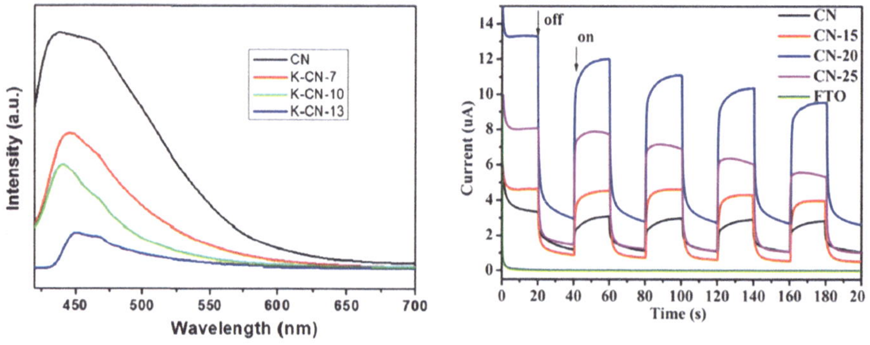

Fig. 4.5 **a** Photoluminescence spectra of k doped GCN (Reproduced with permission from ref. [85]. Copyright 2018, Elsevier) and **b** Transient photocurrent plots of **a** N doped GCN. Reproduced with permission from ref. [84]. Copyright 2016, Elsevier Science Ltd.

to improve the life span of charge carriers. Heterojunction is basically the contact of interfaces between two semiconductors with different band gaps. Formation of GCN heterostructures can significantly be supressed the recombination rate by effective spatial separation of charge carriers across the heterojunction interface. Heterojunction not only enhances the lifetime of charge carriers, but also the light absorption range, band positions and improved surface reactions. These augmented properties finally lead to better photocatalytic performance of pristine GCN. Construction of heterojunctions is considered as most feasible and effective strategy to increase the photocatalytic activity to many folds [89–92].

However, not every material can form a heterojunction with the pristine GCN. The bandgap and band position of both the semiconductor plays a vital role in the formation of coupling hybridization across the interface [93]. The difference in chemical potential between both the semiconductors created band bending in the interface. This bending leads to the movement of charge carriers in opposite directions creating a built in electric field. This effect helps in limiting the recombination by enhancing the spatial separation of electrons and holes [94, 95]. The difference between the lattice spacing of two semiconductor creates a lattice mismatch which can further leads to formation of defects in semiconductor. These defect help in trapping the electrons and thus reduces the diffusion of photoproduced charges. Thus formation of heterostructure was anticipated to be an effective way to upscale the photocatalytic activity of pristine GCN. Various attempts was made to couple pristine GCN with different materials such as CdS, ZnO, TiO_2, BiOI etc. [96–100]. Some of the novel works on GCN based heterostructures are discussed below.

Li et al. fabricated GCN coupled TiO_2 composites via a simple sol–gel process for augmented photodegradation of methyl orange (MO) under UV and visible light irradiation. The XRD patterns for the composites with different weight percentage of GCN are shown in Fig. 4.6a. The peaks of GCN, TiO_2 and the composites coincided well with each other for every prepared composite sample. This indicated the formation of heterojunction between both the nanomaterials. The prepared composite was tasted to perform better photocatalytic activity than the pristine counter parts. TiO_2, owing to its higher bandgap is a UV light active material, whereas the composite was able to generate photoproduced electrons and holes under visible light irradiation. This suggested that the red shifting in the absorption range of TiO_2 is due to the formation of heterojunction with GCN. Subsequently, TiO_2 was unable to degrade MO under visible light whereas the composite showed maximum of ~80% MO degradation owing to better light absorption and spatial separation of photoproduced electron and hole pairs. [101]. A 3D GCN /TiO_2 photocatalyst composite was prepared by Sheng et al. from TiO_2, melamine and ciyanuric acid using a one-step calcination method. From the HRTEM images shown in Fig. 4.6b, a good contact between GCN and TiO_2 can be observed. The thin GCN nanosheets were coated well over the surface of TiO_2. The TiO_2 nanoparticle showed a clear and regular lattice fringe of width 0.35 nm. The photocatalytic activity of the pristine samples and composite were compared by degrading methylene blue (MB) and phenol. The composite exhibited better charge transfer and spatial separation. This leads to the

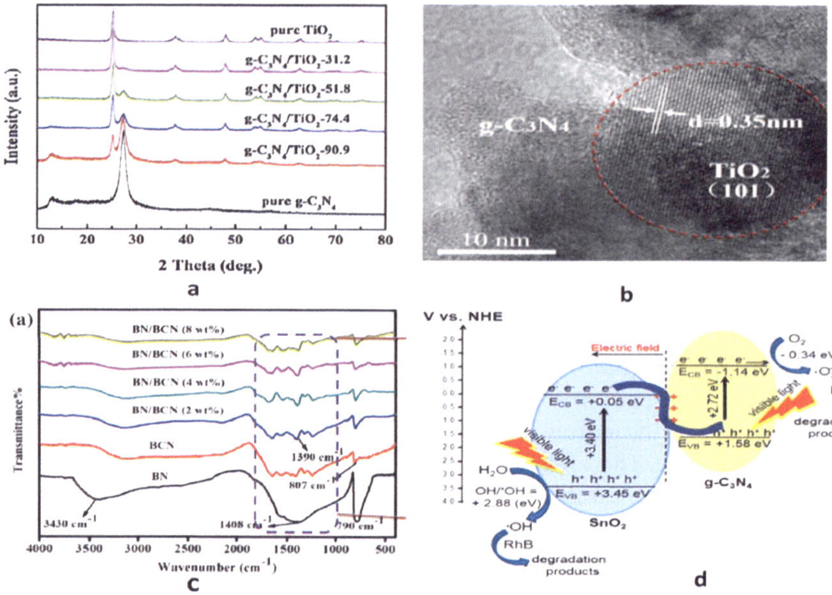

Fig. 4.6 **a** XRD plots for GCN, TiO_2 and GCN/TiO_2 composites (Reproduced with permission from ref. [101]. Copyright 2016, Elsevier), **b** HRTEM images 3D GCN/TiO_2 composite (Reproduced with permission from ref. [102]. Copyright 2019, Elsevier), **c** FTIR spectra of BN, BCN and BN/BCN composites (Reproduced with permission from ref. [86]. Copyright 2020, Elsevier), **d** Schematic illustration of S-scheme charge transfer path between SnO_2 and GCN (Reproduced with permission from ref. [106]. Copyright 2022, Elsevier)

generation of more amount of hydroxyl (·OH) radicals and thus shows better photocatalytic activity. The composite shows 4 and 4.5 times enhanced degradation of MB and phenol degradation respectively [102].

Construction of p–n heterojunction is also another remarkable method to enhance the photogenerated charge separation of GCN. Tian et al. fabricated a Cu_2O/GCN p-n heterojunction by depositing Cu_2O over GCN surface. Owing to the narrow band gap energy of Cu_2O, the optical absorption of GCN was enhanced in the visible region of solar spectrum. Thus, an augmented photocatalytic performance was observed in case of Cu_2O/GCN p-n heterojunction [103]. In another work. Acharya et al. synthesized a boron nitride/B-doped-g-C_3N_4 (BN/BCN) hetertojunction for the degradation of tetracyclin hydrochloride (TCH) in aqueous environment. They prepared the composites with different weight ratios of BN via an in-situ growth process. From the FTIR peaks presented in Fig. 4.6c, the strong bonding within BN and BCN can be easily predicted. The in-plane B-N stretching band at 1408 cm^{-1} got shifted to a higher frequency of 1390 cm^{-1}. With the increase in amount of BN, the intensity of peak formed at 1390 cm^{-1} for the composite increased. These observations from the FTIR spectrum confirmed the formation of heterojunction between BN and BCN. The photocatalytic TCH degradation ability of the as prepared composite was higher than the pristine counterparts. They calculated the rate constant (k) value for the degradation of TCH via different samples. The sample with 4wt% of BN shows maximum activity under sunlight. The k value for BCN sample was 9×10^{-3} min^{-1} whereas for BN/BCN (4 wt%), the value was increased to 34×10^{-1}. These values interpreted the enhancement in photocatalytic activity of GCN through construction of nano-heterojunction with Z-scheme charge transfer mechanism [86].

Fu et al. reported a 2D-2D WO_3/GCN step scheme (S- scheme) hetrostructured composite fabricated through electrostatic self-assembly method. The band positions and composition of the prepared composite was measured by density functional theory (DFT). From the computational calculations, they discovered that the charge transfer across the heterojunction is affected by the difference in work function and fermi levels of the participating semiconductors. The calculated work function of WO_3 was 6.23 eV whereas that for GCN was 4.18 eV. As WO_3 possess a higher work function value, the electrons starts flowing from the surface of GCN to WO_3 until a fermi level equilibrium was established. Due to this charge transfer process, a built in electric field was produced at the heterojunction interface between GCN and WO_3. This helps in enhancing the transfer and separation of charge carrier across the heterojunction interface eventually leading to augmented photocatalytic activity. In this case, the composite performed 1.7 times better photocatalytic H_2 production than that of pristine GCN [104].

Another S scheme based GCN /Bi_2MoO_6 heterostructure was constructed by Zhen et al. The water dispersed Bi_2MoO_6 mixed with GCN was treated under different calcination temperature to form the composite. The composite was proved to be more effective than pristine GCN in photodegradation of phenol under visible light irradiation [105]. Recently, Van et al. reported a SnO_2/GCN system from annealing mixture of SnO_2 and GCN for the photodegradation of Rhodamine B (RhB). The suggested S-scheme charge transfer path between both the nanomaterials is depicted in Fig. 4.6d. From the figure, it can be seen that the electrons and holes produced on the CB of SnO_2 and VB of GCN respectively get recombined due to equivalent band positions and fermi level alignments. The unwanted charge carriers got recombined to give a better separation of useful charge carriers. This mechanism escalates the photodegradation of RhB under visible light irradiation [106].

4.4 Environmental Remediation Through Photocatalytic Technology

4.4.1 Photocatalytic Degradation of Organic Compounds, Dyes and Antibiotics

The rising concentration of organic compounds, dyes and antibiotics in aqueous environment has received great attention in recent years [107]. The photocatalytic degradation of these pollutants using GCN based materials has been found advantageous over the conventional treatment methods. It was observed that modified GCN photocatalysts exhibited enhanced phocatalytic performance for pollutant degradation in comparison to pristine one. Pattnaik et al. reported that about 95% of Congo red (CR) was degraded by exfoliated GCN whereas the bulk GCN degraded only 81%. The enhanced photoactivity of exfoliated GCN might be due to its porous nature. The suggested mechanism of degradation was shown in Fig. 4.7. According to it, $^{\cdot}O_2{}^{-}$ and $^{\cdot}OH$ are the prime reactive species which were confirmed from the scavenging tests for the enhanced degradation of CR. These are produced as per the following equations [64].

$$e - \text{GCN} \underset{\rightarrow}{hv} e^{-}_{\text{CB}} + h^{+}_{\text{VB}} \tag{4.13}$$

$$e^{-}_{\text{CB}} + O_2 \rightarrow {}^{\cdot}O_2^{-} \tag{4.14}$$

$$e^{-}_{\text{CB}} + {}^{\cdot}O_2^{-} + 2h^{+}_{\text{VB}} \rightarrow H_2O_2 \tag{4.15}$$

$$e^{-}_{\text{CB}} + H_2O_2 \rightarrow {}^{\cdot}OH + OH^{-} \tag{4.16}$$

$$\text{CR} + {}^{\cdot}OH \rightarrow H_2O + CO_2 \tag{4.17}$$

$$\text{CR} + {}^{\cdot}OH \rightarrow H_2O + CO_2 \tag{4.18}$$

Or.

The overall reaction may be presented as

$$\text{CR} + {}^{\cdot}OH + {}^{\cdot}O_2^{-} \rightarrow H_2O + CO_2 \tag{4.19}$$

In another study, Pattnaik et al. observed an enhanced photodegradation of ciprofloxacin (CPN) over exfoliated GCN under solar light irradiation. They observed that 78% of 20 ppm CPN was degraded in 1 h by $1 gL^{-1}$ of exfoliated g-C_3N_4. The augmented performance of exfoliated GCN might be attributed to its extremely large surface area and low recombination rate of photogenerated charge carriers [63].

Harraz and co-authors demonstrated the superior performance of Au doped GCN in the photodegradation of methylene blue (MB). The photodegradation rate of the 1%Au/g-C_3N_4 nanocomposite was found to be 2.69 times higher than that of bare GCN [108]. Yan et al. reported that B doped GCN (BCN) exhibited promoted RhB decomposition [109]. Recently, Acharya et al. demonstrated the enhanced performance of exfoliated BCN (eBCN) over BCN in photocatalytic removal of CPN. The degradation efficiency of eBCN was about 1.2 times more than that of bulk BCN. The increased degradation rate was ascribed to the excellent photoinduced charge carriers' separation which was confirmed from the high photocurrent density of 524 mA cm^{-2} displayed by e-BCN [66].

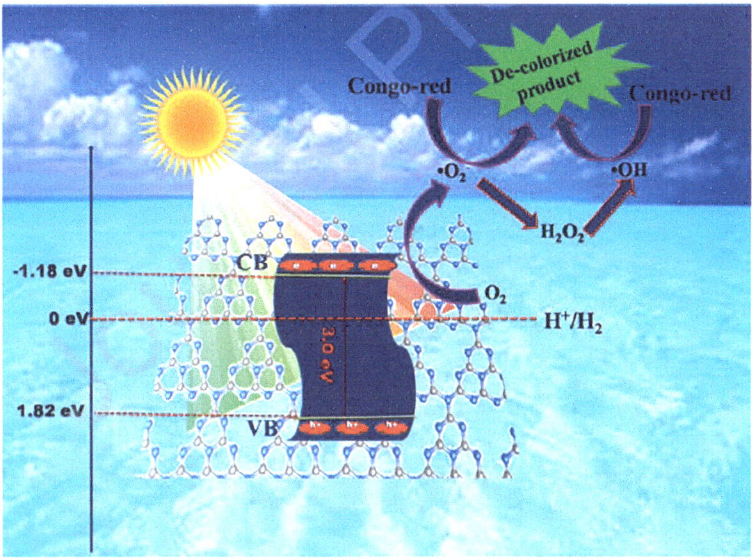

Fig. 4.7 Pictorial presentation of reaction mechanism for degradation of CR over e- GCN under sun light. Reproduced with permission from ref. [64]. Copyright 2019, Elsevier Science Ltd.

Wu group have developed 2D-2D TiO_2/GCN heterostructured nanocomposite for the removal of methyl orange (MO). The face-to-face heterojunction exhibited an improved flat potential, and decreased band gap energy with adequate change in band positions which are found advantageous for promoting charge mobility across the interface, lowering charge transport distance and time. These cumulatively enhanced the photodegradation rate (0.189 min^{-1}) of MO under UV–visible light [110]. Wang et al. constructed a robust GCN/Nb_2O_5 heterojunction through a direct electrospinning route followed by calcination. They observed that niobium pentoxide nanofibers (Nb_2O_5 NFs) were tightly bound onto GCN nanosheets which facilitated a promoted charge carriers' transfer. As a result, the heterostructured photocatalyst showed a remarkable performance in the degradation of phenol and RhB under visible light irradiation. Charge carriers trapping experiments revealed that holes (h^+s) and superoxide radical anions ($\cdot O_2^-$s) were the principal reactive species in the degradation process. The suggested mechanism for organic pollutants' photodegradation is depicted in Fig. 4.8 [111].

Acharya et al. have proposed a type-II p-n heterojunction with Z-scheme charge transfer mechanism for BN/BCN composite photocatalyst in which p-type BCN nanosheets were well covered with n-type born nitride (BN) particles. The heterostructured photocatalyst containing 4 wt. % of BN performed 88% tetracycline (TCH) degradation under solar light irradiation at time period of 60 min. superoxide ($^{\cdot}O_2^-$) and hydroxyl (OH^-) radicals were found as the major reactive species for TCH removal. The suggested mechanism of TCH degradation is illustrated in Fig. 4.9 [86].

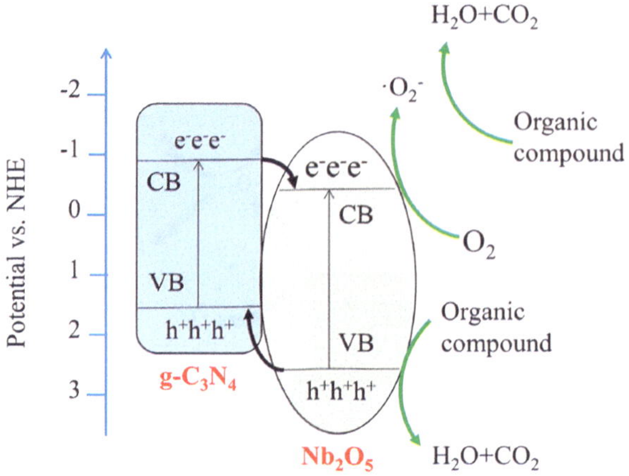

Fig. 4.8 Schematic diagram of photogenerated charges in GCN/Nb_2O_5 heterojunction (Reproduced with permission from ref. [111]. Copyright 2021, Nature)

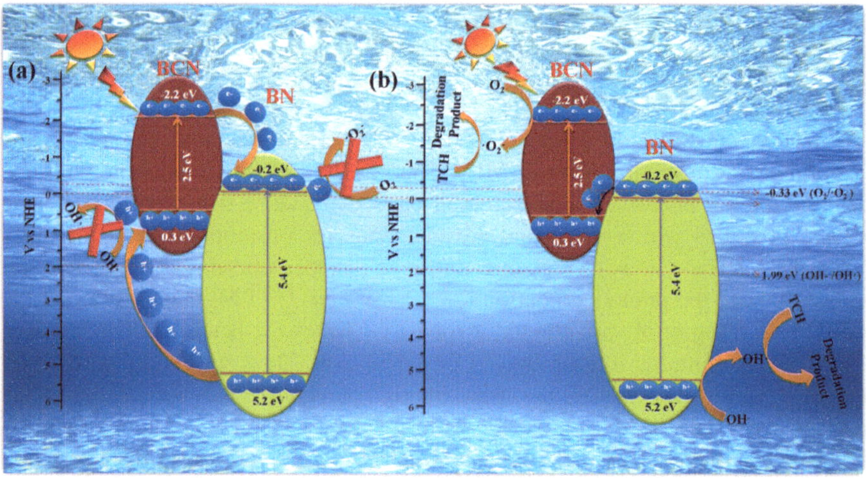

Fig. 4.9 Scheme showing charge transfer mechanism for TCH degradation over BN/BCN composite. Reproduced with permission from ref. [86]. Copyright 2020, Elsevier Science Ltd.

4.4.2 Photocatalytic CO_2 Conversion

The prime product of fossil fuels' combustion is CO_2, which causes greenhouse effect when released into the atmosphere. Conversion of it into selective hydrocarbon-based energy fuels and chemicals is considered as an effective method for CO_2 recycling which is identical to nature's carbon cycle [112–114]. Different types of gaseous/liquid products can be obtained from CO_2 through multiple electron transfer mechanisms as it is the highest oxidation state of carbon. Various reduction products of CO_2 and their standard reduction potential of formation was enumerated in Table 4.1 [42]. It was evident that these are located near the H_2O reduction potential. Therefore, the materials used for solar water splitting can also be used for CO_2 reduction [115]. The main demerits in CO_2 conversion are its high chemical stability and inertness owing to its linear structure with high C = O double bond energy (>750 kJ mol^{-1}) [116, 117]. However, photocatalytic reduction of CO_2 into green solar fuels like methanol (CH_3OH), methane (CH_4), formaldehyde (HCHO) and formic acid (HCOOH) has been recognized as a promising technology as CO_2 level can be imitated in the atmosphere with simultaneous production of useful chemicals and solar fuels [118–120]. The first CO_2 reduction by PEC into CH_3OH, HCOOH, and HCHO using p-GaP photocathode and carbon anode was carried out by Halmann [121]. Since then numerous studies have been reported on photocatalytic CO_2 conversion into value added products [122–124].

Photocatalytic CO_2 degradationtion process mostly relies on the ability of the catalyst to adsorb chemical rection intermediates over its surface. To do the same, the prepared photocatalyst had to have enhanced surface properties and adequate number of surface active sites than its pristine counter parts. Surface properties $SnFe_2O_4$ (SFO)/GCN Z-scheme heterojunction was compared with that of SFO and GCN. From the BET surface area analysis, the specific surface area of $SnFe_2O_4$, GCN and the composite was found to be 11.42, 15.11, and 44.41m^2/g respectively. The incorporation of GCN into the ferrite system enhances the specific surface area to a great extent. These analysis depicted that enhanced surface area and porous structure of the composite helps in better adsorption of CO_2 and other reaction intermediated over the surface active sites. The CO_2 adsorption curves for SFO, CN, and SFO-CN

Table 4.1 Various reduction products of CO_2 their standard reduction potential of formation with equation

Sl no	Name of the product formed	Standard reduction potential (E_0) versus NHE at pH 7	Chemical equation for formation
1	CH_4	−0.24	$CO_2 + 8\,h^+ + 8e^- \rightarrow CH_4 + 2H_2O$
2	CH_3OH	−0.38	$CO_2 + 6\,h^+ + 6e^- \rightarrow CH_3OH + H_2O$
3	HCHO	−0.48	$CO_2 + 4\,h^+ + 4e^- \rightarrow HCHO + H_2O$
4	CO	−0.52	$CO_2 + 2\,h^+ + 2e^- \rightarrow CO + H_2O$
5	HCOOH	−0.61	$CO_2 + 2\,h^+ + 2e^- \rightarrow HCOOH + H_2O$
6	C_2H_4	0. 06	$2CO_2 + 12\,h^+ + 12e^- \rightarrow C_2H_4 + 4H_2O$
7	C_2H_5OH	0.08	$2CO_2 + 12\,h^+ + 12e^- \rightarrow C_2H_5OH + 3H_2O$

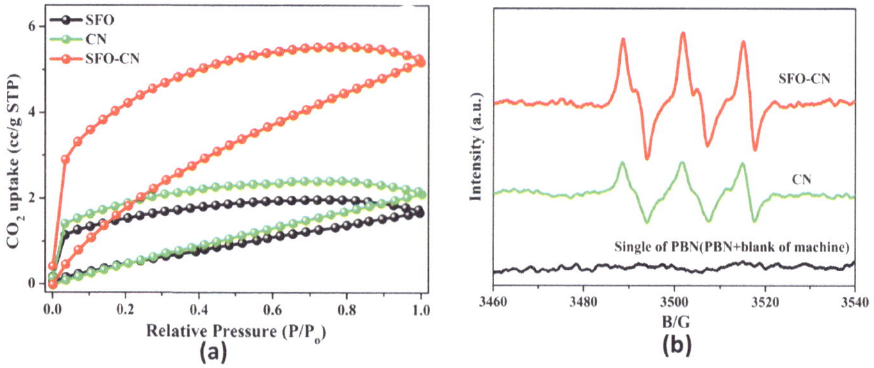

Fig. 4.10 **a** CO_2 adsorption curves **b** the profiles of ESR signal of the radicals in the reaction. Reproduced with permission from ref. [125]. Copyright 2020, Elsevier Science Ltd.

were displayed in Fig. 4.10a. At P/P0 = 1.0, the CO_2 adsorption ability of SFO, CN, and SFO-CN is 1.67, 2.11, and 5.19ccg^{-1} respectively. Figure 4.11a indicated that the SFO-CN with larger pore volume and surface area possessed greater CO_2 adsorption ability [125].

Generally, CO_2 absorbed in the surface active sites is reacted with hydrogen ions and electrons to produce $^{\cdot}CO^{2-}$, which is considered as a key intermediate for the production of CO [126, 127]. The strong signal intensity of the $^{\cdot}CO^{2-}$ radicals in the Electron spin resonance (ESR) spectra of SFO-CN observed in Fig. 4.10b. For SFO-CN demonstrated that release of more number of $^{\cdot}CO^{2-}$ species for efficient reduction CO_2 to CO. In fact, the $SnFe_2O_4$/GCN composite showed 2.2 times higher CO evolution than the pristine CN. The CO evolution value for the CO_2 photodegradation for the composite and pristine GCN was 7.56 and 3.45 μmol/g/h respectively [125].

Li et al. investigated kinetically selective CO production over GCN/Bi_2WO_6. The composite photocatalyst with Z-Scheme heterostructure exhibited an enhanced CO production rate of 5.19 μmol g^{-1} h^{-1} under visible light illumination [128]. According to Sonowal et al., the coupling of GCN quantum dots (GCNQDs) with Zr (IV) MOF resulted a binary composite which yielded methanol of 386 μmol.h^{-1}.g^{-1} through photocatalytic reduction of CO_2. On the other hand the pristine MOF produced only 66 μmol.h^{-1}.g^{-1}. The superior performance of the composite is due to the synergistic effect of GCNQDs, which makes photoinduced electrons available abundantly on the surface of the composite photocatalyst by prominent separation of charge carriers. These electrons selectively reduce CO_2 to methanol [129].

4.4.3 Photocatalytic Removal of Heavy Metal Ions

Heavy metal ions such as Cr (VI), U (VI), As (III), Hg (II) etc. are acutely toxic and are major pollutants in industrial effluents. These cannot be easily degraded in to harmless products unlike organic pollutants and get accumulated in living organisms to cause serious health hazards. Therefore, it has recently been a major challenge to decontaminate these noxious metal ions in order to protect the human health and the environment in particular [130]. Several technologies including solvent extraction, adsorption, chemical precipitation, electrocoagulation, photocatalysis, oxidation/reduction, ultra-filtration, reverse osmosis etc. have been developed for the elimination of these toxic metal ions from aqueous environment [131]. Among these, photocatalysis is considered as a well acclaimed process owing to a number of advantages like conversion of metal ions into nontoxic forms, use of abundantly available solar energy, reusability and low cost [132, 133]. Photocatalytic detoxification of these metal ions was depicted in the following sections.

4.4.3.1 Photocatalytic Reduction of Cr (VI)

As a corrosion inhibitor chromium is widely applied in various industries like dye synthesis, leather tanning, refractories, electroplating, alloy and steel manufacturing. The effluents of these industries containing Cr (VI) usually range up to 500 ppm. The improper disposal and accidental leakage at industry sites also contributes to the Cr (VI) contamination. Besides this, natural oxidation of the Cr (III) of ophiolitic rocks and ultramafic derived soils produce some amount of Cr (VI). It is acutely toxic, carcinogenic, teratogenic and mutagenic in biological systems. The nonbiodegradable and bioaccumulative properties of Cr (VI) allow it to pass through the cell membrane. Then, these are reduced to Cr (III) with the help of reactive oxygen species (ROS) causing intracellular damage. The so formed Cr (III) ions form stable coordination complexes with proteins and nucleic acids. This results in genotoxic damage and other forms of toxicity [9].

Cr (VI) exists as different oxy anionic species in aqueous medium depending on the concentration and pH of the solution. For instance, the dominating species is H_2CrO_4 at pH below 1.0 while $HCrO^{4-}$ and $Cr_2O_7{}^{2-}$ are the predominating species between the pH 1.0 and 6.0. $CrO_4{}^{2-}$ is the stable species above pH 6.0. These oxyanionic species experience repulsion from soil particles carrying negatively charges. As a result, these are passed easily into the aquatic environment and pose threat to aquatic bodies as well as downstream users. Owing to the hazardous impact, the maximum permitted Cr (VI) limit in surface and drinking water was mandated as 0.1 and $0.05 mgL^{-1}$ respectively by the USEPA [134]. In order to minimize the detrimental impact of Cr (VI) on downstream users and aquatic lives, it is essentially required to reduce the concentration below the permissible level before its discharge into the aquatic environment. Conversion of Cr (VI) to Cr (III) is a suitable approach for minimizing Cr toxicity as the latter is about 1000 times less toxic and is an essential nutrient for living organisms. Moreover, it is highly stable, less mobile in aqueous environment and can easily be precipitated as $Cr(OH)_3$ making the separation process easier [4]. Although, chemical precipitation, electrochemical reduction, sulphide precipitation have been adopted for detoxification of Cr (VI), large number of bottlenecks in these processes restrict their applications. On the other hand, photocatalytic reduction Cr (VI) into Cr (III) is considered as an effective, inexpensive and environmentally benevolent technique [33]. Therefore, several researches have been highlighted on these aspects.

Wang et al. prepared Br-doped GCN in which N atoms are substituted by Br atoms without destroying the original GCN framework. The incorporation of Br widened the light response to 800 nm along with the enlargement in surface area and promotion in charge carriers' separation efficiency. As a result Br-GCN experienced an enhanced Cr (VI) reduction rate which was found two times more than that obtained by bare GCN [135]. A CoS_2/g-C_3N_4-rGO hybrid nanocomposite was synthesized by a simple one-pot solvothermal technique. The heterostructured composite reduced 99.8% of Cr (VI) within 120 min at pH = 2 under visible light irradiation. Further, the photoreduction efficiency for the composite was more than 98% after 5 consecutive cycles. The enhanced performance might be attributed to increase in visible light utilization, high surface area and prominent electron–hole pairs separation which are due to formation a heterostructure [136].

4.4.3.2 Photocatalytic U (VI) Reduction

Uranium appears as U (0), U(III), U(IV) and U(VI) states among which U(VI) soluble and U(IV) is sparingly soluble [137]. It is extensively used in nuclear industries. The insufficient processing of spent fuel releases large amount of uranium into the environment. It has been reported that uranium emissions in excess leads to serious environmental pollution. Various methods like adsorption, ion exchange, evaporation and ultrafiltration are used for the treatment of U(VI) [138]. These processes have certain shortcomings due to which a green, efficient, and highly selective technology should be developed for remediation of U (VI). Photocatalytic reduction of soluble U (VI) to insoluble U (IV) can be regarded as advantageous technique. It is predicted that photoreduction of U (VI) takes place in two different ways. First is reduction of U (VI) through single-electron reduction process as shown in Eq. 4.20, U (VI) so formed in the above step is further reduced to U(IV) according to Eq. 4.21 [139].

$$\mathrm{U(VI)} + e^{-}_{\mathrm{CB}} \rightarrow \mathrm{U(V)} \tag{4.20}$$

$$\mathrm{U(V)} + e^{-}_{\mathrm{CB}} \rightarrow \mathrm{U(IV)} \tag{4.21}$$

In the second case, U(VI) is reduced directly to U (IV) by two electrons in acidic medium as presented in Eq. 4.22 [140].

$$\mathrm{UO_2^{2+}} + 4\mathrm{H^+} + e^{-}_{\mathrm{CB}} \rightarrow \mathrm{U^{4+}} + 2\mathrm{H_2O} \tag{4.22}$$

It was reported that the standard reduction potentials of $UO_2{}^{2+}/U^{4+}$, $UO_2{}^{2+}/UO_2$ and U_4O_9/UO_2 redox couple are 0.267, 0.411 and 0.456 V, respectively [141]. This

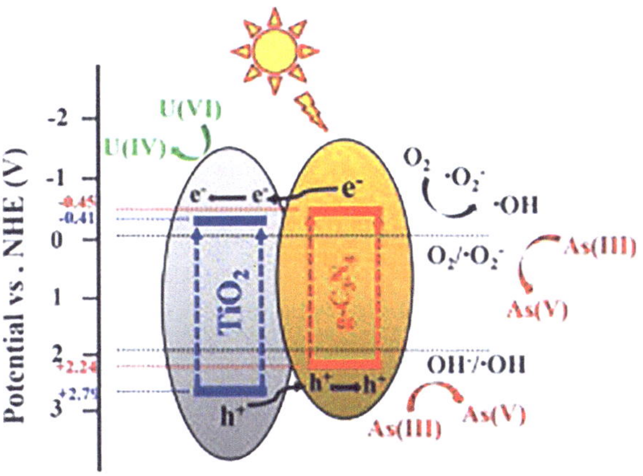

Fig. 4.11 Mechanism of simultaneous U (VI) photoreduction and As (III) photooxidation over GCN/TiO_2 heterostructured photocatalysts. Reproduced with permission from ref. [144]. Copyright 2018, Elsevier Science Ltd.

suggested that GCN based photocatalysts with more negative CB potentials can easily reduce U(VI). In 2016, Lu et al. synthesized boron doped GCN (BCN) for photocatalytic reduction of U (VI) under visible light irradiation in presence of methanol as a hole scavenging agent. The photogenrated electrons (e^-_{CB}s) are used to reduce U (VI) to insoluble U (IV) whereas the photogenerated holes (h^+_{VB} s) oxidize methanol to CO_2 and H_2O [142]. Chen et al. reported that g-C_3N_4/graphene oxide nanosheets (GCN/GO) exhibited 100% U (VI) removal efficiency after five cycles [143]. Jiang et al. investigated the simultaneous reduction of U (VI) and oxidation of As (III) over GCN/TiO_2 heterostructured photocatalysts under simulated solar light illumination. They observed that maximum of 83% UO_2^{2+} is reduced in 240 min. As per the proposed mechanism shown in Fig. 4.11, O2˙⁻, ·OH and h_{VB}^+ were the main active species for As (III) oxidation and U (VI) was reduced with the help of e_{CB}^- [144].

4.4.3.3 Photocatalytic Oxidation of as (III)

The two most stable state of Arsenic is As (V) and As (III). As (V) exist as $H_2AsO_4^-$ and $HAsO_4^{2-}$ species while H_3AsO_3 is the common species of As (III) at neutral pH [145, 146]. It has been reported that toxic effect of As (III) is about 26–60 times more than that of As (V) [147]. Therefore, the World Health Organization (WHO) has mandated 10 mg L^{-1} as the maximum permissible limit of arsenic in drinking water [148]. Water pollution caused by As (III) has been a challenge due to its non-biodegradability as well as accumulative property in food chain [149]. Although several methods including ion exchange, adsorption, chemical oxidation and precipitation are developed, chemical oxidation technique is considered as an effective approach. Oxidizing agents like O_3, MnO_2, Cl_2, $NaClO_7$ and MnOOH are efficiently used to remove As (III). However, these consume more energy and cause secondary pollution. On the other hand, photocatalysis is relatively cheaper and environmentally friendly for the detoxification of As (III) [150, 151]. Ouyang et al. synthesized boron-doped black TiO_2/GCN nanocomposite by sol−gel process followed by in situ decomposition-thermal polymerization technique. The obtained composite exhibited a promoted As (III) photo-oxidation to As (V). Electron spin resonance (ESR) and the radical scavenging experiments evidenced that hydroxyl radicals, superoxide anions and holes synergistically participated to oxidize As (III) [152]. A Mineral-based bentonite/GCN 2D-2D composite was prepared by a self-assembly process. The incorporation of bentonite resulted in enhanced surface area and pronounced excitons' separation. As a result, the prepared photocatalyst showed three times higher photooxidation of As (III) than that exhibited bare GCN. Additionally, high efficiency in As (III) photooxidation was also displayed in a wide pH range from 3 to 8.5. It was observed from ESR and radical trapping tests that $^{\cdot-}O_2$ species is the prime reactive species for the oxidation of As (III) [153].

4.4.3.4 Photocatalytic Reduction of Hg (II)

Various industries like rubber processing, chloralkali, oil refining, paint, fertilizer manufacturing etc. releases Hg to the aquatic environment in form of effluents. It is converted to methyl mercury by bacteria. This organic mercury being non-biodegradable and bio- accumulative in nature, easily enters into the food chain [154]. As it is acutely toxic, introduction of this deadly pollutant in form Hg (II) causes serious detrimental impact on important organs like brain, liver, reproductive system etc. [155]. Therefore, Hg (II) is being received great attention as an acute pollutant for the last several decades and US-EPA has fixed ~2 μg/L as its maximum permissible limit in drinking water [156]. Therefore, adsorption, electrodeposition, co-precipitation and heterogeneous photocatalytic reduction are some of approaches that are practiced for removal of Hg (II) ions [157]. However, photocatalytic reduction of Hg (II) is recently gaining importance owing to its several advantageous features like use of abundantly available water resources and solar energy [158]. Alshaikh et al. reported that $CuAl_2O_4$/g-C_3N_4 nanostructured photocatalysts prepared by MCM-41 and F-127 assisted solution process, exhibited enhanced photoactivity towards reduction of Hg (II). The highest Hg (II) photoreduction rate of 189.4 μmol min^{-1} was observed for 3.0 wt% $CuAl_2O_4$-decorated GCN whereas pristine GCN and $CuAl_2O_4$ displayed a poor performance of 21.95 and 12.95 μmol min^{-1} respectively. The enhanced rate of Hg (II) photoreduction was attributed to formation of a robust heterojunction which minimized the charge carriers' recombination rate and reduction in band gap energy [159]. Mohamed and Ismail constructed a p-n heterojunction between mesoporous $BiVO_4$ NPs and GCNnanosheets through a simple soft and hard template-assisted approach. The synthesized nanocomposites possessed enlarged surgace area with improved crystanility. They proposed Z-Scheme charge transfer route for the photocatalytic reduction of Hg (II) as presented in Fig. 4.12 Maximum of 100% Hg (II) photodegradation was observed by 1.2% $BiVO_4$/GCN nanocomposite in 60 min. The observed reduction efficiency was found to be 3.9 time more than $BiVO_4$ NPs and 4.5 times larger than that of pristine GCN [160].

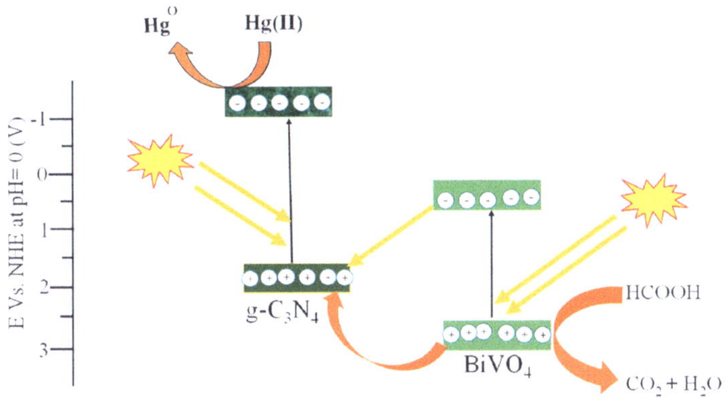

Fig. 4.12 Photocatalytic reduction of Hg (II) over mesoporous $BiVO_4$/GCN nanocomposites through Z-scheme charge transfer process. Reproduced with permission from ref. [160]. Copyright 2021, Elsevier Science Ltd.

4.5 Conclusion and Outlook

In summary, the present chapter highlights the basic mechanism for photocatalytic degradation of pollutants and modification strategies of GCN for enhanced photocatalytic pollutant degradation. Obviously, GCN has proven to be a promising candidate for modification and the modified forms have shown excellent performance for environmental remediation. Therefore, the growth of modified GCN based photocatalysts will be accelerated in near future to eradicate pollutants from our environment. Although, considerable progress has been achieved in the developments of GCN based photocatalysts, there are still many challenges to rationally design such a potential photocatalytic system towards better environmental pollution abatement. In order to address this challenge, further researches may be carried out in following directions.

Textural properties of a photoctalyst are crucial for an optimized optical and electronic structure and hence improved photocatalytic performance. Modification in the textural properties of GCN is essentially required to obtain an augmented photoactivity. For instant, ultrathin GCN nanosheets possess high specific surface area which facilitate to exhibit high photoactivity. Therefore, ultrathin GCN nanosheets have a great potential for the construction of advanced photocatalytic materials. However, GCN usually exhibit the poor specific surface area far below the theoretical value (2500 $m^2\ g^{-1}$). The band gap energy of GCN based materials needs further narrowing in order to harness energy from complete solar spectrum. Moreover, for efficient charge separation at the interface of the heterojunction, the close contact between GCN and the coupled semiconductor is required. This can be achieved through rational design in the preparation process. The surface property of GCN is required to be improved for better adsorption of reactants and construction of heterostructured photocatalysts. Fine tuning of the surface properties at the atomic level might be a successful approach in achieving these aspects.

Acknowledgements The authors express sincere thanks to the management of S'O'A (Deemed to be university), Bhubaneswar for encouraging to carry out the present work.

References

1. R. Acharya, K. Parida, A review on TiO_2/g–C_3N_4 visible-light-responsive photocatalysts for sustainable energy generation and environmental remediation. J. Environ. Chem. Eng. **8**, 103896 (2020)
2. C. Zhai, M. Zhu, Y. Lu, F. Ren, C. Wang, Y. Du, P. Yang, Reduced graphene oxide modified highly ordered TiO_2 nanotube arrays photoelectrode with enhanced photoelectrocatalytic performance under visible-light irradiation. Phys. Chem. Chem. Phys. **16**(28), 14800 (2014)
3. V.K. Gupta, R. Kumar, A. Nayak, T.A. Saleh, M.A. Barakat, Adsorptive removal of dyes from aqueous solution onto carbon nanotubes: a review. Adv. Colloid Interf. Sci. **193**, 24–34 (2013)
4. Z. Jiang, H. Sun, T. Wang, B. Wang, W. Wei, H. Li, S. Yuan, T. An, H. Zhao, J. Yu, P.K. Wong, Nature-based catalyst for visible-light-driven photocatalytic CO_2 reduction, energy environ. Sci. **11**, 2382–2389 (2018)
5. R. Acharya, A. Lenka, K. Parida, Magnetite modified amino group based polymer nanocomposites towards efficient adsorptive detoxification of aqueous Cr (VI): a review. J. Mol. Liq. **337**, 116487 (2021)
6. T.X.H. Le, R. Esmilaire, M. Drobek, M. Bechelany, C. Vallicari, D.L. Nguyen, A. Julbe, S. Tingry, M. Cretin, Design of a novel fuel cell-Fenton system: a smart approach to zero energy depollution. J. Mater. Chem. A. **4**, 17686–17693 (2016)
7. R. Acharya, K. Parida, A review on adsorptive remediation of Cr (VI) by magnetic iron oxides and their modified form, 2020. Biointerf. Res. Appl. Chem. **10**, 5266–5272 (2020)
8. R. Acharya, K. Parida, Valorization of agricultural wastes as low-cost adsorbents towards efficient removal of aqueous Cr(VI), in: Shahid-ul-Islam, Shalla, A.H., Khan, S.A. (Eds.), *Handbook of Biomass Valorization for Industrial Applications*, John Wiley & Sons, pp. 507–530 (2022)
9. R. Acharya, S. Martha, K.M. Parida, Remediation of Cr (VI) using clay minerals, biomasses and industrial wastes as adsorbents, in Shahid-ul-Islam (ed.) *Advanced Materials for Wastewater Treatment*, Scrivener Publishing LLC, pp. 129–170 (2017)
10. R. Acharya, B. Naik, K.M. Parida, Adsorption of Cr (VI) and textile dyes on to mesoporous silica, titanate nanotubes and layer double hydroxides, in: Shahid ul-Islam, B.S. Butola (eds.) *Nanomaterials in the Wet Processing of Textiles*, John Wiley Scrivener USA, pp. 219–260 (2018)
11. M.V. Sofianou, M. Tassi, V. Psycharis, N. Boukos, S. Thanos, T. Vaimakis, J. Yu, C. Trapalis, Solvothermal synthesis and photocatalytic performance of Mn4+-doped anatase nanoplates with exposed 001 facets. Appl. Catal. B Environ. **162**, 27 (2015)
12. T.X.H. Le, M. Bechelany, J. Champavert, M. Cretin, A highly active based graphene cathode for the electro-fenton reaction. RSC Adv. **5**(53), 42536 (2015)
13. D.B. Miklos, C. Remy, M. Jekel, K.G. Linden, J.E. Drewes, U. Hubner, Evaluation of advanced oxidation processes for water and wastewater treatment—a critical review. Water Res. **139**, 118–131 (2018)
14. X. Jiang, M. Manawan, T. Feng, R. Qian, T. Zhao, G. Zhou, F. Kong, Q. Wang, S. Dai, J.H. Pan, Anatase and rutile in evonik aeroxide P25: heterojunctioned or individual nanoparticles? Catal. Today **300**(2018), 12–17 (2018)

15. S. Mansingh, S. Sultana, R. Acharya, M.K. Ghosh, K. Parida, Efficient Photon conversion via double charge dynamics CeO_2-$BiFeO_3$ p-n heterojunction photocatalyst promising toward N_2 fixation and Phenol-Cr (VI) detoxification. Inorg. Chem. **59**, 3856–3873 (2020)
16. S. Mansingh, R. Acharya, S. Martha, K.M. Parida, Pyrochlore $Ce_2Zr_2O_7$ decorated over rGO: a photocatalyst that proves to be efficient towards the reduction of 4-nitrophenol and degradation of ciprofloxacin under visible light. Phys. Chem. Chem. Phys. **20**, 9872–9885 (2018)
17. S. Mishra, R. Acharya, Photocatalytic applications of graphene based semiconductor composites: a review. Mater. Today Proceed. **35**(2021), 164–169 (2021)
18. R. Acharya, B. Naik, K.M. Parida, Cr(VI) remediation from aqueous environment through modified-TiO_2-mediated photocatalytic reduction, Beilstein. J. Nanotechnol. **9**, 1448–1470 (2018)
19. A. Fujishima, K. Honda, Electrochemical photolysis of water at a semiconductor electrode. Nature **238**(5358), 37 (1972)
20. J.H. Carey, J. Lawrence, H.M. Tosine, Photodechlorination of PCB's in the presence of titanium dioxide in aqueous suspensions Bull. Environ. Contam. Toxicol. **16**, 697–701 (1976)
21. M. Nasr, C. Eid, R. Habchi, P. Miele, M. Bechelany, Recent progress on TiO_2 nanomaterials for photocatalytic applications. Chem Sus Chem **11**, 3023–3047 (2018)
22. W. Zhang, H. He, H. Li, L. Duan, L. Zu, Y. Zhai, W. Li, L. Wang, H. Fu, D. Zhao, Visible-light responsive TiO_2-based materials for efficient solar energy utilization. Adv. Energy Mater. **11**, 2003303 (2021)
23. X.C. Wang, K. Maeda, A. Thomas, K. Takanabe, G. Xin, J.M. Carlsson, K. Domen, M. Antonietti, A metal-free polymeric photocatalyst for hydrogen production from water under visible light. Nat. Mater. **8**, 76–80 (2009)
24. G. Liao, Y. Gong, L. Zhang, H. Gao, G.-J. Yang, B. Fang, Semiconductor polymeric graphitic carbon nitride photocatalysts: the '"holy grail"' for the photocatalytic hydrogen evolution reaction under visible light, Energy Environ. Sci. **12**, 2080–2147 (2019)
25. S. Pati, R. Acharya, An overview on g-C_3N_4 as a robust photocatalyst towards the sustainable generation of H_2 energy. Mater. Today Proceed. **35**, 175–178 (2021)
26. S. Zhang, P. Gu, R. Ma, C. Luo, T. Wen, G. Zhao, W. Cheng, X. Wang, Recent developments in fabrication and structure regulation of visible-light driven g-C_3N_4-based photocatalysts towards water purification: a critical review. Catal. Today **335**, 65–77 (2019)
27. J. Wen, J. Xie, X. Chen, X. Li, A review on g-C_3N_4-based photocatalysts. Appl. Surf. Sci. **391**, 72–123 (2017)
28. W.-J. Ong, L.-L. Tan, Y.H. Ng, S.T. Yong, S.-P. Chai, Graphitic carbon nitride (g-C_3N_4)-based photocatalysts for artificial photosynthesis and environmental remediation: are we a step closer to achieving sustainability? Chem. Rev. **116**, 7159–7329 (2016)
29. Q. Zhang, X. Liu, M. Chaker, D. Ma, Advancing graphitic carbon nitride-based photocatalysts toward broadband solar energy harvesting. ACS Materials Lett. **3**, 663–697 (2021)
30. S. Sahoo, R. Acharya, An overview on recent developments in synthesis and molecular level structure of visible-light responsive g-C_3N_4 photocatalyst towards environmental remediation. Mater. Today Proceed. **35**, 150–155 (2021)
31. L. Acharya, B.P. Mishra, S.P. Pattnaik, R. Acharya, K. Parida, Incorporating nitrogen vacancies in exfoliated B-doped g-C_3N_4 towards improved photocatalytic ciprofloxacin degradation and hydrogen evolution. New. J. Chem. **46**, 3493–3503 (2022)
32. T. Su, Q. Shao, Z. Qin, Z. Guo, B. Wu, Role of interfaces in two-dimensional photocatalyst for water splitting. ACS Catal. **8**, 2253–2276 (2018)
33. R. Acharya, S. Pati, K. Parida, A review on visible light driven spinel ferrite-g-C_3N_4 photocatalytic systems with enhanced solar light utilization. J. Mol. Liq. **357**, 119105 (2022)
34. S. Yin, J. Han, T. Zhou, R. Xu, Recent progress in g-C_3N_4 based low cost photocatalytic system: activity enhancement and emerging applications. Catal. Sci. Technol. **5**, 5048–5061 (2015)
35. J. Fu, J. Yu, C. Jiang, B. Cheng, g-C_3N_4-based heterostructured photocatalysts. Adv. Energy Mater. **8**, 1701503 (2018)

36. J. Yang, H. Wang, L. Jiang, H. Yu, Y. Zhao, H. Chen, X. Yuan, L. Liang, H. Li, Z. Wu, Defective polymeric carbon nitride: Fabrications, photocatalytic applications and perspectives. Chem. Eng. J. **427**, 130991 (2022)
37. A. Naseri, M. Samadi, A. Pourjavadi, A.Z. Moshfegh, S. Ramakrishna, Graphitic carbon nitride (g-C_3N_4)-based photocatalysts for solar hydrogen generation: recent advances and future development directions. J. Mater. Chem. A. **5**, 23406 (2017)
38. X. Li, J. Yu, M. Jaroniec, Hierarchical photocatalysts. Chem. Soc. Rev. **45**, 2603–2636 (2016)
39. X. Li, J. Yu, J. Low, Y. Fang, J. Xiao, X. Chen, Engineering heterogeneous semiconductors for solar water splitting. J. Mater. Chem. A. **3**, 2485–2534 (2015)
40. R. Acharya, L. Acharya, K. Parida, $BiFeO_3$-based materials for augmented photoactivity, Khursheed Ahmad and Waseem Raza (eds.) Perovskite Materials for Energy and Environmental Applications, John Wiley & Sons, Inc. p167–217 (2022)
41. S. Borthakur, L. Saikia, $ZnFe_2O_4$@g-C_3N_4 nanocomposites: an efficient catalyst for Fenton-like photodegradation of environmentally pollutant Rhodamine B. J. Environ. Chem. Eng. **7**, 103035 (2019)
42. S. Kim, K.H. Kim, C. Oh, K. Zhang, J.H. Park, Artificial photosynthesis for high-value-added chemicals: Old material, new opportunity. Carbon Energy **4**, 21–44 (2022)
43. W. Zhang, L. Li, Y. Gao, D. Zhang, Graphitic carbon nitride-based materials for photocatalytic reduction of U(VI), New. J. Chem. **44**, 19961–19976 (2020)
44. Y. Wang, Y. Zhang, T.C. Zhang, G. Xiang, C. Wang, S. Yuan, Removal of trace arsenite through simultaneous photocatalytic oxidation and adsorption by magnetic Fe_3O_4@PpPDA@TiO_2 core–shell nanoparticles. ACS Appl. Nano Mater. **3**, 8495–8504 (2020)
45. J. Feng, L. Peng, C. Wu, X. Sun, S. Hu, C. Lin, J. Dai, J. Yang, Y. Xie, Giant moisture responsiveness of VS_2 ultrathin nanosheets for novel touchless positioning interface. Adv. Mater. **24**, 1969–1974 (2012)
46. G. Eda, T. Fujita, H. Yamaguchi, D. Voiry, M.W. Chen, M. Chhowalla, Coherent atomic and electronic heterostructures of single-layer MoS_2. ACS Nano **22**, 7311 (2012)
47. Y. Omomo, T. Sasaki, L.Z. Wang, M. Watanabe, Redoxable nanosheet crystallites of MnO_2 derived via delamination of a layered manganese oxide. J. Am. Chem. Soc. **125**, 3568 (2003)
48. S. Ithurria, M.D. Tessier, B. Mahler, R.P.S.M. Lobo, B. Dubertret, A.L. Efros, Colloidal nanoplatelets with two-dimensional electronic structure. Nat. Mater. **10**, 936 (2011)
49. A.K. Geim, K.S. Novoselov, The rise of graphene. Nat. Mater. **6**, 183–191 (2007)
50. H. Zhang, Ultrathin two-dimensional nanomaterials. ACS Nano **9**, 9451 (2015)
51. P. Xia, B. Zhu, J. Yu, S. Cao, M. Jaronie, Ultra-thin nanosheet assemblies of graphitic carbon nitride for enhanced photocatalytic CO_2 reduction. J. Mater. Chem. A. **5**, 3230 (2017)
52. J. Yang, Z. Chen, N. Mao, D. An, B.D. Wang, Fahlman, Ultrathin g-C_3N_4 nanosheets with an extended visible-light-responsive range for significant enhancement of photocatalysis. RSC Adv. **7**, 2333 (2017)
53. P. Niu, L. Zhang, G. Liu, H.-M. Cheng, Graphene-like carbon nitride nanosheets for improved photocatalytic activities. Adv. Funct. Mater. **22**, 4763 (2012)
54. F. Dong, Y. Li, Z. Wang, W.-K. Ho, Enhanced visible light photocatalytic activity and oxidation ability of porous graphene-like g-C_3N_4 nanosheets via thermal exfoliation. Appl. Surf. Sci. **358**, 393–403 (2015)
55. K. Schwinghammer, M.B. Mesch, V. Duppel, C. Ziegler, J. Senker, B.V. Lotsch, Crystalline carbon nitride nanosheets for improved visible-light hydrogen evolution. J. Am. Chem. Soc. **136**, 1730 (2014)
56. V. Nicolosi, M. Chhowalla, M.G. Kanatzidis, M.S. Strano, J.N. Coleman, Liquid exfoliation of layered materials. Science **340**, 1226419 (2013)
57. J.N. Coleman, M. Lotya, A. O'Neill, S.D. Bergin, P.J. King, U. Khan, K. Young, A. Gaucher, S. De, R.J. Smith, I.V. Shvets, S.K. Arora, G. Stanton, H.Y. Kim, K. Lee, G.T. Kim, G.S. Duesberg, T. Hallam, J.J. Boland, J.J. Wang, J.F. Donegan, J.C. Grunlan, G. Moriarty, A. Shmeliov, R.J. Nicholls, J.M. Perkins, E.M. Grieveson, K. Theuwissen, D.W. McComb, P.D. Nellist, V. Nicolosi, Two-dimensional nanosheets produced by liquid exfoliation of layered materials science **331**, 568 (2011)

58. S. Yang, Y. Gong, J. Zhang, L. Zhan, L. Ma, Z. Fang, R. Vajtai, X. Wang, P.M. Ajayan, Exfoliated graphitic carbon nitride nanosheets as efficient catalysts for hydrogen evolution under visible light. Adv. Mater. **25**, 2452–2456 (2013)
59. D.R. Dreyer, S. Park, C.W. Bielawski, R.S. Ruoff, The chemistry of graphene oxide. Chem. Soc. Rev. **39**, 228 (2010)
60. J.H. Lee, M.J. Park, S.J. Yoo, J.H. Jang, H.J. Kim, S.W. Nam, C.W. Yoon, J.Y. Kim, A highly active and durable Co–N–C electrocatalyst synthesized using exfoliated graphitic carbon nitride nanosheet Nanoscale **7**, 10334 (2015)
61. J. Xu, L. Zhang, R. Shi, Y. Zhu, Chemical exfoliation of graphitic carbon nitride for efficient heterogeneous photocatalysis. J. Mater. Chem. A. **1**, 14766 (2013)
62. F. Cheng, H. Wang, X. Dong, The amphoteric properties of g-C_3N_4 nanosheets and fabrication of their relevant heterostructure photocatalysts by an electrostatic re-assembly route. Chem. Commun. **51**, 7176 (2015)
63. S.P. Pattnaik, A. Behera, S. Martha, R. Acharya, K. Parida, Facile synthesis of exfoliated graphitic carbon nitride for photocatalytic Degradation of ciprofloxacin under solar irradiation. J. Mater. Sci. **54**, 5726–5742 (2019)
64. S.P. Pattnaik, A. Behera, R. Acharya, K. Parida, Green exfoliation of graphitic carbon nitride towards decolourization of congo-red under solar irradiation. J. Environ. Chem. Eng. **7**, 103456 (2019)
65. Y. Fang, I.S. Merenkov, X. Li, J. Xu, S. Lin, M.L. Kosinova, X. Wang, Vertically aligned 2D carbon doped boron nitride nanofilms for photoelectrochemical water oxidation. J. Mat. Chem. **8**, 13059–13064 (2020)
66. L. Acharya, S.P. Pattnaik, A. Behera, R. Acharya, K. Parida, Exfoliated boron nitride (e-BN) tailored exfoliated graphitic carbon nitride (e-CN): an improved visible light mediated photocatalytic approach towards TCH degradation and H2 evolution. Inorg. Chem. **60**, 5021–5033 (2021)
67. B.P. Mishra, K. Parida, Orienting Z scheme charge transfer in graphitic carbon nitride-based systems for photocatalytic energy and environmental applications. J. Mat. Chem. A. **9**, 10039–10080 (2021)
68. S. Patnaik, D.P. Sahoo, K. Parida, Recent advances in anion doped g-C_3N_4 photocatalysts: a review. Carbon **172**, 682–711 (2021)
69. M.S. Khan, F. Zhang, M. Osada, S.S. Mao, S. Shen, Graphitic carbon nitride-based low-dimensional heterostructures for photocatalytic applications. Solar RRL **4**, 1900435 (2020)
70. X. Dong, F. Cheng, Recent development in exfoliated two-dimensional g-C_3N_4 nanosheets for photocatalytic applications. J. Mat. Chem. A. **3**, 23642–23652 (2015)
71. L. Ke, P. Li, X. Wu, S. Jiang, M. Luo, Y. Liu, Z. Le, C. Sun, S. Song, Graphene-like sulfur-doped g-C_3N_4 for photocatalytic reduction elimination of UO_2^{2+} under visible Light. Appl. Catal. B: Environmental. **205**, 319–326 (2017)
72. X. Liu, R. Ma, L. Zhuang, B. Hu, J. Chen, X. Liu, X. Wang, Recent developments of doped g-C_3N_4 photocatalysts for the degradation of organic pollutants. Crit. Rev. Environ. Sci. Technol. **51**, 751–790 (2021)
73. L. Jiang, X. Yuan, Y. Pan, J. Liang, G. Zeng, Z. Wu, H. Wang, Doping of graphitic carbon nitride for photocatalysis: a reveiw. Appl. Catal. B: Environmental **217**, 388–406 (2017)
74. Y. Zhou, W. Lv, B. Zhu, F. Tong, J. Pan, J. Bai, Q. Zhou, H. Qin, Template-free one-step synthesis of g-C_3N_4 nanosheets with simultaneous porous network and S-doping for remarkable visible-light-driven hydrogen evolution. ACS Sustain. Chem. Eng. **7**, 5801–5807 (2019)
75. L. Acharya, G. Swain, B.P. Mishra, R. Acharya, K. Parida, Development of $MgIn_2S_4$ microflower-embedded exfoliated B-Doped g-C_3N_4 nanosheets: p–n heterojunction photocatalysts toward photocatalytic water reduction and H_2O_2 production under visible-light irradiation. ACS Appl. Energy Mater. **5**, 2838–2852 (2022)
76. F. He, Z. Wang, Y. Li, S. Peng, B. Liu, The nonmetal modulation of composition and morphology of g-C_3N_4 -based photocatalysts. Appl. Catal. B: Environ. **269**, 118828 (2020)

77. S. Tonda, S. Kumar, S. Kandula, V. Shanker, Fe-doped and-mediated graphitic carbon nitride nanosheets for enhanced photocatalytic performance under natural sunlight. J. Mat. Chem. A. **2**, 6772–6780 (2014)
78. J. Gao, Y. Wang, S. Zhou, W. Lin, Y. Kong, A facile one-step synthesis of Fe-doped g-C_3N_4 nanosheets and their improved visible-light photocatalytic performance. ChemCatChem **9**, 1708–1715 (2017)
79. S. Hu, F. Li, Z. Fan, F. Wang, Y. Zhao, Z. Lv, Band gap-tunable potassium doped graphitic carbon nitride with enhanced mineralization ability. Dalton Trans. **44**, 1084–1092 (2015)
80. P. Chen, P. Xing, Z. Chen, H. Lin, Y. He, Rapid and energy-efficient preparation of boron doped g-C_3N_4 with excellent performance in photocatalytic H_2-evolution. Int. J. Hydrog. Energy. **43**, 19984–19989 (2018)
81. Y.R. Lin, G.V.C. Dizon, K. Yamada, C.Y. Liu, A. Venault, H.Y. Lin, M. Yoshida, C. Hu, Sulfur-doped g-C_3N_4 nanosheets for photocatalysis: Z-scheme water splitting and decreased biofouling. J. Colloid Interface Sci. **567**, 202–212 (2020)
82. Y. Deng, L. Tang, G. Zeng, Z. Zhu, M. Yan, Y. Zhou, J. Wang, Y. Liu, J. Wang, Insight into highly efficient simultaneous photocatalytic removal of Cr (VI) and 2, 4-diclorophenol under visible light irradiation by phosphorus doped porous ultrathin g-C_3N_4 nanosheets from aqueous media: performance and reaction mechanism. Appl. Catal. B: Environ. **203**, 343–354 (2017)
83. J. Gao, J. Wang, X. Qian, Y. Dong, H. Xu, R. Song, C. Yan, H. Zhu, Q. Zhong, G. Qian, J. Yao, One-pot synthesis of copper-doped graphitic carbon nitride nanosheet by heating Cu–melamine supramolecular network and its enhanced visible-light-driven photocatalysis. J. Solid State Chem. **228**, 60–64 (2015)
84. Y. Zhou, L. Zhang, W. Huang, Q. Kong, X. Fan, M. Wang, J. Shi, N-doped graphitic carbon-incorporated g-C_3N_4 for remarkably enhanced potocatalytic H_2 evolution under visible light. Carbon **99**, 111–117 (2016)
85. Y. Wang, S. Zhao, W. Zhang, J. Fang, Y. Zhou, S. Yuan, One-pot synthesis of K-doped g-C_3N_4 nanosheets with enhanced photocatalytic hydrogen production under visible-light irradiation. Appl. Surf. Sci. **440**, 258–265 (2018)
86. L. Acharya, S. Nayak, S.P. Pattnaik, R. Acharya, K. Parida, Resurrection of boron nitride in p-n type-II boron nitride/B-doped-g-C_3N_4 nanocomposite during solid-state Z-scheme charge transfer path for the degradation of tetracycline hydrochloride. J. Colloid Interface Sci. **566**, 211–223 (2020)
87. J. Zhang, S. Hu, Y. Wang, A convenient method to prepare a novel alkali metal sodium doped carbon nitride photocatalyst with a tunable band structure. RSC adv. **4**, 62912–62919 (2014)
88. G. Mamba, A.K. Mishra, Graphitic carbon nitride (g-C_3N_4) nanocomposites: a new and exciting generation of visible light driven photocatalysts for environmental pollution remediation. Appl. Catal. B. **198**, 347 (2016)
89. N. Tian, H.W. Huang, X. Du, F. Dong, Y.H. Zhang, , Rational nanostructure design of graphitic carbon nitride for photocatalytic applications. J. Mater. Chem. A **7**, 11584–11612 (2019)
90. Q.L. Xu, L.Y. Zhang, J.G. Yu, S. Wageh, A.A. Al-Ghamdi, M. Jaroniec, Direct Z-scheme photocatalysts: principles, synthesis, and applications. Mater. Today **21**, 1042–1063 (2018)
91. P. Zhou, J.G. Yu, M. Jaroniec, All-solid-state Z-scheme photocatalytic systems. Adv. Mater. **26**, 4920–4935 (2014)
92. H. Yang, A short review on heterojunction photocatalysts: Carrier transfer behavior and photocatalytic mechanisms. Mater. Res. Bull. **142**, 111406 (2021)
93. M. Shim, M. McDaniel, N. Oh, Prospects for strained type-II nanorod heterostructures. J. Phys. Chem. Lett. **2**, 2722–2727 (2011)
94. H. McDaniel, P.E. Heil, C.L. Tsai, K.K. Kim, M. Shim, Integration of type II nanorod heterostructures into photovoltaics. ACS Nano **5**, 7677–7683 (2011)
95. H. Wang, L. Zhang, Z. Chen, J. Hu, S. Li, Z. Wang, J. Liu, X. Wang, Semiconductor heterojunction photocatalysts: design, construction, and photocatalytic performances. Chem. Soc. Rev. **43**, 5234 (2014)

96. Z. Zhao, Y. Sun, F. Dong, Graphitic carbon nitride based nanocomposites: a review. Nanoscale **7**, 15–37 (2015)
97. X. Li, M. Edelmannová, P. Huo, K. Kočí, Fabrication of highly stable CdS/g-C3N4 composite for enhanced photocatalytic degradation of RhB and reduction of CO_2. J Mater Sci **55**, 3299–3313 (2020)
98. D.R. Paul, S. Gautam, P. Panchal, S.P. Nehra, P. Choudhary, A. Sharma, ZnO-Modified g-C_3N_4: a potential photocatalyst for environmental application. ACS Omega **5**, 3828–3838 (2020)
99. I. Ali, J.-O. Kim, Optimization of photocatalytic performance of a gC_3N_4–TiO2 nanocomposite for phenol degradation in visible light. Mat. Chem. Phys. **261**, 124246 (2021)
100. R. He, K. Cheng, Z. Wei, S. Zhang, D. Xu, Room-temperature in situ fabrication and enhanced photocatalytic activity of direct Z-scheme BiOI/g-C_3N_4 photocatalyst. Appl. Surf. Sci. **465**, 964–972 (2019)
101. J. Li, Y. Liu, C. Chen, Fabrication of gC_3N_4/TiO_2 composite photocatalyst with extended absorption wavelength range and enhanced photocatalytic performance. J. Photochem. Photobiol. A : Chem. **317**, 151–160 (2016)
102. T. Sheng, Z. Wei, H. Miao, W. Yao, H. Li, Y. Zhu, Enhanced organic pollutant photodegradation via adsorption/photocatalysis synergy using a 3D gC_3N_4/TiO_2 free separation photocatalyst. Chem. Eng. J. **370**, 287–294 (2019)
103. Y. Tian, B. Chang, J. Fu, B. Zhou, J. Liu, F. Xi, X. Dong, J., Graphitic carbon nitride/Cu_2O heterojunctions: Preparation, characterization, and enhanced photocatalytic activity under visible light. Solid State Chem. **212**, 1 (2014)
104. J. Fu, Q. Xu, J. Low, C. Jiang, J. Yu, Ultrathin 2D/2D WO_3/g-C_3N_4 step-scheme H_2-production photocatalyst. Appl. Catal. B: Environ. **243**, 556–565 (2019)
105. Y. Zhen, C. Yang, F. Fu, H. Shen, W. Xue, C. Gu, J. Feng, Y. Zhang, Y. Liang, Photocatalytic performance and mechanism insights of S-scheme g-C_3N_4/Bi_2MoO_6 heterostructure in phenol degradation and hydrogen evolution reaction under visible light. Phys. Chem. Chem. Phys. **22**, 26278–26288 (2020)
106. K.N. Van, H.T. Huu, V.N.N. Thi, T.L.L. Thi, D.H. Truong, Ts.T. Truong, N.N. Dao, V. Vo, D.L. Tran, Y. Vasseghian, Facile construction of S-scheme SnO_2/g-C_3N_4 photocatalyst for improved photoactivity. Chemosphere **289**, 133120 (2022)
107. S. Mishra, R. Acharya, K. Parida, Spinel-ferrite-decorated graphene-based nanocomposites for enhanced photocatalytic detoxification of organic dyes in aqueous medium: a review. Water **15**(1), 81 (2023)
108. M. Faisala, M. Jalalaha, F.A. Harraza, A.M. El-Tonic, A. Khan, M.S. Al-Assiri, Au nanoparticles-doped g-C3N4 nanocomposites for enhanced photocatalytic performance under visible light illumination. Ceram. Int. **46**, 22090–22101 (2020)
109. S.C. Yan, Z.S. Li, Z.G. Zou, Photodegradation of rhodamine B and methylorange over boron-doped g-C_3N_4 under visible light irradiation. Langmuir **26**, 3894–3901 (2010)
110. W. Gu, F. Lu, C. Wang, S. Kuga, L.-Z. Wu, Y. Huang, M. Wu, Face-to-Face interfacial assembly of ultrathin g-C_3N_4 and anatase TiO_2 nanosheets for enhanced solar photocatalytic activity, ACS Appl. Mater. Interfaces. **9**, 28674–28684 (2017)
111. L. Wang, Y. Li, P. Han, Electrospinning preparation of g-C_3N_4/Nb_2O_5 nanofibers heterojunction for enhanced photocatalytic degradation of organic pollutants in water. Sci. Rep. **11**, 22950 (2021)
112. V. Kumaravel, J. Bartlett, S.C. Pillai, Photoelectrochemical conversion of carbon dioxide (CO_2) into fuels and value-added products. ACS Energy Lett. **5**, 486–519 (2020)
113. I. Masood ul Hasan, L. Peng, J. Mao, R. He, Y. Wang, J. Fu, N. Xu, J. Qiao, Carbon-based metalfree catalysts for electrochemical CO_2 reduction: activity, selectivity, and stability. Carbon Energy **3**, 24–49
114. X. Chang, T. Wang, J. Gong, CO_2 photo-reduction: insights into CO_2 activation and reaction on surfaces of photocatalysts. Energy Environ. Sci. **9**, 2177–2196 (2016)
115. Y. Shioya, K. Ikeue, M. Ogawa, M. Anpo, synthesis of transparent Ti-containing mesoporous silica thin film materials and their unique photocatalytic activity for the reduction of CO_2 with H_2O. Appl Catal A-Gen. **254**, 251–259 (2003)

116. A. Hasani, M.A. Teklagne, H.H. Do, S.H. Hong, Q.V. Le, S.H. Ahn, S.Y. Kim, Graphene-based catalysts for electrochemical carbon dioxide reduction. Carbon Energy. **2**, 158–175 (2020)
117. D. Chen, Y. Wang, D. Liu, H. Liu, C. Qian, H. He, J. Yang, Surface composition dominates the electrocatalytic reduction of CO_2 on ultrafine CuPd nanoalloys. Carbon Energy. **2**, 443–451 (2020)
118. E.V. Kondratenko, G. Mul, J. Baltrusaitis, G.O. Larrazábal, J. Pérez-Ramírez, Status and perspectives of CO_2 conversion into fuels and chemicals bycatalytic, photocatalytic and electrocatalytic processes. Energy Environ. Sci. **6**, 3112–3135 (2013)
119. W. Dai, H. Xu, J. Yu, X. Hu, X. Luo, X. Tu, L. Yang, Photocatalytic reduction of CO_2 into methanol and ethanol over conducting polymers modified Bi_2WO_6 microspheres under visible light. Appl. Surf. Sci. **356**, 173–180 (2015)
120. S.W. Jo, B.S. Kwak, K.M. Kim, J.Y. Do, N.-K. Park, S.O. Ryu, H.-J. Ryu, J.-I. Baek, M. Kang, Effectively CO_2 photoreduction to CH_4 by the synergistic effects of Ca and Ti on Ca-loaded TiSiMCM-41 mesoporous photocatalytic systems. Appl. Surf. Sci. **355**, 891–901 (2015)
121. M. Halmann, Photoelectrochemical reduction of aqueous carbon dioxide on p-type gallium phosphide in liquid junction solar cells. Nature **275**, 115–116 (1978)
122. W. Jiang, X. Yin, F. Xin, Y. Bi, Y. Liu, X. Li, Preparation of $CdIn_2S_4$ microspheres and application for photocatalytic reduction of carbondioxide. Appl. Surf. Sci. **288**, 138–142 (2014)
123. Y.-X. Pan, Z.-Q. Sun, H.-P. Cong, Y.-L. Men, S. Xin, J. Song, S.-H. Yu, Photocatalytic CO_2 reduction highly enhanced by oxygen vacancies on Pt-nanoparticle-dispersed gallium oxide. Nano Res. **9**, 1689–1700 (2016)
124. P. Li, H. Xu, L. Liu, T. Kako, N. Umezawa, H. Abe, J. Ye, Constructing cubic-orthorhombic surface-phase junctions of $NaNbO_3$ towards significant enhancement of CO_2 photoreduction. J. Mater. Chem. A. **2**, 5606–5609 (2014)
125. Y. Jia, H. Ma, W. Zhang, G. Zhu, W. Yang, N. Son, M. Kang, C. Liu, Z-scheme $SnFe_2O_4$-graphitic carbon nitride: reusable, magnetic catalysts for enhanced photocatalytic CO_2 reduction. Chem. Eng. J. **383**, 123172 (2020)
126. X. Li, J. Yu, M. Jaroniec, X. Chen, Cocatalysts for selective photoreduction of CO_2 into solar fuels. Chem. Rev. **119**, 3962–4179 (2019)
127. Q. Zhang, C.F. Lin, B.Y. Chen, T. Ouyang, C.T. Chang, Deciphering visible light photoreductive conversion of CO_2 to formic acid and methanol using waste prepared material. Environ. Sci. Technol. **49**, 2405–2417 (2015)
128. M. Li, L. Zhang, X. Fan, Y. Zhou, M. Wu, J. Shi, Highly selective CO_2 photoreduction to CO over $g\text{-}C_3N_4/Bi_2WO_6$ composites under visible light. J Mater Chem A. **3**, 5189–5196 (2015)
129. K. Sonowal, N. Nandal, P. Basyach, L. Kalita, S.L. Jain, L. Saikia, Photocatalytic reduction of CO_2 to methanol using Zr (IV)-based MOF composite with $g\text{-}C_3N_4$ quantum dots under visible light irradiation. J. CO2 Utiliz **57**, 101905 (2022)
130. Y. Zou, X. Wang, A. Khan, P. Wang, Y. Liu, A. Alsaedi, T. Hayat, X. Wang, Environmental remediation and application of nanoscale zero-valent iron and its composites for the removal of heavy metal ions: a review environ. Sci. Technol. **50**, 7290–7304 (2016)
131. C. Chen, Z. Chai, X. Wang, Metal–organic framework-based materials: superior adsorbents for the capture of toxic and radioactive metal ions. Chem. Soc. Rev. **47**, 2322–2356 (2018)
132. Z. Zhang, C. Liu, Z. Dong, Y. Dai, G. Xiong, Y. Liu, Y. Wang, Y. Wang, Y. Liu, synthesis of flower-like $MoS_2/g\text{-}C_3N_4$ nanosheet heterojunctions with enhanced photocatalytic reduction activity of uranium(VI) Appl. Surf. Sci. **520**, 146352 (2020)
133. R. Acharya, P. Pani, Visible light susceptible doped TiO2 photocatalytic systems: an overview. Mater. Today: Proc. **67**, 1276–1282 (2022)
134. S.P. Tripathy, S. Subudhi, R. Acharya, R. Acharya, M. Das, K.M. Parida, Adsorptive removal of Cr (VI) onto UiO-66-NH_2 and its determination by radioanalytical techniques. J. Radioanal. Nucl. Chem. **322**, 983–992 (2019)
135. M. Wang, Y. Zeng, G. Dong, C. Wang, Br doping of $g\text{-}C_3N_4$ towards enhanced photocatalytic performance in Cr (VI) reduction, Chinese. J. Catal. **41**, 1498–1510 (2020)

136. Y. Wang, S. Bao, Y. Liu, W. Yang, Y. Yu, M. Feng, K. Li, Efficient photocatalytic reduction of Cr(VI) in aqueous solution over CoS_2/g-C_3N_4-rGO nanocomposites under visible light. Appl. Surf. Sci. **510**, 145495 (2020)
137. Y.H. Wang, M. Frutschi, E. Suvorova, V. Phrommavanh, M. Descostes, A.A.A. Osman, G. Geipel, L.R. Bernier-, Mobile uranium(IV)-bearing colloids in a mining-impacted wetland. Nat. Commun. **4**, 2942–2950 (2013)
138. Z. Huang, Z. Li, Q. Wu, L. Zheng, L. Zhou, Z. Chai, X. Wang, W. Shi, Simultaneous elimination of cationic uranium(vi) and anionic rhenium(vii) by graphene oxide–poly(ethyleneimine) macrostructures: a batch, XPS, EXAFS, and DFT combined study, Environ. Sci. Nano. **5**, 2077–2087 (2018)
139. S. Lee, U. Kang, G. Piao, S. Kim, D.S. Han, H. Park, Homogeneous photoconversion of seawater uranium using copper and iron mixed-oxide semiconductor electrodes. Appl. Catal. B. **207**, 35–41 (2017)
140. G. Wang, J. Zhen, L. Zhou, F. Wu, N. Deng, Adsorption and photocatalytic reduction of U (VI) in aqueous TiO_2 suspensions enhanced with sodium formate. J. Radioanal. Nucl. Chem. **304**, 579–585 (2015)
141. S. Tripathi, R. Bose, A. Roy, S. Nair, S. Ravishankar, Synthesis of hollow nanotubes of Zn_2SiO_4 or SiO_2: mechanistic understanding and uranium adsorption behaviour. ACS Appl. Mater. Interfaces. **7**, 26430–26436.
142. C. Lu, R. Chen, X. Wu, M. Fan, Y. Liu, Z. Le, S. Jiang, S. Song, Boron doped g-C_3N_4 with enhanced photocatalytic UO_2^{2+} reduction performance. Appl. Surf. Sci. **360**, 1016–1022 (2016)
143. C. Lu, P. Zhang, S. Jiang, X. Wu, S. Song, M. Zhu, Z. Lou, Z. Li, F. Liu, Y. Liu, Y. Wang, Z. Le, Photocatalytic reduction elimination of UO_2^{2+} pollutant under visible light with metal-free sulfur doped g-C_3N_4 photocatalyst. Appl. Catal. B. **200**, 378–385 (2017)
144. X. Jiang, Q. Xing, X. Luo, F. Li, J. Zou, S. Liu, X. Lia, X. Wang, Simultaneous photoreduction of Uranium(VI) and photooxidation of Arsenic(III) in aqueous solution over g-C_3N_4/TiO_2 heterostructured catalysts under simulated sunlight irradiation. Appl. Catal. B. **228**, 29–38 (2018)
145. Y. Wang, X.M. Liu, Q.I. Chen, T.C. Zhang, L. Ouyang, S.J. Yuan, Simultaneous photocatalytic oxidation and adsorption for efficient As(III) removal by magnetic BiOI/γ-Fe_2O_3 core-shell nanoparticles. Mater. Today Chem. **24**, 100823 (2022)
146. M.N. Maga˜na, A.E. Gonz´alez, L.M. Ix, S.C. Díaz, R. G´omez, Improved photocatalytic oxidation of arsenic (III) with WO_3/TiO_2 nanomaterials synthesized by the sol-gel method. J. Environ. Manag. **282**, 111602 (2021)
147. H. Eslami, M.H. Ehrampoush, A. Esmaeili, A.A. Ebrahimi, M.H. Salmani, M.T. Ghaneian, H. Falahzadeh, Efficient photocatalytic oxidation of arsenite from contaminated water by Fe_2O_3-Mn_2O_3 nanocomposite under UVA radiation and process optimization with experimental design. Chemosphere **207**, 303–312 (2018)
148. W. Ali, N. Mushtaq, T. Javed, H. Zhang, K. Ali, A. Rasool, A. Farooqi, Vertical mixing with return irrigation water the cause of arsenic enrichment in groundwater of district Larkana Sindh, Pakistan. Environ. Pollut. **245**, 77–88 (2019)
149. A.S. Maghsoudi, S. Hassani, K. Mirnia, M. Abdollahi, Recent advances in nanotechnology-based biosensors development for detection of arsenic, lead, mercury, and cadmium. Int. J. Nanomed. **16**, 803–832 (2021)
150. S.-Y. Yu, Y. Liu, H.-T. Ren, Z.-Y. Liu, X. Han, Importance of the ligand-to-metal charge transfer (LMCT) pathway in the photocatalytic oxidation of arsenite by TiO_2. Phys. Chem. Chem. Phys. **24**, 13661–13670 (2022)
151. M. Xiao, R. Li, J. Yin, J. Yang, X. Hu, H. Xiao, W. Wang, T. Yang, Enhanced photocatalytic oxidation of As(III) by TiO_2 modified with Fe_3O_4 through Ti-O-Fe interface bonds. Colloids Surf., A **651**, 129678 (2022)
152. L. Ouyang, Y. Zhang, Y. Wang, X. Wang, S. Yuan, Insights into the adsorption and photocatalytic oxidation behaviors of boron-doped TiO_2/gC_3N_4 nanocomposites toward As(III) in aqueous solution. Ind. Eng. Chem. Res. **60**, 7003–7013 (2021)

153. C. Wang, Y. Dai, X. Fu, X. Lu, J. Zhang, A novel layer-layer crossed structure of bentonite/g-C_3N_4 for enhanced photocatalytic oxidation of arsenic (III) in a wide pH range. Surf. Interf. **26**, 101365 (2021)
154. X. Wang, S.O. Pehkonen, A.K. Ray, Photocatalytic reduction of Hg(II) on two commercial TiO_2 catalysts. Electrochim. Acta **49**, 1435–1444 (2004)
155. K. Sundseth, J. Pacyna, E. Pacyna, N. Pirrone, R. Thorne, Global sources and pathways of mercury in the context of human health. Int. J. Environ. Res. Publ. Health **14**, 105–119 (2017)
156. Y. Fu, J. Jiang, Z. Chen, S. Ying, J. Wang, J. Hu, Rapid and selective removal of Hg (II) ions and high catalytic performance of the spent adsorbent based on functionalized mesoporous silica/poly (m-aminothiophenol) nanocomposite. J. Mol. Liq. **286**, 110746 (2019)
157. Y. Li, M. Xia, F. An, N. Ma, X. Jiang, S. Zhu, D. Wang, J. Ma, Superior removal of Hg (II) ions from wastewater using hierarchically porous, functionalized carbon. J. Hazard. Mater. **371**, 33–41 (2019)
158. G.G. Lenzi, C.V.B. Fávero, L.M.S. Colpini, H. Bernabe, M.L. Baesso, S. Specchia, O.A.A. Santos, Photocatalytic reduction of Hg(II) on TiO_2 and Ag/TiO_2 prepared by the sol–gel and impregnation methods. Desalination **270**, 241–247 (2011)
159. H. Alshaikh, A. Shawky, L.S. Roselin, Promoted visible-light photocatalytic reduction of Hg^{2+} over $CuAl_2O_4$-decorated g-C_3N_4 nanoheterojunctions synthesized by solution process. J. Env. Chem. Eng. **9**, 106778 (2021)
160. R.M. Mohamed, A.A. Ismail, Mesoporous $BiVO_4$/2D-g-C_3N_4 heterostructures for superior visible light-driven photocatalytic reduction of Hg (II) ions. Ceram. Intern. **47**, 26063–26073 (2021)

Chapter 5
Antibacterial Properties of Two-Dimensional Nanomaterials

Elishba Noor, Usman Liaqat, Waqas Qamar Zaman, Sabir Hussain, Asif Shahzad, Kashif Rasool, Zaeem Bin Babar, and Waheed Miran

Abstract Many bacterial species have developed the ability to tolerate multiple drugs, and hence there is a severe threat in treating infectious diseases. Since the conventional drugs are becoming largely ineffective, there is an urgent need to find novel antibacterial strategies. To this end, the development of nanomaterials, mainly two-dimensional (2D) nanomaterials, has emerged as a new class of antimicrobial agents that demonstrate strong antimicrobial properties and are less susceptible to bacterial resistance. Various mechanisms associated with the antibacterial activity of 2D nanomaterials have been exhibited, including physical or mechanical damage, the release of controlled drug/metallic ions, multi-mode synergistic antibacterial activity, oxidative stress, and photothermal/photodynamic effects. In addition to the detailed mechanisms involved in antibacterial activities, this chapter discusses various types of 2D nanomaterials used for antibacterial activities. For better understanding, the 2D nanomaterials are classified into carbon and non-carbon based for antibacterial activities. The present challenges and possible future directions for the development of 2D nanomaterials with advanced antimicrobial properties are also discussed.

E. Noor · U. Liaqat · W. Miran (✉)
School of Chemical and Materials Engineering, National University of Sciences and Technology, Islamabad 44000, Pakistan
e-mail: waheed.miran@scme.nust.edu.pk

W. Q. Zaman · Z. B. Babar
Institute of Environmental Sciences and Engineering (IESE), School of Civil and Environmental Engineering (SCEE), National University of Sciences and Technology (NUST), Sector H-12, Islamabad, Pakistan

S. Hussain
Department of Space Science, Institute of Space Technology, Islamabad 44000, Pakistan

A. Shahzad
Department of Materials Science and Engineering, Uppsala University, Box 534, 75121 Uppsala, Sweden

K. Rasool
Qatar Environment and Energy Research Institute, Hamad Bin Khalifa University (HBKU), Qatar Foundation, 5824 Doha, Qatar

N. Kumar et al. (eds.), *Two-Dimensional Materials for Environmental Applications*, Springer Series in Materials Science 332, https://doi.org/10.1007/978-3-031-28756-5_5

5.1 Introduction

Bacterial infections are a global health concern as they cause millions of deaths annually [1, 2]. To overcome this concern, different classes of antibiotics have been discovered/synthesized, which resulted in saving countless lives through effective control of lethal bacterial infections [3]. The antibacterial efficiency of conventional drugs is known to be reduced when biofilms are formed [4]. Further, the alarming misuse and abuse of conventional antibiotics are leading to an immense increase in antibiotic resistance, "superbugs" that are fastly emerging as a global emergency [5, 6]. The consequences of multidrug resistance are alarming as there is an estimation of losing more than ten million lives yearly by 2050, coupled with substantial economic losses [7]. Therefore, alternative technologies for multidrug resistance bacteria are in high demand to alleviate this pressing challenge.

Given that nanotechnology offers humans the means for manipulation, controlling, and processing at atomic and molecular levels, recent advancements in antibacterial nanotechnologies enable various opportunities to effectively remove drug-resistant bacteria via distinct mechanisms [8, 9]. Additionally, since nanomaterials are biocompatible and physiochemically stable, delivery of naturally and synthetically produced antibacterial compounds via nanomaterials is envisioned as a practical strategy [10, 11]. Compared to conventional antibacterial drugs, nanomaterials are considered to have lower chances of developing bacterial resistance because their membrane permeability is better, they are largely benign and they have several bactericidal actions [12]. Among different nanomaterials, 2D nanomaterials have a large surface area and can be surface functionalized to achieve required interactions with bacterial membranes that can enhance the antibacterial effect [13]. Further, even a lower dose of bactericidal agents based on 2D nanomaterials compared to conventional antibiotics can be used to help overcome multidrug resistance problems and lower unfavourable side effects [14]. To this end, various mechanisms have been discovered/identified for antibacterial activities by 2D nanomaterials, such as bacterial membrane damage by physical contact, oxidative stress, light-mediated bactericidal effect, controlled drug/metallic ions releasing, and synergistic effects of 2D nanomaterials [15, 16].

Considerable progress in developing 2D nanomaterial-based antibacterial agents involving carbon and non-carbon-based 2D nanomaterial have been achieved. Among carbon-based 2D nanomaterials, graphene, graphitic carbon nitride (g-C3N4), and transition metal carbides (MXenes) have dominant antibacterial properties [17]. In contrast, transition metal dichalcogenides (TMDCs), transition metal oxides (TMOs), black phosphorus (BP) and layered double hydroxides (LDHs) have been studied under non-carbon based 2D nanomaterials class for antibacterial properties [18]. A systematic summary of the recent research advancement in such 2D nanomaterials in bactericidal activities will be useful for further research progress in the field.

Here, advancements in 2D nanomaterial-based antibacterial agents are summarized. Various antibacterial mechanisms discussed in the literature are briefly outlined. Further, 2D nanomaterials are classified into carbon and non-carbon based, and their antibacterial properties are presented. Finally, the challenges in the progress of the field, major obstacles offered in real clinical systems, and future perspectives are discussed. This chapter will not only help in understanding the properties and antibacterial mechanisms of 2D nanomaterials but also encourage many scientists to bring this field to practicality by addressing research gaps in the near future.

5.2 Antibacterial Mechanisms by 2D Nanomaterials

Nanomaterials, unlike traditional antibiotics, have good membrane permeability, benign biocompatibility, and potential for various antibacterial activities; therefore, they are less likely to activate bacterial resistance [19, 20]. The pathway for the antimicrobial activity followed by nanomaterials depends on their shape, size, functionality and core of the material [21]. There has been vast progress in the synthesis of nanomaterials of different dimensionalities, i.e., 0D (zero dimensional), 1D (one dimensional), 2D (two dimensional) and 3D (dimensional). Dimensionality changes the intrinsic property of materials hence causing their antibacterial mechanism to differ as well [22]. 2D nanomaterials are preferred over 0D and 1D because of their large surface area, easy surface functionalisation, degradability in human bodies and low dose requirement [23, 24, 25]. These nanomaterials provide antibiotic-free antibacterial strategies via physical contact destruction, generation of reactive oxygen species (ROS), photo-induced antibacterial effects, controlled drug release, synergistic antibacterial effects [18, 26]. Here, we briefly summarized the mechanisms involved in antibacterial actions by 2D nanomaterials.

5.2.1 Membrane Damage by Physical Contact

Conventional antibiotics mainly target bacterial cell membranes through their chemical digestion or by preventing the formation of a new one [27]. But over time, the bacterial cell envelope has evolved, becoming a physical barrier to antibacterial drugs. Also, the teichoic acid in gram-positive bacteria and the phosphate group in the lipid bilayer of the outer layer of gram-negative render polarity to bacterial surfaces, compromising the activity of hydrophobic antimicrobials across the membranes (Fig. 5.1) [21].

The nanomaterials in contact with bacterial cell membranes disrupt it because of their unique physicochemical properties like geometry, size, surface charge, surface topology and hydrophobicity [29]. For instance, physical rupture of the cell membrane can be achieved by using 2D graphene sheets having sharp edges disrupting the lipid bilayer when aligned perpendicular to the bacteria surface (Fig. 5.2) [30]. Non-carbon-based 2D nanomaterials, such as nanosheets of WS_2, MoO_3, $BiSe_3$, and RuO_2, damage the cell membrane drastically when vertically

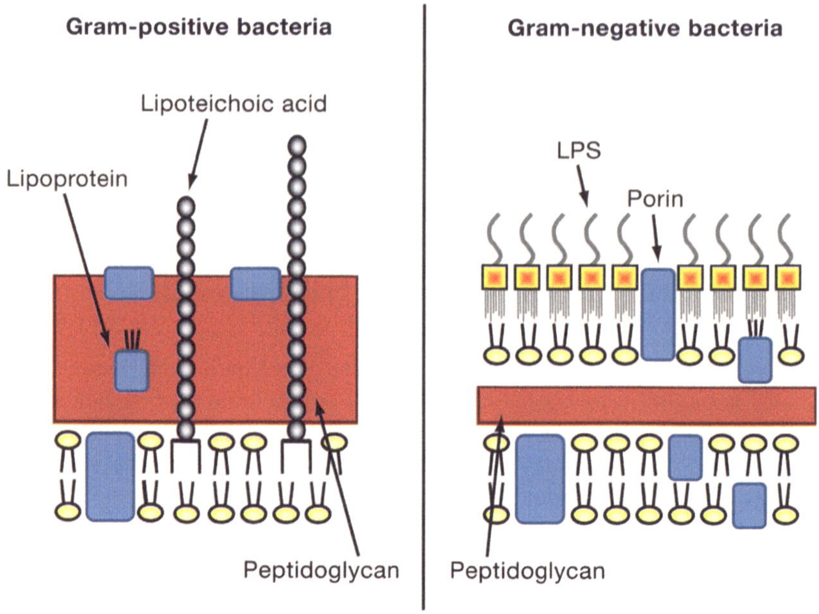

Fig. 5.1 Structure of cell wall of gram-positive and gram-negative bacteria. Reproduced with permission from Akira et al. [28]. Copyright 2006, Elsevier

aligned against the membrane [31, 32]. Compared to graphene-based, the results of theoretical simulation predict that it is not evident whether the physicochemical properties of non-carbon-based nanomaterials affect the bactericidal activity or not and needs to be explored more [18].

5.2.2 *Bacterial Inactivation by Oxidative Stress*

The bacterial activity can be inhibited by the generation of oxidative stress, which is mainly of two types, i.e., ROS-dependent and ROS-independent.

5.2.2.1 ROS-Dependent Oxidative Stress

Chemical change to the bacterial membrane can be caused by the excessive accumulation of ROS leading to oxidative stress. ROS are the by-product generated as a result of oxidative metabolic processes that lead to cell lysis by deteriorating the essential cellular components, preventing cell differentiation, its survival and ultimately leading to cell death [34, 35]. The typical ROS generated by 2D nanomaterials includes hydroxyl radicals ($^{\cdot}OH$), hydrogen peroxide (H_2O_2), singlet oxygen ($^1O_2$)

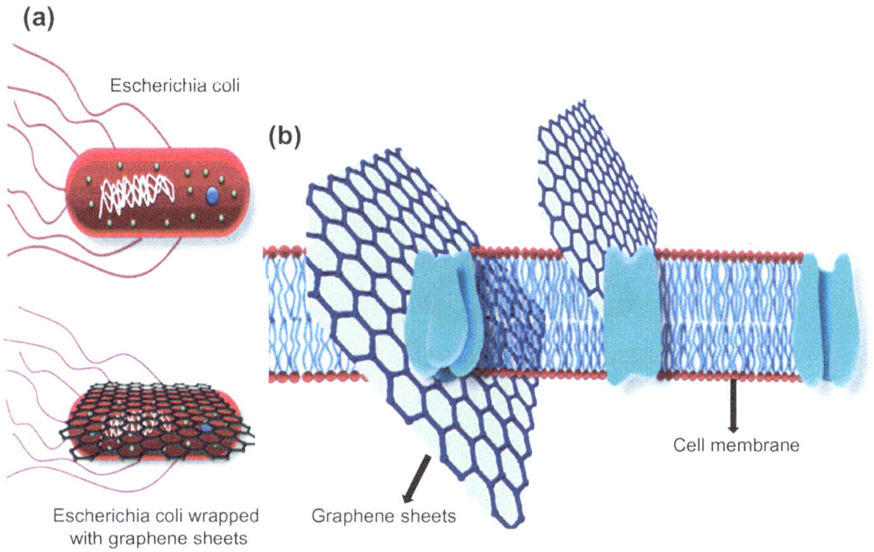

Fig. 5.2 **a** Schematic showing *E.coli* and graphene sheets encapsulating *E.coli*. **b** Graphene Sheets enabled physical damage to cell membrane by rupturing the cell membrane. Reproduced with permission from Singh et al. [33]. Copyright 2014, Royal Society of Chemistry

and superoxide anion radicals ($^{\cdot}O_2^-$). These oxidative species, in particular, react with thiols in protein, resulting in the deactivation of membrane receptors (Fig. 5.3) [36]. Various derivatives of graphene could generate ROS to enhance the bactericidal effect. For example, graphene oxide (GO) and reduced graphene oxide (rGO) can kill the bacteria such as *Xanthomonas oryzae pv. Oryzae* and *Pseudomonas aeruginosa* through production of ROS [37, 38, 39]. Recently, 2D nanomaterials beyond graphene (NBG) have been found to be effective in the antibacterial mechanism through ROS generation, such as nanosheets of 2D MoS_2 that selectively aim the target cell instead of control cell, which makes them beneficial for their analysis in further biomedical applications [40].

5.2.2.2 ROS-Independent Oxidative Stress

Even though ROS-induced oxidative stress has a prominent role in antibacterial activity, different studies have shown that various nanomaterials also induce ROS-independent oxidative stress. In this process, the cellular organelles and components such as lipids, proteins and DNA are oxidized directly through nanomaterials without any production of ROS [41, 42]. Various shreds of evidence have been provided regarding bactericidal activity through ROS-independent oxidative stress of carbon-based, such as GO and rGO, as well as non-carbon based, including chemically exfoliated MoS_2 nanosheets killing the *E.coli* in a very short period [43].

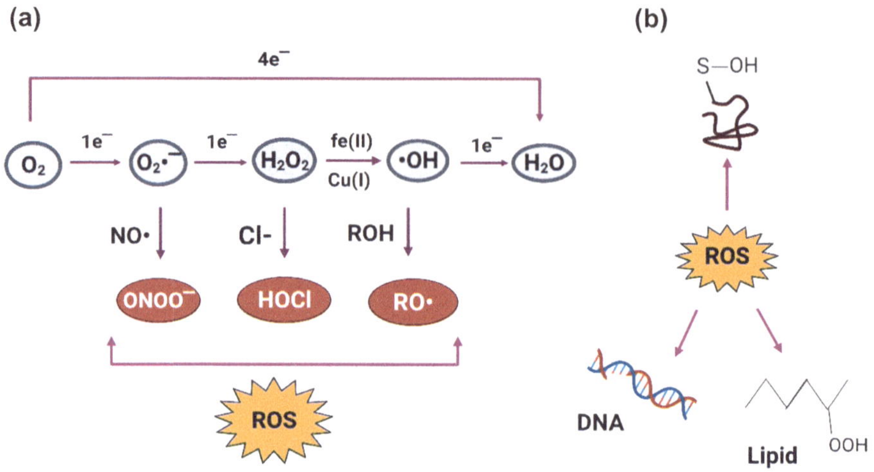

Fig. 5.3 **a** General scheme for ROS production; **b** Damaging effects of ROS to Biomacromolecules. Constructed by using Biorenders.com with permission

5.2.3 Light-Mediated Bactericidal Effects

Over the past years, light-induced antibacterial mechanisms have been given more attention because of their particular advantages such as selective treatment of target, insignificant side effects and non-invasiveness. In these processes, the photoactive agents of nanomaterials are excited by irradiation through light and converted into heat or ROS [18]. The different photo-induced mechanism includes photocatalytic activity (PCT), photothermal activity (PTT) and photodynamic therapy (PDT), which are further discussed below.

5.2.3.1 Photothermal Activity

Studies have shown that bacteria, at a temperature of more than 55 °C, are killed, leading to the disintegration of heat shock proteins. Still, this method is only limited to food production and preservation. So, new strategies have been explored to generate heat with high biocompatibility and effectiveness. For this purpose, the photothermal effect has been reported in which near-infrared (NIR) light is irradiated to convert light to heat. NIR (λ = 700–1400 nm) easily penetrates the mammalian cells and increase the temperature of the targeted area leading to the inactivation of a bacterial cell by denaturation of enzymes while causing negligible damage to mammalian tissue cells [44, 45].

Among carbon-based 2D nanomaterials, reduced rGO is widely used in PTT due to increased availability and good electrical conductance as compared to graphene oxide

(GO). In contrast, NIR photothermal nanoagents in the case of non-carbon-based 2D nanomaterials include MoS_2, black phosphorus (BP), and Sb_2S_3 nanosheets.

5.2.3.2 Photocatalytic/Photodynamic Therapy (PCT/PDT)

It is established that ROS generation induced through light leads to distinct oxidative stress in a bacterial cell. Usually, light-mediated production of ROS is achieved in two ways: photocatalytic and photo-dynamic processes.

In PCT, the nanomaterials act as catalysts and convert the cellular components and structures into toxic species such as ROS by separating the holes (h^+) and electrons (e^-) in the valence and conduction band on irradiation of light. Typically, light with a wavelength from ultraviolet (UV) to visible (VIS) causes the generation of oxidative species hence leading to cell lysis [46] (Fig. 5.4). There are mainly three steps of photocatalytic antibacterial mechanism, i.e. (i) 2D nanomaterials having large surface area, then attach with bacterial surface offer more active centres for catalytic activity; (ii) the production of charge-separated species on irradiation of light; (iii) the interaction of activated charged sites with neighbouring oxygen and water molecules to produce oxidative species (ROS) [47]. Among various photocatalyst, GO/g-CN heterojunction, and bismuth oxybromide (BiOBr) nanosheets have attracted tremendous attention in efficiently killing bacteria [48, 49].

Photodynamic therapy (PDT) is another bactericidal mechanism employed to avoid the use of antibiotics. In this process upon irradiation of light the photosensitizers in nanomaterials react when the frequency of light matches with the absorption frequency of photosensitisers. This produces oxidative species such as $OH^{\cdot}$ radicals, superoxides, $^1O_2$ (singlet oxygen) etc., that disrupt the cell membrane and DNA molecules [50]. Recent studies have shown that research is now more focused on graphene-based materials [51], MoS_2 [52], and black phosphorus (BP) nanosheets for ROS generation.

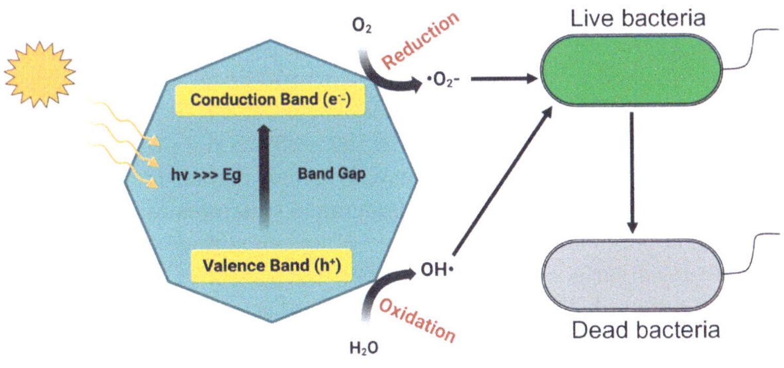

Fig. 5.4 General mechanism of photocatalytic impact of 2D nanomaterials on bacteria. Constructed by using Biorenders.com with permission

5.2.3.3 Synergistic Effects of 2D Nanomaterials

Although the antibacterial mechanisms mentioned above are widely accepted against conventional antibiotics, they also have some limitations. For example, in the physical contact mechanism, the antibacterial activity depends on the surface area of nanomaterials, and their antibacterial efficiency considerably decreases with time. In the oxidative stress mechanism, the antibiotic-free activity mainly depends on the surface area, conductivity, and size of nanomaterials. Even though the light-mediated antibacterial mechanism is widely accepted but targeted lysis of bacteria, oxygen deficiency in infected areas and photocatalytic efficiency are the factors that limit the antibacterial activity. So, for better results, current research is more inclined towards multi-mode synergistic antibacterial mechanisms rather than single-mode. Summary of various synergistic antibacterial mechanisms includes synergy of: (i) PTT with PDT [53, 54, 55] [54], (ii) PTT with oxidative stress [42], (iii) PTT with catalytic activity [56, 42], [57, 58], and (iv) chemotherapy with PDT [59]. Rapid antibacterial activity is shown by chitosan-assisted MoS_2 hybrid coated over Ti material surface, giving synergistic effects of PTT and PDT under dual light irradiation (660 nm visible light and 808 nm NIR) (Fig. 5.5). This hybrid has increased bactericidal effects towards *E. coli* and *S. aureus* [60].

5.3 Carbon-Based 2D Nanomaterials for Antibacterial Activities

Since bacterial resistance to presently offered antibiotics highlights the strong need for new antibacterial drug agents [61, 62], nanomaterial with antibacterial characteristics can be utilized for various antibacterial practices. Various classes of nanomaterials together with medicinal drug polypeptides, metal nanoparticles (NPs), semiconductor NPs, compound nanostructures, and carbon-based nanomaterials are studied each beneath in vitro/vivo circumstances to manage and fight microorganism infection [63]. Many characteristic properties of carbon-based nano-scaled 2D materials, such as electronic conduction, hydrophobicity, metric linear unit sharp edges, and the creation of aerophilic strain make them ideal candidates for achieving disinfectant properties. It is necessary to notice that the antibacterial properties of carbon-containing 2D nanomaterials like graphene, graphitic carbon nitride, and MXene could offer unique and numerous physical/chemical structures like nanosheet form, kind of delamination, nanosheets size, range of layers, surface chemistry, range of defects on the layer-surface, and accumulation [64]. A brief detail of commonly employed carbon-based 2D nanomaterials for antibacterial purposes is provided below.

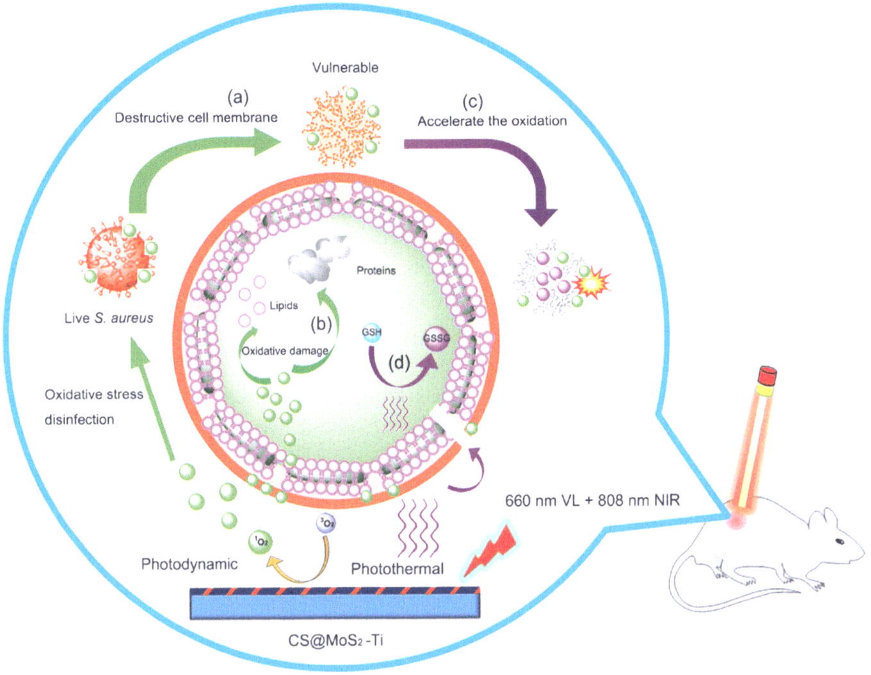

Fig. 5.5 Schematic showing the synergistic photodynamic and photothermal antibacterial activity via dual lights irradiation using CS@MoS2 hybrid coating made on the surface of Ti material. Reproduced with permission from Feng et al. [53]. Copyright 2018, Wiley Online Library

5.3.1 Antibacterial Performance of GO

Graphene and associated 2D nanomaterial, including GO have a huge interest in studies concerning their primary houses and capacity applications. Given the capacity of massive usage, studies on the ecological effect of this magnificent nanomaterial have all started to boom rapidly. Graphene is made of carbon particles and prepared in a hexagonal honeycomb arrangement because of the sp^2 hybridization. In this particular framework, the p-s orbital of carbon atom in graphene is perpendicular to the layers' direction, which creates a significant polyatomic π link through the whole layer, generated in electrical, thermal, visual, and mechanical characteristics, and a spacious particular surface area [65]. Graphene can also be developed through chemical-based vaporization and thermal stripping techniques. GO is fabricated by oxidization using the Hummers method. Furthermore, in photocatalysis applications, graphene is being practiced due to its raised electrical conductivity (106 S m^{-1}). In graphene-based nanocomposites, it performs an excellent role in receiving or transporting electrons, enhancing catalytic efficiency. Currently, the leading means used to improve the antibacterial characteristics of graphene-oriented materials are:

tuning the size and number of nanosheet surface modifications, creating a hetero junction design, and developing a composite method [66].

Up to now, multiple practices for graphene synthesis have been developed and may be into two groups: physical and chemical synthesis procedures. Physical techniques include the exfoliation of stacked graphite sheets into graphene sheets through van der Waals force interference operating mechanical exfoliation or key liquid-phase exfoliation. The chemical techniques include the fabrication of a variety of graphene-comprising materials through chemical-based reactions, e.g., chemical vapor deposition (CVD) method, epitaxial growth of sheets, chemical reduction, etc. The graphene family has inherent antibacterial properties and antibacterial materials such as GO surface decoration with silver, metal ions/oxides/sulfides NPs, polymers, antibiotics, and enzyme functionalization [65] (Fig. 5.6).

Graphene-associated nanomaterials are dynamic, and their potential ecotoxicological impact has been inspected in a range of recent investigations using several aquatic and terrestrial microorganisms. Bacteria is an excellent example to be used as model microbes in examining the ecotoxicity of developed 2D nanomaterials, as they are the fundamental unit of the food chain and are ever-present in water, soil, and sediment. Many scientific studies have examined the bacterial toxicity of graphene-associated nanomaterials and their toxicity to wastewater precipitate microorganisms [67]. The toxicity of microbes has been found to associate with the physiology of the investigated creatures and the physical/chemical characteristics of GO, such as inorganic reduction, structure, and size. Numerous possible applications resulted from the advancement in GO antimicrobial process have been suggested, especially the use of GO for the manufacturing of biofouling-resistant water filtration membranes.

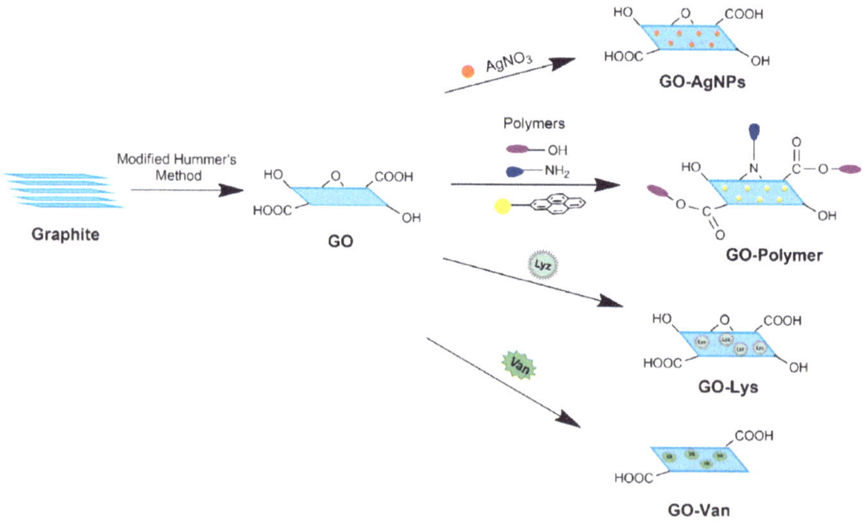

Fig. 5.6 Schematic representation for the preparation of graphene-based nanocomposites. Reproduced with permission from Kumar et al. [65]. Copyright 2019, MDPI

Lately, the prospective anti-bacterial usages of graphene and graphene-derived nanomaterials have driven significant importance, though it is still debated how GO acts as an antibacterial agent. However, the antibacterial properties of graphene or GO are still limited. Also, nanocomposite materials such as polymer-coated nanoparticles, diamond/carbon nanocomposites, and metal oxide nanoparticles composites with graphene/GO/r-GO are being used in therapeutic trials as an agent [68]. Furthermore, the antibacterial activities of GO can be studied by operating scanning electron microscopy (SEM), atomic force microscopy (AFM), and transmission electron microscopy (TEM). In addition, GO has been examined on the molecular level to improve its antibacterial characteristics [69].

5.3.2 Antibacterial Properties of Graphitic Carbon Nitride

The polymeric material carbon nitrides are composed of carbon and nitrogen. These are derived from carbon by exchanging carbon atoms with nitrogen, and become the most exciting candidates for numerous applications. For instance, other carbon components and carbon nitride materials have an extended historical past because in 1834, Berzelius developed materials with the name of "melon" (linear polymers of connected tri-s-triazines) through secondary nitrogen [70, 71]. However, the possible significance of this material was not thoroughly identified till the new investigation, practically because it was chemically inert, insoluble in acidic, neutral, or basic solvents. So, the protected composition of graphitic carbon nitride (g-C_3N_4), an interesting earth-abundant material that can be viewed, contains an identifiable two-dimensional design, exceptional chemical reliability, and a tunable framework electronic. Pure g-C_3N_4 has low activity and is photocatalytic due to the fast recombination of photogenerated electron–hole pairs. g-C_3N_4 material might work as a famous prospect for linking with several practical materials to boost efficiency because of its unique electronic structure.

The earlier ground-breaking work has activated significant fascination with the fabrication, customization, and use of g-C_3N_4-designed materials [71]. The most common precursors used in the chemical synthesis of g-C_3N_4 are activated nitrogen-abundant and oxygen-free compounds with C-N preboned structures, such as triazine and heptazine derivatives. Many of these precursors are volatile, difficult to obtain, and/or extremely explosive. Due to their low thermodynamic stability and the fact that they require surface defects for catalysis, single-phase sp^3-hybridized carbon nitrides are difficult to synthesize. It also seems that defective materials are much more valuable than perfect ones. Thus, forming g-C_3N_4 with defects is an appealing topic, as soon as the material is likely found in catalysis and antibacterial activities [72, 73, 71]. The g-CN_4 and the composites comprised of g-C_3N_4 are potential catalysts with outstanding physical and chemical characteristics and hold substantial potential in antibacterial activities. Nonetheless, due to the narrow absorption of visible light and easy charges' recombination, the antibacterial activities of bulk g-C_3N_4 are few. Subsequently, several g-C_3N_4-derived materials have been fabricated with high surface areas, enhanced electron–hole pair separation capabilities, and prolonged visible light absorption ranges, adding a synergistic effect to improve their antibacterial activities.

Various approaches have been deployed to enhance the antibacterial efficiency of g-C_3N_4, involving functionalization with noble metal nanoparticles, such as gold and silver nanoparticles [74, 75]. Surface plasmon resonance of nanoparticles improves performance that is antibacterial by expanding the spectral absorption range and advancing the synthesis of photogenerated carriers in g-C_3N_4. Further, noble metal nanoparticles act as electron traps, catching free e− , and so suppressing the recombination of photogenerated carriers. [76] grew Ag/g-C_3N_4 nanocomposites by merging thermal polymerization with photo-assisted reduction. An interactive antibacterial proficiency was attained with excellent sterilization activity of the Ag/g-C_3N_4 nanocomposite compared to pure g-C_3N_4. Particularly, the handling of noble Ag nanoparticles from the g-C_3N_4 nanosheets considerably boosts the light that is a visible region as a result of the surface plasmon resonance effect of these Ag nanoparticles together with charge transfer amongst the Ag, additionally, the g-C_3N_4 nanosheets. Similarly, the strong surface that is unique resonance absorption of gold nanoparticles covers an array of spectra, like the visible and near-infrared light (NIR) (Fig. 5.7) [77].

5.3.3 Antibacterial Properties of MXene

MXenes, a novel class of 2D nanomaterials, developed from a family of transition metal carbides/nitrides or carbonitrides represented by the general formula $M_{n+1}X_nT_x$ (n = 1–3), where M represents an early transition metal group, X is for carbon, and/or nitrogen and T_x are functional groups (–OH, –O–, and/or –F) present on the surface of 2D MXene layers. To date, many types of MXenes have been reported counting $Ti_3C_2T_x$, Ti_2CT_x, and $Nb_4C_3T_x$ [79, 80, 81]. Ti_3CNT_x, many more are being synthesized or theoretically anticipated based on the available MAX phase precursor. MXene is two dimensional, stratified transition metal inorganic compound or chemical compound carved to get a bedded structure (Anasori and Gogotsi, 2019, [82]. MXenes don't solely have the characteristics of an outsized specific extent and high conduction (9.88 × one hundred and five S m^{-1}), like graphene, but conjointly the benefits of the versatile adjustment of parts and manageable minimum Nano layer thickness. Mainly, the titanium-MXenes, including Ti_2CT_x and $Ti_3C_2T_x$, are the most potential candidates for antibacterial activities because of the elemental abundance and non-toxic decay of products. So far, only limited studies have explored the antibacterial applications of MXenes; however, an encouraging number of scientific research is in progress. The antibacterial behaviors of $Ti_3C_2T_x$ were examined for *Escherichia coli* (E. coli) and *Bacillus subtilis* (B. subtilis) as model Gram (+) and Gram (−) bacteria, respectively, in aqueous suspension [83, 64].

Rasool et al. [83] examined as well as compared the anti-bacterial application of MAX phase (Ti_3AlC_2), single-layered exfoliated (S-$Ti_3C_2T_x$), and few-layered

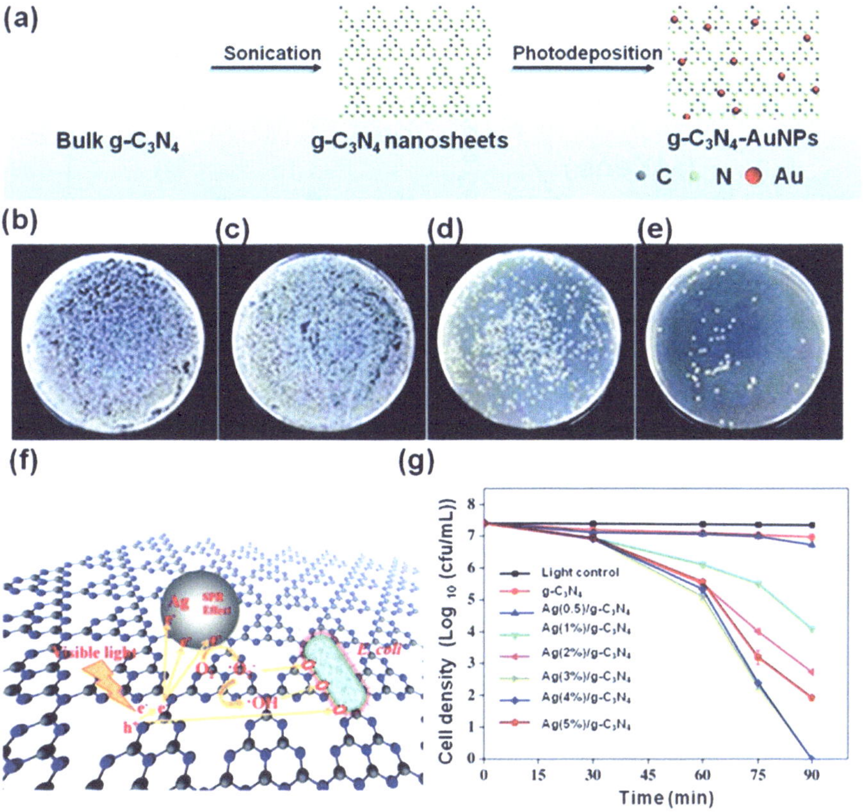

Fig. 5.7 **a** Fabrication of g-C_3N_4-Au nanocomposite, **b** *E. coli* bacteria + control sample, **c** *E. coli* + g-C_3N_4-1.0%, and **d** *E. coli* + g-C_3N_4-1.0% Au + 10 min irradiation. **e** *E. coli* + g-C_3N_4-1.0% Au + 20 min irradiation, Reproduced with permission from [78]. Copyright 2019, American Chemical Society, (f) the mechanism of *E. coli* inactivation in the presence of Ag/g-C_3N_4 under visible light, and (g) disinfection efficiencies of *E. coli*. Reproduced with permission from Dai et al. [77]. Copyright 2017, Elsevier

(F-$Ti_3C_2T_x$ that is exfoliated MXene sheets over the *B. subtilis* and *E. coli* bacteria. The arrangement of the antimicrobial activity towards both the bacterial strains had been revealed as S-$Ti_3C_2T_x$ $\gg$ F-$Ti_3C_2T_x$ > Ti_3AlC_2, showing the importance of the role of the MXene depth in identifying specific antimicrobial impact [83]. Moreover, the bacterial stability research and optical density growth curves demonstrated dose-dependent bactericidal characteristics of MXenes for the inspected strains. They even compared the antimicrobial activity of $Ti_3C_2T_x$ with GO (well-recognized 2D material for antimicrobial properties) and revealed enhanced antibacterial activity of MXenes than GO. This indicates the huge potential of 2D materials, namely MXenes, for efficient antimicrobic applications [84] (Fig. 5.8).

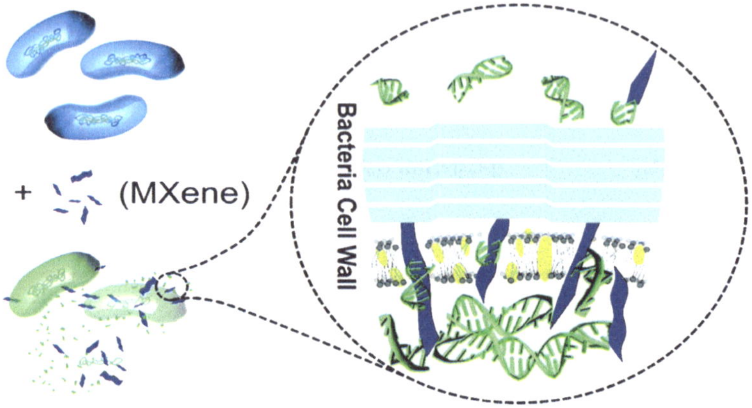

Fig. 5.8 Suggested antibacterial procedure of activity to MXene nanosheets. Reproduced with permission from Arabi Shamsabadi et al. [84]. Copyright 2018, American Chemical Society

The fluorescence imaging and flow cytometry investigated by Shamsabadi et al. (2018) revealed that the antimicrobial activity of the MXene nanosheets is size and exposure time-depending [84]. The MXene sheets, along with smaller sizes, are discovered to be more damaging in the direction of the bacterial cells, especially when subjected for a much longer time. Dealing with the bacterial inoculum with 0.09–4.40 μm $Ti_3C_2T_x$ MXene nanosheets for 8 h without light exposure takes to >90% bacterial killing [85].

The antibacterial synergy between MXenes and metal nanostructures has been investigated to design additional effective MXene-based antibacterial formulations. In this particular framework, Pandey et al. [86] described an Ag NP-$Ti_3C_2T_x$ MXene-based antifouling nanocomposite membrane for water refinement methods [86]. MXene-based composites with polymers, such as MXenes with exceptional physical characteristics, hydrophilic surfaces, and metallic conductivity, increase polymers mechanical and thermal properties [82]. Single-layer MXenes have a larger accessible surface area and are more compatible with polymers than multi-layered MXenes. As a result, MXenes are often delaminated before being combined with polymers. Utilizing the lactate dehydrogenase release assay and various microscopic techniques, such as SEM and TEM, process of antibacterial activity for MXenes are investigated. This investigation revealed that microbial inactivation by MXenes is caused by cell membrane dislocation and responsive ROS production. Rasool et al. also reported the synthesis of a micrometer-thick $Ti_3C_2T_x$ membrane which showed antibacterial activity against widespread waterborne bacteria that supports their prospective use as anti-biofouling membranes in water and wastewater treatment procedures [69].

5.4 Non-carbon-Based 2D Nanomaterials for Antibacterial Activities

Although 1-D nanomaterials have extensive applications, they are known to have low surface areas and respective loadings. Additionally, they possess singular photothermal functionality limiting their utility in clinical therapeutics. Conversely, 2-D nanomaterials possess extensive surface areas and complex structures in the form of sheets and layers. High surface areas of 2-D nanomaterials owing to sheet-like structures, are significantly photo-reactive and facilitate photodynamic and photothermal therapies even for non-carbon based 2D nanomaterials. Therefore, these distinctive properties render 2-D nanomaterials with extraordinary electronic and physio-chemical features. A wide range of non-carbon-based 2-D nanomaterials was developed and investigated for antibacterial applications, including transition metal-based dichalcogenides, metal oxides, and metal hydroxide [18, 87] due to their inherent large surface areas facilitates which promote their application in a range of therapeutics [88]. Non-carbon-based nanomaterials (i.e., Transition metal dichalcogenides (TMDs)) are relatively more functional in drug delivery, phototherapic agents for cancerous cells and bacteria, and their utilization as healing agents for bacterial wounds [89, 90, 91, 92, 43, 93, 58, 94] in comparison with their bulk configuration by optimizing their physicochemical characteristics such as pH, specific surface modifications and functionalization, etc., to increase their abatement efficacy. In the following section, the antibacterial potential of various non-carbon-based 2-D materials is discussed in detail. The mechanisms involved in antibacterial action and recent advancements in the practical application of TMD and TMD-based heterostructures are also explained.

Initially, in 2014, Yang and colleagues [95] investigated the application of 2-D chemically exfoliated MoS_2 as an antibacterial agent. They proposed that its antimicrobial action might be due to membrane cleaving and subsequent oxidation processes. In recent years, scientists and researchers have grown their research interests to investigate the development and application of modified 2-D MoS_2-based nanomaterials for bacterial abatement [89, 90, 91, 86, 43, 93, 58, 94]. Several studies have reported that modified MoS_2 nanosheets with different chemical moieties or combinations of MoS_2 with nanoparticles, such as gold, silver, etc., were effectively employed in photothermal, photocatalytic, and chemical disinfection systems.

Pandit and coworkers [43] also reported that chemically modified MoS_2 possessed a potent antibacterial action against Gram-positive and negative bacterial species. Furthermore, the antibacterial features of exfoliated MoS_2 were found to be significant against Methicillin-Resistant *Staphylococcus Aureus* (MRSA) and *Pseudomonas aeruginosa* with a minimum inhibitory concentration (MIC) value of 1.88 ppm. The net positive and negative charges on the surfaces of modified MoS_2 and bacteria were responsible for high antibacterial action. Consequently, it exhibited high efficacy in killing bacteria. Additionally, it was shown that the antibacterial activity was tuned by changing the hydrophobicity of positively charged MoS_2.

In a recent study [89], an antibacterial agent was developed by attaching melittin antimicrobial peptide (AMP) with MoS_2 via synchronic photothermal and photodynamic therapies and AMP processing and consequently achieved a 100% multidrug resistance against bacteria. It demonstrated a 100% ability to kill bacterial species of MRSA, *E. coli*, and *Klebsiella pneumoniae*. Furthermore, it was observed that approximately 28% of multidrug-resistant bacteria were killed by employing a MoS_2-based nano-platform. Moreover, 20% of bacteria were solely killed by AMP. Also, 100% of bacteria were killed using NIR light and nanoplate form. This 100% synergic killing is comprised of two steps. Initially, pores were made by AMP. Then, these pores help to diffuse heat and promote oxidation during photodynamic and photothermal therapies.

In another study, Roy and colleagues [93] developed nanosheets based on chitosan and MoS_2 and investigated their antibacterial properties. They found that the developed nanosheets effectively treated Gram-negative and -positive bacteria. Additionally, the MIC and MBC values of these nanosheets against *E. coli* and *S. aureus* were found to be 30 and 60 $\mu g\ mL^{-1}$, 90 and 120 $\mu g\ mL^{-1}$, respectively. A high concentration of nanosheets was required for a significant antibacterial action owing to the presence of a peptidoglycan layer around the Gram-positive *S. aureus*. The primary mechanisms involved in antibacterial activity were membrane destruction, deactivating metabolism, and oxidation. Thus, MoS_2 nanosheets seemed to be effective as antibacterial plating and wound bandages for prospective applications in bioscience, bioengineering, and ecology.

Cao and coworkers [90] developed a poly-(dimethyl diallyl ammonium chloride) (PDDA)-Ag^+-Cys-MoS_2 heterostructure as an antimicrobial agent by infusing cysteine (Cys) modified molybdenum disulphide (MoS_2) with a minimal concentration of Ag^+ ions and subsequent coating with positively charged polyelectrolyte and investigated it for treating Gram-negative *E. coli* and Gram-positive *S. aureus*. The infected wound model was utilized. Initially, the back of the mice was slit and MRSA was injected (1×10^6). Then, the mice were categorized into seven sections depending on ministrations with PBS buffer, $AgNO_3$, silver nanoparticles (AgNPs), PDDA, Cys-MoS_2, Ag^+-Cys- MoS_2, and PDDA-Ag^+-Cys- MoS_2. It was observed that after three days of ministrations, the fragments of the epidermal layer were found in control groups, and PDDA-Ag^+-Cys-MoS_2 demonstrated the best healing and antibacterial activities (Fig. 5.9).

For the case of photothermal disinfection, innovative 2-D photocatalysts based on plasma and semiconductors have also been investigated. They exhibited extraordinary antibacterial features. Recently, Plasmonic MoO_3@x–Ag was studied as an effective antibacterial agent which absorbed near-infrared (NIR) light with significant efficacy. In addition to its antibacterial action, it also converted absorbed light in to heat, which was then transferred to Ag that emitted Ag ions to augment further the bacterial disinfection [96].

In another study, relatively non-hazardous and economic nanosheets (NSs) based on extremely thin antimony selenide (Sb_2Se_3) were developed and investigated as highly effective 2-D nano-agents for antibacterial properties. It was using the

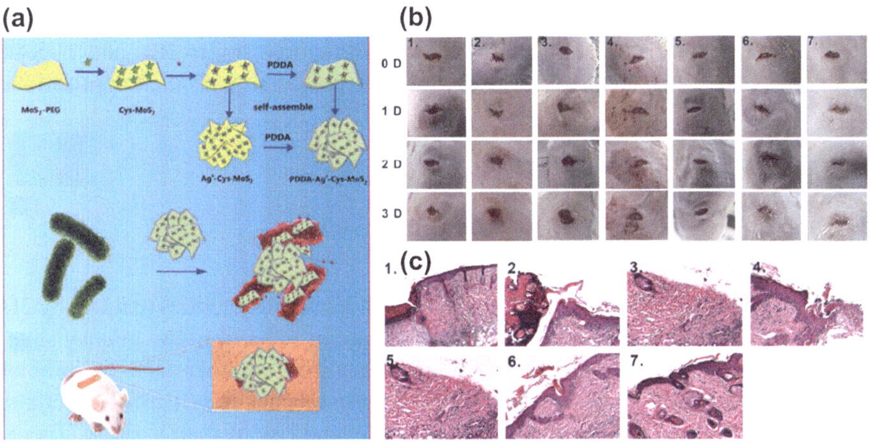

Fig. 5.9 **a** Synthesis of poly(dimethyldiallylammonium) chloride (PDDA)-Ag^+-Cys-MoS_2, **b** Pictures of mice's wound taken at 0–3 days for seven categories of ministrations of antibacterial agent, and **c** Pictures of specific parts of stained skin tissues for seven categories. The seven categories were comprised of PBS buffer, $AgNO_3$, silver nanoparticles (AgNPs), PDDA, Cys-MoS_2, Ag^+-Cys-MoS_2, and PDDA-Ag^+-Cys-MoS_2. Reproduced with permission from Begum et al. [14]. Copyright 2017, American Chemical Society

liquid exfoliation method, comprised of cyro-pretreatment and poly(vinyl pyrrolidone) (PVP)-based exfoliation. On treatment with bacteria, the PVP-based Sb_2Se_3 nanoparticles showed high antibacterial effectivity. It was due to the production of reactive oxygen and serrated fragmentation of the membrane aided by the attrition between bacteria and Sb_2Se_3 NSs. In the presence of IR radiations, owing to effective IR absorption and simultaneous photothermal conversion, Sb_2Se_3 NSs efficiently inhibited and completely abated the Gram-negative Escherichia coli, Gram-positive methicillin-resistant Staphylococcus aureus, and wild bacteria from natural water samples in only 5 min. The corresponding concentration of Sb_2Se_3 NSs was approximately 300 μM. Therefore, Sb_2Se_3-based nanosheets seemed to possess promising antibacterial properties for multiple species of bacteria [97].

Brookite, rutile, and anatase are the three common phase structures of TiO_2, but anatase is the most widely utilized photocatalyst [98]. To further investigate its photocatalytic capability, Arnab et al. synthesized MoS_2-TiO_2 nanomaterial with appreciable antibacterial action [99]. Also, Ma et al. developed Fe_3O_4–TiO_2 nanosheets using lamellar reverse micelles and solvothermal process for effective antibacterial properties, which exhibited excellent antibacterial action for *E. Coli* under solar radiations [76].

The use of BPs as 2D antimicrobial nanoagents is also getting attention. For instance, an antimicrobial agent was synthesized by incorporating the titanium aminobenzenesulfanato complexes (Ti-SA4) onto BPs. A substantial loading capacity of 43% was attained owing to strong P–Ti coordination between Ti-SA4 and BPs. This resulted in enhanced stability of BPs against oxidation. The Ti-SA4@BPs showed strong antibacterial activity in comparison to BPs and Ti-SA4 alone. Ti-SA4@BPs is envisioned to have a huge potential in clinical applications [100].

5.5 Challenges and Future Perspectives

Although there has been a dramatic increase in antibacterial activities realized by 2D nanomaterials recently, but their practical applications are still in the infancy state. The major hurdles involved their competition with current commercial antimicrobial peptides, scalability, impurities and heterogeneity in synthesized 2D nanomaterials, precision to position 2D nanomaterials in bacteria, and risk of their toxicity. Therefore, efforts are required to be dedicated to the large-scale production of 2D nanomaterials with high purity and homogeneity, as well as good selectivity for bacterial cells while maintaining low cytotoxicity for human cells.

Preparing and purifying 2D nanomaterials of high quality is essential for ensuring consistent results, as low reproducibility might be a problem in large-scale production. It's also crucial to ensure that efficient processes for making 2D nanomaterials can be scaled up successfully. Attempts have been made to decrease the production costs and environmental burdens, i.e., it is critical to developing green and economical methods for producing high-quality 2D nanomaterials in large quantities. Further efforts are required for an in-depth understanding of the process for 2D nanomaterials synthesis, simplify preparation and purification methods, lower the amounts of precursors and catalysts, and working with milder conditions.

Before being employed in clinical therapy, 2D nanomaterials must thoroughly evaluate their biosafety. 2D nanomaterials have demonstrated a wide range of biomedical applications in addition to being used as an antibacterial agent. However, the clinical use of these nanomaterials is dependent on their biosafety. There is still room for improvement regarding the biosafety of 2D nanomaterials because there is little data on their dosage, incubation period, and surface modification-related biosafety.

Utilizing computer modelling is a new dimension in improving the antibacterial properties of 2D nanomaterials. Computers can already replicate acts that humans cannot execute as science and technology continue to improve quickly. They discovered, which is a useful addition to the experiment, that the size and orientation of 2D nanomaterials influence how cells and nanomaterials interact. We may combine theoretical calculation with computer modelling to further study the unresolved antibacterial mechanisms.

Despite the difficulties mentioned above, 2D nanomaterials still hold promise for antibacterial applications because of the widespread use of 2D nanomaterials in a variety of fields; their synthesis and purification technologies are becoming more advanced; the availability of biodegradable 2D nanomaterials can reduce their toxicity in antimicrobial applications with regard to both human health and the environment; other biomedical applications.

Acknowledgements We acknowledges School of Chemical and Materials Engineering (SCME), National University of Sciences and Technology, Sector H-12, Islamabad, Pakistan for their support.

Funding Information This work received no external funding.

References

1. R.E. Baker, A.S. Mahmud, I.F. Miller, M. Rajeev, F. Rasambainarivo, B.L. Rice, S. Takahashi, A.J. Tatem, C.E. Wagner, L.-F. Wang, A. Wesolowski, C.J.E. Metcalf, Infectious disease in an era of global change. Nat. Rev. Microbiol. **20**, 193–205 (2022)
2. L. Giraud, A. Tourrette, E. Flahaut, Carbon nanomaterials-based polymer-matrix nanocomposites for antimicrobial applications: a review. Carbon **182**, 463–483 (2021)
3. M.I. Hutchings, A.W. Truman, B. Wilkinson, Antibiotics: past, present and future. Curr. Opin. Microbiol. **51**, 72–80 (2019)
4. M.E. Olson, H. Ceri, D.W. Morck, A.G. Buret, R.R. Read, Biofilm bacteria: formation and comparative susceptibility to antibiotics. Can. J. Vet. Res. **66**, 86–92 (2002)
5. V. Nizet, Stopping superbugs, maintaining the microbiota. Sci. Transl. Med. **7**, 295ed8–295ed8 (2015)
6. C.L. Ventola, The antibiotic resistance crisis: part 1: causes and threats. P t **40**, 277–283 (2015)
7. M.E. de Kraker, A.J. Stewardson, S. Harbarth, Will 10 million people die a year due to antimicrobial resistance by 2050? PLoS Med. **13**, e1002184 (2016)
8. E. Sánchez-López, D. Gomes, G. Esteruelas, L. Bonilla, A.L. Lopez-Machado, R. Galindo, A. Cano, M. Espina, M. Ettcheto, A. Camins, A.M. Silva, A. Durazzo, A. Santini, M.L. Garcia, E.B. Souto, Metal-based nanoparticles as antimicrobial agents: an overview. Nanomaterials (Basel) **10** (2020)
9. S. Sharmin, M.M. Rahaman, C. Sarkar, O. Atolani, M.T. Islam, O.S. Adeyemi, Nanoparticles as antimicrobial and antiviral agents: a literature-based perspective study. Heliyon **7**, e06456 (2021)
10. M.A. Gatoo, S. Naseem, M.Y. Arfat, A.M. Dar, K. Qasim, S. Zubair, Physicochemical properties of nanomaterials: implication in associated toxic manifestations. Biomed. Res. Int. **2014**, 498420 (2014)
11. B. Jia, X. Du, W. Wang, Y. Qu, X. Liu, M. Zhao, W. Li, Y.-Q. Li, Nanophysical antimicrobial strategies: a rational deployment of nanomaterials and physical stimulations in combating bacterial infections. Adv. Sci. **9**, 2105252 (2022)
12. L. Wang, C. Hu, L. Shao, The antimicrobial activity of nanoparticles: present situation and prospects for the future. Int J Nanomed. **12**, 1227–1249 (2017)
13. T. Seifi, A.R. Kamali, Anti-pathogenic activity of graphene nanomaterials: a review. Colloids Surf., B **199**, 111509 (2021)
14. S. Begum, A. Pramanik, D. Davis, S. Patibandla, K. Gates, Y. Gao, P.C. Ray, 2D and heterostructure nanomaterial based strategies for combating drug-resistant bacteria. ACS Omega **5**, 3116–3130 (2020)
15. B. Anasori, Y. Gogotsi, *Introduction to 2D Transition Metal Carbides and Nitrides (MXenes)*. 2D Metal Carbides and Nitrides (MXenes) (Springer, 2019)
16. M. Sathishkumar, S. Geethalakshmi, M. Saroja, M. Venkatachalam, P. Gowthaman, Chapter Three—Antimicrobial activities of biosynthesized nanomaterials, in *Comprehensive Analytical Chemistry*, ed. S.K. Verma, A.K. Das (Elsevier, 2021)

17. Q. Xin, H. Shah, A. Nawaz, W. Xie, M.Z. Akram, A. Batool, L. Tian, S.U. Jan, R. Boddula, B. Guo, Q. Liu, J.R. Gong, Antibacterial carbon-based nanomaterials. Adv. Mater. **31**, 1804838 (2019)
18. L. Mei, S. Zhu, W. Yin, C. Chen, G. Nie, Z. Gu, Y. Zhao, Two-dimensional nanomaterials beyond graphene for antibacterial applications: current progress and future perspectives. Theranostics **10**, 757 (2020)
19. Y. Li, X. Liu, L. Tan, Z. Cui, D. Jing, X. Yang, Y. Liang, Z. Li, S. Zhu, Y. Zheng, Eradicating multidrug-resistant bacteria rapidly using a multi functional g-C_3N_4@ Bi_2S_3 nanorod heterojunction with or without antibiotics. Adv. Func. Mater. **29**, 1900946 (2019)
20. Y. Wang, Y. Jin, W. Chen, J. Wang, H. Chen, L. Sun, X. Li, J. Ji, Q. Yu, L. Shen, Construction of nanomaterials with targeting phototherapy properties to inhibit resistant bacteria and biofilm infections. Chem. Eng. J. **358**, 74–90 (2019)
21. A. Gupta, S. Mumtaz, C.-H. Li, I. Hussain, V.M. Rotello, Combatting antibiotic-resistant bacteria using nanomaterials. Chem. Soc. Rev. **48**, 415–427 (2019)
22. X.-L. Hu, Y. Shang, K.-C. Yan, A.C. Sedgwick, H.-Q. Gan, G.-R. Chen, X.-P. He, T.D. James, D. Chen, Low-dimensional nanomaterials for antibacterial applications. J. Mater. Chem. B **9**, 3640–3661 (2021)
23. N. Beyth, Y. Houri-Haddad, A. Domb, W. Khan, R. Hazan, Alternative antimicrobial approach: nano-antimicrobial materials. Evid.-Based Complement. Alternat. Med. (2015)
24. L. Shi, J. Chen, L. Teng, L. Wang, G. Zhu, S. Liu, Z. Luo, X. Shi, Y. Wang, L. Ren, The antibacterial applications of graphene and its derivatives. Small **12**, 4165–4184 (2016)
25. L.-S. Wang, A. Gupta, V.M. Rotello, Nanomaterials for the treatment of bacterial biofilms. ACS Infect. Dis. **2**, 3–4 (2016)
26. Y. Wang, Y. Yang, Y. Shi, H. Song, C. Yu, Antibiotic-free antibacterial strategies enabled by nanomaterials: progress and perspectives. Adv. Mater. **32**, 1904106 (2020)
27. K. Lohner, New strategies for novel antibiotics: peptides targeting bacterial cell membranes. Gen. Physiol. Biophys. **28**, 105–116 (2009)
28. S. Akira, S. Uematsu, O. Takeuchi, Pathogen recognition and innate immunity. Cell **124**, 783–801 (2006)
29. E.P. Ivanova, J. Hasan, H.K. Webb, G. Gervinskas, S. Juodkazis, V.K. Truong, A.H. Wu, R.N. Lamb, V.A. Baulin, G.S. Watson, Bactericidal activity of black silicon. Nat. Commun. **4**, 1–7 (2013)
30. X. Lu, X. Feng, J.R. Werber, C. Chu, I. Zucker, J.-H. Kim, C.O. Osuji, M. Elimelech, Enhanced antibacterial activity through the controlled alignment of graphene oxide nanosheets. Proc. Natl. Acad. Sci. **114**, E9793–E9801 (2017)
31. K. Krishnamoorthy, M. Veerapandian, K. Yun, S.J. Kim, New function of molybdenum trioxide nanoplates: toxicity towards pathogenic bacteria through membrane stress. Colloids Surf., B **112**, 521–524 (2013)
32. R. Kurapati, K. Kostarelos, M. Prato, A. Bianco, Biomedical uses for 2D materials beyond graphene: current advances and challenges ahead. Adv. Mater. **28**, 6052–6074 (2016)
33. P. Singh, P. Shandilya, P. Raizada, A. Sudhaik, A. Rahmani-Sani, A. Hosseini-Bandegharaei, Review on various strategies for enhancing photocatalytic activity of graphene based nanocomposites for water purification. Arab. J. Chem. **13**, 3498–3520 (2020)
34. J.A. Lemire, J.J. Harrison, R.J. Turner, Antimicrobial activity of metals: mechanisms, molecular targets and applications. Nat. Rev. Microbiol. **11**, 371–384 (2013)
35. M.Y. Memar, R. Ghotaslou, M. Samiei, K. Adibkia, Antimicrobial use of reactive oxygen therapy: current insights. Infect. Drug Resist. **11**, 567 (2018)
36. Z. Shaw, S. Kuriakose, S. Cheeseman, M.D. Dickey, J. Genzer, A.J. Christofferson, R.J. Crawford, C.F. McConville, J. Chapman, V.K. Truong, Antipathogenic properties and applications of low-dimensional materials. Nat. Commun. **12**, 1–19 (2021)
37. J. Chen, X. Wang, H. Han, A new function of graphene oxide emerges: inactivating phytopathogenic bacterium Xanthomonas oryzae pv. Oryzae. J. Nanoparticle Res. **15**, 1–14 (2013)

38. S. Gurunathan, J.W. Han, A.A. Dayem, V. Eppakayala, J.-H. Kim, Oxidative stress-mediated antibacterial activity of graphene oxide and reduced graphene oxide in Pseudomonas aeruginosa. Int. J. Nanomed. **7**, 5901 (2012)
39. F. Perreault, A.F. De Faria, S. Nejati, M. Elimelech, Antimicrobial properties of graphene oxide nanosheets: why size matters. ACS Nano **9**, 7226–7236 (2015)
40. D.K. Ji, Y. Zhang, Y. Zang, J. Li, G.R. Chen, X.P. He, H. Tian, Targeted intracellular production of reactive oxygen species by a 2D molybdenum disulfide glycosheet. Adv. Mater. **28**, 9356–9363 (2016)
41. G.S. Bang, S. Cho, N. Son, G.W. Shim, B.-K. Cho, S.-Y. Choi, DNA-assisted exfoliation of tungsten dichalcogenides and their antibacterial effect. ACS Appl. Mater. Interf. **8**, 1943–1950 (2016)
42. Z. Yuan, B. Tao, Y. He, J. Liu, C. Lin, X. Shen, Y. Ding, Y. Yu, C. Mu, P. Liu, Biocompatible MoS_2/PDA-RGD coating on titanium implant with antibacterial property via intrinsic ROS-independent oxidative stress and NIR irradiation. Biomaterials **217**, 119290 (2019)
43. S. Pandit, S. Karunakaran, S.K. Boda, B. Basu, M. De, High antibacterial activity of functionalized chemically exfoliated MoS_2. ACS Appl. Mater. Interf. **8**, 31567–31573 (2016)
44. D. Jaque, L. Martínez Maestro, B. del Rosal, P. Haro-Gonzalez, A. Benayas, J.L. Plaza, E. Martín Rodríguez, J. García Solém, Nanoparticles for photothermal therapies. Nanoscale **6**, 9494–9530 (2014)
45. W.-Y. Pan, C.-C. Huang, T.-T. Lin, H.-Y. Hu, W.-C. Lin, M.-J. Li, H.-W. Sung, Synergistic antibacterial effects of localized heat and oxidative stress caused by hydroxyl radicals mediated by graphene/iron oxide-based nanocomposites. Nanomed.: Nanotech. Biol. Med. **12**, 431–438 (2016)
46. Y. Liu, X. Zeng, X. Hu, J. Hu, X. Zhang, Two-dimensional nanomaterials for photocatalytic water disinfection: recent progress and future challenges. J. Chem. Technol. Biotechnol. **94**, 22–37 (2019)
47. C. Liu, D. Kong, P.-C. Hsu, H. Yuan, H.-W. Lee, Y. Liu, H. Wang, S. Wang, K. Yan, D. Lin, Rapid water disinfection using vertically aligned MoS 2 nanofilms and visible light. Nat. Nanotechnol. **11**, 1098–1104 (2016)
48. L. Sun, T. Du, C. Hu, J. Chen, J. Lu, Z. Lu, H. Han, Antibacterial activity of graphene oxide/g-C_3N_4 composite through photocatalytic disinfection under visible light. ACS Sustain. Chem. Eng. **5**, 8693–8701 (2017)
49. D. Wu, B. Wang, W. Wang, T. An, G. Li, T.W. Ng, H.Y. Yip, C. Xiong, H.K. Lee, P.K. Wong, Visible-light-driven BiOBr nanosheets for highly facet-dependent photocatalytic inactivation of Escherichia coli. J. Mater. Chem. A **3**, 15148–15155 (2015)
50. Z. Xu, Y. Gao, S. Meng, B. Yang, L. Pang, C. Wang, T. Liu, Mechanism and in vivo evaluation: photodynamic antibacterial chemotherapy of lysine-porphyrin conjugate. Front. Microbiol. **7**, 242 (2016)
51. B.Z. Ristic, M.M. Milenkovic, I.R. Dakic, B.M. Todorovic-Markovic, M.S. Milosavljevic, M.D. Budimir, V.G. Paunovic, M.D. Dramicanin, Z.M. Markovic, V.S. Trajkovic, Photodynamic antibacterial effect of graphene quantum dots. Biomaterials **35**, 4428–4435 (2014)
52. W. Liu, Y. Feng, H. Tang, H. Yuan, S. He, S. Miao, Immobilization of silver nanocrystals on carbon nanotubes using ultra-thin molybdenum sulfide sacrificial layers for antibacterial photocatalysis in visible light. Carbon **96**, 303–310 (2016)
53. Z. Feng, X. Liu, L. Tan, Z. Cui, X. Yang, Z. Li, Y. Zheng, K.W.K. Yeung, S. Wu, Electrophoretic deposited stable chitosan@ MoS_2 coating with rapid in situ bacteria-killing ability under dual-light irradiation. Small **14**, 1704347 (2018)
54. Y. Li, X. Liu, L. Tan, Z. Cui, X. Yang, Y. Zheng, K.W.K. Yeung, P.K. Chu, S. Wu, Rapid sterilization and accelerated wound healing using Zn^{2+} and graphene oxide modified g-C_3N_4 under dual light irradiation. Adv. Func. Mater. **28**, 1800299 (2018)
55. X. Zhang, C. Zhang, Y. Yang, X. Huang, R. Hang, X. Yao, Light-assisted rapid sterilization by a hydrogel incorporated with Ag_3PO_4/MoS_2 composites for efficient wound disinfection. Chem. Eng. J. **374**, 596–604 (2019)

56. Y. Lin, D. Han, Y. Li, L. Tan, X. Liu, Z. Cui, X. Yang, Z. Li, Y. Liang, S. Zhu, Ag_2S@ WS2 heterostructure for rapid bacteria-killing using near-infrared light. ACS Sustain. Chem. Eng. **7**, 14982–14990 (2019)
57. B. Li, L. Tan, X. Liu, Z. Li, Z. Cui, Y. Liang, S. Zhu, X. Yang, K.W.K. Yeung, S. Wu, Superimposed surface plasma resonance effect enhanced the near-infrared photocatalytic activity of Au@ Bi_2WO_6 coating for rapid bacterial killing. J. Hazard. Mater. **380**, 120818 (2019)
58. W. Yin, J. Yu, F. Lv, L. Yan, L.R. Zheng, Z. Gu, Y. Zhao, Functionalized nano-MoS_2 with peroxidase catalytic and near-infrared photothermal activities for safe and synergetic wound antibacterial applications. ACS Nano **10**, 11000–11011 (2016)
59. C. Zhang, D.-F. Hu, J.-W. Xu, M.-Q. Ma, H. Xing, K. Yao, J. Ji, Z.-K. Xu, Polyphenol-assisted exfoliation of transition metal dichalcogenides into nanosheets as photothermal nanocarriers for enhanced antibiofilm activity. ACS Nano **12**, 12347–12356 (2018)
60. M. Li, L. Li, K. Su, X. Liu, T. Zhang, Y. Liang, D. Jing, X. Yang, D. Zheng, Z. Cui, Highly effective and noninvasive near-infrared eradication of a Staphylococcus aureus biofilm on implants by a photoresponsive coating within 20 min. Adv. Sci. **6**, 1900599 (2019)
61. A.M. Jastrzębska, E. Karwowska, T. Wojciechowski, W. Ziemkowska, A. Rozmysłowska, L. Chlubny, A. Olszyna, The atomic structure of Ti_2C and Ti_3C_2 MXenes is responsible for their antibacterial activity toward E. coli bacteria. J. Mater. Eng. Perform. **28**, 1272–1277 (2019)
62. B. Li, Y. Luo, Y. Zheng, X. Liu, L. Tan, S. Wu, Two-dimensional antibacterial materials. Progr. Mater. Sci. 100976 (2019)
63. L. Wang, C. Hu, L. Shao, The antimicrobial activity of nanoparticles: present situation and prospects for the future. Int. J. Nanomed. **12**, 1227 (2017)
64. K. Rasool, R.P. Pandey, P.A. Rasheed, S. Buczek, Y. Gogotsi, K.A. Mahmoud, Water treatment and environmental remediation applications of two-dimensional metal carbides (MXenes). Mater. Today **30**, 80–102 (2019)
65. P. Kumar, P. Huo, R. Zhang, B. Liu, Antibacterial properties of graphene-based nanomaterials. Nanomaterials **9**, 737 (2019)
66. N. Dwivedi, C. Dhand, P. Kumar, A. Srivastava, Emergent 2D materials for combating infectious diseases: the potential of MXenes and MXene–graphene composites to fight against pandemics. Mater. Adv. **2**, 2892–2905 (2021)
67. S. Yaragalla, K.B. Bhavitha, A. Athanassiou, A review on graphene based materials and their antimicrobial properties. Coatings **11**, 1197 (2021)
68. W.-C. Hou, P.-L. Lee, Y.-C. Chou, Y.-S. Wang, Antibacterial property of graphene oxide: the role of phototransformation. Environ. Sci. Nano **4**, 647–657 (2017)
69. K. Rasool, K.A. Mahmoud, D.J. Johnson, M. Helal, G.R. Berdiyorov, Y. Gogotsi, Efficient antibacterial membrane based on two-dimensional Ti_3C_2Tx (MXene) nanosheets. Sci. Rep. **7**, 1598 (2017)
70. M.L. Cohen, Structural, electronic and optical properties of carbon nitride. Mater. Sci. Eng., A **209**, 1–4 (1996)
71. J. Zhu, P. Xiao, H. Li, S.A. Carabineiro, Graphitic carbon nitride: synthesis, properties, and applications in catalysis. ACS Appl. Mater. Interfaces. **6**, 16449–16465 (2014)
72. L. Maya, D.R. Cole, E.W. Hagaman, Carbon–nitrogen pyrolyzates: attempted preparation of carbon nitride. J. Am. Ceram. Soc. **74**, 1686–1688 (1991)
73. D.C. Nesting, J.V. Badding, High-pressure synthesis of sp2-bonded carbon nitrides. Chem. Mater. **8**, 1535–1539 (1996)
74. S. Kang, Y. Fang, Y. Huang, L.-F. Cui, Y. Wang, H. Qin, Y. Zhang, X. Li, Y. Wang, Critical influence of g-C_3N_4 self-assembly coating on the photocatalytic activity and stability of Ag/AgCl microspheres under visible light. Appl. Catal. B **168**, 472–482 (2015)
75. S. Patnaik, D.P. Sahoo, K. Parida, An overview on Ag modified g-C_3N_4 based nanostructured materials for energy and environmental applications. Renew. Sustain. Energy Rev. **82**, 1297–1312 (2018)
76. S. Ma, S. Zhan, Y. Jia, Q. Zhou, Superior antibacterial activity of Fe_3O_4-TiO_2 nanosheets under solar light. ACS Appl. Mater. Interf. **7**, 21875–21883 (2015)

77. K. Yan, C. Mu, L. Meng, Z. Fei, P.J. Dyson, Recent advances in graphite carbon nitride-based nanocomposites: structure, antibacterial properties and synergies. Nanoscale Adv. **3**, 3708–3729 (2021)
78. J. Dai, J. Song, Y. Qiu, J. Wei, Z. Hong, L. Li, H. Yang, Gold nanoparticle-decorated g-C3N4 nanosheets for controlled generation of reactive oxygen species upon 670 nm laser illumination. ACS Appl. Mater. Interf. **11**, 10589–10596 (2019)
79. M. Naguib, V.N. Mochalin, M.W. Barsoum, Y. Gogotsi, 25th anniversary article: MXenes: a new family of two-dimensional materials. Adv. Mater. **26**, 992–1005 (2014)
80. A. Shahzad, M. Nawaz, M. Moztahida, J. Jang, K. Tahir, J. Kim, Y. Lim, V.S. Vassiliadis, S.H. Woo, D.S. Lee, Ti_3C_2Tx MXene core-shell spheres for ultrahigh removal of mercuric ions. Chem. Eng. J. **368**, 400–408 (2019)
81. A. Shahzad, K. Rasool, M. Nawaz, W. Miran, J. Jang, M. Moztahida, K.A. Mahmoud, D.S. Lee, Heterostructural TiO_2/Ti_3C_2Tx (MXene) for photocatalytic degradation of antiepileptic drug carbamazepine. Chem. Eng. J. **349**, 748–755 (2018)
82. X. Zhan, C. Si, J. Zhou, Z. Sun, MXene and MXene-based composites: synthesis, properties and environment-related applications. Nanoscale Horizons **5**, 235–258 (2020)
83. K. Rasool, M. Helal, A. Ali, C.E. Ren, Y. Gogotsi, K.A. Mahmoud, Antibacterial activity of Ti_3C_2T x MXene. ACS Nano **10**, 3674–3684 (2016)
84. A. Arabi Shamsabadi, G.M. Sharifian, B. Anasori, M. Soroush, Antimicrobial mode-of-action of colloidal Ti_3C_2Tx MXene nanosheets. ACS Sustain. Chem. Eng. **6**, 16586–16596
85. G.P. Lim, C.F. Soon, M. Morsin, M.K. Ahmad, N. Nayan, K.S. Tee, Synthesis, characterization and antifungal property of Ti_3C_2Tx MXene nanosheets. Ceram. Int. **46**, 20306–20312 (2020)
86. R. Pandey, K. Rasool, E.M. Vinod, B. Aïssa, Y. Gogotsi, K. Mahmoud, Ultrahigh-flux and fouling-resistant membrane based on layered silver/MXene(Ti_3C_2Tx) nanosheets. J. Mater. Chem. A **6** (2018)
87. H. Moon, J. Bang, S. Hong, G. Kim, J.W. Roh, J. Kim, W. Lee, Strong thermopower enhancement and tunable power factor via semimetal to semiconductor transition in a transition-metal dichalcogenide. ACS Nano **13**, 13317–13324 (2019)
88. X.Y. Wong, A. Sena-Torralba, R. Alvarez-Diduk, K. Muthoosamy, A. Merkoci, Nanomaterials for nanotheranostics: tuning their properties according to disease needs. ACS Nano **14**, 2585–2627 (2020)
89. S. Begum, A. Pramanik, K. Gates, Y. Gao, P.C. Ray, Antimicrobial peptide-conjugated MoS_2-based nanoplatform for multimodal synergistic inactivation of superbugs. ACS Appl. Bio Mater. **2**, 769–776 (2019)
90. F. Cao, E. Ju, Y. Zhang, Z. Wang, C. Liu, W. Li, Y. Huang, K. Dong, J. Ren, X. Qu, An efficient and benign antimicrobial depot based on silver-infused MoS_2. ACS Nano **11**, 4651–4659 (2017)
91. Y. Feng, Q. Chen, Q. Yin, G. Pan, Z. Tu, L. Liu, Reduced graphene oxide functionalized with gold nanostar nanocomposites for synergistically killing bacteria through intrinsic antimicrobial activity and photothermal ablation. ACS Appl. Bio Mater. **2**, 747–756 (2019)
92. S. Karunakaran, S. Pandit, B. Basu, M. De, Simultaneous exfoliation and functionalization of 2H-MoS_2 by thiolated surfactants: applications in enhanced antibacterial activity. J. Am. Chem. Soc. **140**, 12634–12644 (2018)
93. S. Roy, A. Mondal, V. Yadav, A. Sarkar, R. Banerjee, P. Sanpui, A. Jaiswal, Mechanistic insight into the antibacterial activity of chitosan exfoliated MoS_2 nanosheets: membrane damage, metabolic inactivation and oxidative stress. ACS Appl. Bio Mater. **2**, 2738–2755 (2019)
94. H. Zhao, C. Zhang, Y. Wang, W. Chen, P.J.J. Alvarez, Self-damaging aerobic reduction of graphene oxide by Escherichia coli: role of GO-mediated extracellular superoxide formation. Environ. Sci. Technol. **52**, 12783–12791 (2018)
95. X. Yang, J. Li, T. Liang, C. Ma, Y. Zhang, H. Chen, N. Hanagata, H. Su, M. Xu, Antibacterial activity of two-dimensional MoS_2 sheets. Nanoscale **6**, 10126–10133 (2014)
96. W. Zhang, Z. Mou, Y. Wang, Y. Chen, E. Yang, F. Guo, D. Sun, W. Wang, Molybdenum disulfide nanosheets loaded with Chitosan and silver nanoparticles effective antifungal activities: in vitro and in vivo. Mater Sci Eng C Mater Biol Appl **97**, 486–497 (2019)

97. Z. Miao, L. Fan, X. Xie, Y. Ma, J. Xue, T. He, Z. Zha, Liquid exfoliation of atomically thin antimony selenide as an efficient two-dimensional antibacterial nanoagent. ACS Appl. Mater. Interf. **11**, 26664–26673 (2019)
98. B. Wang, M.K.H. Leung, X.-Y. Lu, S.-Y. Chen, Synthesis and photocatalytic activity of boron and fluorine codoped TiO_2 nanosheets with reactive facets. Appl. Energy **112**, 1190–1197 (2013)
99. A. Pal, T.K. Jana, T. Roy, A. Pradhan, R. Maiti, S.M. Choudhury, K. Chatterjee, MoS_2-TiO_2 nanocomposite with excellent adsorption performance and high antibacterial activity. ChemistrySelect **3**, 81–90 (2018)
100. Z. Li, L. Wu, H. Wang, W. Zhou, H. Liu, H. Cui, P. Li, P.K. Chu, X.-F. Yu, Synergistic antibacterial activity of black phosphorus nanosheets modified with titanium aminobenzenesulfanato complexes. ACS Appl. Nano Mater. **2**, 1202–1209 (2019)
101. K. Rasool, K.A. Mahmoud, D.J. Johnson, M. Helal, G.R. Berdiyorov, Y. Gogotsi, Efficient antibacterial membrane based on two-dimensional Ti_3C_2Tx (MXene) nanosheets. Sci. Rep. **7**, 1–11 (2017)

Chapter 6
Graphene-Based Photocatalysts for the Elimination of Pollutants in Water

Ajay Kumar, Irwing Ramirez, Priyanka Chaudhary, Kuldeep Kumar, Munish Sharma, and Rohit Rana

Abstract Water is an essential part of living things as the life cycle, and water cycle are inextricably linked. The development of industrialization caused the water to become hazardous by emitting dyes, pollutants, organic wastes and other harmful chemicals into the water bodies. It's a big challenge to treat wastewater for human health and to preserve the ecosystem. In recent years, graphene-based photocatalytic materials have gained much attention to eliminate liquid pollutants. This is due to the tremendous optical, electrical and physicochemical properties of this unique two-dimensional material. On integration with other semiconducting, metallic or polymeric materials, graphene remarkably boosts the photocatalytic activity of materials toward contaminants destruction. The high charge carrier mobility, high surface area and excellent mechanical strength of graphene-based nanocomposites make them suitable for photocatalytic applications. This book chapter will focus on the recent significant advances in developing graphene-based photocatalytic materials. The principle of photocatalysis, the basic properties of graphene and the mechanism of how the photocatalytic efficiency against the removal of the liquid pollutant can be enhanced when coupled with graphene has been discussed in this book chapter. Furthermore, current challenges and future recommendations for developing graphene-based photocatalysts are also discussed.

A. Kumar (✉) · I. Ramirez
School of Engineering and Innovation, The Open University, Milton Keynes MK7 6AA, UK
e-mail: ajay.kumar@open.ac.uk

P. Chaudhary
School of Basic Sciences, Advanced Materials Research Center, Indian Institute of Technology Mandi, Kamand, Mandi, Himachal Pradesh 175075, India

K. Kumar · M. Sharma
Department of Physics, School of Basic and Applied Sciences, Maharaja Agrasen University, Solan, Himachal Pradesh 174103, India

R. Rana
Natural Product Chemistry and Process Development Division, CSIR-Institute of Himalayan Bioresource Technology, Palampur, Himachal Pradesh 176061, India

N. Kumar et al. (eds.), *Two-Dimensional Materials for Environmental Applications*, Springer Series in Materials Science 332, https://doi.org/10.1007/978-3-031-28756-5_6

6.1 Introduction

The demand for fresh water has increased manyfold in the last decade due to increased world's population. According to the report published by United Nations in 2018 on World Water Development, there will be a one-third increase in the demand for drinkable water by 2050 [1]. As clean water is primarily required for various industries and pharmaceutical companies, and households. After use, the release of the water into the environment without treatment adversely affects aquatic as well as human life. Thus, it is highly desirable to treat wastewater using a suitable technique so that it can be reused which can also solve the problem of water scarcity.

To remove the impurities and harmful water pollutants from the waste water, various techniques have been developed with the technological advancements [2]. The waste water can be treated in three effective ways as follows.

(i) Biological treatment,
(ii) Physical treatment, and
(iii) chemical treatment.

Biological treatment of water involves the use of various microbes to degrade the biodegradable water pollutants through biological process. This type of treatment degrade the water pollutant and produce the water, CO_2 and other chemicals. The challenge in this type of water treatment lies in requirement of large area, slow treatment process and bring water to its natural appearance of clean water. Furthermore, physical treatment of water involves physically filter the water using filtration method, sedimentation etc. which again put forward the challenge of slow treating process. It is appreciable in the physical treatment method that no chemicals are used while treating the waste water with physical methods. An another way is chemical method which involves treatment of water through chemical reaction. Oxidation, precipitation, and adsorption are examples of chemical wastewater treatment.

The photocatalytic treatment of wastewater using nanomaterials has gained a lot of attention due to its potential to solve environmental issues, the biggest challenge of the twenty-first century [3, 4]. Many nanomaterials show inherent confinement quantum effects, high surface-to-volume ratio, and large surface area which are favourable to increase the photocatalytic reaction.

Various fundamental issues related to photocatalysts that still need to be resolved before implementing them for large-scale industrial applications. For example, high cost, low efficiency, mismatch of the band gap, recombination of charge carriers and low absorption of solar energy [5]. On absorption of light photons having energy greater than the band gap of the material, the electrons and holes get separated and reside in the conduction band (CB) and valence band (VB), respectively (Fig. 6.1). These separated charge carriers (electrons and holes) migrate towards the surface-active sites wherein they combine with the adsorbed water molecules or moisture to generate highly reactive species such as $OH^{\cdot}$, $O_2^{-\cdot}$, $OOH^{\cdot}$, etc. Further, these highly reactive species react with the pollutant species to mineralize them into CO_2 and water. One could improve the photocatalytic performance of the nanomaterials by addressing the following three key points [6, 7];

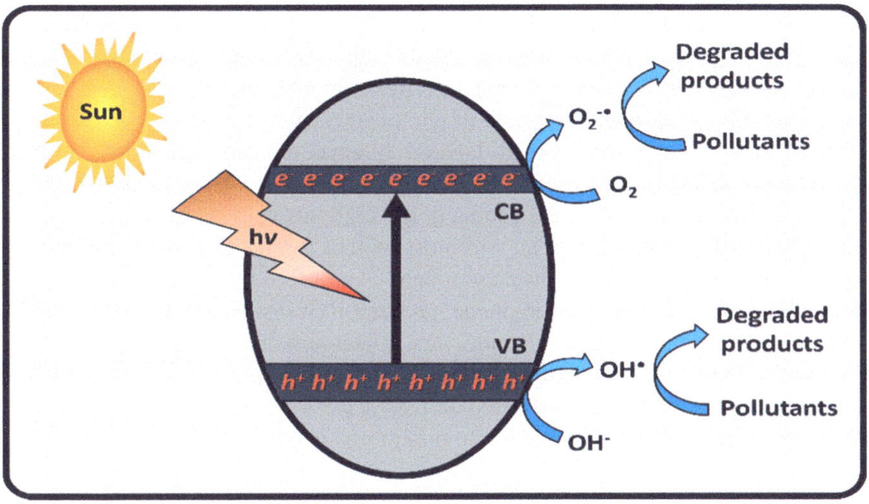

Fig. 6.1 Photoexcitation of the semiconducting materials and subsequent oxidation and reduction reactions for generation of radical species

(1) Absorption of the broad solar spectrum,
(2) Effective separation of the charge carriers, and
(3) Facilitated movement of the separated charge carriers towards the surface-active sites.

Various attempts have been made to improve the efficiency of the photocatalysts such as doping with cationic or anionic species [8], coupling with the metal nanoparticles (NPs) [9] or other semiconductor photocatalysts [10], exfoliation [11] and surface modifications to improve the porosity, surface area, and active facets of the photocatalysts [12, 13]. Despite such modifications at the lab scale, the commercialization of such photocatalytic subsystems for wastewater treatment needs to achieve a certain criterion in terms of cost, efficiency and environmental impact is yet to be realized [14]. Among various two-dimensional support materials, graphene-based materials are the most reliable due to their low cost, high stability and excellent efficiency.

6.2 Basic Properties of Graphene-Based Materials

The carbonaceous materials exhibit unique structural, electrical, and redox properties and tend to add an attractive feature to photocatalysis. The carbonaceous and graphene-based photocatalytic materials have attracted immense attention due to the unique flexible sp^2 hybridized carbon network [15, 16]. Graphene-based materials show promising applications in various fields, such as energy storage [17], biosensors [18], photocatalysis [19], antibacterial and antimicrobial activity [20] etc.

The graphene-based nanomaterials showed excellent electron mobility and electrical conductivity, high optical transparency, and good chemical and thermal stability. Due to the promising electronic and optical properties of graphene-based nanocomposites they are highly demanding in the field of photocatalysis.

Graphene is a two-dimensional layered honeycomb-like structural material formed by sp^2 hybridized carbon atoms. Graphene is a fascinating material with a large surface area, good electrical conductivity, high intrinsic electron mobility and optical transmittance [21, 22]. The two most widely studied derivates of graphene are graphene oxide (GO) and reduced graphene oxide (RGO) [23]. Furthermore, the modified graphene derivatives (graphene, GO and RGO) with other materials such as metals or non-metals or semiconductors, metal–organic frameworks (MOF's) aerogels, etc., offer an outstanding platform for photocatalytic applications. It was found that the modified graphene-based derivative exhibits better photocatalytic and adsorption properties as compared to the pristine graphene [24].

6.3 Synthesis and Properties of Graphene-Based Composites

There are several chemical and physical methods for designing and developing graphene-based composites. Although the chemical vapour deposition and physical exfoliation methods can produce high-quality graphene materials, they show poor dispersion and contact angle due to insufficient functional groups making them less suited for photocatalytic applications [25, 26]. The chemical oxidation of graphene modifies its surface with many oxygen functionalities (hydroxyl, epoxides, carboxyl groups) responsible for increased interlayer distance, thereby allowing better solubility and adsorption of the organic pollutants [26]. In GO, the hydroxyls and epoxides groups are extensively occupied on the basal planes, while the aldehyde, ketone and carboxylic groups are mainly located on the edge of GO [27]. The oxygenated functional groups of GO affect its mechanical, electronic, and electrochemical properties and improve its dispersion in solvents. However, the high dispersion of GO in both aqueous and non-aqueous solvents is challenging during its recovery after the treatment process. Also, the presence of a large number of oxygen functionalities in the GO dimmish its electrical conductivity, which is inappropriate for photocatalytic activity.

To overcome this drawback of GO, further reduction of the oxygen functionalities was made by various techniques. The RGO exhibits similar properties as pristine graphene with restored hexagonal honeycomb-like structure and electrical conductivity due to abundant conjugated π-bonds that were not there in GO [28]. However, with the reduction of oxygen functionalities in RGO, there will be a reduction in the repulsive forces between the sheets, causing lower interlayer distance [29]. The aggregations or restacking due to strong π–π interactions could reduce some potential active sites of RGO nanosheets. However, to prevent the aggregation of RGO nanosheets various strategies have been used such as surface modification, composite formation, in-situ reduction etc.

Muiva and coworkers prepared multilayered graphene/zinc oxide (MLG/ZnO) nanocomposite using ex-situ casting of ZnO and MLG nanopowders [30]. The MLG wasobtained via a green route from cellulose extracted from corn husk. The extracted cellulose was pyrolyzed at 800 °C for 2 h under a nitrogen atmosphere (Fig. 6.2). The carbonized corn husk (CH-C) was mixed homogeneously with KOH and pyrolyzed again in a nitrogen atmosphere to activate the carbon material (CH-CA). Finally, the activated material was treated with H_2SO_4 to etch out intercalated K^+ ions resulting in highly porous and activated MLG nanosheets. Furthermore, the prepared nanocomposite was used for the photocatalytic degradation of congo red (CR) and rhodamine B (RhB) under natural sunlight irradiations.

Recently, our group has synthesized a ternary nanocomposite of 2D-2D GCN/RGO heterojunction decorated with Au nanostars (NST) using a hydrothermal synthesis route (Fig. 6.3) [31]. A chemical reduction method was used for the synthesis of the Au NST. The mixture of the Au NST, protonated GCN and GO with negative surface charge was treated hydrothermally to synthesize the composite with strong electrostatic interactions. Further, a series of nanocomposites were synthesized by varying the weight ratio as 1, 2, 3, and 5 wt%. It was found that the

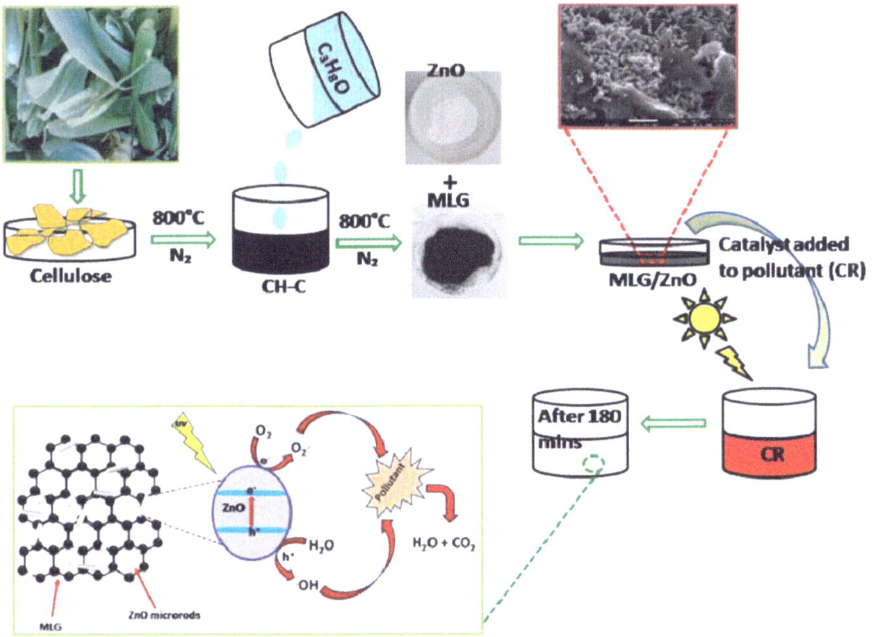

Fig. 6.2 Schematic presentation for the synthesis of multilayered graphene/ZnO nanocomposite and its photocatalytic activity for the degradation of organic pollutants. Reproduced with permission from Ref. [30] Copyright 2022, Elsevier Science Ltd.

optimized catalyst ACG2 showed the highest efficiency for the degradation of the methylene blue (MB) and tetracycline (TC) dyes. Similarly, other ternary nanocomposites synthesized using the hydrothermal synthesis route have been reported in the literature for the degradation of organic pollutants such as $RGO/ZnO/MoS_2$ and $CNT/Zno/MoS_2$ [32], Ag_3VO_4/Cu-MOF/RGO [33], $AgFeO_2$-graphene/$Cu_2(BTC)_3$ MOF [34], CdS/RGO/BiOI [35], and GCN/graphene/TiO_2 [36].

Furthermore, another synthesis strategy (chemical reflux method) was adopted for the fabrication of RGO nanocomposite with Fe_3O_4@polypyrrole [37]. The free radical polymerization of the pyrrole monomer using ammonium persulfate as an initiator in an acidic solution of GO, resulted in the formation of polypyrrole/RGO (PPy/RGO) nanohybrid. Further, the chemical reflux method was used for the coating of Fe_3O_4 nanorods onto the PPy/RGO nanohybrid as shown in Fig. 6.4a. The XRD patterns show the presence of the characteristic peaks of GO and further the appearance of a broad peak at 24.5° in the PPy/RGO hybrid indicated the successful reduction of GO into RGO (Fig. 6.4b). Similarly, the change in the intensity ratio of D-band due to the defect-filled nature of graphitic carbon at 1347 cm^{-1} and G-band due to

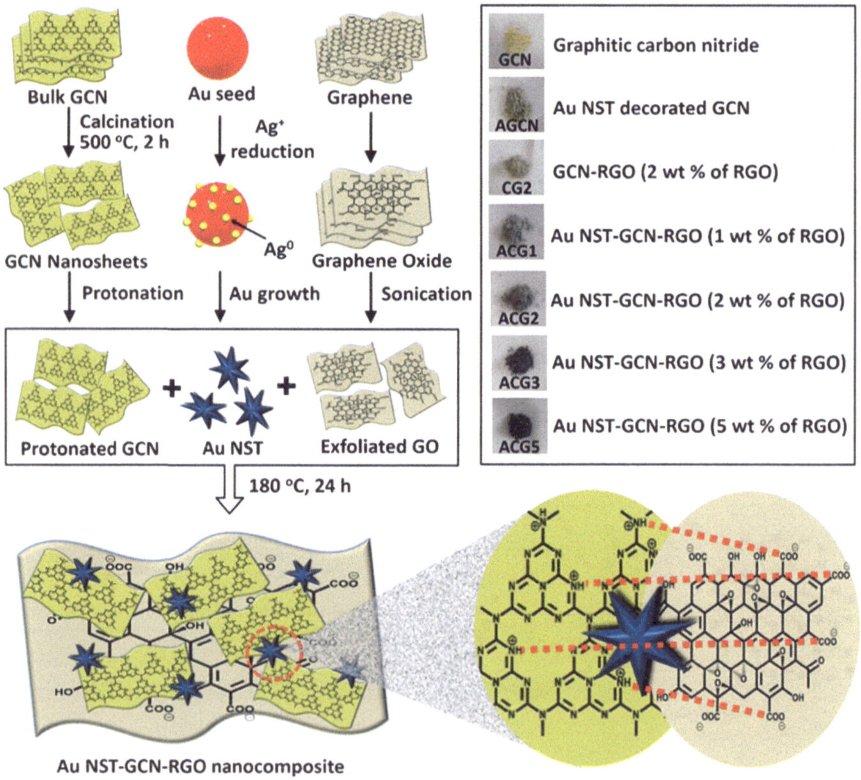

Fig. 6.3 Schematic representation for the synthesis of Au NST/GCN/RGO nanocomposites. Reproduced with permission from Ref. [31]. Copyright 2021, Royal Society of Chemistry

stretching vibration modes in graphitic carbon sheet at 1582 cm^{-1} in Raman spectra indicate a reduction of GO into RGO as shown in Fig. 6.4c. The downward shift of the D and G bands in the representative nanocomposite, Fe_3O_4@PPy/RGO signifies the strong interactions between Fe_3O_4 and PPy/RGO hybrids.

In another report, a quaternary nanocomposite consisting of $NaYF_4$:Yb/Er@CdS core—shell nanostructures decorated with Au NPs supported on RGO nanosheets has been synthesized using the multistep hydrothermal method for the removal of the pharmaceutical pollutant, ciprofloxacin (CIP) [9]. The rational design and development of the catalyst showed the efficient absorption of the broadband solar spectrum (visible and NIR regions). The carbonaceous 2D material, RGO, not only acts as the support for the NPs but plays multiple roles, such as electron acceptor and transporter. In addition, the RGO nanosheet adsorbs the pollutant molecules onto the

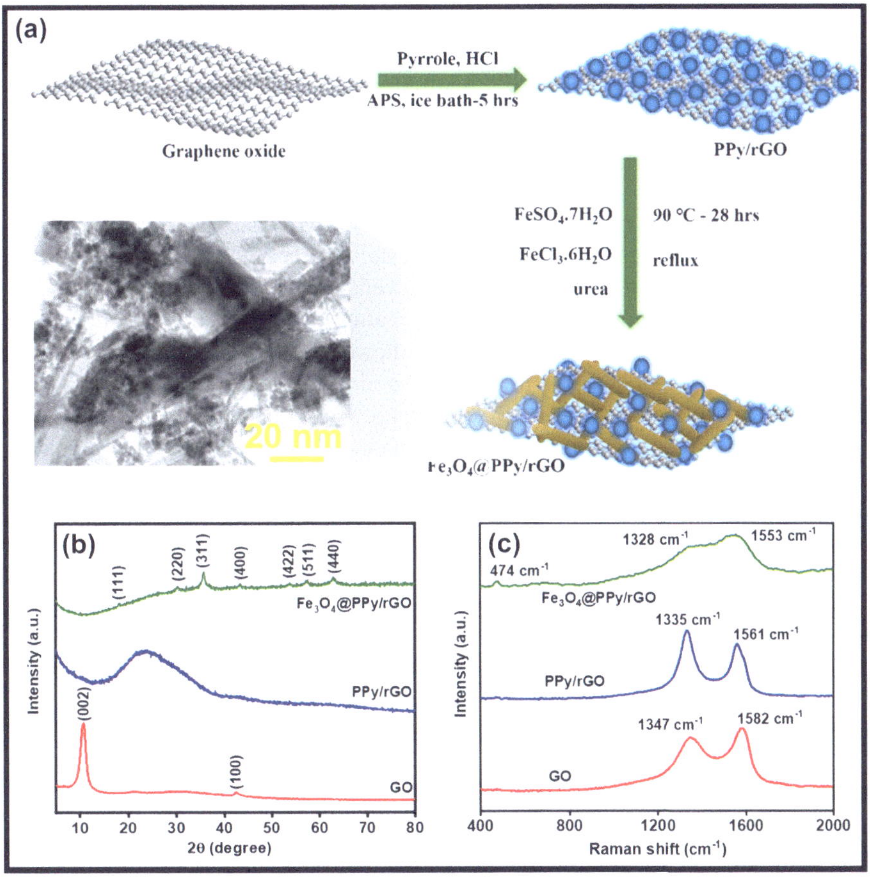

Fig. 6.4 **a** Schematic presentation of the synthesis of Fe_3O_4@PPy/RGO nanocomposite, **b** XRD patterns and **c** Raman spectra of GO, PPy/RGO and Fe_3O_4@PPy/RGO nanocomposite. Reproduced with permission from Ref. [37]. Copyright 2021, Elsevier Science Ltd.

surface, thereby facilitating their photocatalytic degradation. Other quaternary photocatalytic materials are also reported in the literature for the removal of the organic and inorganic contaminants from the wastewater, for example, $Ag_2CrO_4/TiO_2/Au/RGO$ hybrid biofilm [38], Carbon QDs/RGO-S@GCN/B@GCN [39], etc.

6.4 Photocatalytic Degradation of Organic Pollutants

Various intermediates and radical species are generated after photoexciting the graphene-based photocatalysts. These species are responsible for destroying both organic and inorganic components of pollutants in wastewater. Table 6.1 summarizes the recent reports on the photocatalytic degradation of organic pollutants using graphene-based nanocomposites. Peng et al., have synthesized 2D-2D GCN/GO nanocomposite using a hydrothermal synthesis route through π-π stacking interactions [40]. Further, the Ag NPs were uniformly deposited onto GCN/GO nanocomposite through the *in-situ* reduction method. As-synthesized nanocomposite was used for surface-enhanced Raman scattering (SERS) detection of biomolecules (adenine) and adsorption and photocatalytic degradation of organic pollutants (methylene blue).

A TiO_2/GO nanocomposite synthesized using the hydrothermal method was used to treat mineral processing wastewater photocatalytically [41]. The UV–Vis diffuse reflectance spectroscopy (DRS) analysis showedan extraordinary improvement in the absorption of the TiO_2 in the visible region after loading 18% of GO (Fig. 6.5a). Furthermore, the photocurrent measurements showed the higher photocurrent density of the 18% TiO_2/G nanocomposite as compared to the bare graphene nanosheets indicating the efficient separation of the charge carriers for better photocatalytic performance (Fig. 6.5b). The representative nanocomposite showed a degradation efficiency of 97.03% in 190 min under visible light irradiations (Fig. 6.5c). The improved photocatalytic performance of the catalyst was attributed to the huge specific surface area, low electron–hole suppression rate and extended absorption in the visible range. The bare GO shows a high rate of removal of the dye for an initial 30 min that could possibly be due to the adsorption process. However, the removal rate decreases significantly with further increases in time. Figure 6.5d depicts the EPR spectra suggesting that the higher intensity in 18% TiO_2/G nanocomposite under visible light irradiations is due to high oxygen vacancies in the samples. The EPR analysis showed that the 18% TiO_2/G photocatalytic material can be easily excited under visible light to generate electrons responsible for its high photocatalytic performance.

In another report, Graphene-based ZnCr layered double hydroxide (LDH) nanocomposites were utilized for the sonophotocatalytic performances for the degradation of rifampicin dye [42]. For this purpose, the ZnCr LDH and its nanocomposite with RGO were synthesized using a facile co-precipitation method., The maximum degradation efficiency of 87.3%, was achieved for rifampicin dye within 60 min of visible light irradiation. The high photocatalytic performance of the ZnCr

Table 6.1 Summary of graphene-based photocatalysts for degradation of pollutants in wastewater

Photocatalysts	Light source	Pollutants concentration	Reaction time	Year (Ref.)
Ag NPs/GCN/GO nanocomposite	Visible light (405 nm)	RhB (1×10^{-6} M)	60 min	(2022) [40]
Corn husk multi-layered graphene/ZnO nanocomposite	Natural sunlight	RhB, CR and DOX (1.25×10^{-5} mol L^{-1})	RhB (86%), CR (100%) and DOX (90%) in 180 min	(2022) [30]
Indium hydroxide/Nitrogen-doped graphene aerogels	High-pressure mercury lamp (400 W)	*p*-NP (20 mg L^{-1})	98% in 180 min	(2022) [44]
B doped RGO supported bismuth ferrite	300 W Xenon lamp	TC and RhB (10 mg L^{-1})	TC (86.7%) and RhB (99.4%) in 90 min	(2022) [45]
RGO-TiO_2 (photocatalysis linked with Fenton process)	shortwave light (UV254), black light (UV365), UV LED torch (UV395), visible light (400–700 nm) and sunlight	PA (200 mg L^{-1})	100% in 24 min (UV254), 22 min (UV365), 19 min (UV395), 18 min (VIS), 15 min (sunlight)	(2022) [46]
3D graphene/$Cd_{0.5}Zn_{0.5}S$ hybrids	300 W Xenon lamp ($\lambda >$ 420 nm)	CIP (5 mg L^{-1})	95% removal in 15 min	(2022) [47]
ZnCr LDH/RGO nanocomposite	LED lamp (visible light)	RIF (15 mg L^{-1})	87.3% removal in 60 min	(2022) [42]
TiO_2/graphene nanocomposite	300 W Xenon lamp	PBX (20 mg L^{-1})	97.03% in 130 min	(2022) [41]
Graphene- AgNPs@GCN	Visible light	MB (5 mg L^{-1})	88.5% in 140 min	(2022) [48]
$FeTiO_3$/GO nanocomposite	150 W Xenon lamp	Phenol (50 mg L^{-1})	100% in 150 min	(2021) [49]
$ZnSnO_3$/graphene aerogel	Simulated sunlight	CIP (100 mg L^{-1})	100% in 120 min	(2021) [43]
Fe_3O_4 NRs-polypyrrole/RGO	Two 250 W tungsten-halogen lamps	ACP (1 mg L^{-1})	84% in 120 min	(2021) [37]

(continued)

Table 6.1 (continued)

Photocatalysts	Light source	Pollutants concentration	Reaction time	Year (Ref.)
RGO/Zno/MoS_2 ternary nanocomposite	Visible light (100 W)	Aniline (80 mg L^{-1})	100% in 120 min	(2020) [32]
RGO-Bi_2WO_6	85 W CFL bulbs (450–650 nm)	LVX (80 mg L^{-1})	74.3% in 120 min	(2020) [50]
Ag_3VO_4/Cu-MOF/RGO	Visible light	AB92 (10 mg L^{-1})	80% in 120 min	(2020) [33]
$AgFeO_2$-graphene/$Cu_2(BTC)_3$ MOF	Natural sunlight	AMC and DCF (5 mg L^{-1})	AMC (92.9%) and DCF (91.4%) in 150 min	(2020) [34]
Carbon QDs/RGO-S@GCN/B@GCN	300 W Xe lamp (450 mW cm^{-2}, 380–780 nm) and direct sunlight	CMP (10 mg L^{-1})	Visible light 90% in 120 min) and sunlight (99.2% in 90 min)	(2020) [39]
Ag_2CrO_4/TiO_2/Au/RGO biofilm	125 W Hg lamp	MB (10 mg L^{-1})	97% in 52 min	(2020) [38]
RGO-N-ZnO	500 W tungsten halogen lamp	MB (10 mg L^{-1})	98.5% in 120 min	(2020) [51]
GO/TiO_2	500 W Xenon lamp	MB (5 mg L^{-1})	92% in 60 min	(2020) [52]
N doped graphene/TiO_2	300 W mercury lamp	BPA (20 mg L^{-1})	100% in 60 min	(2020) [53]
GO-derived C-doped $SrTiO_3$	500 W Xenon lamp	MB (8 mg L^{-1}), MO (3.3 mg L^{-1}), RhB (4.8 mg L^{-1}), phenol (9.4 mg L^{-1})	Over 95% of MO, MB, and RhB, and 70% of phenol and BPA in 3 h	(2020) [54]
Lanthanum-substituted Zn spinel ferrite @RGO	UV lamp (0.6 mW cm^{-2})	RhB (30 mg L^{-1})	100% in 60 min	(2020) [55]
GCN/graphene/TiO_2	300 W Xenon lamp (300 mW cm^{-2})	TC (15 mg L^{-1}), CIP (3 mg L^{-1}), BPA (5 mg L^{-1}) and RhB (20 mg L^{-1})	TC (83.5%) in 80 min, CIP (61.7%) in 60 min, BPA (79.5%) in 70 min and RhB (98.0%) in 50 min	(2020) [36]

(continued)

Table 6.1 (continued)

Photocatalysts	Light source	Pollutants concentration	Reaction time	Year (Ref.)
Bi_2S_3-$BiVO_4$ graphene aerogel	300 W Xe lamp ($\lambda > 420$ nm)	Cr (VI) (50 mg L^{-1}) and BPA (10 mg L^{-1})	99% removal of Cr (VI) and BPA in 120 min	(2020) [56]
GO/ZnO-Ag nanocomposite	Natural sunlight	MB (3.13×10^{-5} M)	100% in 40 min using GO/ZnO-Ag	(2020) [57]
CuO-TiO_2/RGO	UV light (312 nm)	MO (10 mg L^{-1})	99% in 90 min	(2019) [58]
SnO_2 aerogel/rGO nanocomposite	40 W UV lamp (370 nm)	MO (1×10^{-5} M)	84% in 60 min	(2019) [59]
GO/TiO_2	Hg lamp	RhB (10 mg L^{-1}) and AG25 (40 mg L^{-1})	100% in 120 min (RhB) and 97% in 180 min (AG25)	(2019) [60]
Graphene based NiAlFe LDH	500 W Xe lamp	CIP (10 mg L^{-1})	93% in 120 min	(2019) [61]
SnS_2/RGO	65 W CFL lamp ($\lambda > 400$ nm)	RBB and RBR (10 ppm)	99.7% (RBB) and 97% (RBR) in 150 min	(2019) [62]
Porous graphene/ZnO	Natural sunlight	MO (13 mg L^{-1})	100% in 150 min	(2019) [63]
CdS/RGO/BiOI	500 W Xe lamp ($\lambda > 420$ nm)	MDHB (20 mg L^{-1})	~80% in 240 min	(2019) [35]
Ordered Porous TiO_2-Pt/RGO	300 W Xe lamp ($\lambda > 420$ nm)	MO (20 mg L^{-1})	82% in 120 min	(2019) [64]
RGO-ZnTe	Solar simulator	TC (10 mg L^{-1})	~70% in 45 min	(2019) [65]

AB92 = acid blue 92; ACP = acetaminophen; AG25 = acid green 25; AMC = amoxicillin; BPA = bisphenol A; CMP = Chloramphenicol; CIP = ciprofloxacin; CR = congored; DCF = diclofenac; DOX = doxycycline; GCN = graphitic carbon nitride; GO = graphene oxide; LDH = layered double hydroxide; LVX = levofloxacin; MB = methylene blue; MDBH = methyl 3,5-dichloro-4-hydroxybenzoate; MO = methyl orange; NPs = nanoparticles; *p*-NP = p-nitrophenol; PA = picric acid; PBX = potassium butyl xanthate;RBB = remazol brilliant blue; RBR = remazol brilliant red; RGO = reduced graphene oxide; RhB = rhodamine B; RIF = rifampicin; TC = tetracycline

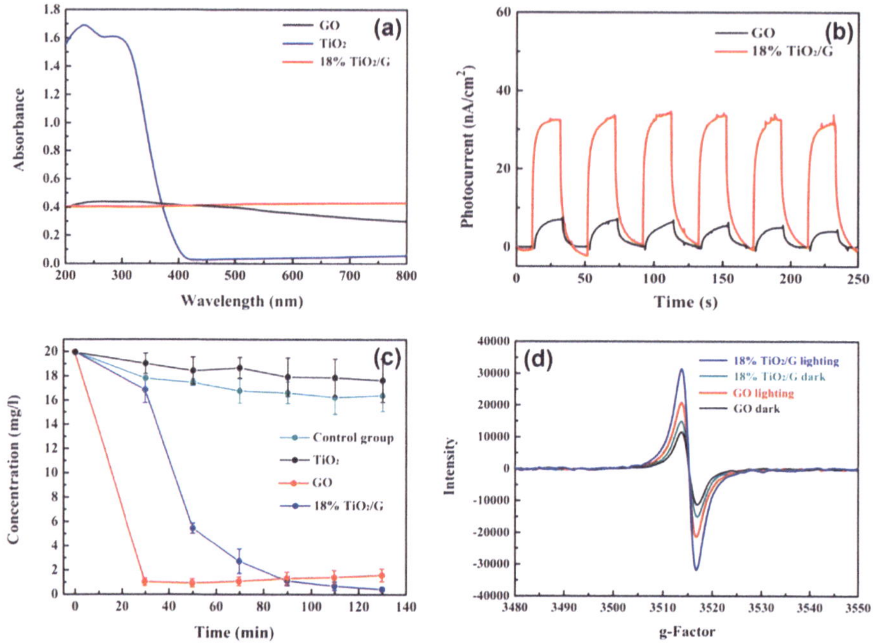

Fig. 6.5 **a** UV–VIS-DRS spectra, **b** photocurrent-time (I-t) plot, **c** photocatalytic activity under visible light and **d** EPR spectra of GO and 18% TiO_2/G nanocomposite. Reproduced with permission from Ref. [41] Copyright 2022, Elsevier Science Ltd.

LDH/RGO nanocomposite was attributed to the enhancement in the production of active hydroxyl ($OH^{\cdot}$) radical species under visible light irradiations. The radical scavenger studies confirm that the $OH^{\cdot}$ radical plays the most significant role in degrading the dye molecule. Furthermore, the ultrasonication process benefits the process by preventing the agglomeration of the catalyst in solution, improving mass transport and increasing the availability of the active sites by continuously cleaning of the catalyst surface.

A macroscopic $ZnSnO_3$/graphene aerogel structure has been synthesized that showed excellent adsorption and degradation of the CIP from wastewater under visible light irradiations [43]. The catalyst synthesized with an optimized mass ratio (1:2) of the $ZnSn(OH)_6$ to GO, denoted as ZGA-4, exhibits the 100% removal efficiency of the pollutant. The improved photocatalytic performance of the catalyst was ascribed to the optimized interfacial and electronic band structures responsible for the separated photogenerated charge carriers and immensely produced active species such as $OH^{\cdot}$ and $O_2^{-\cdot}$ radicals, as shown in Fig. 6.6.

The graphene-based materials play a major role in enhancing the photocatalytic activity of the semiconducting materials for the degradation of organic pollutants. Various intermediates and radical species are generated after photoexciting the graphene-based photocatalysts. These species are responsible for destroying both

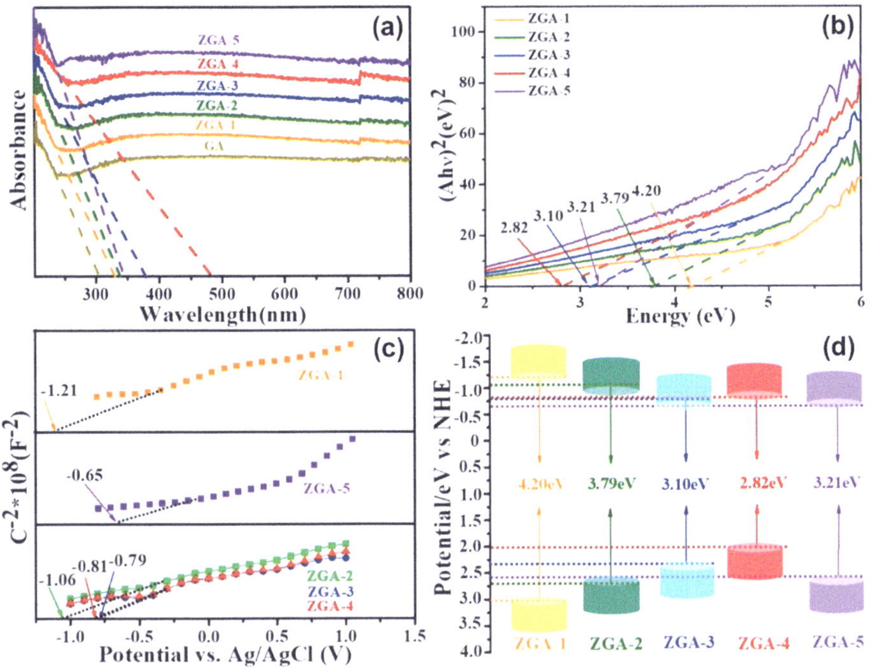

Fig. 6.6 **a** UV–VIS diffuse reflectance spectra; **b** Tauc's plots; **c** Mott-Schottky diagram and **d** band gaps of the ZGA nanocomposites. Reproduced with permission from Ref. [43] Copyright 2021, Elsevier Science Ltd.

organic and inorganic components of pollutants in wastewater. Table 6.1 summarizes the recent reports on the photocatalytic degradation of organic pollutants using graphene-based nanocomposites.

6.5 Summary and Future Perspective

In summary, graphene-based materials open new opportunities in the advanced photocatalysis process. The outstanding physical and chemical properties of graphene-based photocatalysts such as large surface area, high electron conductivity, abundant surface-active sites, high mechanical strength and good adsorption and reusability properties make it the most reliable candidate in the field of photocatalysis. This chapter discusses various strategies, such as surface modification, doping, and multijunction formation, to improve the photocatalytic performance of graphene-based materials. Coupling the two-dimensional carbonaceous materials (graphene, GO and RGO) with other nanomaterials like metal oxides and metal nanoparticles exhibit synergistic improvement in the photocatalytic efficiency and stability of the catalyst. Also, the optimized amount of graphene-based materials in the catalyst tunes the bandgap to enhance the light absorption of the catalyst. From the above discussion of the literature on binary, ternary, and quaternary composites of graphene-based materials, it can be concluded that the appropriate modification could result from highly efficient materials for adsorption and photocatalytic treatment of wastewater. The production of cost-effective, photostable and efficient graphene-based materials at a large scale is most desirable and challenging as well. Thus, more work needs to be done by researchers to ensure the commercialization of such materials for large-scale photocatalytic applications. There are extensive opportunities in the research area to extend the perspective of graphene-based materials for wastewater treatment.

Acknowledgements Ajay Kumar and Irwing Ramirez acknowledge Waste2Fresh grant under Horizon 2020 program. The authors thank The Open University for granting financial resources, access to facilities, and training to conduct this work.

References

1. U. Water, *2018 UN World Water Development Report, Nature-based Solutions for Water* (2018)
2. A. Saravanan et al., Effective water/wastewater treatment methodologies for toxic pollutants removal: processes and applications towards sustainable development. Chemosphere **280**, 130595 (2021)
3. F.E. Osterloh, Inorganic materials as catalysts for photochemical splitting of water. Chem. Mater. **20**(1), 35–54 (2008)
4. S. Jamaly, A. Giwa, S.W. Hasan, Recent improvements in oily wastewater treatment: progress, challenges, and future opportunities. J. Environ. Sci. **37**, 15–30 (2015)
5. S.M. Gupta, M. Tripathi, A review of TiO_2 nanoparticles. Chinese Sci. Bull. **56**(16), 1639–1657
6. C. Belver et al., Semiconductor Photocatalysis for Water Purification, in *Nanoscale Materials in Water Purification*. (Elsevier, 2019), pp.581–651
7. Y. Xu et al., Rate enhancement and rate inhibition of phenol degradation over irradiated anatase and rutile TiO_2 on the addition of NaF: new insight into the mechanism. J. Phys. Chem. C **111**(51), 19024–19032 (2007)
8. A. Kudo et al., Effects of doping of metal cations on morphology, activity, and visible light response of photocatalysts. Chem. Phys. **339**(1–3), 104–110 (2007)
9. A. Kumar et al., Rational design and development of lanthanide-doped $NaYF_4$@ CdS–Au–RGO as quaternary plasmonic photocatalysts for harnessing visible–near-infrared broadband spectrum. ACS Appl. Mater. Interf. **10**(18), 15565–15581 (2018)
10. R. Marschall, Semiconductor composites: strategies for enhancing charge carrier separation to improve photocatalytic activity. Adv. Func. Mater. **24**(17), 2421–2440 (2014)
11. P. Qiu et al., Fabrication of an exfoliated graphitic carbon nitride as a highly active visible light photocatalyst. J. Mater. Chem. A **3**(48), 24237–24244 (2015)
12. H. Park et al., Surface modification of TiO_2 photocatalyst for environmental applications. J. Photochem. Photobiol., C **15**, 1–20 (2013)
13. M. Li et al., Facet, junction and electric field engineering of bismuth-based materials for photocatalysis. ChemCatChem **10**(20), 4477–4496 (2018)
14. M. Kitano et al., Recent developments in titanium oxide-based photocatalysts. Appl. Catal. A **325**(1), 1–14 (2007)
15. T. Wehling et al., Molecular doping of graphene. Nano Lett. **8**(1), 173–177 (2008)

16. P.V. Kamat, Graphene-based nanoassemblies for energy conversion. J. Phys. Chem. Lett. **2**(3), 242–251 (2011)
17. R. Kumar, et al., Laser processing of graphene and related materials for energy storage: state of the art and future prospects. Progr. Energy Combust. Sci. 100981 (2022)
18. C. Deepa, L. Rajeshkumar, M. Ramesh, Preparation, synthesis, properties and characterization of graphene-based 2D nano-materials for biosensors and bioelectronics. J. Mater. Res. Technol. (2022)
19. M. Sajna, et al., An overview of graphene-based 2D/3D nanostructures for photocatalytic applications. Top. Catal. 1–25 (2022)
20. G. Lekshmi et al., Multifunctional oil-produced reduced graphene oxide–silver oxide composites with photocatalytic, antioxidant, and antibacterial activities. J. Colloid Interf. Sci. **608**, 294–305 (2022)
21. S. Guo, S. Dong, Graphene nanosheet: synthesis, molecular engineering, thin film, hybrids, and energy and analytical applications. Chem. Soc. Rev. **40**(5), 2644–2672 (2011)
22. C. Zhu, D. Du, Y. Lin, Graphene and graphene-like 2D materials for optical biosensing and bioimaging: a review. 2D Mater. **2**(3), 032004 (2015)
23. Y. Ai et al., Insights into the adsorption mechanism and dynamic behavior of tetracycline antibiotics on reduced graphene oxide (RGO) and graphene oxide (GO) materials. Environ. Sci. Nano **6**(11), 3336–3348 (2019)
24. A. Gulati, R. Kakkar, Graphene-based adsorbents for water remediation by removal of organic pollutants: theoretical and experimental insights. Chem. Eng. Res. Des. **153**, 21–36 (2020)
25. X. Cui et al., Liquid-phase exfoliation, functionalization and applications of graphene. Nanoscale **3**(5), 2118–2126 (2011)
26. H.M. Hegab, L. Zou, Graphene oxide-assisted membranes: fabrication and potential applications in desalination and water purification. J. Membr. Sci. **484**, 95–106 (2015)
27. X.J. Lee et al., Review on graphene and its derivatives: synthesis methods and potential industrial implementation. J. Taiwan Inst. Chem. Eng. **98**, 163–180 (2019)
28. S. Gupta et al., Emerging trends in the syntheses of heterocycles using graphene-based carbocatalysts: an update. Top. Curr. Chem. **377**(3), 1–62 (2019)
29. C.M. Park et al., Potential utility of graphene-based nano spinel ferrites as adsorbent and photocatalyst for removing organic/inorganic contaminants from aqueous solutions: a mini review. Chemosphere **221**, 392–402 (2019)
30. D.P. Sebuso et al., Corn husk multilayered graphene/ZnO nanocomposite materials with enhanced photocatalytic activity for organic dyes and doxycycline degradation. Mater. Res. Bull. **151**, 111800 (2022)
31. A. Kumar et al., Plasmon induced hot electron generation in two dimensional carbonaceous nanosheets decorated with Au nanostars: enhanced photocatalytic activity under visible light. Mater. Chem. Front. **5**(3), 1448–1467 (2021)
32. P. Ghasemipour et al., Developing the ternary ZnO doped MoS_2 nanostructures grafted on CNT and reduced graphene oxide (RGO) for photocatalytic degradation of aniline. Sci. Rep. **10**(1), 1–16 (2020)
33. E. Akbarzadeh et al., Preparation and characterization of novel Ag_3VO_4/Cu-MOF/rGO heterojunction for photocatalytic degradation of organic pollutants. Mater. Res. Bull. **121**, 110621 (2020)
34. E.M. El-Fawal, S.A. Younis, T. Zaki, Designing $AgFeO_2$-graphene/Cu_2 $(BTC)_3$ MOF heterojunction photocatalysts for enhanced treatment of pharmaceutical wastewater under sunlight. J. Photochem. Photobiol. A: Chem. **401**, 112746 (2020)
35. N. Ma et al., In situ synthesis of a cadmium sulfide/reduced graphene oxide/bismuth Z-scheme oxyiodide system for enhanced photocatalytic performance in chlorinated paraben degradation. Chem. Eng. J. **359**, 530–541 (2019)
36. Z. Wu et al., MXene Ti_3C_2 derived Z–scheme photocatalyst of graphene layers anchored TiO_2/g–C_3N_4 for visible light photocatalytic degradation of refractory organic pollutants. Chem. Eng. J. **394**, 124921 (2020)

37. R.S. Kumar et al., Fe_3O_4 nanorods decorated on polypyrrole/reduced graphene oxide for electrochemical detection of dopamine and photocatalytic degradation of acetaminophen. Appl. Surf. Sci. **556**, 149765 (2021)
38. R. Biswas et al., Novel green approach for fabrication of $Ag_2CrO_4/TiO_2/Au/r\text{-}GO$ hybrid biofilm for visible light-driven photocatalytic performance. J. Phys. Chem. C **124**(5), 3373–3388 (2020)
39. A. Kumar et al., Carbon quantum dots and reduced graphene oxide modified self-assembled $S@C_3N_4/B@C_3N_4$ metal-free nano-photocatalyst for high performance degradation of chloramphenicol. J. Mol. Liq. **300**, 112356 (2020)
40. G.-Z. Peng et al., Photocatalytic degradation and reusable SERS detection by Ag nanoparticles immobilized on g-C_3N_4/graphene oxide nanosheets. Surf. Coat. Technol. **435**, 128212 (2022)
41. M. Jiang et al., Photocatalytic degradation of xanthate in flotation plant tailings by TiO_2/graphene nanocomposites. Chem. Eng. J. **431**, 134104 (2022)
42. T.S. Rad et al., Graphene-based ZnCr layered double hydroxide nanocomposites as bactericidal agents with high sonophotocatalytic performances for degradation of rifampicin. Chemosphere **286**, 131740 (2022)
43. S. Dong et al., Interfacial and electronic band structure optimization for the adsorption and visible-light photocatalytic activity of macroscopic $ZnSnO_3$/graphene aerogel. Compos. B Eng. **215**, 108765 (2021)
44. Y. Jiang, J. Zhang, R. Balasubramanian, Nitrogen-doped graphene aerogels with rational indium hydroxide decoration for highly efficient photocatalytic of p-nitrophenol. J. Environ. Chem. Eng. **10**(1), 107125 (2022)
45. R. Preetha et al., Promoting photocatalytic interaction of boron doped reduced graphene oxide supported $BiFeO_3$ nanocomposite for visible-light-induced organic pollutant degradation. J. Alloy. Compd. **904**, 164038 (2022)
46. B. Usharani, V. Manivannan, Enhanced photocatalytic activity of reduced graphene oxide-TiO_2 nanocomposite for picric acid degradation. Inorg. Chem. Commun. **142**, 109660 (2022)
47. G. Zeng et al., Highly efficient photocatalytic degradation of the emerging pollutant ciprofloxacin via the rational design of a magnetic interfacial junction of mangosteen peel waste-derived 3D graphene hybrid material. Environ. Sci. Nano **9**(4), 1298–1314 (2022)
48. C. Chen et al., Graphene nanofiltration membrane intercalated with AgNP@g-C_3N_4 for efficient water purification and photocatalytic self-cleaning performance. Chem. Eng. J. **441**, 136089 (2022)
49. M. Moradi et al., Ultrasound-assisted synthesis of $FeTiO_3$/GO nanocomposite for photocatalytic degradation of phenol under visible light irradiation. Sep. Purif. Technol. **261**, 118274 (2021)
50. M. Arya et al., Hydrothermal synthesis of rGO-Bi_2WO_6 heterostructure for the photocatalytic degradation of levofloxacin. Opt. Mater. **107**, 110126 (2020)
51. M. Suresh, A. Sivasamy, Fabrication of graphene nanosheets decorated by nitrogen-doped ZnO nanoparticles with enhanced visible photocatalytic activity for the degradation of Methylene Blue dye. J. Mol. Liq. **317**, 114112 (2020)
52. T.A. Kurniawan et al., Functionalizing TiO_2 with graphene oxide for enhancing photocatalytic degradation of methylene blue (MB) in contaminated wastewater. J. Environ. Manage. **270**, 110871 (2020)
53. Y. Zhao et al., Enhanced activation of peroxymonosulfate by nitrogen-doped graphene/TiO_2 under photo-assistance for organic pollutants degradation: Insight into N doping mechanism. Chemosphere **244**, 125526 (2020)
54. C.-W. Chang, C. Hu, Graphene oxide-derived carbon-doped $SrTiO_3$ for highly efficient photocatalytic degradation of organic pollutants under visible light irradiation. Chem. Eng. J. **383**, 123116 (2020)
55. B.-M. Jun et al., Accelerated photocatalytic degradation of organic pollutants over carbonate-rich lanthanum-substituted zinc spinel ferrite assembled reduced graphene oxide by ultraviolet (UV)-activated persulfate. Chem. Eng. J. **393**, 124733 (2020)

56. Q. Liang et al., Simultaneous Cr (VI) reduction and bisphenol A degradation by a 3D Z-scheme Bi_2S_3-$BiVO_4$ graphene aerogel under visible light. Chem. Eng. J. **384**, 123256 (2020)
57. N.A. Al-Rawashdeh, O. Allabadi, M.T. Aljarrah, Photocatalytic activity of graphene oxide/zinc oxide nanocomposites with embedded metal nanoparticles for the degradation of organic dyes. ACS Omega **5**(43), 28046–28055 (2020)
58. S.G. Babu et al., Synergistic effect of sono-photocatalytic process for the degradation of organic pollutants using CuO–TiO_2/rGO. Ultrason. Sonochem. **50**, 218–223 (2019)
59. T. Kim et al., Facile synthesis of SnO_2 aerogel/reduced graphene oxide nanocomposites via in situ annealing for the photocatalytic degradation of methyl orange. Nanomaterials **9**(3), 358 (2019)
60. M. Adly, S.M. El-Dafrawy, S. El-Hakam, Application of nanostructured graphene oxide/titanium dioxide composites for photocatalytic degradation of rhodamine B and acid green 25 dyes. J. Mater. Res. Technol. **8**(6), 5610–5622 (2019)
61. J. Liang et al., Constructing high-efficiency photocatalyst for degrading ciprofloxacin: three-dimensional visible light driven graphene based NiAlFe LDH. J. Colloid Interf. Sci. **540**, 237–246 (2019)
62. L. Dashairya et al., SnS_2/RGO based nanocomposite for efficient photocatalytic degradation of toxic industrial dyes under visible-light irradiation. J. Alloy. Compd. **774**, 625–636 (2019)
63. L. Wang et al., Enhanced photocatalytic degradation of methyl orange by porous graphene/ZnO nanocomposite. Environ. Pollut. **249**, 801–811 (2019)
64. J. Huo, C. Yuan, Y. Wang, Nanocomposites of three-dimensionally ordered porous TiO_2 decorated with Pt and reduced graphene oxide for the visible-light photocatalytic degradation of waterborne pollutants. ACS Appl. Nano Mater. **2**(5), 2713–2724 (2019)
65. K. Chakraborty, T. Pal, S. Ghosh, RGO-ZnTe: a graphene based composite for tetracycline degradation and their synergistic effect. ACS ACS Appl. Nano Mater. **1**(7), 3137–3144 (2018)

Chapter 7
Adsorptive Removal of Pollutants Using Graphene-based Materials for Water Purification

Lesego Tabea Temane, Jonathan Tersur Orasugh, and Suprakas Sinha Ray

Abstract Adsorption methods have been employed for pollution control as well as cleanup all around the world. Composite materials have been the most suitable candidates for high-standard adsorption systems. So also, when merged with graphene or its derivatives, they become very effective candidates for adsorbing environmental contaminants found in water. The combined effect of graphene oxide, as well as its engineered material nanostructures (hybrids, composites, etc.), has also been shown to significantly contribute to the adsorption of heavy metals, toxic organic chemicals (colorants, diverse volatile organic compounds (VOCs), pesticides, chemical fertilizer, drugs), as well as other suspended particles pollutants of water, particularly industrial effluents. The broad surfaces of graphene oxide's derivatives and nanocomposites are bonded with a variety of reactionary oxygen-containing functionalities, giving them exceptional stability and adsorption efficiency in an aqueous environment. This also enables them to be recycled for numerous adsorption–desorption cycles. The present chapter discusses all of these graphene-based materials, their adsorption phenomena, and their application to water to cleanse and purify it.

7.1 Introduction

Environmentalists are now more worried about wastewater management due to the rapid growth in the global population and ongoing technological development. Millions of tons of harmful inorganic (heavy metal ions) and organic (dyes, polycyclic aromatic hydrocarbons, etc.) compounds are present in the wastewater produced daily by a variety of industries; including tannery, fertilizer, paper manufacturing, domestic dwellings, etc., [1, 2].

L. T. Temane · J. T. Orasugh (✉) · S. S. Ray
DST-CSIR National Centre for Nanostructured Materials, DSI-CSIR Nanotechnology Innovation Centre, Council for Scientific and Industrial Research, Pretoria 0001, South Africa
e-mail: calunivjto@gmail.com

J. T. Orasugh · S. S. Ray
Department of Chemical Sciences, University of Johannesburg, Doornfontein, Johannesburg 2028, South Africa

N. Kumar et al. (eds.), *Two-Dimensional Materials for Environmental Applications*, Springer Series in Materials Science 332, https://doi.org/10.1007/978-3-031-28756-5_7

There are two categories of harmful water pollutants; inorganic and organic, within the inorganic pollutants; heavy metals, such as Cu, Ag, Hg, Fe, Zn, Cd, Pb, Sn, Cr, Mn, Co, Al, Ni, and As, are among the metals and metalloids in the category of trace elements with an elemental density greater than or equal to 4 ± 1 g/cm^3. The concentrations beyond the recommended World Health Organization (WHO) limits are considered harmful to the soil and aquatic systems. These pollutants' strong reactivity and tendency to form complexes are controlled by the design of their electronic atom shells, leading to favorable physiological and biochemical activities [1]. Moreover, these metals are stable, with a high rate of movement and solubility in water, improper management of metal-polluted wastewater discharges could lead to various adverse implications on both the environment and human health. Additionally, as plants ingest these metals, they can infiltrate the food chains of humans and animals and severely impact their health and essential functions [3, 4]

On the other hand, it has been noticed that anthropogenic processes of heavy metals go beyond the natural fluxes for several elements. Fossil fuels release nickel, mercury, vanadium, selenium, and tin, as well as other critical anthropogenic sources such as smelting, which releases arsenic, copper, and zinc; pesticides, that discharges arsenic, as well as automotive exhaust; which releases lead.

Due to the frequent production of goods to satisfy the desires of the vast population, People's actions have proven to add more to wastewater [4]. Organic wastewater pollutants are mostly produced by the food industry, pharmaceuticals, cosmetics, textile, dye, leather, and pesticide businesses, the processing of vegetable oils, oil and gas production companies, and home sewage [5, 6].

Both organic (such as colors, medications, personal care items, PAHs, phenols, herbicides, and pesticides) and inorganic (such as heavy metals and earth rare metals) contaminants can cause cancer and build up in the body over time. As a result, wastewater treatment is a pressing issue since it negatively influences on people, animals, and the environment [2]. Table 7.1 presents certain pollutants, their sources, and the risks they pose.

Toxic contaminants have been removed from wastewater using a variety of treatment technologies, including adsorption [8], advanced oxidation process (AOPs) [9], membrane separation [10], reverse osmosis [11, 12], chemical precipitation [13], ion-exchange [14], electro-chemical treatment [15], and biological treatments [16–18]. Among these technologies, adsorption (Fig. 7.1) is considered the most promising treatment method because of its high efficiency, lack of sophistication, affordability, and viability for large-scale implementation. As a result, research into creating improved adsorbent materials has gained attention during the past few decades [17]. Furthermore, it does not produce hazardous contaminants during the treatment process, which would result in secondary pollution [2]. Adsorption is known to be a physicochemical procedure wherein the adsorbate material transfers itself to the surface of the adsorbent material as a liquid or gas. Four stages make up the adsorption phenomenon: bulk, film/external mass diffusion, pore/intraparticle diffusion, combination, and finally, physical/chemical reactions. Here, bulk diffusion is significantly reduced by even distributed dispersion of the adsorbent or adsorbate [7].

Table 7.1 Harmful metals, dyes, pesticides, and germs found in water, as well as their negative effects. Reproduced with permission from Orasugh and Ray [7], Copyright 2022, Springer Nature

Category	Contaminant	Origin/source	Side effect(s)
Dyes	Basic Red-9	Industrial wastes	Carcinogenic, dermatitis, teratogenicity, histopathological effects
	Congo Red	Industrial wastes	May be carcinogenic, dermatitis, irritant to the eye, skin, mucus membrane, mutagenic, etc
	Crystal Violet	Industrial wastes	Lethal to renal and respiratory system develops chemical cystitis, etc
	Malachite Green	Industrial wastes	Extremely carcinogenic, histopathological effect, mitotic poisoning, mutagenic, teratogenic, etc
	Methyl Blue	Industrial wastes	Causes hemolytic anemia, breast cancer, hyperbilirubinemia, respiratory challenges
	Rhodamine-8	Industrial wastes	Eyes irritant, carcinogenic to animals (both marine and land)
Heavy metal ions and oxides	As	Pesticides, chemical wastes, Bi-products from mines	Enzyme-inhibitor, Carcinogenic
	Be	Nuclear plants, space productions, coal plants, etc	Poisonous, cancer-causing
	B	Industrial bi-products, cleansing agents, etc	Deadly to florae and algae
	Cd	Industrial ejection, metal electroplating, Ni–Cd from batteries, etc	One of the origins of high blood pressure, kidney disorder, anemia, illness, etc
	Cr	Metal electroplating industries, mining, tanning, etc	Cancer-causing

(continued)

Table 7.1 (continued)

Category	Contaminant	Origin/source	Side effect(s)
	Cu	Metal electroplating industries, mining, etc	Deadly to florae and algae
	Fluorine (F)	Natural ecological sources, manufacturing bi-products	Causes bone impairment, dappled teeth
	Pb	Drains, crude oil, coal, mining, etc	Causes anemia, kidney malfunction, Neurological disarrays
	Mn	Mining and manufacturing bi-product, the bacteriological act of Mn minerals of low-slung pH	Deadly to florae
	Hg	Mining and manufacturing bi-product, coal, etc	Extremely deadly in its form as CH_3Hg^-, Hg^{2+}
	Molybdenum (Mo)	Natural and manufacturing surplus	Noxious to faunae
	Selenium (Se)	Natural and coal sources	Poisonous
	Zn	Metal coating companies, manufacturing waste	Poisonous to ecological plants
Insecticides/pesticides	Atrazine	Agricultural weed control sprays	Cause endocrine disruption, hormonal disorder, fetotoxicity, skin irritant, an increased relative risk of ovarian neoplasia, can induce mammary tumors in rats, immunotoxicity, behavioral alterations
	Cyromazine	Agricultural weed control sprays	Causes serious irritation to the skin and eye irritation. It can also cause respiratory aggravation. It is harmful to oceanic life, presenting lasting impacts

(continued)

Table 7.1 (continued)

Category	Contaminant	Origin/source	Side effect(s)
	Hexazinone	Agricultural weed control sprays	The nervous system, immune system, endocrine function, development or reproduction, and carcinogenicity or mutagenicity
	Metribuzin	Agricultural weed control sprays	Central nervous system damage, kidney and liver damage, narcosis and labored breathing, deficiency or excess of the human growth hormone
	Prometon	Agricultural weed control sprays	Discreetly exasperating to the skin, eyes, and respiratory expanse, liver, and heart of animals
	Prometryn	Agricultural weed control sprays	Nausea, Sore throat, Respiratory, depression, Muscle weakness
	Simazine	Agricultural weed control sprays	Trouble in strolling, quake, seizures, loss of motion, cyanosis, eased back breath, miosis, stomach agony, looseness of the bowels, impeded adrenal capacity, lower food admission, higher water consumption, incoordination, quakes, shortcoming, harm to the testicles, kidneys, liver, and thyroid, unsettling influences in sperm creation, and gene mutations

(continued)

Table 7.1 (continued)

Category	Contaminant	Origin/source	Side effect(s)
	Desethylterbutilazine	Agricultural weed control sprays	Hormonal disorders, fetotoxicity, skin irritant, and an increased relative risk of ovarian neoplasia can induce mammary tumors in rats, immunotoxicity, behavioral alterations
	Terbuthylazine	Agricultural weed control sprays	Mildly to moderately irritating to the eyes and slightly irritating to the skin
Waterborne bacteria	Bacillus cereus		Vomiting, nausea, diarrhea, mucus
	Enterococcus faecalis		Shortness of breath, fever, bleeding gums, chest pain, diarrhea, fatigue, stiff neck, and abdominal pain
	Escherichia coli		Abdominal pain, watery diarrhea, intestinal toxin damage, nausea, vomiting, fever, and fatigue
	Pseudomonas aeruginosa		Dermatitis, gastrointestinal infection, soft tissue contamination, urinary expanse contagion, bone and joint infection
	Salmonella typhimurium		Diarrhea, vomiting, abdominal cramps, and fever
	Staphylococcus		Bloodstream, bone, and joint infections, pneumonia, scalded skin syndrome, toxic shock syndrome, etc

(continued)

Table 7.1 (continued)

Category	Contaminant	Origin/source	Side effect(s)
	Shigella flexneri		Dysentery, drug resistance, joint pains, painful urination, virulence, and eye irritation
	Vibrio		Abdominal cramps, fever, nausea, decreased blood pressure, skin lesions, etc

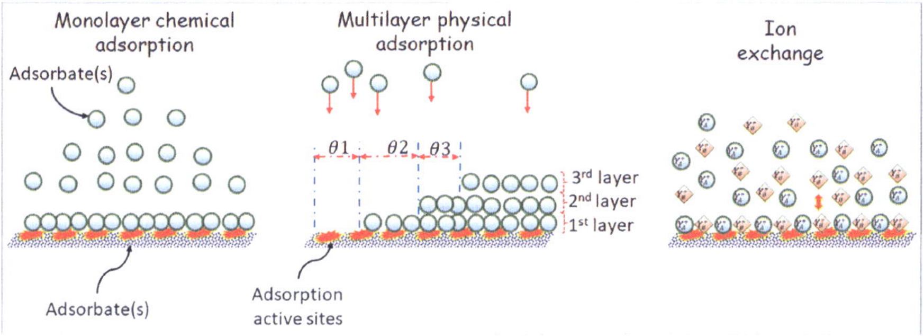

Fig. 7.1 An illustration of the adsorption process

Researchers have studied many newly discovered adsorbents in recent years for the exclusion of dangerous contaminants from aqueous systems, including carbon-based materials [19, 20], clay minerals [21], polymers [22], agricultural wastes [23, 24], zeolites [24, 25], etc., to remove micro/nano pollutants from wastewater, most of these adsorbents lack selectivity for the target contaminants or have low adsorption and desorption effectiveness.

Amongst carbon-based materials, graphene nanostructures have recently expanded their use in environmental remediation as a new brand of adsorbent for cleaning wastewater. It has been considered the most promising adsorbent for arresting a variety of organic and inorganic pollutants due to its excellent delocalized p-electron systems, a significant number of active sites, and high theoretical surface area. It also has unique mechanical and physicochemical properties, including exceptional mechanical strength, electrical conductivity, good thermal stability, large aspect ratio, flexibility, as well as negligible thickness [1].

7.2 History of Graphene and Its Derivatives

Graphene is a sheet of carbon atoms organized in a single layer to create a closely bound hexagonal honeycomb lattice. Graphene is a carbon allotrope with sp^2-bonded atoms. Graphite is produced when layers of graphene are piled on top of one another. Carbon, the fourth most plentiful element in the universe, is also the next most prevalent component of the human body. All known forms of life on Earth have been attributed to carbon [26].

In graphene, the extent of the carbon–carbon bond is around 0.142 nm. Its layer height is said to be only 0.33 nm. It is the thinnest material currently renowned and among the most powerful. Graphene is almost entirely translucent. Even the tiniest atom of helium cannot fit through its structure due to its extreme density [27]. The development of new technologies, as well as the advancement of existing ones, can both benefit from the use of graphene [26].

In 1859, Benjamine Collins Brodie first identified the extremely lamellar structure of thermally reduced graphite oxide by thermally reducing graphite oxide by the reaction of graphite with potassium chlorate and fuming nitric acid, a solution of crystallite graphene oxide (GO) was created. Later on, this GO was weaved into paper. In 1919, Kohlschutter and Haenni finished a preliminary investigation into the characteristics of this graphene oxide paper [27].

In 1916, powder diffraction was used to identify the structure of graphene, and in 1924, single-crystal X-ray diffraction analysis was used to discover the internal structure of graphene. Wallace originally examined the graphene theory in 1947, and this work served as a crucial foundation for understanding the electrical characteristics of three-dimensional (3D) graphite. The evolving massless Dirac equation was first identified by Semenoff, DiVincenzo, and Mele [26, 28, 29].

The earliest TEM photographs of few-layer graphite were published in 1948 by Ruess and Vogt. Following this, Ubbelohde and Lewis extracted a single atom of graphite. They discovered that the basal-plane conductivity of graphite intercalation mixtures was surprisingly higher than that of the original graphite [27].

The first thorough investigation of few-layer graphite with decreased GO monolayer flakes was done by Boehm and colleagues in 1962. Although no layered graphite smaller than 50–100 layers had been effectively created before 2004, very thin films of graphite had been produced by mechanical exfoliation as early as 1990. To begin with, atomically thin graphitic films were produced using exfoliation procedures that are akin to sketching techniques. Later, multilayer samples with a thickness of 10 nm could be produced [26].

It is important to note that Dr. Bor Jang, who has the first graphene patent, has done much research on the material. He holds more than forty patents covering the manufacturing and use of graphene, including the first patent for single layer graphene in 2002 as well as the first patents for single layer graphene reinforced metal, carbon, glass, as well as ceramic-matrix composites but also polymer composites. But because Dr. Jang hardly ever produced any scholarly papers, he is essentially unknown in the academic community [27].

Graphene production-related patent applications were initially announced in October 2002 and were first approved in 2006. After two years of attempting, Geim and Novoselov finally isolated single-atom-thick crystallites from bulk graphite. These crystallites were then deposited onto thin silicon dioxide (SiO_2) or a silicon wafer, from which the graphene was electrically isolated. The cleavage method allowed for the first detection of the anomalous quantum Hall effect in graphene. Theoretically, Berry's phase of the massless Dirac fermion was predicted and theoretically presented and published in Nature in 2005, providing direct evidence of graphene. Kim and Zhang, as well as Geim's group, reported on this outcome [26].

In 2010 Novoselov and Geim were awarded the Nobel Prize in Physics for their groundbreaking studies on graphene. A very large-scale research initiative with a graphene Flagship and 150 partner organizations was started in 2013 and was financed by the European Commission with a total budget of €1 billion. Graphene commercialization moved forward quickly after commercial-scale fabrication of the material was shown. A recent advancement is that an integrated graphene electronics device was developed commercially in 2017, 13 years after graphene was first created in a lab. San Diego's Nanomedical Diagnostics markets this item to pharmaceutical developers [26].

7.3 Various Methods for the Formulation of Graphene

There are two primary approaches for formulating graphene: top-down (destruction) and bottom-up (building), as illustrated in Fig. 7.2. The top-down techniques, such as mechanical exfoliation [30], arc discharge [31], oxidative exfoliation-reduction [32], liquid-phase exfoliation (LPE) [33], and unzipping of carbon nanotubes (CNT) [34], typically separate and delaminate the layers of graphite into single-, bi-, as well as few-layer graphene. Essentially, these processes destroy bigger precursors like graphite and other carbon-based ones for the formation of nanoscale graphene. Certain top-down processes are known to be extremely scalable and provide high-quality products. Still, they struggle to produce goods with consistent qualities, have low yields, and rely heavily on the finite graphite raw material [35].

On the other hand, bottom-up techniques use carbon sources besides graphite to create graphene and its derivatives. These processes start with atomically small precursors and turn them into graphene compounds. Chemical vapor deposition (CVD) [36], epitaxial growth [37], substrate-free gas-phase synthesis (SFGP) [38], template route [39], and complete organic synthesis [40], are some of the bottom-up techniques. Although the bottom-up methods result in graphene products with a vast surface area and nearly no defects, they frequently demand expensive manufacturing costs.

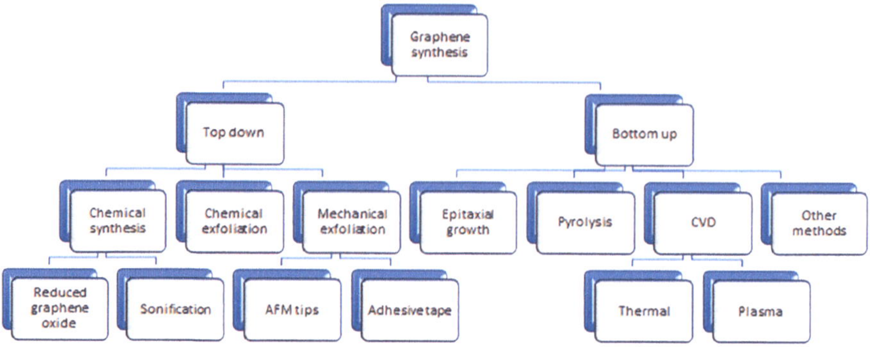

Fig. 7.2 Schematic representation of the different methods of graphene synthesis

7.3.1 Mechanical Exfoliation

Depending on directional approaches, such as the vectors of normal force and shear force, such as the vectors of normal force and shear force mechanical exfoliation can be divided into different categories [35].

7.3.1.1 Micromechanical Cleavage

Amongst normal force approaches is the Scotch-tape method (Fig. 7.3), which is often referred to as micro-mechanical exfoliation, was invented by Novoselov and Geim in 2004 and uses an adhesive tape to separate graphene from graphite crystals. Many layers of graphene separate from graphite during the first peel. The graphene is continually pulled from the adhesive tape to obtain mono, bi, or a few layers of graphene. Finally, the adhesive is eliminated in a solvent like acetone after the tape has been applied to a substrate [27].

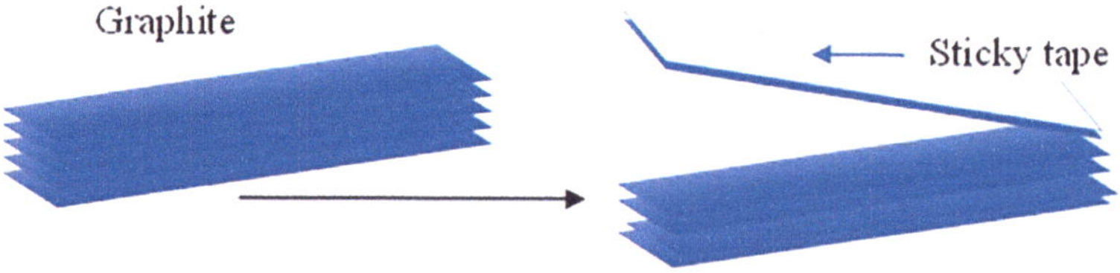

Fig. 7.3 A schematic for the Scotch-tape-based micromechanical cleavage of producing graphene

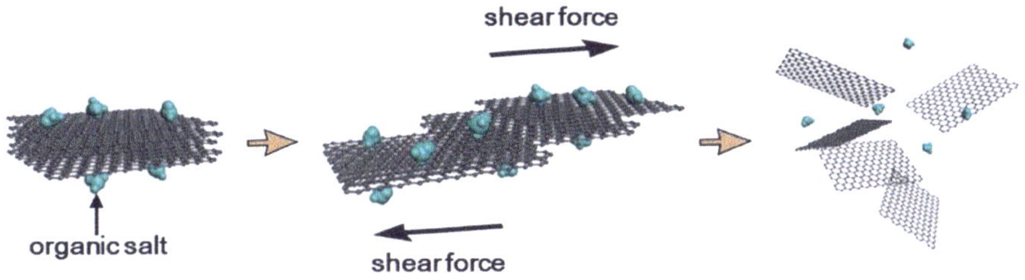

Fig. 7.4 Liquid-phase exfoliation of graphite in the presence of sodium citrate, potassium citrate, and sodium tartrate. Reproduced with permission from Liang et al. [42]. Copyright 2021, Elsevier Science Ltd.

7.3.1.2 Sonification

Using an ultrasonic approach to separate individual layers allows for the successful exfoliation of graphite in liquid conditions. Bath sonication (BS) and tip sonication (TS) are the two different types of sonication methods used to prepare monolayer or few-layer graphene sheets by exfoliating graphene; they have both been used simultaneously or separately [33]. Figure 7.4 depicts the liquid-phase exfoliation (LPE) process.

Ultrasonic exfoliation typically follows the following steps: Step 1–dispersion of graphite in a solvent, step 2–exfoliation, and step 3–purification. By chemically wetting down graphite and then ultrasonically processing it in organic solvents, graphene flakes can be created without surfactants. The process of exfoliation during ultrasonication is induced by shear forces and cavitation, which is the growth as well as the collapse of micrometer-sized bubbles/voids in liquids because of pressure variations. The attractive forces between the sheets of graphene must be balanced after exfoliation by the solvent-graphene interfacial interaction. The best solvents for dispersing graphene the kind that reduce the interfacial tension [$mN\ m^{-1}$] between the liquid and the graphene flakes or the force that exists between the two [41].

7.3.1.3 Ball Milling

Graphene flakes can also be created through lateral exfoliation of graphite using a shear force. A valuable method for creating shear force is ball milling. It is a mechanical method of separating graphene layers from raw graphite. However, the crystalline structural architecture of the graphene sheets is susceptible to destruction in the event of normal-force-induced exfoliation brought on by the contact of balls with the sheets' surface during milling. In this case, fragmentation rather than exfoliation takes place [43]. There are two potential mechanisms at work in the majority of ball milling equipment that produce the exfoliation and fragmentation effects. The main one is shearing force, which is considered to be a superior mechanical method of exfoliation. This approach is very desirable for producing large-scale graphene flakes. The additional one is caused by the balls colliding or making vertical impacts as they roll. Large flakes can be broken up into smaller ones in this fashion, and occasionally crystalline structures can be destroyed and transformed into amorphous or non-equilibrium phases. Therefore, it is anticipated that obtaining high-quality and large-sized graphene will decrease the secondary effect [44].

7.3.1.4 Fluid Dynamics

Because graphite particles may move with the liquid in the fluid dynamics, they can be repeatedly exfoliated at various positions. This property makes it fundamentally distinct from sonication and ball milling, making it a potentially effective method for producing graphene at scale. There are two levels of fluid dynamics: mild and intense [44]. The fluid dynamics process has three categories: vortex fluidics, pressure-driven fluid dynamics, and mixer-driven fluid dynamics.

7.3.1.5 Vortex Fluidics Devices (VFD)

The VFD exfoliation method produces defect-free graphene with minimal energy. The schematic diagram for a VFD is shown in Fig. 7.5 [45]. VFD is seen in the cross-section in Fig. 7.6. A tube with one end closed and the other open makes up a VFD. Strong shear stress produced by the tube's high-speed rotation results in forming a thin film in the liquid. The thin film's volume is barely a few milliliters. The shear fluidic film can be adjusted by altering the tube's velocity, alignment, and other operating variables (Fig. 7.6) [46]. According to fluid dynamics, a fast-spinning fluid creates Stewartson/Ekman layers, which are boundary and shear layers with regard to the spinning direction. The liquid surface has a downward flow direction, while the tube surface has an upward flow direction. Chen et al. [47] exfoliated graphene by shearing vortex fluidic sheets using NMP as the solvent. Centrifugal force will cause a coating to first form on the tube wall from dispersed graphite particles. The tube wall will experience partial lifting and sliding of the graphite layers. By now, the centrifugal force of the moving graphite particles against the wall will start a shear-triggered motion along the tube [45, 47]. However, using the continuous flow mode of vortex fluidics, Wahid et al. [48] have successfully created a graphene hybrid in water. The advantage of the continuous flow mode is that another liquid may be added to the VFD [45], increasing the shear force in the film as a result of the frictional force created as the fluid spins around the tube. In the beginning, as shown in Fig. 7.5, the restricted mode converts graphite into multi-layer graphene in water. The liquid in the continuous flow mode in the VFD is then injected with multi-layer graphene generated in the restricted mode. After more recirculation, the combination product known as VFD employing the Taylor-Couette flow reactor was established by Tran and colleagues for the high-yield synthesis of bulk few-layer graphene [49]. The graphite and solvent mixture is pumped into the reactor, shredded between a rotating inner and stationary outer cylinder. Thus, graphite flakes will be peeled by

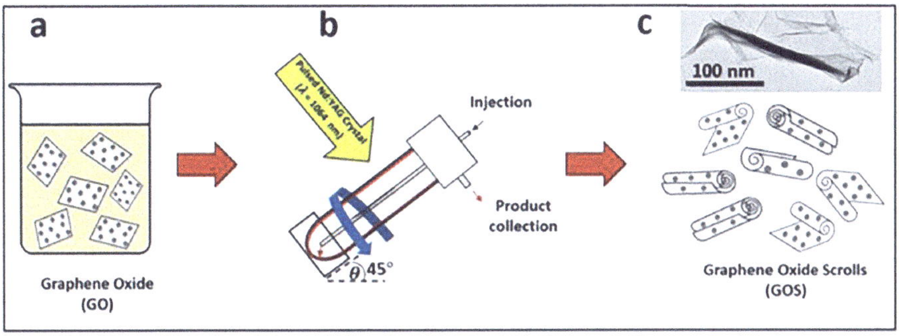

Fig. 7.5 Schematic illustration of the experimental procedure for fabricating GOS from GO sheets. **a** Solvated GO sheets before processing in the VFD. **b** Schematic of the experimental set up for the vortex fluidic device (VFD) and Nd: YAG pulsed laser irradiation (operating at 1064 nm with the optimised power at 250 mJ) and rotational speed at 4 k rpm. **c** GOS after processing in the VFD, inset is TEM image for GOS. (A colour version of this figure can be viewed online.) Reproduced with permission from Alharbi et al. [45]. Copyright 2018, Elsevier Science Ltd.

the powerful shear force and torque produced by the vortex flow. With fewer flaws, this method can effectively produce few-layer graphene. Only 2.5 mm separates the moving and fixed cylinders, which restricts the capacity [49].

7.3.1.6 Pressure-Driven Fluid Dynamics

Large-scale graphene production is accomplished using the pressure-driven fluid dynamics (PFD) technique. PFD exfoliates graphene via several flow channels. Complex fluid dynamics in the flow channels can be caused by the pressure difference between the intake and outflow. As a result of the PFD device's graphite particles' ability to move together with the fluid in the flow channel [50], the flow channel can be millimeter- or micrometer-sized. The volume of the compression and expansion tubes of PFD devices can be altered, and the important features of the fluid dynamics in the PFD include cavitation, pressure release, viscous shear stress, turbulence, and collision. These fluid dynamics properties produce normal and shear forces that will operate as a driving force for exfoliation. For instance, normal force from cavitation and pressure discharge might result in exfoliation. Moreover, Graphite self-exfoliates to form single- or few-layer graphene due to viscous and Reynolds shear stresses produced by velocity gradient and turbulence, respectively.

7.3.1.7 Mixer-Driven Fluid Dynamics.

A prospective substitute for the sonication method for exfoliating graphene is mixer-driven fluid dynamics (MFD). Poorly linked nanoparticle separation using MFD is efficient and straightforward. A rotor/stator blender offered for sale can be used to

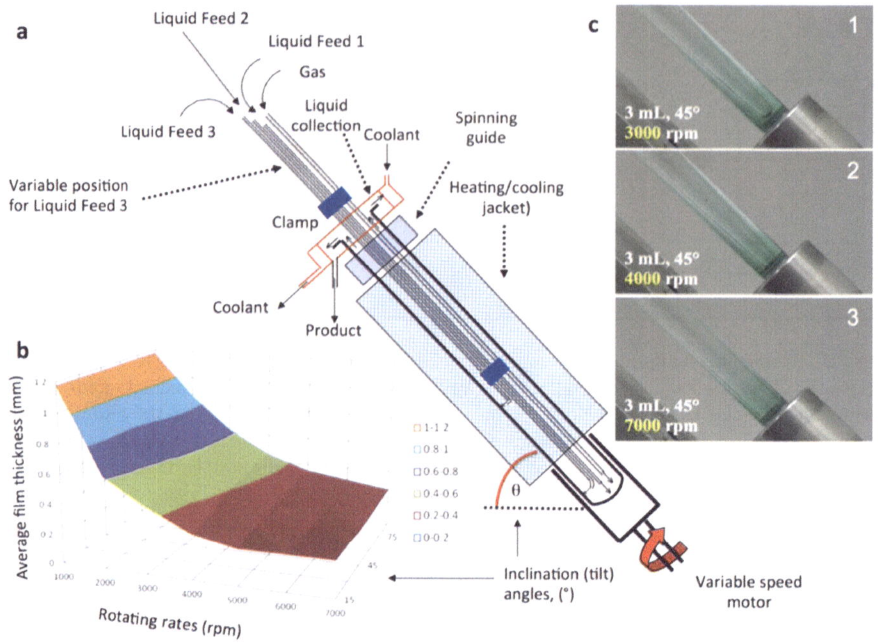

Fig. 7.6 **a** components of the device. **b** Average film thickness (mm) versus tilt angle ө. **c** Photographs showing the film of liquid developed for different speeds. Reproduced with permission from Yasmin et al. [46]. Copyright 2013, Nature Publishing Group

set up an MFD system. The key exfoliating component of the mixer is its rotor-based head [50, 51]; if necessary, the rotor's diameter can be adjusted. Researchers have also employed the MFD approach to exfoliate graphite in the N-methyl pyrrolidone (NMP) dispersion media [51].

7.3.1.8 Hydrodynamic-Assisted Exfoliation Method

This approach is comparatively new and has not been widely adopted for the preparation of graphene. In their study, Zhang et al. demonstrated this innovative approach when they synthesized outstanding quality graphene with only a few layers, as depicted in Fig. 7.7 [52]. This method can be scaled to commercial quantities to produce graphene nanosheets for advanced applications.

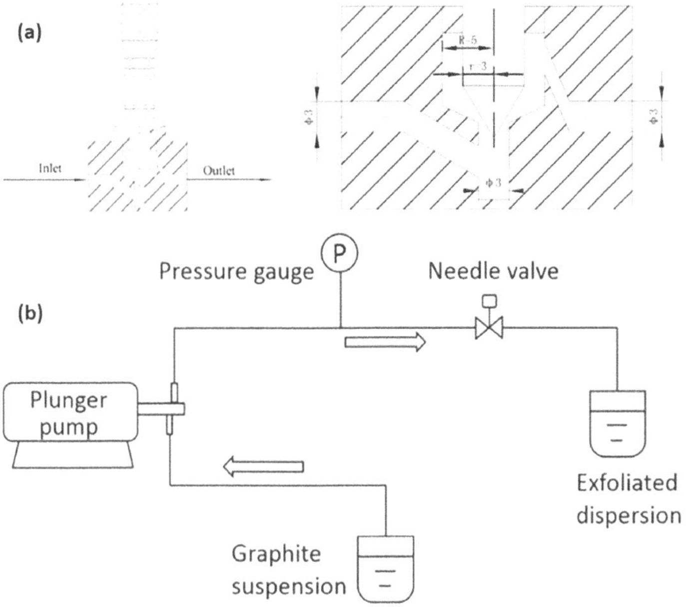

Fig. 7.7 **a** A schematic view of the needle valve and **b** Schematic diagram of the hydrodynamic-assisted exfoliation process. Reproduced with permission from Zhang et al. [52]. Copyright 2018, Springer Nature

7.3.2 Chemical Exfoliation

Chemical exfoliation is another well-known process for producing graphene, similar to mechanical exfoliation. Alkali metals are intercalated with the graphite structure through the process of chemical exfoliation, which isolates few-layer graphene distributed in solution. Alkali metals are the elements in the periodic table that can readily combine with graphite to form intercalated structures at different stoichiometric ratios. One of the main benefits of alkali metals is that they easily fit in the interlayer spacing due to their smaller ionic radii than the graphite interlayer gap [53].

The first chemical exfoliation of few-layer graphite, later known as "graphene," utilizing potassium (K) as the intercalating agent creating alkali metal, was described by Kaner and colleagues in 2003 [54]. Potassium (K) forms a KC_8 intercalated compound when interacting with graphite at 200 °C under an inert helium environment (less than 1 ppm H_2O and O_2).

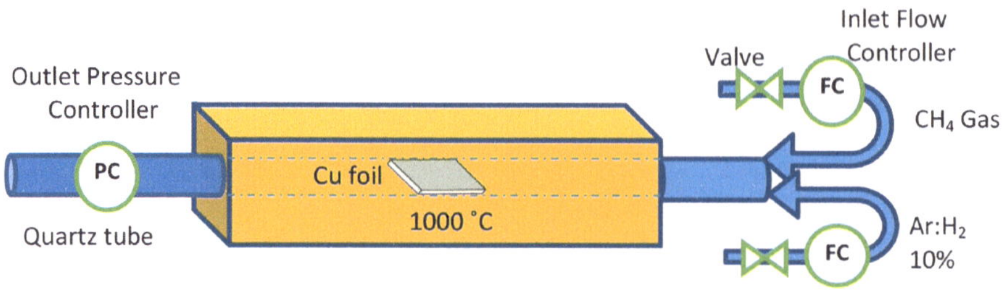

Fig. 7.8 Chemical vapor deposition (CVD) process. Reproduced with permission from Moosa et al. [57] Copyright 2021, TUBITAK

7.3.2.1 Thermal Chemical Vapor Deposition Process

The synthesis of graphene on an industrial scale can be done easily using the CVD process. To manufacture graphene from highly volatile carbon sources using the CVD process (Fig. 7.8), a high-temperature, high-vacuum, and chemically inert environment are necessary. Common carbon precursors utilized in the manufacture of graphene include methane, methanol, ethane, ethylene, ethanol, acetylene, polymers, and scrap plastic. To produce extremely crystalline graphite coatings on Ni surfaces, the first CVD technique, known as thermal CVD, was introduced in 1966 [50]. The preparation costs of this method are relatively high (cost: 1–2 USD cm^2; retail price: 10 USD cm^2) despite their outstanding quality and low defect density [55, 56].

Transition metals, such as crystalline Ni [58], Cu [59, 60], or CuNi alloys, as well as Ru [61], Pt [62], Co [63], Ti [64], Re [65], etc. are frequently used as CVD substrates. Crystalline Ni, Cu, or Ni-Cu alloys are extremely inexpensive compared to other substrates and produce high monolayer graphene coverage [66].

7.3.2.2 Plasma-Enhanced CVD

In the Plasma-enhanced CVD (PECVD), plasma is created in a void chamber, and then thin films are deposited on the substrate surface. It also includes the chemical interaction between the reacting gases. RF (AC frequency), microwave, and inductive coupling are all used in the IN PECVD system (electrical currents produced by electromagnetic induction). It is possible to produce graphene without a catalyst at a relatively low temperature, making it more practical for large-scale industrial applications [67]. Even though gas-phase precursor materials are used, it is expensive. Graphene sheets have also been synthesized in this manner [68]. PECVD is used to create monolayer and thin layers of graphene on various substrates, including Si, SiO_2, $Al_2\ O_3$, Mo, Zr, Ti, Hf, Nb, W, Ta, and 304 stainless-steel.

7.3.3 Epitaxial Growth

One of the most lauded processes for producing graphene is epitaxial heat growth on a single crystalline silicon carbide (SiC) surface. The word "epitaxy" is Greek in origin; the prefix "epi" means "over" or "upon," while the word "taxis" means "order" or "arrangement." Epitaxial growth occurs when a single crystalline layer is deposited on a single crystalline substrate, producing an epitaxial film. In this case, the fundamental process is an annealed surface made of a SiC sample, the desorption of atoms from it. Because carbon has a far lower vapor pressure than silicon, the silicon atoms desorb at high temperatures, leaving the carbon atoms left to create graphitic layers, commonly known as few-layer graphene (FLG). The outcome is the same when using various heating techniques like resistive heating or e-beam heating. In order to prevent contamination, the heating process is conducted in a vacuum. To liberate enough carbon atoms for the creation of one graphene layer, approximately three bilayers of SiC are required, and this is derived from the molar densities. Thus, single-crystalline SiC substrates, it fabricates high-crystalline graphene. Depending on the substrate, there are two general epitaxial growth processes: homo-epitaxial and hetero-epitaxial development. A homo-epitaxial layer has a film placed on a substrate made of the same material. In contrast, a hetero-epitaxial layer has a film and substrate of different materials [69].

7.3.4 Pyrolysis

In a bottom-up approach, the chemical synthesis of graphene was carried out using the Solvo thermal technique. In a closed vessel, the molar ratio of sodium to ethanol throughout this heat reaction was 1:1. By pyrolyzing sodium ethoxide with sonication, graphene sheets may be smoothly separated. This resulted in the production of graphene sheets up to 10 μm in size. The resulting sheet's Raman spectroscopy displayed a wide D-band, G-band, and the IG/ID intensity ratio of 1.16, all indicative of faulty graphene. The advantages of this method included easy fabrication at low temperatures of high-purity, functionalized graphene. Despite this, graphene had numerous flaws, making it's quality unsuitable [70]. The thermal breakdown of silicon carbide (SiC) is another key method used to create graphene. High temperatures cause Silicon to desorb, leaving carbon remaining to create a few layers of graphene [71]. The continuous manufacture of graphene films in a mm scale at a temperature of 750 °C on a thin film of nickel plated on SiC substrate has significantly improved this method [72]. This method's advantage is the continuous generation of graphene films across the entire SiC-coated surface. But large-scale production of graphene is not possible using this process. The thermal breakdown of ethylene occurs at 1000 K using a similar strategy. The ability to create high-purity graphene mono-layers is a benefit of this synthesis technique [73].

7.3.5 Unzipping Method

Using chemical and plasma-etched techniques, a carbon nanotube (CNT) is unzipped. The term "graphene nano ribbon" (GNR) refers to a narrow, long graphene strip with sharp edges. The width of the nanotube affects how an electronic state changes from semimetal to semiconductor [74]. If the initial nanotube has multiple walls or only one, multi-layer or single-layer graphene will result. The precursor nanotubes' diameter determines the breadth of the nanoribbons that are thus generated. Lithium (Li) with ammonia creates multi-walled CNTs (MWNTs) (NH_3). The production of graphene and derivatives from CNT has been studied using a variety of unzipping techniques, some of which are depicted in Fig. 7.9: chemical assault [75], plasma etching [76], intercalation and exfoliation [77], and metal-catalyzed cutting. H_2SO_4 treatment and $KMnO_4$ oxidation are the first steps in the chemical unzipping of CNT in order to break the axial-directional C–C bond [75]. The axial-directional C–C bond of the CNT is first broken by H_2SO_4 treatment, followed by $KMnO_4$ oxidation [75]. This approach produced a low yield of graphene while destroying a sizable amount of precursor [78]. Before chemical unzipping, MWCNT was intercalated with a chemical medium (oxalic acid) to increase yield. Oxalic acid molecules can easily intercalate in the lattice space between MWCNT layers because their size (0.38 nm) is consistent with the MWCNT interlayer spacing (0.34 nm) [79].

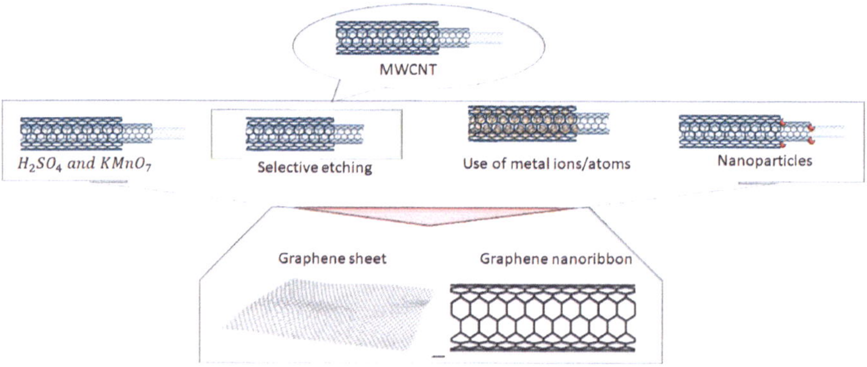

Fig. 7.9 **a** MWCNT unzipping process treatment using $KMnO_4$ and H_2SO_4 for GNR synthesis, **b** MWCNT etching by moderately embedding it within a polymeric membrane with Ar plasma-assisted synthesis, **c** Peeling off the graphene sheet involves inserting alkali-metal atoms between the concentric cylinders of MWCNT. Peeling off the graphene sheet involves inserting alkali-metal atoms between the concentric cylinders of MWCNT. Peeling off the graphene sheet involves inserting alkali-metal atoms between the concentric cylinders of MWCNT, as reported in Cano-Marquez [77], and **d** Catalytic metal nanoparticles acting as scissors to cut along the MWCNT length

7.3.6 Arc Discharge Method

Carbon nanoparticles have been produced via arc discharge, which occurs when electricity is run across two electrodes of graphite [80]. Using this method, it is possible to generate controlled amounts of graphene sheets [80]. The experimental process for disintegrating 2D multi-layer graphene sheets (MLGs) utilizing toluene droplets in an arc discharge approach has been explored [81]. The connection between toluene and water, or the encapsulation of the 2D MLGs into toluene droplets, was reported to have demonstrated a wrinkly exterior of the generated granules, having a diameter of about 600 nm [81]. Additionally, the spherical granule presented a hollow structure of intersecting ridges over a thin, consistent layer. By arc discharge in a moderate-pressure hydrogen atmosphere, graphene flakes have also been created without the use of any catalytic reactions [82]. Utilizing a variety of buffer gases, layer graphene has been produced using the arc discharge method [82]. These authors' atomic force microscope (AFM) result showed that the thickness of graphene synthesized using arc discharge was around 0.7–1 nm matching 2–3 layers as reported in Fig. 7.10. Graphene sheets cannot roll up because the hydrogen gas in buffer gas can break the bonds holding carbon atoms together. In a mixed H_2/He environment, Rao et al. found that graphite rods can be used as electrode materials with potentials between 100 and 150 A to create graphene nanosheets [83]. Then, rather than using a mixed H_2/He environment, Wang et al. devised an improved arc discharge method for generating graphene nanosheets in air [84]. They found that low pressure favored the manufacture of other carbon nanomaterials, such as carbon nano-horns and nanospheres whereas high pressure promoted the formation of graphene nanosheets.

7.4 Graphene-Based Materials

7.4.1 Go

GO is a graphene-based material with a lamellar structure made up of layers of oxidized carbon from the initial graphene lattice. Recent research has found that each GO layer, in addition to the carbon backbone, contains many oxygen functions, making each layer a multifunctional network [85]. The Lerf–Klinowski (LK) model [86, 87], which is the most widely used, predicts that GO is made up of two randomly dispersed domains: (a) pure graphene with sp^2–hybridized carbon atoms, and (b) domains containing sp^3–hybridized and oxidized carbon. Compared to graphene, GO has a greater number of oxygen-containing functional groups on its sheet layer, complicating its structure and affecting its characteristics.

The carboxyl and carbonyl groups were introduced near the single layer's edge. In contrast, the hydroxyl and epoxy groups were arbitrarily distributed over the single GO layer. An oxidized aliphatic six-membered ring region and an unoxidized benzene

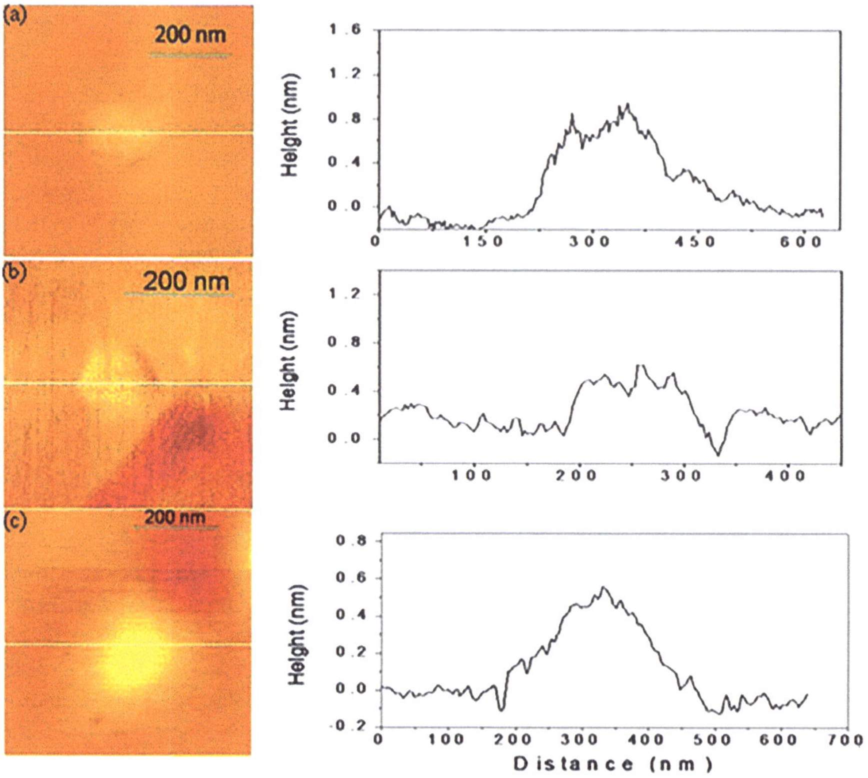

Fig. 7.10 Noncontact mode AFM images of **a, b** pure (HG) and **c** boron-doped (BG) graphenes with height profiles. Reproduced with permission from Subrahmanyam et al. [82]. Copyright 2009, American Chemical Society

ring region were both present in GO. The degree of oxidation and random distribution determines the relative sizes of these two regions. The L-K structural model, however, ignores the influence of raw graphene, oxidants, and oxidation processes in favor of a set of assumptions. Using scanning electron microscopy (SEM), Erickson and colleagues investigated GO nano-plates. They discovered that the material not only has an oxidized zone with high disorder and an unoxidized graphene region but also has hole defects brought on by over oxidation and lamellar peeling [88]. Consequently, various models have been put out in recent years, including the dynamic structure model (DSM) [89, 90] and the binary structure model [91].

7.4.1.1 Formulation of GO

Strong acids (nitric and/or sulfur) and potassium chlorate are used in the documented procedures for making graphene oxide (by Brodie, Staudenmaier, and Hoffman)

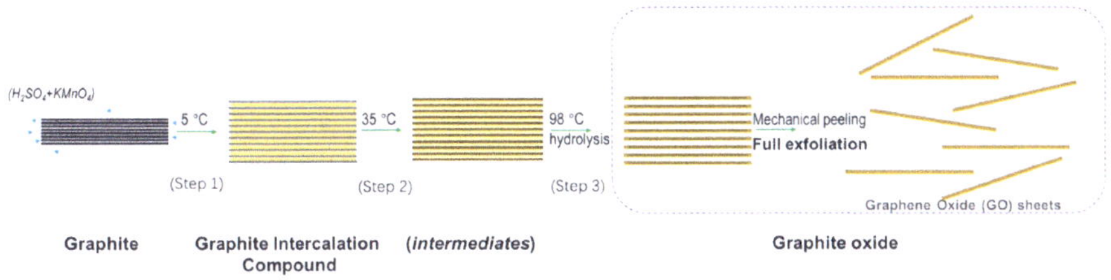

Fig. 7.11 A chemical representation of the Hummers-Offeman method for preparing GO. Reproduced with permission from Sun [93]. Copyright 2019, Elsevier Science Ltd

[92]. The Hummers method from 1958 is frequently used to delaminate and oxidize graphite because it is more convenient than the techniques developed by Brodie and his early followers. This method is based on a mixture of sulfuric acid and potassium permanganate (Fig. 7.11, the first two steps). The entire process is divided into three steps: a time for the intercalation of graphite and a period for the simultaneous/following oxidization of the aforementioned graphite intercalation compounds (GICs); next, to obtain homogeneous GO solution (Fig. 7.11, the third step), GO then hydrolyzes and exfoliates. Hummers' oxidation and exfoliation of graphite provide the foundation for contemporary advancements in GO preparation. Notably, many groups favor using a short burst of high-intensity ultrasound to disassemble graphite oxide's stacked structure into GO sheets [93].

7.4.2 Molecular Skeleton of Single-Layer GO

It is vital to define and offer a description of GO (Fig. 7.12) based on current research techniques before moving on to a study of its qualities. The monolayer of carbon atoms that makes up GO contains both (substantially) sp^2- and (partially) sp^3-hybridized carbon atoms that have functional groups that contain oxygen in both the basal (hydroxyl and epoxy) and edge planes (carboxyl, carbonyl). Sp^2-hybridized carbon atoms can be considered unoxidized regions, while sp^3-hybridized carbon atoms can be considered oxidized regions with covalently bound to oxygen-containing functional groups. By directly visualizing lattice atoms and topological defects in a GO monolayer using high-resolution transmission electron microscope (HRTEM), it was determined that these sp^3-hybridized carbon clusters are situated marginally above or beneath the plane of sp^2-hybridized carbon atoms [94–96].

Using a platinum coil similar to Fig. 7.13a, samples were exposed utilizing heating chips for TEM (DENS) to suspend the GO over the vacuum for improved contrast and atomic resolution imaging; slits were created in the silicon nitride thin film using a focused ion beam, as illustrated in Fig. 7.13b. Although monolayer portions are evident, GO has wrinkles and looks clumped uniformly across many samples, as illustrated in Fig. 7.13c, d. the existence of an amorphous component on the surface of the GO, which reacts rapidly to the beam's energy, prevents direct imaging of the

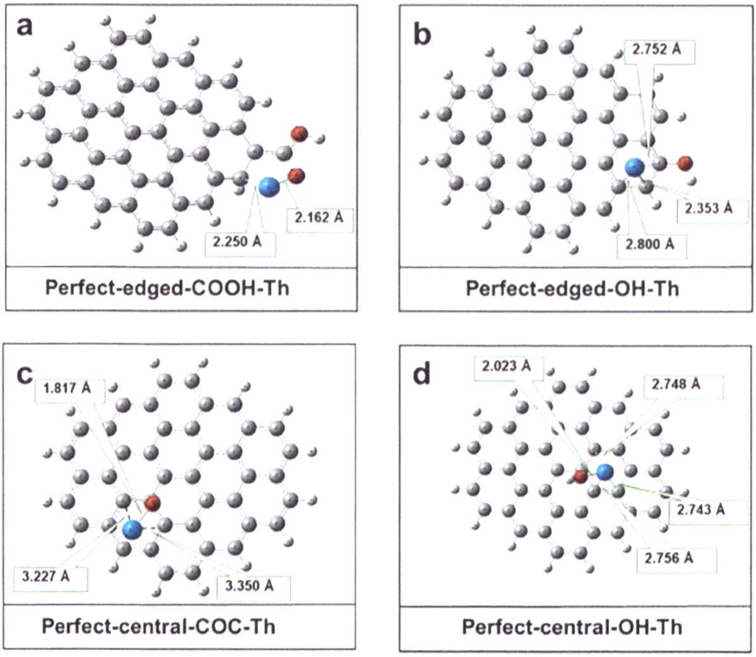

Fig. 7.12 Different structures of the oxygen functional group on the perfect graphene surface. **a** Perfect-edged-OH-Th^{4+}, **b** perfect-edged -COOH-Th^{4+}, **c** perfect-central-COC-Th^{4+}, and **d** perfect-central-OH-Th^{4+}. Reproduced with permission from Gao et al. [97]. Copyright 2022, Elsevier Science Ltd

hexagonal lattice structure at ambient temperature and in the high vacuum of the TEM. When the specimen is heated to 500 °C, it becomes more crystalline, and we can see the atomic structure of GO. To improve contrast, a section of GO was depicted with false color in Fig. 7.13a. Regions of disorder that divide the nanocrystallites from one another are apparent, as are the borders of the (folded over bilayer, high contrast) structure. The closed edges' great contrast distinguishes them from heavier atoms that would have contaminated the sample. At a temperature of 700 °C, Fig. 7.13b depicts the atomic structure of a patch of monolayer GO with enough clarity to show imperfections (pentagons and heptagons) in the hexagonal sp^2 lattice [98].

It is possible to use spectroscopic methods of analysis, such as solid-state nuclear magnetic resonance (NMR), Raman spectroscopy, X-ray photoelectron spectroscopy (XPS), and Fourier transform infrared (FTIR) spectroscopy to determine the chemical composition of GO and oxygen-containing functional groups. FTIR spectroscopy enables one to determine the existence of sorbed water, carboxyl and hydroxyl groups (broadband at 3000–3800 cm^{-1}), aliphatic groups (2700–2980 cm^{-1}), carbonyl groups (1650–1750 cm^{-1}), C = C (1500–1630 cm^{-1}), hydroxyl groups (1300–1480 cm^{-1}), and epoxy groups (1000–1280 cm^{-1}) in GO (Fig. 7.14c) [99].

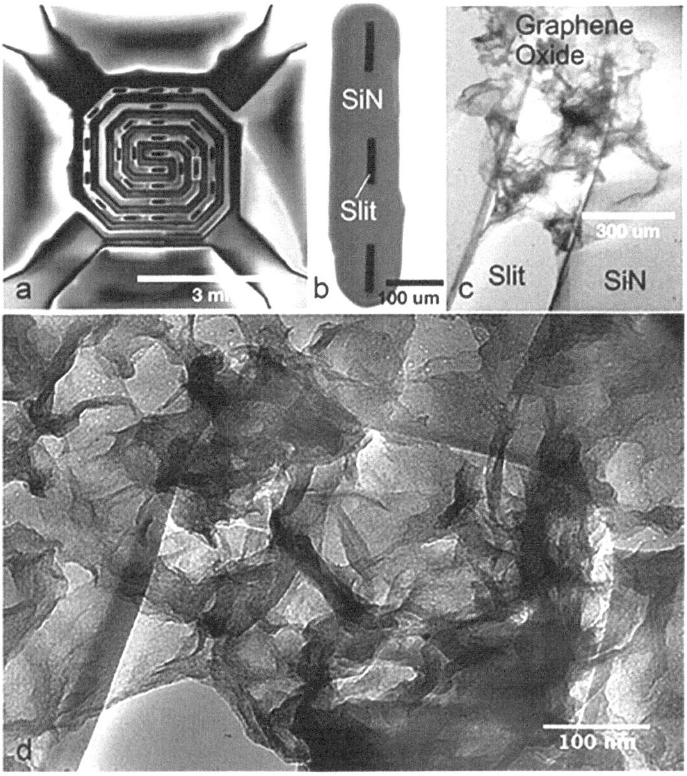

Fig. 7.13 Low-voltage, high-resolution TEM was used to directly observe the structure of graphene oxide. A heating chip containing a platinum coil **a** was used to study the temperature dependence of the structure. A slit was introduced into the thin silicon nitride membrane using a focused ion beam **b** in order to obtain suspended GO. Low-magnification images **c,d** show the macroscopic morphology of graphene oxide: clumped flakes with monolayer regions. Reproduced with permission from Dave et al. [98]. Copyright 2016, American Chemical Society

The NMR spectra of GO show that its basic lines are a signal at about 60 ppm attributable to carbon atoms coupled to epoxy functional groups, a signal related to carbon atoms coupled to hydroxyl functional groups at around 70 ppm, and a signal attributable to sp2 carbon at about 130 ppm. Lactone, carbonyl, and ketone groups are responsible for three further weak signals at 101, 167, and 191 ppm [100].

The characteristics of carbon and oxygen bonding in all of their forms, including unoxidized carbon atoms (sp^2 carbon), C–O–C, C = O, and COOH, could be revealed by the XPS technique (Fig. 7.15). Sp^2 carbon atoms (286.79 eV) and atoms of carbon joined to a hydroxyl (C–OH, 286.93 eV), epoxy (COC, 286.55 eV), carbonyl (C = O, 288.64 eV), OH (532.80 eV), CO–OH (532.10 eV), and carboxyl groups (COOH, 290.54 eV) are responsible for five different chemical components that make up the C1 s signal of GO, according to reports [101].

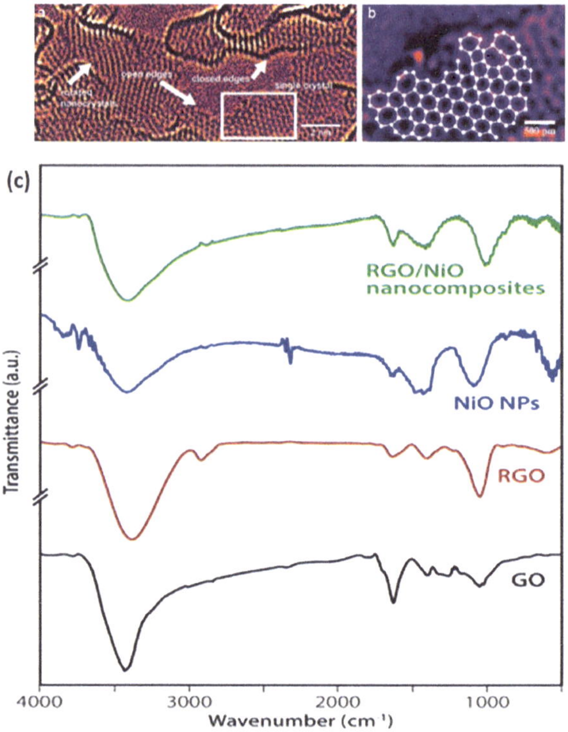

Fig. 7.14 Atomic structure can be resolved using in situ heating with AC-TEM at 700 °C with features shown in **a**. Sufficient resolution is achieved to identify defects in the bond structure **b**. In **b**, not all of the atomic structure could be accurately resolved, likely due to structural changes occurring during image acquisition, and only the directly resolved atomic structure is indicated by white dots and lines [98], and **c** FT-IR spectrum of GO, RGO, NiO NPs, and RGO/NiO nanocomposites. Reproduced with permission from Sadhukhan et al. [99]. Copyright 2022, Elsevier Science Ltd

All sp^2-hybridized oxidized carbon possesses a large G peak at around 1594 cm^{-1} in its Raman spectra, as does GO [99]. The spectrum also exhibits D, G, and D + G bands at 1334, 1594, and 2928 cm^{-1}, respectively, caused by structural flaws created by functional groups containing oxygen on the basal carbon plane [99].

The enormous number of structural flaws, the non-stoichiometric arrangement, and the wide range of functional groups containing oxygen, and inconsistent GO preparation procedures have led to a significant dispute regarding the chemical architecture of GOs. So, throughout the past 80 years, many structural (chemical) models of GiO and/or GO have been proposed [86, 89, 102–105], as summarized in Fig. 7.16. The highly regarded models for GO structure include the Hofmann, Ruess, Scholz-Boehm, Nakajima-Matsuo, Lerf-Klinowski, Dekany, and Ajayan models [7].

7.4.3 Reduced GO

GO was first just thought to be a means of producing cheap, mass quantities of graphene. The complete restoration of the structural characteristics of graphene was thus the most crucial objective of the transformation GO. At high levels of oxidation, the structure and chemical composition of GO are unstable. Due to the high surface

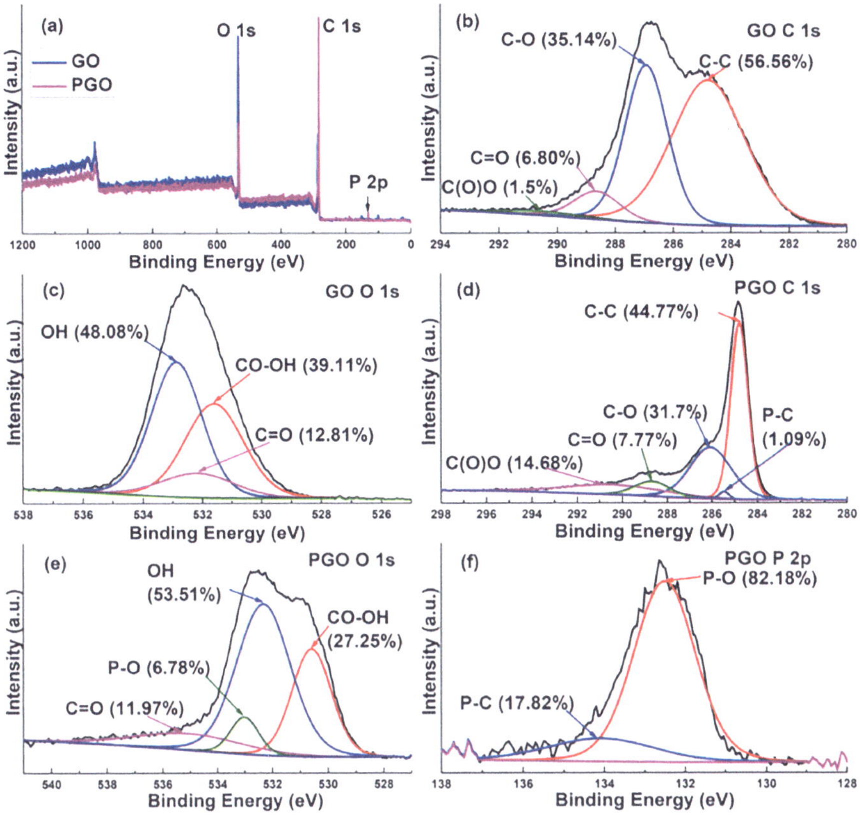

Fig. 7.15 **a** XPS survey spectrum of GO and PGO; **b** C 1 s; **c** O 1 s of GO; **d** C 1 s; **e** O 1 s; and **f** P 2p of PGO. Reproduced with permission from Qiyu et al. [101]. Copyright 2021, Elsevier Science Ltd

reactivity of the GO particle, the large portion of functional groups containing oxygen can be readily removed from the particle's basal plane. Several studies provide the widest selection of techniques for GO restoration [106–108]. The transformational techniques currently in use GO to graphene might be split into three groups: physical (thermal), chemical, and hybrid, where both physical and chemical factors are simultaneously at play.

The heating of GO either directly or by irradiation (microwave, infrared, visible, or ultraviolet (UV)) in a vacuum, inert or reducing atmosphere is a frequent characteristic of thermal techniques. The conversion of GO into rGO (reduced GO) occurs across a wide temperature range. Thermal reduction has the apparent advantage of simultaneously removing oxygen-containing groups and reestablishing the structure of GO through thermal annealing. It encourages the restoration of oxidation deficiencies of the GO basal plane afloat of sp^3-sp^2 rehybridization of carbon atoms [92].

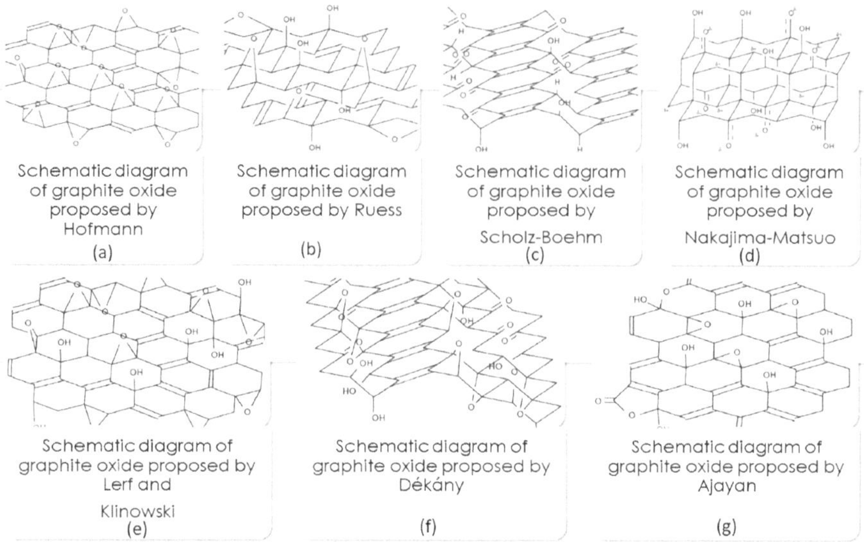

Fig. 7.16 GiO structural models: **a** Hofmann model, **b** Ruess model, **c** Scholz-Boehm model, **d** Nakajima-Mastsuo model, **e** Lerf-Klinowski model, **f** Dekany model, and **g** Ajayan model. Reproduced with permission from Orasugh and Ray [7]. Copyright 2022, Springer Nature

Different reactants are used in chemical reduction. Hydrazine hydrate is a frequently used reagent for the reduction of GO. Fe powder, sodium borohydride, hydroquinone, hexamethylenetetramine, hydroiodic acid, sodium, and potassium alkaline solutions are a few commonly used reactants that successfully restore GO to graphene. It could be done at a moderate or even room temperature in a liquid medium or vapor atmosphere [107]. Methods for catalyst-assisted photoreduction could potentially be categorized as chemical processes. For instance, under UV irradiation, the inclusion of titanium dioxide particles efficiently aids in reducing GO in an aqueous medium.

The solvothermal approach, which combines treatment with chemicals and heat under supercritical circumstances, can augment the extent of reduction attained by chemical methods [109]. The same hybrid strategy enables better electric conductance nitrogen doping of decreased GO. Under gentle UV irradiation in an argon environment, it was shown that ULGO platelets might be perforated under control. With this technique, carboxyl groups can simultaneously modify the margins of holes [110].

7.5 Adsorption and Its Kinetics, Isotherms, as Well as Thermodynamics

The mass transfer process of adsorption transfers contaminants from the liquid phase to the solid adsorbent. Because of its many benefits, including its straightforward design, affordable, low maintenance requirements, and high efficiency, Adsorption is among the most employed methods in water and wastewater treatment. The results of the adsorption kinetic study are utilized to determine the mass transfer mechanisms, adsorption rate, and performance of the applied adsorbent. Designing adsorption systems requires an understanding of the adsorption kinetics. The outward diffusion is the initial step. Adsorbate is transferred through the liquid film surrounding the adsorbent in this step. The concentration differential between the surface of the adsorbent and the bulk solution is what drives external diffusion. Internal diffusion comes next. Internal diffusion refers to the adsorbate's diffusion within the pores of the adsorbent. The third stage entails the adsorption of the adsorbate into the adsorbent's active sites [111]. The following sections explain numerous adsorption kinetic models, isotherms, and thermodynamics.

7.5.1 Adsorption Isotherms

Adsorption investigations are essential for establishing an adsorbent's adsorption capacity and propensity. As a result, many different isotherm equations have been produced, some of which have a scientific foundation and others more empirical in character.

7.5.1.1 Langmuir Isotherm Model (LM)

When studying the theoretical effects of gases adhering to solid surfaces, Langmuir (1918) viewed sorption as a chemical process [112]. To put it simply, the hyperbolic form of the Langmuir isotherm equation is:

$$q_e = q_{max} \frac{K_L C_e}{1 + K_L C_e} \tag{7.1}$$

where C_e represents the equilibrium adsorbate molar concentration in solution and q_e is the equilibrium weight-based adsorption capacity, q_{max} is the theoretical maximum adsorption capacity by weight, and K_L represents the equilibrium constant of the adsorption reaction with a 0unit of l mol^{-1} (mol L^{-1}).

The following presumptions are used to generate the Langmuir isotherm [113, 114]. Their corresponding plots are;

$$\frac{C_e}{q_e}\ vs\ C_e \tag{7.2}$$

$$\frac{1}{q_e}\ vs\ \frac{1}{C_e} \tag{7.3}$$

$$q_e\ vs\ \frac{q_e}{bC_e} \tag{7.4}$$

$$\frac{C_e}{C_e}\ vs\ q_e \tag{7.5}$$

This model and the Freundlich isotherm model (FRIM) [115] are the most adopted models in the study of dye adsorption kinetic.

7.5.1.2 Redlich–Peterson Isotherm Model (R-PIM)

Freundlich and Langmuir's isotherms are combined to create the Redlich-Peterson model, which has three elements. The empirical equation of the model, which that can be used for both homogeneous and heterogeneous systems of adsorption, has connections between concentrations in the denominator and numerator that are exponential and linear, respectively., that demonstrate the equilibrium of adsorption over a wide range of concentration [116].

$$q_e = K_{rp}\frac{C_e}{1 + \alpha_{rp}C_e^{\beta}} \tag{7.6}$$

where K_{rp} and α_{rp} are the Redlich-Peterson constants and β is essentially an exponent, ranging between 0 and 1. Equation 7.13 becomes the Langmuir isotherm equation if β is equal to 1. A Freundlich-type equation can approximate the Redlich-Peterson isotherm equation when the value of the term $\alpha_{rp}C_e^{\beta}$ is significantly larger than one [113].

7.5.1.3 Khan Isotherm Model (KM)

To represent the adsorption of aromatics by activated carbon, Khan et al. [117] suggested an isotherm equation, which can be expressed in the shortened version shown below [117]:

$$q_e = q_{max}\frac{b_k C_e}{(1 + b_k C_e)^{n_k}} \tag{7.7}$$

where the two constants b_K and n_K are present. The combined property of the Langmuir and Freundlich isotherm equations is shown in the Khan isotherm equation, much like the Sips and Redlich-Peterson isotherm equations. Equation 7.1, for instance, reduces to the Langmuir isotherm if n_K is equal to 1. In contrast, Eq. 7.1 can be modified to a Freundlich-type isotherm when the value of the term $b_K C_e$ is significantly more than unity.

7.5.1.4 Toth Isotherm Model (TM)

Equation 7.8, which describes adsorption on heterogeneous adsorbent surfaces, is an isotherm equation that was developed using the potential theory [118]:

$$q_e = q_{max} \frac{b_T C_e}{\left(1 + (b_T C_e)^{1/n_T}\right)^{n_T}} \tag{7.8}$$

where there are two constants, b_T and n_T. The Tóth isotherm equation failed to capture the feature of the Freundlich-type adsorption, whereas Eq. 7.8 reduces to the Langmuir-type isotherm equation if $n_T = 1$.

7.5.1.5 Radke-Prausnitz Isotherm Model (R-PIM)

The Radke-Prausnitz isotherm is frequently used to describe adsorption systems because it has several unique characteristics at low adsorbate concentrations. The Radke-Prausnitz isotherm is described by the following phrase [119]:

$$q_e = a_R b_R \frac{C_e^{\alpha_R}}{a_R + b_R C_e^{\alpha_R - 1}} \tag{7.9}$$

where a_R, b_R, and α_R are all used as constants.

7.5.1.6 Dubinin-Radushkevich Isotherm Model (D-RIM)

The adsorption is thought to be proportional to the volume of adsorbent pores, according to the Dubinin-Radushkevich (DR) isotherm model. This model is suited to heterogeneous surfaces and considers the pore structure in the adsorbent. The DR isotherm model describes physical or chemical adsorption processes [120, 121]. The matching equation is as follows:

$$q_e = q_{DR} e^{\left(-K_{DR} \varepsilon^2\right)} \tag{7.10}$$

where K_{DR} is the constant of the adsorption energy, which is connected to mean adsorption energy (E), which is further defined as $E = 1/\sqrt{2K_{DR}}$, ε is the Polanyi potential, and q_{DR} is the Dubinin-Radushkevich constant reflecting the putative monolayer saturation capacity [113].

$$\text{Thus, } \varepsilon = \text{RT}\ln\left(1 + \frac{1}{C_e}\right) \tag{7.11}$$

7.5.1.7 Frumkin Isotherm Model (FIM)

The species interaction with adsorbed species was considered when developing the Frumkin isotherm equation. The Frumkin isotherm can be expressed as follows [113, 122]:

$$\frac{\theta}{1-\theta}e^{(-f\theta)} = K_F C_e \tag{7.12}$$

where is the adsorbent surface's coverage percentage ($0 \leq \theta \leq 1$). Equation 7.12 can be further linearized by replacing it with its definition formula to provide the following result:

$$\ln\left[\left(\frac{q_e}{q_F - q_e}\right)\frac{1}{C_e}\right] = \ln K_F + 2f\frac{q_e}{q_f} \tag{7.13}$$

Given by the Dubinin-Radushkevich isotherm equation (q_{DR}), q_F is the theoretical monolayer saturation capacity, f is the interaction coefficient, and K_F is the equilibrium constant. Equation 7.20 simplifies to the Langmuir-type isotherm if $f = 0$, i.e., Adsorbate species do not interact with one another. The adsorption of colors on activated carbon made from discarded apricots has been studied using Eq. 1.13 [113].

7.5.1.8 Flory–Huggins Isotherm Model (F-HIM)

The Flory–Huggins's solution theory offers an efficient and straightforward mathematical model for the thermodynamics of polymer mixtures. Entropy and enthalpy are combined to form the dissolution process, which might then be described by the Flory–Huggins equation [113, 123, 124]:

$$\frac{\theta}{n_{FH}(1-\theta)^{n_{FH}}}e^{(-2n_{FH}\alpha_{FH}\theta)} = K_{FH}\,C_e \tag{7.14}$$

where n_{FH} is constant, FH is an efficient constant showing the correlation between adsorbed molecules, and K_{FH} is the so-called equilibrium constant determined by the Flory–Huggins isotherm [113, 125]. Equation 7.14 becomes the Frumkin isotherm, as can be observed if $n_{FH} = 1$.

7.5.1.9 Bet Isotherm Model

The BET equation, created in 1938 by Brunauer, Emmett, and Teller, describes multimolecular adsorption [113, 126]:

$$q_e = \frac{BC_e q^0}{(C_s - C_e)[1 + (B - 1)C_e/C_s]} \tag{7.15}$$

q^0 is the quantity of solute adsorbed per unit weight of adsorbent in producing a whole monolayer on the surface, B is a constant pertaining to the energy of interaction with the surface, and C_s is the solute's saturation concentration. According to the BET model, several adsorbate molecule layers are thought to form at the surface.

The BET model also assumes that a given layer does not have to form fully before the beginning of the following layers. As a result, there will be different types of surfaces involved in the equilibrium condition depending on how many layers of molecules are present on each surface site. Adsorption from solutions with the added supposition that layers subsequent to the first have equivalent adsorption energies [113].

7.5.1.10 Temkin Isotherm Model (TIM)

This isotherm was initially created by Temkin and Pyzhev (1940) and is founded on idea that the heat of adsorption will drop linearly with an increase in adsorbent coverage [127, 128]:

$$q_e = \frac{RT}{b_t}\ln(a_t C_e) \tag{7.16}$$

where R is the gas constant, T is the absolute temperature in Kelvin, b_t is an adsorption heat constant, and a_t is the Temkin isotherm constant [113].

7.6 Factors Influencing the Removal of Pharmaceuticals on GBAs

Adsorption is a process influenced by several governing factors that ultimately decide the final adsorption capacity. These factors encompass surface functional groups, sorption kinetics, temperature, pH, adsorbent dosage, initial adsorbate concentration, agitation/stirring/shaking, etc. as discussed herewith in this subsection. This is also the fate of graphene-based adsorbents (GBAs) for the adsorptive exclusion of pollutants from wastewater.

7.6.1 Surface Functional Groups

The surface functionalities present upon the GBAs surface (though not neglecting pore physisorption) are the main factors governing all other elements that have an impact on the adsorption mechanism of these adsorbents. It is also vital to remember that the adsorbate surface functionalities also play an important role in this respect. For effective adsorption of contaminants from wastewater, it is expected that the surface functionalities on the adsorbent surface as well as the adsorbate, should possess opposite charges to enhance better electrostatic attraction between both [129]. Numerous of these preferred materials fall under the category of graphene nanocomposites with different organic compounds, where it is well demonstrated that the existence of functional groups proficient in physically as well as chemically engaging with the contaminant molecules is a key factor in enabling the adsorption process [129]. To this end, Bhattacharyya et al. [129], in their report, demonstrated MB dye adsorption on the surface of graphene oxide/activated carbon through its surface functional groups, as presented in Fig. 7.17. It is well established that surface functionalities present in GBAs are very important to the adsorptive removal of contaminants [129–144]. This implies that the design and preparation of multi-functionalized GBA materials will go a long way in solving the challenge of selective adsorptive removal of contaminants from wastewater.

7.6.2 Effect of pH

The variation in the pH of wastewater has been shown to greatly affect the adsorptive removal of contaminants within the wastewater affecting the adsorbents' surface functionalities or charge. This phenomenon is, in most cases, different for diverse pollutants due to their surface functionalities interaction within the liquor. The solid surface's net charge reaches zero at the point of zero charges, or PZC. According to the pH of the medium, various functionalities on the GO and GO-based adsorbents surfaces could go through protonation and de-protonation processes. De-protonation

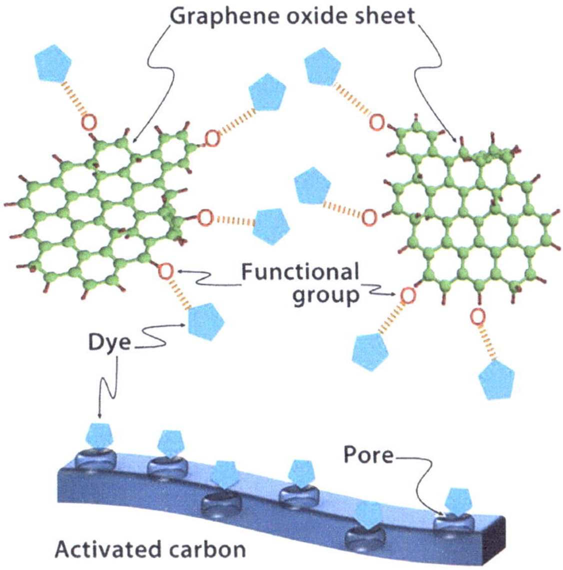

Fig. 7.17 The schematic diagram for the mechanism of adsorption. Reproduced with copyright permission from Bhattacharyya et al. [129]. Copyright 2021, Elsevier Science Ltd

of surface functional groups occurs as the medium pH rises over the PZC, making the surface more negative and enhancing the electrostatic interaction with the positively charged contaminants. However, protonation takes place when the pH decreases below the PZC, positively charging the surface to increase the electrostatic attraction for negatively charged contaminants.

Researchers have demonstrated in their work that changing the pH of Basic Fuchsin (BF) as well as Malachite Green (MG) dye solutions within the examined range did not affect their ability to adsorb material [145]. Their findings were supported by revelations reported by another group who concluded that MG adsorption is caused by electrostatic forces as well as other interacting forces, like hydrophobic interactions, because they did not see appreciable variations in MG adsorption with variation in the dye solution pH. For MB dye, it was found that the adsorptive capacity increases steadily as the pH rises [146]. This tendency was first noted when investigating how the initial pH affected the adsorption of MB by materials using GBA [147]. For the NB, however, the effect of the pH alteration was more apparent when the value was raised from 6 to 8. The molecular charge of the adsorbate and the adsorbent charge may be connected. Being a cationic dye, MB will compete with a large amount of H^+ for adsorption sites in acidic medium. As a result, the $-OH$ and $-COOH$ groups on the surface of the Agar-GO hydrogel protonate to generate $-OH_2^+$, causing electrostatic repulsion with the cationic dye and lowers the adsorption capacity. When the pH rises and the hydrogel's $-OH$ and $-COOH$ groups deprotonate, creating -COO- and -O-, the H^+ competition will nevertheless become less relevant. Consequently, the electrostatic pull will rise and the adsorption capacity will be enhanced [147]. Consequently, the authors used pH = 8 in subsequent testing for solutions of methylene blue (MB) and Nile blue A

(NB-A), while tests on malachite green and basic fuchsin were conducted without first altering the pH.

This kind of observation has also been supported elsewhere by Bhattacharyya et al. [129, 138] the authors demonstrated using activated carbon@GO composite for the adsorptive MB removal that the adsorption percentage was only about 40% when the pH is very low, or when the environment is acidic. This is so that the carboxylic acid groups, primarily from GO and present on the nanocomposite adsorbent, are not changed into their equivalent negatively charged carboxylate ions when the pH is very low. Thus, surface adsorption mostly occurs via the tiny pores on the AC surface. This is still the only adsorption mechanism at low pH levels, and as a result, the adsorption percentage remains very low. The acidic carboxylic groups begin to change into negatively charged carboxylate ions, generating a negative ion charge crib upon the adsorbents surface as the pH moves toward a basic medium. Therefore, as the pH is raised, two different kinds of adsorption begin to occur: surface adsorption, which is aided by the surface pores of the AC, and electrostatic attraction, occurring between the positively charged MB dye and the negatively charged carboxylate ions/groups. Higher pH values make both adsorption modes significant, and as pH rises, so does the adsorption percentage, which continuously rises until it attains >90% in a relatively basic media (pH 9). Similar observations with regard to cationic dye adsorption have been supported by other works of literature [129, 138, 146, 148].

Regarding the adsorption of toxic/heavy metals, reports have it that the adsorption process characteristic of GBAs at diverse pH slightly deviates diverse heavy metals [149]. Using MGO (PZC = 7.1), the impact of pH (4–10) on the adsorptive As (V) and As (III) speciation has been investigated (Yoon, 2017). As (V) species are found in negatively charged species beyond pH 2.2, whilst As (III) species are found in nonionic forms below pH 9.2, it is hypothesized that pH has significantly impacts on speciation. Due to the enhanced contests among the negatively charged As(V) species and the hydroxide ions (OH-) and increased negative charges on the adsorbent surface sites, As(III) removal percentage by MGO would not be significantly affected by a change in pH up to 9.1, but As(V) removal percentage does slightly decrease with a rise in the pH [150]. As opposed to the parent Cu^{2+} ions, which had a lower affinity for the Fe_3O_4@GO/MnO_x nanocomposite adsorbent, the hydrolyzed species of $Cu(OH)^+$, $Cu_2(OH)_2^+$, and other hydroxyl species were produced when the pH of the Cu^{2+} solution was raised [151]. However, an excessive rise in solution pH over the PZC of the adsorbent may result in precipitations of heavy metal ions in alkaline media, which would reduce the effectiveness of adsorption because the active sites of the adsorbent would be blocked. For instance, it has been shown that the precipitation of $Zn(OH)_2$ is mostly responsible for the steady decline in the removal percentage of Zn(II) upon MGO beyond pH 8.0 [152].

Another group of researchers likewise found, as shown in Fig. 7.18a, that only 60–65% of the dye was removed at pH 4–6, supporting the theory that the adsorption mechanism is mainly based on surface phenomena [138]. More than 80% of the color was absorbed when the pH was raised to 8. They attributed this to the de-protonation of the adsorbent's surface functionalities at elevated pH levels, which caused a negative charge buildup upon the surface of the adsorbent, which attracted the positively charged dye molecules. Nearly 89–90% of the dye was adsorbed at a pH of 9, beyond which there was no further increase in adsorption. The movement of dyes may be hampered in extremely alkaline environments. They concluded that the medium is acidic at lower pH values, the carboxylic functionalities maintain their form, while no anions are generated. It is reasonable to infer that the adsorption occurring at lesser pH is nearly exclusively surface adsorption since electrostatic pull does not impact the adsorption process in the absence of negatively charged groups. On the other hand, the carboxylic acid groups change into carboxylate anions at elevated pH levels when the adsorption liquor is basic, leading to a connection between the negatively charged groups and the positively charged entities of the cationic dye. Only a minor portion of the adsorbate is adsorbed upon the adsorbent surface at elevated pH levels; the majority occurs through electrostatic bonding [138]. Hence, it appeared that a pH of 9 would be ideal for effective adsorption.

It can be resolved that dependent on the surface charge of the GBA used but also the surface charge of the adsorbate to be removed, the pH of the adsorption liquor most favor the creation of opposite surface charges on the surface of the adsorbent as well as the adsorbate for effective adsorption process to occur.

7.6.3 Temperature Effect

The influence of temperature on the adsorptive removal of pollutants by GBA materials has also been studied by researchers to a great extent within recent decades [129, 138, 148, 153–155].

Results from adsorption experiments for the adsorptive removal of MB were evaluated at various experimental temperatures (20–50 °C). The MB% removed augmented from 88.35% to about 98%, and it was clear from the data that there was a simultaneous rise in temperature from 20 to 40 °C. However, the amount removed marginally decreased at 45 °C to 96.64% and did not significantly change as the temperature rose. An ideal temperature rise can increase porosity, consequently, overall adsorbent pore volume. As a result, the dye molecules would have been more likely to diffuse into the GO@CNC pores and stick to the outside surface. It has also been hypothesized that this process was endothermic by the temperature-dependent rise in the percentage of MB elimination. The physical links and interactions amongst the active sites for adsorption upon the molecules of the adsorbent and adsorbate might, however, become weaker at extremely elevated temperatures (in this example, 50 °C and beyond) [153].

According to Bhattacharyya et al., temperature is one of the key deciding factors for the adsorption process. The adsorbent content was kept at 35 mg/L, and the authors looked at variations in the adsorption percentage versus temperature, as shown in Fig. 7.17d below [138]. They discovered that whereas adsorption is minimal at low temperatures, it becomes more significant as the temperature rises. The improved dye molecule diffusion rate and the decreased solution viscosity of the adsorbent particles may contribute to the temperature-dependent improvement in the adsorption

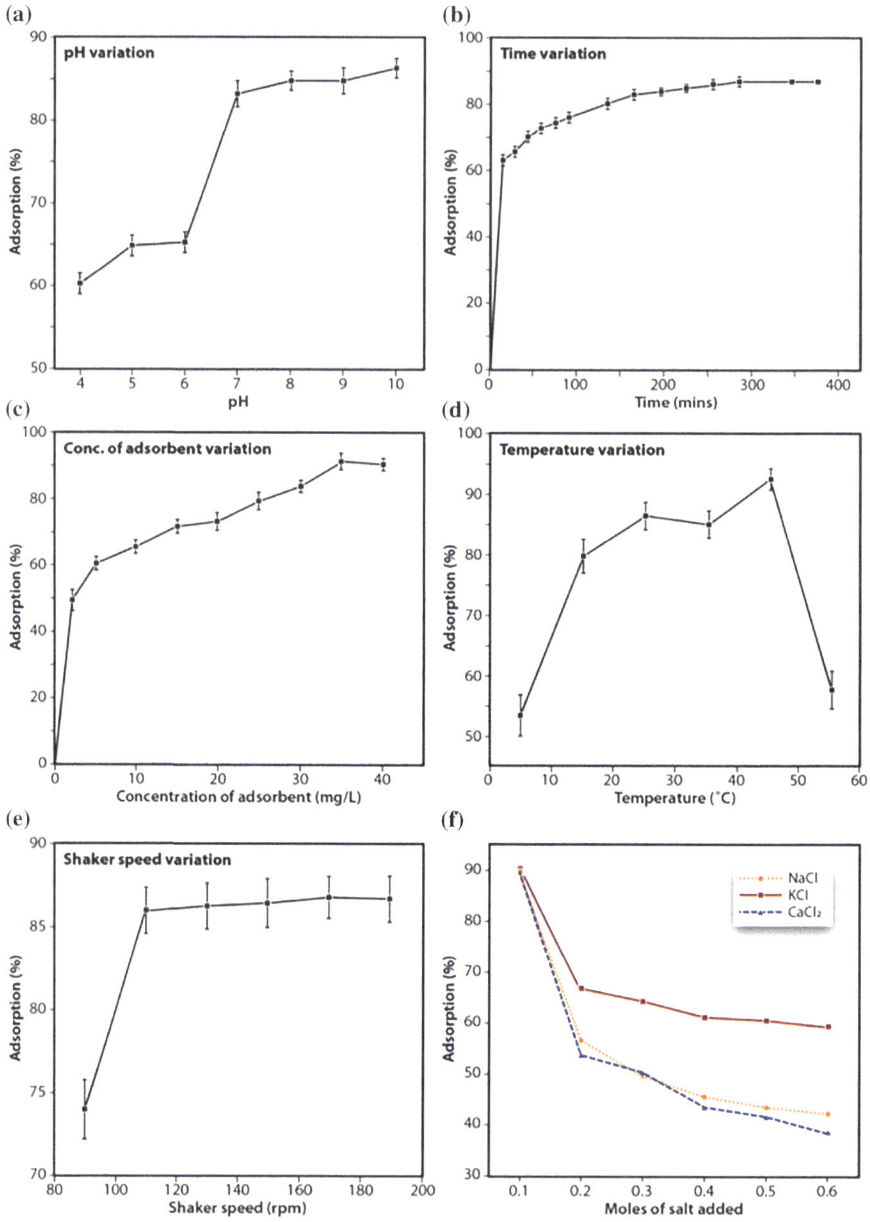

Fig. 7.18 Using GO-glu-Starch bio-composite: **a** Effect of pH on MB adsorption, **b** Effect of time on MB adsorption, **c** Effect of adsorbent concentration on MB adsorption, **d** Effect of temperature variation on MB adsorption, **e** Effect of shaker speed variation on MB adsorption, and **f** Effect of addition of salt concentration on MB adsorption. Reproduced with permission from Bhattacharyya et al. [138]. Copyright 2018, Elsevier Science Ltd

rate. Maximum adsorption efficiency occurs at 45 °C, the ideal temperature when both the surface and electrostatic attraction work efficiently. The solute molecules' higher kinetic energy decreases the electrostatic interaction between the dye and the adsorbent composite as the temperature increases. Hence, certain dye molecules become dislodged from the adsorbent, which causes the percentage of adsorption to drop at 55 °C. The nature of adsorption, in this case, changes to primarily surface adsorption [138]. This phenomenon has been supported by other reports [129, 148, 153]. Figure 7.17a–f presents most of the effect of other factors (pH, adsorbent concentration, temperature, shaker speed variation, and addition of salt) affect the adsorption of MB on GBA.

So, we can predictively conclude from the available literature that the adsorption behavior of GBAs is better at temperatures not greater than 45 °C.

7.6.4 Effect of Contact Time

The time of contact's impact or the time needed to eliminate different contaminants from wastewater upon GBA material is a crucial element in its adsorption process but also the mechanism's prediction. The elimination of pollutants often begins quickly, slows slowly, and eventually stops once equilibrium is reached [149]. This phenomenon occurs because the adsorbent initially has many free adsorption sites on its surface. The adsorption operation slowed down with time and eventually reached saturation as the adsorption sites became occupied along with a decline in their number. For instance, MB has been reported elsewhere to display 75% removal using MrGO (manganese ferrite functionalized rGO) within the first 10 min of contact, but equilibrium wasn't attained until 60 min had passed [156]. Corresponding to this, the absorption of Pb(II) by TEPA-GO/$MnFe_2O_4$, as reported by Xu et al., was said to be fast for the initial 20 min before slowing down until equilibrium was attained after 90 min [157]. This behavioral trend by GBAs has been supported by other researchers [154, 157–160].

7.6.5 Initial Concentration of Adsorbate

The preliminary amount of pollutant influences the adsorption process. Nevertheless, the proportion of pollutant removal declines as the preliminary amount of the contaminant rises. A larger starting concentration of the pollutant may help the adsorption process since, normally, the equilibrium adsorption capacity increases with pollutant concentration [149]. The higher likelihood of collision between contaminant ions and adsorbent particles may explain why the absorption capacity of GBAs increases as contaminant concentration rises. Additionally, the fact that the contaminant's concentration gradient is relatively high and all of GO's surface active sites are unoccupied at the beginning may contribute to the difference in the extent of adsorption [149].

Since initial dye molecule concentrations vary between industries, it is crucial to assess the ability of adsorbents to remove various amounts [161]. This author looked at how the initial concentration of the methyl violet dye impacted the effectiveness of adsorption using various adsorbers. The adsorption yields utilizing MMT, MMT/$CoFe_2O_4$, and MMT/GO/$CoFe_2O_4$ declined from 96.43, 97.18, and 98.65 to 28.55, 29.65, and 32.41%, respectively, when the starting dye concentration was raised from 10 to 300 mg/L. All dye molecules attach to and interact with the adsorbent surface in low quantities, which may explain why the adsorption process performs worse when the initial concentration of pollutants increases. With high adsorbate concentrations, part of the dye molecules remains in the aqueous solution because the MV molecules cannot completely adhere to the adsorbent surface [161].

With regards to available literature, it has been proven over and over that increase in adsorbate concentration, be it dyes, toxic metallic ions, radionuclides, pharmaceutical residues, pesticides, etc., beyond the threshold results directly in decline in the adsorption of the contaminant(s) [129, 148, 153–155, 160].

7.6.6 Effect of Different Coexisting Competing Ions

The presence of different ionic groups within the adsorption liquor is of great importance in the adsorption process as it directly correlates to the pH of the liquor, which in turn influences the selective adsorption process.

To demonstrate the effect of salt concentration (affecting several coexisting competing ions) on the effectiveness of MB adsorption, the authors created solutions of conventional inorganic salts with various molar concentrations while holding all other variables constant. The base is added, forming carboxyl and negative hydroxyl ions from the carboxylic acid and hydroxyl groups. Positively charged MB molecules then attach to the negative ions, removing the MB dyes. The "ionic environment" may gradually improve the shielding of the charge of cationic dye molecules due to the addition of salt, which lowers the rate of adsorption. The formation of outer sphere complexes is demonstrated here by ionic-strength-dependent adsorption [138]. The percentage of adsorption reduces with the inclusion of KCl and NaCl salt, where Na^+ and K^+ ions are generated. The decline is more pronounced in the presence of Na^+ ions than K^+ ions because their sizes are smaller. Due to the presence of alkaline hydroxyl ions within the liquor, $CaCl_2$ also forms $Ca(OH)_2$, which precipitates in addition to opposing the sites for adsorption via the ions. Since the hydroxyl functionalities are employed to create the precipitated calcium hydroxide in the case of $CaCl_2$, this occurrence may also be partially to blame for the decrease in adsorption. This finding shows that the primary driving factor behind adsorption is the electrostatic interaction between the cationic dye molecules and the anionic adsorbent [129, 138]. Figure 7.18f demonstrates the effect of contending ions on the adsorptive performance of the adsorbent on MB (dye). The authors proved in their report that the addition of salts that created an ion-contending environment for the dye or adsorbent led to reduced adsorptive removal of MB.

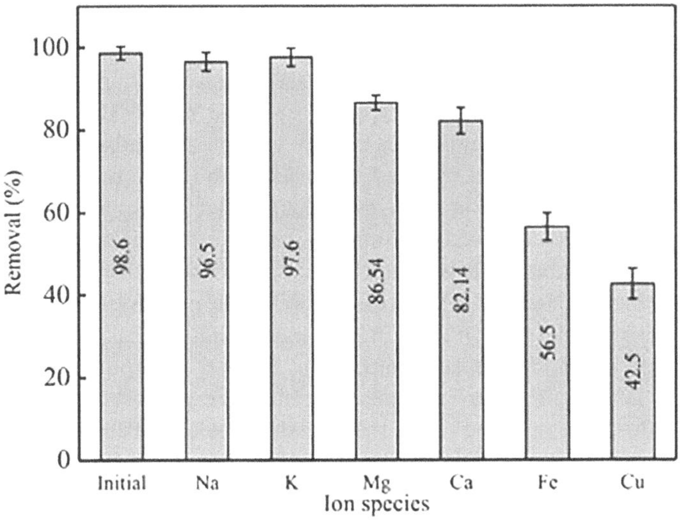

Fig. 7.19 Effect of different competitive ions on uranium removal rate ($C_0 = 50$ ppm, pH = 5, T = 25 °C, P = 101 kPa, E = −1.2 V, t = 200 min). Reproduced with copyright permission from Cao et al. [162], Elsevier, 2022

In actual applications, contending ions are unavoidable: a group of academics looked at how various ions affected the electro sorption of uranium upon GO@NC-700. The six competing ions that have been chosen are Fe^{2+}, K^+, Ca^{2+}, Na^+, Mg^{2+}, as well as Cu^{2+} (Fig. 7.19). The impact of each ion's pairing with the contaminant at the same concentration on uranium exclusion was studied. The six ions were grouped into three groups based on their impact on uranium removal effectiveness. The alkali metals Na^+ and K^+ did not considerably impact the removal rate of uranium. The removal rate of uranium was greatly influenced by the alkaline earth metals Mg^{2+} and Ca^{2+}. After including Mg^{2+} and Ca^{2+}, the removal efficiency decreased to 86.5% and 82.1% as compared to the uranium solution in the blank. Whenever the charge stayed the same, ions with a lower hydration radius were much more dominating. Thus the selectivity of electro-sorption to ions remained highly dependent upon the charge of the ion [162]. Contrary to alkali metals, transition metals (Fe^{2+}, Cu^{2+}) can perform redox reactions within this site as well as generate precipitation to prevent the uranium adsorption, which has a significant impact on the removal rate of uranium [162].

7.6.7 *Effect of Agitation*

The effect of agitation, though only studied by a few researchers, is critical regarding GBA material's rate of adsorption and its effectiveness, as it enhances the transportation of the adsorbate to the adsorbent and vice-versa.

At a temperature of 45 °C along with other optimal adsorption conditions, Bhattacharyya et al. investigated the impact of shaker (agitation rate) speed on the adsorption process (Fig. 7.18e) [138]. They found that the adsorption percentage remains low at a relatively low speed of 90 rpm because the dye (MB) molecules could not approach the adsorbent molecules closely enough. The adsorption percentage improved with increased shaker speed up to 110 rpm as the adsorbate molecules could get in close juxtaposition to the essential architectural elements of the GBA. They explained the occurrence by pointing out that, in part, as agitation speed was increased, the rate of dye diffusion from the liquid bulk towards the boundary liquid layer next to biocomposite GBA increased due to the improvement of turbulence, leading to enhanced adsorption [37]. In this case, the adsorption process is aided by surface adsorption and electrostatic attraction. Beyond 110 rpm, where the pollutants and the adsorbent are properly near together, the proportion of adsorption is unaffected. This is mostly due to the robustness of the electrostatic bond, which is unaffected by an increase in rotational speed [138]. The report by these authors has been substantiated in their other report(s) [129].

7.6.8 *Effect of Adsorbent Dosage*

The effect of increasing or decreasing the adsorbent amount on the adsorption of wastewater contaminants is vital as it directs the researchers to decide the optimum amount of adsorbent to be utilized for effective wastewater remediation without wastage, thereby making the process economical. This factor may not present the same trend for the adsorption behavior of graphene-based materials with respect to dyes [148, 153], radionuclides [163, 164], toxic metals [165], pharmaceutical residues [166, 167], etc.

With regards to radionuclides, Lingamdinne et al. discovered an intriguing outcome of the rGONF but also GONF dose impact on Th (IV) along with U(VI) adsorption [163]. As the dosage of rGONF increased from 0.1 to 0.3 g/L, the elimination % of Th(IV) and U(VI) increased quickly until reaching a threshold around 0.3 g/L and 0.4 g/L for Th(IV) and U(VI), respectively. As opposed to that, as the dosage of GONF grew from 0.1 to 0.4 g/L and attained its threshold of 0.5 g/L, the elimination of U(VI) and Th(IV) was quick. However, when the dosage of rGONF and GONF rose, the U(VI) and Th(IV) adsorption capacity per unit mass (q_e) declined (Fig. 7.20) [163]. Their findings were attributed to the intramolecular interactions between adsorbent molecules, which decreased the amount of active sites per unit area of rGONF and GONF surfaces. Though the scenario conversed above may not be the same for all radionuclides removal from wastewater using an adsorption process for graphene-based materials.

No matter the kind of GBA material used for the adsorptive removal of wastewater contaminants, its efficiency should be tested towards the maximum adsorbent amount/dosage required for maximum adsorption capacity.

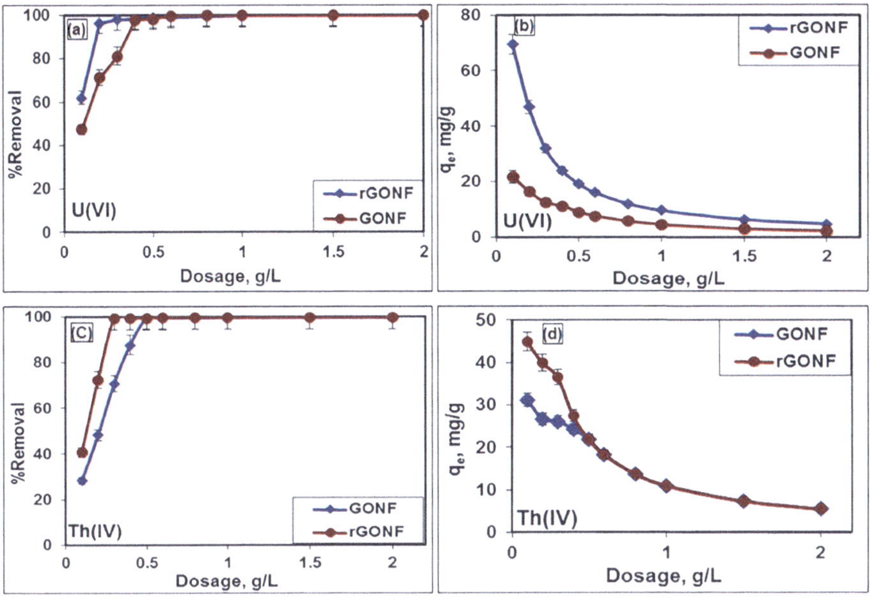

Fig. 7.20 GONF and rGONF dosage effect on the adsorption removal of U(VI) **a, b** and Th(IV) **c, d** at pH 3.5 and room temperature for pre-determined equilibration time. $C_0 = 10$ mg/L. Reproduced with permission from Lingamdinne et al. [163]. Copyright 2017, Elsevier Science Ltd

7.7 Application of GBA Materials for Water Purification

7.7.1 Exclusion of Toxic/Heavy Metals from Contaminated Water Using GO and Its Composites/Hybrids

Owing to their exceptional physicochemical characteristics, including chemical stability, vast surface area, elevated adsorption capacity, as well as recyclability, graphene-based nanomaterials as well as their advanced materials are regarded as auspicious adsorbents for the elimination of toxic metals, dyes, as well as other hazardous contaminants from aqueous environments. The acceptable limits of toxic/heavy metals in water bodies as per WHO and US-EPA are presented in Table 7.2.

The utilization of GBA materials for the exclusion of toxic/heavy metals from wastewater has been broadly explored by researchers. Readers are referred to Table 7.3 below for some instances as per the literature.

Table 7.2 Permissible levels of heavy metal pollutants in water. Reproduced with permission from Adel et al. [149], Elsevier, 2022

Permissible level (mg L^{-1})	WHO	US-EPA
Total dissolved solids	500	500
Manganese	0.01–0.02	0.05
Arsenic	0.010	0.01
Cadmium	0.005	0.005
Chromium (VI)	0.1	0.05
Lead	0.01	0.015
Mercury	0.006	0.002
Copper	2.0	1.3
Iron	0.3	0.3
Zinc	3.0	5.0
Nickel	0.07	0.02
Thallium	0.002	0.002

7.7.2 Removal of Organic Contaminants Using GO/GOs-Based Composites

The application of GO, rGO, and its modified forms, along with its composites/hybrids for the adsorptive exclusion of organic contaminants from wastewater, has been widely explored by researchers in recent decades [176–178]. These organic contaminants may be dyes/pigments [168, 178], pharmaceutical residues [177], organic pesticides [179], etc.

7.7.2.1 Removal of Pharmaceutical Traces from Wastewater

The remains of pharmaceutical drugs released into wastewater bodies from animal/human urine or excretion are all, in most cases, washed into water bodies, thereby creating toxicity to the water. The removal of these harmful pharmaceutical residues is highly recommended in water treatment plants for effective wastewater remediation towards its reuse.

The utilization of graphene, its oxides, as well as its derivatives and composites/hybrids such as has been widely explored by researchers globally towards the adsorptive removal of pharmaceutical residues from wastewater/effluents [160, 180, 181]. It can be postulated with considerations from the available literature that the kinetics and isotherms mostly followed by pharmaceutics adsorbed onto graphene-based materials is Pseudo-second-order kinetic (PSOM) and Langmuir isotherm model (LM), as can be observed from Table 7.4.

Table 7.3 GBAs and their application for toxic metals removal

Formulation	Toxic metals removed	Adsorbent dosage	pH for max. adsorption	Temp (K)	Maximum adsorption capacity	Adsorption isotherm and kinetics		Refs.
						Isotherm	Kinetics	
Polypyrrole functionalized Cobalt oxide Graphene (COPYGO)	Pb (II)	0.01 g/L	5.5	323.5	780.363 mg/g	LM	PSOM	[168]
	Cd (II)	0.01 g/L	6.1	323.5	794.188 mg/g	LM	PSOM	[168]
	CR	0.01 g/L	5.2	323.5	659.056 mg/g	LM	PSOM	[168]
Nitrogen, sulfur-graphene nanoplatelets (NS-GrNPs)	Hg^{2+}	100 mg/L	–	–	987.06 mg/g	–	–	[169]
	Pb^{2+}	100 mg/L	–	–	935.58 mg/g	–	–	[169]
	Cu^{2+}	100 mg/L	–	–	776.2 mg/g	–	–	[169]
GO@lignin	Pb^{2+}	20 mg/L	7.0	298.15	147.88 mg/g	LM	PSOM	[170]
	Ni^{2+}	20 mg/L	7.0	298.15	110.25 mg/g	LM	PSOM	[170]
PVC@PPD@GO	Pb^{2+}	–	6.5	343.15	45 mg/g	–	–	[171]
	Cu^{2+}	–	6.5	343.15	39 mg/g	–	–	[171]
	Ni^{2+}	–	6.5	343.15	37 mg/g	–	–	[171]
	Co^{2+}	–	6.5	343.15	36 mg/g	–	–	[171]
	Zn^{2+}	–	6.5	343.15	34 mg/g	–	–	[171]
	Cd^{2+}	–	6.5	343.15	33 mg/g	–	–	[171]
GO chitosan	Hg^{2+}	50 mg/L	4.5	Room temp. (RT)	324 ± 3.30 mg/g	LM	PSOM	[172]

(continued)

Table 7.3 (continued)

Formulation	Toxic metals removed	Adsorbent dosage	pH for max. adsorption	Temp (K)	Maximum adsorption capacity	Adsorption isotherm and kinetics		Refs.
						Isotherm	Kinetics	
	Cu^{2+}	50 mg/L	4.5	RT	130 ± 2.80 mg/g	LM	PSOM	[172]
GO/montmorillonite (G/M/S)	Cu^{2+}	10 mg/L	6.0	323.15	100.82 mg/g	FM	PSOM	[173]
TrGO	Pb^{2+}	80 mg/L	4.0	298.15	803.84 mg/g	FM	–	[174]
	Cd^{2+}	80 mg/L	4.0	298.15	395.80 mg/g	FM	–	[174]
Magnetite/G-AC-800	Pb^{2+}	0.25 g/L	5.0	RT	153.2 mg/g	LM	PSOM	[175]
Polyampholyte hydrogel adsorbent functionalized with β-cyclodextrin and graphene oxide (P(AA-MMA)/MGO/CA-CD/NH_2)	Cu^{2+}	0.5 g/L	6.0	298	140 mg/g	LM	PSOM	[176]

Table 7.4 GBAs and their application for pharmaceutical traces removal

Formulation	Pharmaceutical traces removed	Adsorbent dosage	pH for max. adsorption	Temp. (K)	Maximum adsorption capacity	Adsorption isotherm and kinetics		Refs.
						Isotherm	Kinetics	
GO	Carbamazepine (CBZ)	1 g/L	2.0	308,15	9.2 mg/g	LM	PSOM	[160]
Graphene	Sulfamethoxazole (SMX)	60 mg/L	8.0	–	210.08 mg/g	LM	PSOM	[180]
	Acetaminophen (ACM)	60 mg/L	6.0	–	56.21 mg/g	LM	PSOM	[180]
GO	Metformin	40 mg/L	6.0	288–303	50.47 mg/g	Freundlich isotherm (FRIM)	PSOM	[166]
GO	Acetaminophen	200 mg	8.0	296–3323	18.06 mg/g	Type IV isotherm with an H3 type hysteresis loop	PSOM	[177]
	Caffeine	200 mg	8.0	77	19.72 mg/g	Type IV isotherm with an H3 type hysteresis loop	PSOM	[177]
	Aspirin	200 mg	12.0	77	13.02 mg/g	Type IV isotherm with an H3 type hysteresis loop	PSOM	[177]
rGO@magnetite (RGO-M)	Ciprofloxacin (CIP)	0.2 g/L	6.2	298	18.22 mg/g	LM	PSOM	[167]
	Norfloxacin (NOR)	0.2 g/L	6.2	298	22.20 mg/g	LM	PSOM	[167]

(continued)

Table 7.4 (continued)

Formulation	Pharmaceutical traces removed	Adsorbent dosage	pH for max. adsorption	Temp. (K)	Maximum adsorption capacity	Adsorption isotherm and kinetics		Refs.
						Isotherm	Kinetics	
GO	Tetracycline	0.181 mg/mL	3.6	298,15	313 mg/g	LM	PSOM	[182]
GO	Trimethoprim (TMP)	15 mg/L			204.08	LM	PSOM	[181]
	Isoniazid (INH)	15 mg/L			13.89	LM	PSOM	[181]

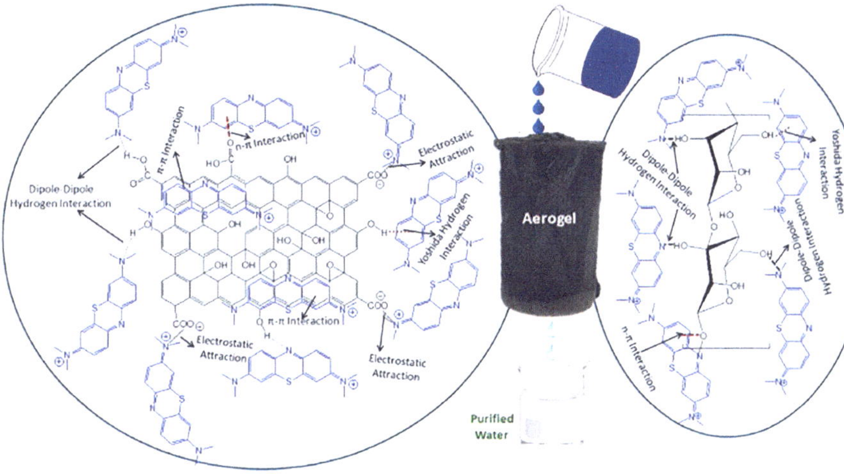

Fig. 7.21 Schematic representation of plausible interactions between active surface sites/chemical functionalities of CGA and MB dye molecules. Reproduced with permission from Joshi et al. [183]. Copyright 2022, Elsevier Science Ltd

7.7.2.2 Removal of Dyes

Adsorptive adsorbent materials must meet critical criteria for commercial feasibility and ecological sustainability to be practical for wastewater cleanup applications, especially for toxic dyes discharged into water bodies from diverse human or industrial activities. As per literature, graphene-based materials as adsorbents have been shown to meet these demands considering their effective adsorptive removal of dyes/pigments from wastewater and their ability to be regenerated easily [168, 176, 178, 183, 184].

It has been proposed that the interactive routes that aid the adsorptive removal of dyes/pigments from wastewater bodies by graphene-based materials are cation-π interactions, n-π, π-π, electrostatic, dipole–dipole, as well as Yoshida hydrogen bonding/linkages as in Fig. 7.21 [183]. As per the available literature within our reach, we have presented diverse GBA materials utilized for the adsorptive removal of numerous dyes by researchers in Table 7.5.

7.7.2.3 Removal of Pesticides

All living things require access to clean water; thus, human activities are strongly tied to water safety [179]. The main sources of pesticides in bodies of water are diffuse and point sources, especially the diffuse source [179].

The use of graphene nanostructured materials aimed at pesticide adsorption has recently gained a lot of interest from researchers [179].

Table 7.5 GBAs and their application for dyes/pigments removal

Formulation	Dyes/pigment removed	Adsorbent dosage	pH for max. adsorption	Temp. (K)	Maximum adsorption capacity	Adsorption isotherm and kinetics		Refs.
						Isotherm	Kinetics	
Polypyrrole functionalized Cobalt oxide Graphene (COPYGO)	MB	0.01 g/L	7.2	308.5	663.018 mg/g	LM	PSOM	[168]
GO chitosan	MB	50 mg/L	7.1	RT	141 ± 6.60 mg/g	LM	PSOM	[172]
	Crystal violet (CV)	50 mg/L	7.1	RT	121 ± 3.50 mg/g	LM	PSOM	[172]
P(AA-MMA)/MGO/CA-CD/NH_2	MB	0.5 g/L	12.0	318	3185 mg/g	LM	PSOM	[176]
	Malachite green (MG)	0.5 g/L	8.0	298	3315 mg/g	LM	PSOM	[176]
	CR	0.5 g/L	6.0	298	1058 mg/g	LM	PSOM	[176]
Montmorillonite-reduced graphene oxide composite aerogel (M@rGO)	MB	-	6.5		450.90 mg/g	LM	PSOM	[178]
Cellulose@GO (CGA)	MB	50 mg/100 mL	-	RT	490 mg/g	LM	PSOM	[183]
polyaniline/dicarboxyl acid cellulose@GO (PANI@DCC@GO)	Reactive brilliant red K-2G	40 mg/150 mL	2.0	308.15	733.3 mg/g	Freundlich (FM)	PSOM	[184]
montmorillonite/graphene oxide/$CoFe_2O_4$ (MMT/GO/$CoFe_2O_4$)	Methyl violet (MV)	1 g/L	9.0	298.15	97.26 mg/g	LM	PSOM	[161]
polysaccharide@GO (PS-GO)	MB	5 mg/L	10.0	298	789 mg/g	LM	PSOM	[185]
CNC: GO	Basic blue 7 (BB7)	0.1 g/L	6.0	RT	891	Redlich-Peterson (RPM)	PSOM	[186]

(continued)

Table 7.5 (continued)

Formulation	Dyes/pigment removed	Adsorbent dosage	pH for max. adsorption	Temp. (K)	Maximum adsorption capacity	Adsorption isotherm and kinetics		Refs.
						Isotherm	Kinetics	
	Reactive orange 122 (RO)	0.1 g/L	6.0	RT	711	RPM	PSOM	[186]
rGO	MG	10 mg/L	8.0	303	279.85	LM	PSOM	[187]
	Rhodamine B (RhB)	0.1 g/L	6.0	RT	560	RPM	PSOM	[186]

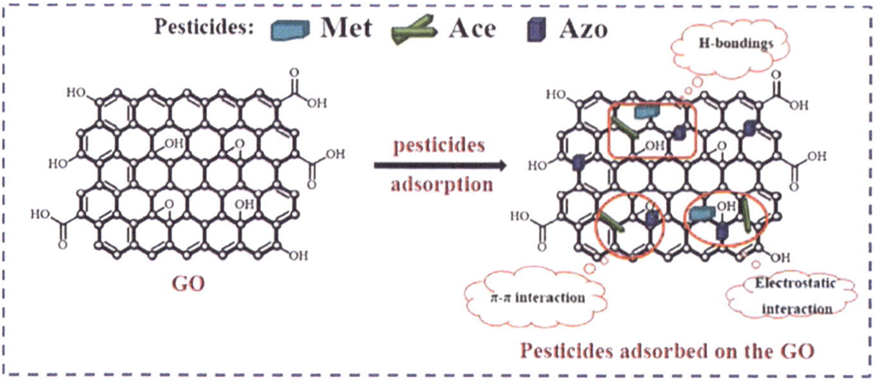

Fig. 7.22 Possible mechanism for the adsorption of pesticide molecules on GO/rGO. Reproduced with permission from Shi et al. [179]. Copyright 2022, Elsevier Science Ltd

The adsorption of pesticides using GBA materials has been postulated to be due to three interfacial/interacting mechanisms, namely [179, 188];

(i) H-bonding interaction.
(ii) Electrostatic interaction as well as.
(iii) π-π interaction.

These interacting mechanisms have been demonstrated by Shi et al. [179], in their study, where they utilized both GO and rGO for the removal of Methomyl (Met), Acetamiprid (Ace), and Azoxystrobin (Azo) (Fig. 7.22). As per the literature, several reports on the use of various GBA materials for adsorptive removal of pesticides from wastewater are presented in Table 7.6.

7.7.3 Removal of Radionuclides

Because exposure of living organisms to radioactive wastewater can result in various health issues due to chemical toxicity and radiation, radioactive waste discharge being a source of worry is currently on the rise. Large amounts of radionuclides, such as radioactive Cs isotopes with long half-lives, ^{134}Cs, ^{135}Cs, and ^{137}Cs, are created during fission and found in effluent from research facilities, nuclear power plants, nuclear weapons testing, and spent fuel reprocessing. As a result, it is crucial to remove Cs from wastewater [191, 192]. Given their strong affinity, good solubility in water, and ease of movement in the environment, it is still challenging to remove these radioactive contaminants from wastewaters effectively [192].

Radionuclides can be removed with the help of ion exchange, membrane separation, chemical precipitation, adsorption, and evaporation, which are very effective processes. Due to its low cost and ease of use, adsorption is regarded as a practical approach. For the treatment of radioactive effluents, many different adsorbents have

Table 7.6 GBAs and their application for pesticides removal

Formulation	Pesticides removed	Adsorbent dosage	pH for max. adsorption	Temp. (K)	Maximum adsorption capacity (mg/g)	Adsorption isotherm and kinetics		Refs.
						Isotherm	Kinetics	
GO	Methomyl (Met)	1 mg/L	12.0	298	106.22	Sips model	PSOM and Elovich models (EVM)	[179]
	Acetamiprid (Ace)	1 mg/L	2.0	298	285.96	Sips model	PSOM and EVM	[179]
	Azoxystrobin (Azo)	1 mg/L	2.0	298	2896.84	Sips model	PSOM and EVM	[179]
rGO	Methomyl (Met)	1 mg/L	12.0	298	96.86	Sips model	PSOM and EVM	[179]
	Acetamiprid (Ace)	1 mg/L	2.0	298	357.65	Sips model	PSOM and EVM	[179]
	Azoxystrobin (Azo)	1 mg/L	2.0	298	2818.04	Sips model	PSOM and EVM	[179]
AS/NZVI/GO	Methomyl	10 mg/L	7.0	RT	59.13	Freundlich isotherm model (FRIM)	PSOM	[188]
	Isoprocarb	10 mg/L	7.0	RT	21.33	FRIM	PSOM	[188]
	Carbaryl	10 mg/L	7.0	RT	61.91	FRIM	PSOM	[188]
Polyacrylic acid-modified graphene oxide (GO-PAA)	Phenol	0.2 mg/L	2.0	298.15	10	LM	–	[189]
GO	Phenol	0.2 mg/L	2.0	298	70	LM	–	[190]

been created. Nevertheless, the currently available adsorbents' efficiency, capacity, and selectivity fall short of the requirements for use. Therefore, the advancement of new GBAs is required to efficiently treat radioactive effluents.

Nuclear as well as mining operations result in the release of a significant amount of radioactive material into the aquatic environment. The aquatic life is endangered as a result [193]. The released fission product enters the food chain through water systems. Human activities, including mineral processing, fertilizer use, and lignite burning in powder plants emit radionuclides. As a result, a successful treatment method is needed to get rid of long-lived radionuclides. Environmental effects from radioactive waste produced during mining and nuclear energy generation are long-lasting [192]. The by-products of fission, ^{90}Sr, ^{137}Cs, ^{235}U, and ^{129}I, can enter the food chain through the water system. Water contaminated with these radionuclides can seep into the soil, be absorbed by plants, and eventually make its way to animals and people. Radionuclides are produced through the processing of ores, the burning of lignite in power plants, and the usage of fertilizers. Freshwater contamination by these radionuclides is a serious issue [192].

In a research study, Sun et al. [165] reported the development of iron-rGO for the adsorptive exclusion of U (VI) [194]. MXene@GO nanocomposites have been explored by Li et al. to removed U(VI) recently. These compounds could be effectively absorbed by their synthesized adsorbent/nanocomposites, which could also be recycled up to five times [165].

Greater focus is being placed on graphene oxide because of its unmatched physicochemical characteristics, including its high specific surface area and abundance of surface functional groups [191, 192]. Table 7.7 presents diverse GBA materials and their application for the removal of radionuclides, along with their property performance.

Sharma et al. reported an innovative novel Magnetite@GO@Chitosan (M@GO@Cs) adsorbent for effectual adsorption of U(VI) from wastewater where the bonding of U(VI) was ascribed to be with the -CONH and -OH functionalities complexations in M@GO@Cs even as shown by the scheme below (Fig. 7.23). They observed that the adsorption isotherm of U(VI) onto the M@GO@Cs obeyed the Langmuir isotherm (LM) while the kinetic was that of PSOM corresponding to other reports for the removal of radionuclides from water [164, 192].

GO-based inverse spinel nickel ferrite (GONF) and reduced GO-based inverse spinel nickel ferrite (rGONF) nanocomposite, which were created by co-precipitating GO with nickel and iron salts in one pot, were used to study the removal of uranium(VI) but also thorium(IV) [163]. According to their findings, rGONF and GONF possess porous surface morphologies with mean particle sizes of 32.16 and 41.41 nm, respectively. The synthesis of ferromagnetic GONF and superparamagnetic rGONF, was validated by tests using the magnetic property measurement system (MPMS). Their research suggested that the PSOM kinetics of adsorption was well suited to the adsorption of Th(IV) and U(VI) [163]. Their adsorption isotherm data demonstrated that the monolayer on the homogenous surface of the GONF and rGONF was responsible for the adsorption of Th(IV) and U(VI). The system temperature rose from 293 towards 333 ± 2 K, increasing the adsorptions of both

Table 7.7 GBAs and their application for radionuclides removal

Formulation	Radionuclide removed	Adsorbent dosage	pH	Temp. (K)	Maximum adsorption capacity	Adsorption isotherm and kinetics		Refs.
						Isotherm	Kinetics	
Prussian blue analog-doped reduced graphene oxide aerogels (RGOA/TPBAs)	Radioactive cesium (Cs^+orCs(I))	0.25 g/L	7.0	298.15	226.98 mg/g	LM	PSOM	[191]
MXene@graphene oxide	U(VI)	0.3–0.5 g/L	5.0	298	1003.5 mg/g	LM	PSOM	[164]
Graphene oxide composite nitrogen-doped carbon (GO@NCs)	U(VI)	50 ppm	5.0	298.15	879.2 mg/g	LM		[162]
Chitosan@graphene oxide	U(VI)	160 mg/L	6.0	303	822 mg/g	LM	PSOM	[196]
M@GO@Cs	U(VI)	15 mg/L	6.0		504 mg/g	LM	PSOM	[195]
GO@nickel ferrite (GONF)	Uranium (VI) (U(VI))	0.3 mg/L	5.0	333 ± 2	135.13 mg/g	LM	PSOM	[163]
	Thorium (IV) (Th(IV))	0.3 mg/L	5.0	333 ± 2	88.49 mg/g	LM	PSOM	[163]
rGONF	U(VI)	0.3 mg/L	5.0	333 ± 2	200 mg/g	LM	PSOM	[163]
	Th(IV)	0.3 mg/L	5.0	333 ± 2	126.68 mg/g	LM	PSOM	[163]
	U(VI)	2 mg/100 ml	6.5	298	1532.35 mg/g	LM	PSOM	[197]
	Eu(III)	2 mg/100 ml	6.5	298	886.44 mg/g	LM	PSOM	[197]
GO/hydroxyapatite (HAP)	U(VI)	0.1 g/L	3.0	Room temp	373.00 mg/g	LM	PSOM	[198]
PVA/GO	Sr(II)	0.5 g/L	7.0	298	20.07 mg/g	LM	PSOM	[199]

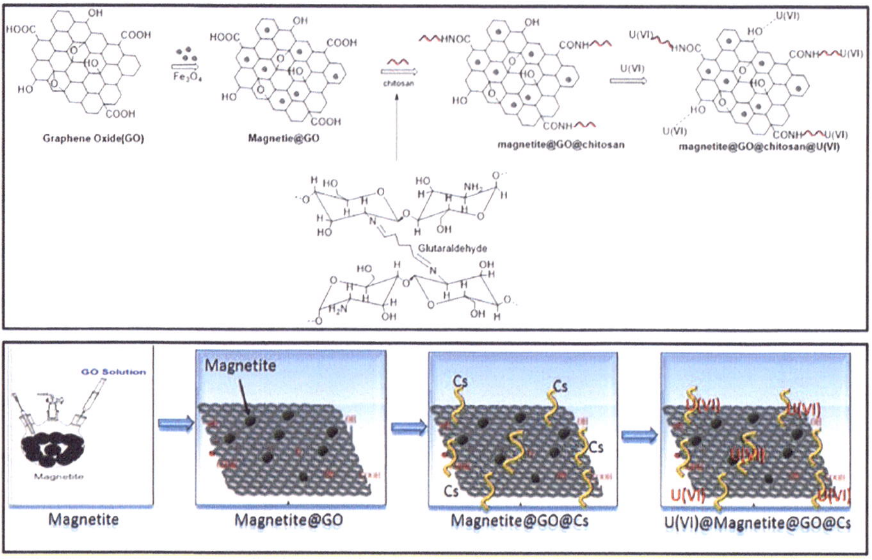

Fig. 7.23 Adsorption of uranium on the surface of M@GO@Cs. Reproduced with permission from Sharma et al. [195]. Copyright 2022, Elsevier Science Ltd

U(VI) and Th(IV). The Th(IV) and U(VI) adsorption onto rGONF and GONF were endothermic, according to the thermodynamic studies. Without suffering a material reduction in adsorption capacity, rGONF and GONF, which an external magnetic field could separate, were recycled and reused up to five times.

7.8 Possible Rejuvenation of GBAs for Reuse

It can be assumed that after desorption-regeneration with a variety of potential eluents, graphene derivatives regain most of their adsorption capacities for phenols and dyes. However, some adsorbents exhibit some loss in adsorption capacity during each regeneration phase due to contaminants that bind strongly and do not desorb. However, a retention capacity between 50 and 95% efficiency has been reportedly attained, as presented in Table 7.8. Some research attributed the decline in efficiency to the wasting of adsorbents. Straightforward desorption practices were sufficient to go through rejuvenation and provide a reuse capacity due to the comparatively modest attraction forces between pollutants and graphene adsorbents, with the bulk of investigations utilizing ethanol and DI water as eluents. It is clear that not all researchers looked into the GO adsorbents' regenerative capabilities, even though this is a crucial aspect of the "recovery and re-use" of adsorbate in a sustainable circular economy idea.

Table 7.8 Recyclability of graphene-based adsorbent for diverse contaminants as per literature

Formulated adsorbent	Adsorbate	Efficiency	Number of cycles	Eluent	Refs.
Ag_3PO_4/rGH	BPA	50%	5	–	[200]
CNC: GO	Basic blue 7 (BB7)	–	2	Water, EtOH, or acetone	[186]
	Reactive orange 122 (RO)	–	2	Water, EtOH, or acetone	[186]
	Rhodamine B (RhB)	–	2	Water, ethanol, or acetone	[186]
3D GO	BPA	90.3%	5	n-hexane	[201]
Bi_2O_3@GO	RhB	80%	7	EtOH	[202]
PS-GO	MB	69.71%	5	EtOH	[185]
Ag_3PO_4/GO	MB	98%	8	3% H_2O_2	[203]
Mag-EC-Tyr/GRO	Phenol	56%	5	Water	[204]
AS/NZVI/GO	Methomyl	>70%	5	0.1 mol/L NaOH solution	[188]
	Isoprocarb	>70%	5	0.1 mol/L NaOH solution	[188]
	Carbaryl	>70%	5	0.1 mol/L NaOH solution	[188]
GO/PPy	Phenol	95%	6	EtOH	[205]
Cu-BDC@GrO	BPA	91%	5	EtOH	[206]
GH-AgBr@rGO	BPA	90%	5	–	[207]

7.9 Current Challenges, Conclusion, and Prospects

7.9.1 Current Challenges

The exploration of chemical exfoliation for the manufacture of nanomaterials based on graphene oxide, its derivatives, or composites/hybrids has been largely observed to be performed via the Hummer method or its modified forms as it is most preferred. Even though experiments require relatively little participation, graphene-based nanomaterials take a long time to prepare. Therefore, in the future, it should be considered to eliminate or substitute most of the chemicals from the process of creating GBAs. It is also responsible enough to take a closer look at ways to reduce the use of chemicals or replace them with less expensive chemicals or better sustainable ones, both of which would help achieve a process that is both adaptable to industries, sustainable, as well as cost-effective.

7.9.2 Conclusions

Graphene composites are one of the most well-investigated adsorbents used for water treatment. This research has assessed the performance of various composites, with certain materials having noteworthy maximum adsorption capacities of over 2700 mg/g. The isotherms, as per literature, have been mostly examined using the Langmuir and Freundlich isotherm models. Additionally, it has been frequently demonstrated that the adsorption rate research had the best fit for the pseudo-second-order kinetic equation. There still exists large room for rGO-based composites, which as per available studies, are modified by organic modification and may be able to remove challenging contaminants from wastewater discharges successfully. The fact that GO and the anionic character of the pollutants, do not reject one another explains this capacity. Two methods using a three-dimensional (3D) adsorbent structure and UV light as a catalyst for photocatalytic degradation can be adopted to enhance the removal efficiencies of diverse graphene-based tested adsorbents.

7.9.3 Prospects

The following are some examples of how the financial, chemical, physical, as well as biological characteristics and attributes of the adsorbate affect the adsorbent's design as well as capacity: Batch, fixed bed, expanded bed, and fluidized bed contacting systems; cost and type of adsorbent raw material feedstock; precise surface area of the finished adsorbent; mean pore diameter, pore volume, and pore size dispersion; surface functional group type and number; ability to recover the adsorbed species for recycling and other uses, as well as the ability to replenish the adsorbent; operating expenses (OPER) and capital expenditures (CAPEX) (OPEX). Because the bulk of the literature did not cover any of these requirements, there are significant gaps in our understanding of the prospective commercial pertinence of these novel improved GO materials. Throughout many instances, the exploration has been centered on the creation of a nanostructured material with a materials science focus that was subsequently evaluated using such a tiny proportion of carefully chosen devices, as opposed to searching for an answer to a specific pollutant challenge and afterwards, from its character traits, modeling a custom adsorbent. Continued studies must create design models as well as simulated studies to forecast precision and accuracy, engage in design optimization, as well as offer economic evaluations of the treatment process using the knowledge from these fundamental studies. Within the near future, it is expected that researchers should/would focus on creating advanced GBAs embedded with a dual or hybrid function of adsorbing contaminants along with

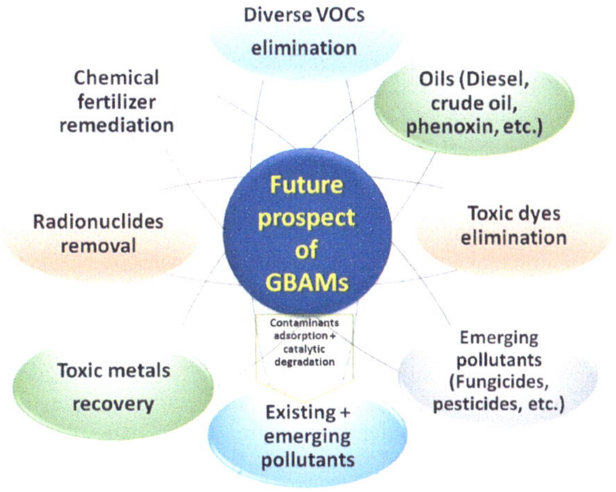

Fig. 7.24 The future prospect of GBA materials in adsorption as well as adsorption plus photocatalysis

their degradation, as depicted in Fig. 7.24. There is also a need to design and engineer effective GBAs seeing studies regarding their exploration for the adsorptive removal of agro-pollutants like fungicides have been rare or limited. Future research studies should also investigate the effect of agitation on the adsorption of contaminants from wastewater because only a handful of literature has considered this factor which significantly affects the adsorption process. The relationship between the factors affecting the adsorption process should also be studied extensively by researchers to optimize the use of all these factors with a single adsorption system, especially in the case of industrial effluent remediation.

Acknowledgements The authors wish to acknowledge the Department of Chemical Sciences, University of Johannesburg, Doornfontein, Johannesburg 2028, South Africa, and DSI-CSIR Nanotechnology Innovation Centre, Council for Scientific and Industrial Research, Pretoria 0001, South Africa.

Declaration The authors have no conflict of interest.

References

1. I. Ali et al., Graphene based adsorbents for remediation of noxious pollutants from wastewater. Environ. Int. **127**, 160–180 (2019)
2. M.D.F. Hossain, N. Akther, Y. Zhou, Recent advancements in graphene adsorbents for wastewater treatment: current status and challenges. Chinese Chem. Lett. **31**(10), 2525–2538 (2020)
3. N. Khatri, S. Tyagi, Influences of natural and anthropogenic factors on surface and groundwater quality in rural and urban areas. Front. Life Sci. **8**(1), 23–39 (2015)
4. V. Masindi, K.L. Muedi, Environmental contamination by heavy metals. Heavy Metals **10**, 115–132 (2018)

5. F.E. Titchou et al., Removal of organic pollutants from wastewater by advanced oxidation processes and its combination with membrane processes. Chem. Eng. Proc. Proc. Intensific. **169**, 108631 (2021)
6. R.R.L. Vidal, J.S. Moraes, Removal of organic pollutants from wastewater using chitosan: a literature review. Int. J. Environ. Sci. Technol. **16**(3), 1741–1754 (2019)
7. J.T. Orasugh, S.S. Ray, Nanocellulose-graphene oxide-based nanocomposite for adsorptive water treatment, in *Functional Polymer Nanocomposites for Wastewater Treatment*. (Springer, 2022), pp.1–53
8. Z.-H. Huang et al., Adsorption of lead (II) ions from aqueous solution on low-temperature exfoliated graphene nanosheets. Langmuir **27**(12), 7558–7562 (2011)
9. J.L. Wang, L.J. Xu, Advanced oxidation processes for wastewater treatment: formation of hydroxyl radical and application. Crit. Rev. Environ. Sci. Technol. **42**(3), 251–325 (2012)
10. G. Ciardelli, L. Corsi, M. Marcucci, Membrane separation for wastewater reuse in the textile industry. Resour. Conserv. Recycl. **31**(2), 189–197 (2001)
11. A. Bodalo-Santoyo et al., Application of reverse osmosis to reduce pollutants present in industrial wastewater. Desalination **155**(2), 101–108 (2003)
12. S.K. Nataraj, K.M. Hosamani, T.M. Aminabhavi, Distillery wastewater treatment by the membrane-based nanofiltration and reverse osmosis processes. Water Res. **40**(12), 2349–2356 (2006)
13. T.R. Harper, N.W. Kingham, Removal of arsenic from wastewater using chemical precipitation methods. Water Environ. Res. **64**(3), 200–203 (1992)
14. R. Bochenek, R. Sitarz, D. Antos, Design of continuous ion exchange process for the wastewater treatment. Chem. Eng. Sci. **66**(23), 6209–6219 (2011)
15. E. Ofir et al., Boron removal from seawater by electro-chemical treatment as part of water desalination. Desalin. Water Treat. **31**(1–3), 102–106 (2011)
16. T. Pickett, J. Sonstegard, B. Bonkoski, Using biology to treat selenium: biologically treating scrubber wastewater can be an attractive alternative to physical-chemical treatment. Power Eng. **110**(11), 140–143 (2006)
17. H. Liu, H. Qiu, Recent advances of 3D graphene-based adsorbents for sample preparation of water pollutants: a review. Chem. Eng. J. **393**, 124691 (2020)
18. M.S. Ali, J.T. Orasugh, D. Chattopadhyay, Bacillus subtilis-based biofilms, in *Application of Biofilms in Applied Microbiology*, ed. by M.P. Shah, (Elsevier, Elsevier, 2022)
19. M.M.-A. Aslam et al., Functionalized carbon nanotubes (Cnts) for water and wastewater treatment: preparation to application. Sustainability **13**(10), 5717 (2021)
20. M. Kumari et al., Transformation of solid plastic waste to activated carbon fibres for wastewater treatment. Chemosphere **294**, 133692 (2022)
21. H. Es-sahbany et al., Adsorption of heavy metal (Cadmium) in synthetic wastewater by the natural clay as a potential adsorbent (Tangier-Tetouan-Al Hoceima-Morocco region). Mater. Today Proc. **45**, 7299–7305 (2021)
22. A. Sarı, T.A. Saleh, M. Tuzen, Development and characterization of polymer-modified vermiculite composite as novel highly-efficient adsorbent for water treatment. Surfaces Interfaces **27**, 101504 (2021)
23. A.A. Al-Gheethi et al., Sustainable approaches for removing Rhodamine B dye using agricultural waste adsorbents: a review. Chemosphere **287**, 132080 (2022)
24. I. Anastopoulos et al., Sunflower-biomass derived adsorbents for toxic/heavy metals removal from (waste) water. J. Mol. Liq. **342**, 117540 (2021)
25. Z.-D. Peng et al., Removal of cadmium from wastewater by magnetic zeolite synthesized from natural, low-grade molybdenum. Sci. Total Environ. **772**, 145355 (2021)
26. T. Sattar, Current review on synthesis, composites and multifunctional properties of graphene. Top. Curr. Chem. **377**(2), 1–45 (2019)
27. H. Shinohara, A. Tiwari, *Graphene: An Introduction to The Fundamentals and Industrial Applications*. (John Wiley & Sons, 2015)
28. D.P. DiVincenzo, E.J. Mele, Self-consistent effective-mass theory for intralayer screening in graphite intercalation compounds. Phys. Rev. B **29**(4), 1685 (1984)

29. A. Geim, Many pioneers in graphene discovery. APS News **19**(1), 4 (2010)
30. Y. Huang et al., Universal mechanical exfoliation of large-area 2D crystals. Nat. Commun. **11**(1), 1–9 (2020)
31. G. Borand, N. Akçamlı, D. Uzunsoy, Structural characterization of graphene nanostructures produced via arc discharge method. Ceram. Int. **47**(6), 8044–8052 (2021)
32. A. Kaur, J. Kaur, R.C. Singh, Tailor made exfoliated reduced graphene oxide nanosheets based on oxidative-exfoliation approach. Fullerenes Nanotubes Carbon Nanostruct. **26**(1), 1–11 (2018)
33. Y. Xu et al., Liquid-phase exfoliation of graphene: an overview on exfoliation media, techniques, and challenges. Nanomaterials **8**(11), 942 (2018)
34. A.A. Silva et al., Graphene sheets produced by carbon nanotubes unzipping and their performance as supercapacitor. Appl. Surf. Sci. **446**, 201–208 (2018)
35. X.J. Lee et al., Review on graphene and its derivatives: synthesis methods and potential industrial implementation. J. Taiwan Inst. Chem. Eng. **98**, 163–180 (2019)
36. M. Saeed et al., Chemical vapour deposition of graphene—synthesis, characterisation, and applications: a review. Molecules **25**(17), 3856 (2020)
37. G. Li et al., Epitaxial growth and physical properties of 2D materials beyond graphene: from monatomic materials to binary compounds. Chem. Soc. Rev. **47**(16), 6073–6100 (2018)
38. Y. Sun, J. Zhang, Strategies for scalable gas-phase preparation of free-standing graphene. CCS Chem. **3**(4), 1058–1077 (2021)
39. Y. Yang et al., Bottom-up fabrication of graphene on silicon/silica substrate via a facile soft-hard template approach. Sci. Rep. **5**(1), 1–7 (2015)
40. V. Singh et al., Graphene based materials: past, present and future. Prog. Mater Sci. **56**(8), 1178–1271 (2011)
41. A. Ciesielski, P. Samorì, Graphene via sonication assisted liquid-phase exfoliation. Chem. Soc. Rev. **43**(1), 381–398 (2014)
42. B. Liang et al., Organic salt-assisted liquid-phase shear exfoliation of expanded graphite into graphene nanosheets. J. Materiom. **7**(6), 1181–1189 (2021)
43. N. Kumar et al., Top-down synthesis of graphene: a comprehensive review. FlatChem **27**, 100224 (2021)
44. M. Yi, Z. Shen, A review on mechanical exfoliation for the scalable production of graphene. J. Mater. Chem. A **3**(22), 11700–11715 (2015)
45. T.M.D. Alharbi et al., Shear stress mediated scrolling of graphene oxide. Carbon **137**, 419–424 (2018)
46. L. Yasmin et al., Optimising a vortex fluidic device for controlling chemical reactivity and selectivity. Sci. Rep. **3**(1), 1–6 (2013)
47. X. Chen, J.F. Dobson, C.L. Raston, Vortex fluidic exfoliation of graphite and boron nitride. Chem. Commun. **48**(31), 3703–3705 (2012)
48. M.H. Wahid et al., Functional multi-layer graphene–algae hybrid material formed using vortex fluidics. Green Chem. **15**(3), 650–655 (2013)
49. T.S. Tran et al., High shear-induced exfoliation of graphite into high quality graphene by Taylor-Couette flow. RSC Adv. **6**(15), 12003–12008 (2016)
50. M.G. Sumdani et al., Recent advances of the graphite exfoliation processes and structural modification of graphene: a review. J. Nanopart. Res. **23**(11), 1–35 (2021)
51. K.R. Paton et al., Scalable production of large quantities of defect-free few-layer graphene by shear exfoliation in liquids. Nat. Mater. **13**(6), 624–630 (2014)
52. Z. Zhang et al., Efficient production of high-quality few-layer graphene using a simple hydrodynamic-assisted exfoliation method. Nanoscale Res Lett **13**(1), 416 (2018)
53. W. Choi, J.-W. Lee, *Graphene: synthesis and applications*. (CRC press, 2011)
54. L.M. Viculis, J.J. Mack, R.B. Kaner, A chemical route to carbon nanoscrolls. Science **299**(5611), 1361 (2003)
55. Z. Ereš, S. Hrabar, Low-cost synthesis of high-quality graphene in do-it-yourself CVD reactor. Automatika **59**(3–4), 254–260 (2018)
56. Z. Zhang et al., Top-down bottom-up graphene synthesis. Nano Futures **3**(4), 042003 (2019)

57. A. Moosa, M. Abed, Graphene preparation and graphite exfoliation. Turk. J. Chem. **45**(3), 493–519 (2021)
58. Y. Lee et al., Wafer-scale synthesis and transfer of graphene films. Nano Lett. **10**(2), 490–493 (2010)
59. X. Li et al., Large-area synthesis of high-quality and uniform graphene films on copper foils. Science **324**(5932), 1312–1314 (2009)
60. X. Li et al., Evolution of graphene growth on Ni and Cu by carbon isotope labeling. Nano Lett. **9**(12), 4268–4272 (2009)
61. S. Jia-Tao et al., Effect of strain on geometric and electronic structures of graphene on a Ru (0001) surface. Chin. Phys. B **18**(7), 3008 (2009)
62. L. Gao et al., Repeated growth and bubbling transfer of graphene with millimetre-size single-crystal grains using platinum. Nat. Commun. **3**(1), 1–7 (2012)
63. D.Y. Usachov et al., Epitaxial B-graphene: large-scale growth and atomic structure. ACS Nano **9**(7), 7314–7322 (2015)
64. B.-J. Park et al., Defect-free graphene synthesized directly at 150 C via chemical vapor deposition with no transfer. ACS Nano **12**(2), 2008–2016 (2018)
65. Y. Yao, C.-P. Wong, Monolayer graphene growth using additional etching process in atmospheric pressure chemical vapor deposition. Carbon **50**(14), 5203–5209 (2012)
66. C.-M. Seah, S.-P. Chai, A.R. Mohamed, Mechanisms of graphene growth by chemical vapour deposition on transition metals. Carbon **70**, 1–21 (2014)
67. N.G. Shang et al., Catalyst-free efficient growth, orientation and biosensing properties of multilayer graphene nanoflake films with sharp edge planes. Adv. Func. Mater. **18**(21), 3506–3514 (2008)
68. A.N. Obraztsov et al., DC discharge plasma studies for nanostructured carbon CVD. Diam. Relat. Mater. **12**(3–7), 917–920 (2003)
69. C. Berger et al., Ultrathin epitaxial graphite: 2D electron gas properties and a route toward graphene-based nanoelectronics. J. Phys. Chem. B **108**(52), 19912–19916 (2004)
70. M. Choucair, P. Thordarson, J.A. Stride, Gram-scale production of graphene based on solvothermal synthesis and sonication. Nat. Nanotechnol. **4**(1), 30–33 (2009)
71. H. Hibino, H. Kageshima, M. Nagase, Graphene growth on silicon carbide. NTT Technical Review (Web). **8**(8) (2010)
72. Z.-Y. Juang et al., Synthesis of graphene on silicon carbide substrates at low temperature. Carbon **47**(8), 2026–2031 (2009)
73. Y. Pan et al., Highly ordered, millimeter-scale, continuous, single-crystalline graphene monolayer formed on Ru (0001). Adv. Mater. **21**(27), 2777–2780 (2009)
74. Z. Chen et al., Graphene nano-ribbon electronics. Phys. E **40**(2), 228–232 (2007)
75. D.V. Kosynkin et al., Longitudinal unzipping of carbon nanotubes to form graphene nanoribbons. Nature **458**(7240), 872–876 (2009)
76. L. Jiao et al., Narrow graphene nanoribbons from carbon nanotubes. Nature **458**(7240), 877–880 (2009)
77. A.G. Cano-Marquez et al., Ex-MWNTs: graphene sheets and ribbons produced by lithium intercalation and exfoliation of carbon nanotubes. Nano Lett. **9**(4), 1527–1533 (2009)
78. J. Liu et al., Graphene oxide nanoribbon as hole extraction layer to enhance efficiency and stability of polymer solar cells. Adv. Mater. **26**(5), 786–790 (2014)
79. W.E. Mahmoud, F.S. Al-Hazmi, G.H. Al-Harbi, Wall by wall controllable unzipping of MWCNTs via intercalation with oxalic acid to produce multilayers graphene oxide ribbon. Chem. Eng. J. **281**, 192–198 (2015)
80. J.T. Orasugh et al., Carbon nanotube and nanofiber reinforced polymer composites, in *Encyclopedia of Materials: Plastics and Polymers*, ed. by M.S.J. Hashmi (Elsevier, Oxford, 2022), pp.837–859
81. S. Kim et al., An aqueous single reactor arc discharge process for the synthesis of graphene nanospheres. Small **11**(38), 5041–5046 (2015)
82. K.S. Subrahmanyam et al., Simple method of preparing graphene flakes by an arc-discharge method. J. Phys. Chem. C **113**(11), 4257–4259 (2009)

83. C.E.E. Rao, A.E. Sood, Graphene: the new two-dimensional nanomaterial. Angew. Chem. Int. Ed. **48**(42):7752–7777
84. Z. Wang et al., Low-cost and large-scale synthesis of graphene nanosheets by arc discharge in air. Nanotechnology **21**(17), 175602 (2010)
85. T. Szabó et al., Evolution of surface functional groups in a series of progressively oxidized graphite oxides. Chem. Mater. **18**(11), 2740–2749 (2006)
86. A. Lerf et al., Structure of graphite oxide revisited. J. Phys. Chem. B **102**(23), 4477–4482 (1998)
87. H. He et al., Solid-state NMR studies of the structure of graphite oxide. J. Phys. Chem. **100**(51), 19954–19958 (1996)
88. K. Erickson et al., Determination of the local chemical structure of graphene oxide and reduced graphene oxide. Adv. Mater. **22**(40), 4467–4472 (2010)
89. A. Dimiev et al., Pristine graphite oxide. J. Am. Chem. Soc. **134**(5), 2815–2822 (2012)
90. S. Eigler, A. Hirsch, Chemistry with graphene and graphene oxide—challenges for synthetic chemists. Angew. Chem. Int. Ed. **53**(30), 7720–7738 (2014)
91. J.P. Rourke et al., The real graphene oxide revealed: stripping the oxidative debris from the graphene-like sheets. Angew. Chem. Int. Ed. **50**(14), 3173–3177 (2011)
92. A.T. Dideikin, A.Y. Vul, Graphene oxide and derivatives: the place in graphene family. Front. Phys. **6**, 149 (2019)
93. L. Sun, Structure and synthesis of graphene oxide. Chin. J. Chem. Eng. **27**(10), 2251–2260 (2019)
94. C. Gómez-Navarro et al., Atomic structure of reduced graphene oxide. Nano Lett. **10**(4), 1144–1148 (2010)
95. D.Y. Kornilov, S.P. Gubin, Graphene oxide: structure, properties, synthesis, and reduction (a review). Russ. J. Inorg. Chem. **65**(13), 1965–1976 (2020)
96. N. Vats et al., Electron microscopy of polyoxometalate ions on graphene by electrospray ion beam deposition. Nanoscale **10**(10), 4952–4961 (2018)
97. Y. Gao et al., Revealing the role of oxygen-containing functional groups on graphene oxide for the highly efficient adsorption of thorium ions. J. Hazard. Mater. **436**, 129148 (2022)
98. S.H. Dave et al., Chemistry and structure of graphene oxide via direct imaging. ACS Nano **10**(8), 7515–7522 (2016)
99. S. Sadhukhan et al., Synthesis of RGO/NiO nanocomposites adopting a green approach and its photocatalytic and antibacterial properties. Mater. Chem. Phys. **247**, 122906 (2020)
100. D. Chen, H. Feng, J. Li, Graphene oxide: preparation, functionalization, and electrochemical applications. Chem. Rev. **112**(11), 6027–6053 (2012)
101. Q. Lian et al., Enhanced adsorption of resorcinol onto phosphate functionalized graphene oxide synthesized via Arbuzov Reaction: a proposed mechanism of hydrogen bonding and π-π interactions. Chemosphere **280**, 130730 (2021)
102. W. Gao et al., New insights into the structure and reduction of graphite oxide. Nat. Chem. **1**(5), 403–408 (2009)
103. T. Nakajima, Y. Matsuo, Formation process and structure of graphite oxide. Carbon **32**(3), 469–475 (1994)
104. G. Ruess, Über das graphitoxyhydroxyd (graphitoxyd). Monatshefte für Chemie und verwandte Teile anderer Wissenschaften **76**(3), 381–417 (1947)
105. W. Scholz, H.P. Boehm, Untersuchungen am graphitoxid. VI. Betrachtungen zur struktur des graphitoxids. Zeitschrift für anorganische und allgemeine Chemie. **369**(3–6), 327–340 (1969)
106. W. Chen, L. Yan, P.R. Bangal, Preparation of graphene by the rapid and mild thermal reduction of graphene oxide induced by microwaves. Carbon **48**(4), 1146–1152 (2010)
107. S. Pei, H.-M. Cheng, The reduction of graphene oxide. Carbon **50**(9), 3210–3228 (2012)
108. K.A.I. Yan et al., Designed CVD growth of graphene via process engineering. Acc. Chem. Res. **46**(10), 2263–2274 (2013)
109. H. Wang et al., Solvothermal reduction of chemically exfoliated graphene sheets. J. Am. Chem. Soc. **131**(29), 9910–9911 (2009)

110. M.K. Rabchinskii et al., Nanoscale perforation of graphene oxide during photoreduction process in the argon atmosphere. J. Phys. Chem. C **120**(49), 28261–28269 (2016)
111. J. Wang, X. Guo, Adsorption kinetic models: physical meanings, applications, and solving methods. J. Hazard. Mater. **390**, 122156 (2020)
112. I. Langmuir, The constitution and fundamental properties of solids and liquids. Part I. Solids. J. Am. Chem. Soc. **38**(11), 2221–2295 (1916)
113. Y. Liu, J. Wang, *Fundamentals and applications of biosorption isotherms, kinetics and thermodynamics*. (Nova Science Publishers, 2009)
114. R. Sips, On the structure of a catalyst surface. J. Chem. Phys. **16**(5), 490–495 (1948)
115. H. Freundlich, Over the adsorption in solution. J. Phys. chem **57**(385471), 1100–1107 (1906)
116. O. Redlich, D.L. Peterson, A useful adsorption isotherm. J. Phys. Chem. **63**(6), 1024–1024 (1959)
117. A.R. Khan, R. Ataullah, A. Al-Haddad, Equilibrium adsorption studies of some aromatic pollutants from dilute aqueous solutions on activated carbon at different temperatures. J. Colloid Interface Sci. **194**(1), 154–165 (1997)
118. J. Toth, A multicomponent isotherm for liquid adsorption. Acta Chim. Acad. Sci. Hung **69**, 311–322 (1971)
119. C.J. Radke, J.M. Prausnitz, Adsorption of organic solutes from dilute aqueous solution of activated carbon. Ind. Eng. Chem. Fundam. **11**(4), 445–451 (1972)
120. M. Dubinin, The potential theory of adsorption of gases and vapors for adsorbents with energetically nonuniform surfaces. Chem. Rev. **60**(2), 235–241 (1960)
121. M.M. Majd, et al., Adsorption isotherm models: A comprehensive and systematic review (2010−2020). Sci. Total Environ. 151334 (2021)
122. T. Grchev et al., Adsorption of polyacrylamide on gold and iron from acidic aqueous solutions. Electrochim. Acta **36**(8), 1315–1323 (1991)
123. P.J. Flory, Thermodynamics of high polymer solutions. J. Chem. Phys. **10**(1), 51–61 (1942)
124. M.L. Huggins, Some properties of solutions of long-chain compounds. J. Phys. Chem. **46**(1), 151–158 (1942)
125. M. Kaisheva, G. Saraivanov, A. Anastopoulos, Adsorption studies of hexadecyltributylphosphonium bromide at the mercury/solution interface through differential capacity measurements. Langmuir **7**(10), 2380–2384 (1991)
126. S. Brunauer, P.H. Emmett, E. Teller, Adsorption of gases in multimolecular layers. J. Am. Chem. Soc. **60**(2), 309–319 (1938)
127. M.J. Temkin, V. Pyzhev, *Recent Modifications to Langmuir Isotherms*. (1940)
128. C. Aharoni, M. Ungarish, *Kinetics of activated chemisorption. Part 2—Theoretical models*. J. Chem. Soc. Faraday Trans. 1 Phys. Chem. Condens. Phases **73**, 456–464 (1977)
129. A. Bhattacharyya, et al., Design of an efficient and selective adsorbent of cationic dye through activated carbon–graphene oxide nanocomposite: study on mechanism and synergy. Mater. Chem. Phys. **260** (2021)
130. L. Abou Chacra et al., Application of graphene nanoplatelets and graphene magnetite for the removal of emulsified oil from produced water. J. Environ. Chem. Eng. **6**(2), 3018–3033 (2018)
131. H. Adamu, P. Dubey, J.A. Anderson, Probing the role of thermally reduced graphene oxide in enhancing performance of TiO2 in photocatalytic phenol removal from aqueous environments. Chem. Eng. J. **284**, 380–388 (2016)
132. M.T. Al-Shemy, A. Al-Sayed, S. Dacrory, Fabrication of sodium alginate/graphene oxide/nanocrystalline cellulose scaffold for methylene blue adsorption: kinetics and thermodynamics study. Sep. Purif. Technol. **290**, 120825 (2022)
133. T.S. Anirudhan, J.R. Deepa, Nano-zinc oxide incorporated graphene oxide/nanocellulose composite for the adsorption and photo catalytic degradation of ciprofloxacin hydrochloride from aqueous solutions. J. Colloid Interface Sci. **490**, 343–356 (2017)
134. N.I.F. Aris et al., Superhydrophilic graphene oxide/electrospun cellulose nanofibre for efficient adsorption of organophosphorus pesticides from environmental samples. R. Soc. Open Sci. **7**(3), 192050 (2020)

135. O. Bagoole et al., Functionalized three-dimensional graphene sponges for highly efficient crude and diesel oil adsorption. Environ. Sci. Pollut. Res. **25**(23), 23091–23105 (2018)
136. S. Bai et al., One-pot solvothermal preparation of magnetic reduced graphene oxide-ferrite hybrids for organic dye removal. Carbon **50**(6), 2337–2346 (2012)
137. U. Baig, M. Faizan, M. Sajid, Effective removal of hazardous pollutants from water and deactivation of water-borne pathogens using multifunctional synthetic adsorbent materials: a review. J. Clean. Prod. 126735 (2021)
138. A. Bhattacharyya et al., Development of an auto-phase separable and reusable graphene oxide-potato starch based cross-linked bio-composite adsorbent for removal of methylene blue dye. Int. J. Biol. Macromol. **116**, 1037–1048 (2018)
139. J. Chen et al., Direct reduction of graphene oxide/nanofibrillated cellulose composite film and its electrical conductivity research. Sci. Rep. **10**(1), 3124 (2020)
140. L. Chen et al., Removal of methylene blue from water by cellulose/graphene oxide fibres. J. Exp. Nanosci. **11**(14), 1156–1170 (2016)
141. X. Chen, et al., Adsorption of heavy metals by graphene oxide/cellulose hydrogel prepared from NaOH/urea aqueous solution. Materials (Basel) **9**(7) (2016)
142. Z. Dong et al., Bio-inspired surface-functionalization of graphene oxide for the adsorption of organic dyes and heavy metal ions with a superhigh capacity. J. Mater. Chem. A **2**(14), 5034–5040 (2014)
143. C. Gao et al., Preparation of reduced graphene oxide aerogel and its adsorption for Pb(II). ACS Omega **5**(17), 9903–9911 (2020)
144. H. Gu, et al., Nanocellulose nanocomposite aerogel towards efficient oil and organic solvent adsorption. Adv. Compos. Hybrid Mater. 1–10 (2021)
145. C.M.B. de Araujo et al., Wastewater treatment using recyclable agar-graphene oxide biocomposite hydrogel in batch and fixed-bed adsorption column: bench experiments and modeling for the selective removal of organics. Colloids Surf. A **639**, 128357 (2022)
146. A.A. Adeyi et al., Simultaneous adsorption of malachite green and methylene blue dyes in a fixed-bed column using poly (acrylonitrile-co-acrylic acid) modified with thiourea. Molecules **25**(11), 2650 (2020)
147. L. Chen et al., High performance agar/graphene oxide composite aerogel for methylene blue removal. Carbohyd. Polym. **155**, 345–353 (2017)
148. A. Zaman et al., Biopolymer-based nanocomposites for removal of hazardous dyes from water bodies, in *Innovations in Environmental Biotechnology*. (Springer, 2022), pp.759–783
149. M. Adel et al., Removal of heavy metals and dyes from wastewater using graphene oxide-based nanomaterials: a critical review. Environ. Nanotechnol. Monitor. Manag. **18**, 100719 (2022)
150. Y. Yoon et al., Synthesis of magnetite/non-oxidative graphene composites and their application for arsenic removal. Sep. Purif. Technol. **178**, 40–48 (2017)
151. H. Zhang et al., Synthesis of KMnO4-treated magnetic graphene oxide nanocomposite (Fe3O4@ GO/MnOx) and its application for removing of Cu2+ ions from aqueous solution. Nanotechnology **29**(13), 135706 (2018)
152. C. Bulin et al., Magnetic graphene oxide nanocomposite: one-pot preparation, adsorption performance and mechanism for aqueous Mn (II) and Zn (II). J. Phys. Chem. Solids **156**, 110130 (2021)
153. A. Zaman et al., Facile one-pot in-situ synthesis of novel graphene oxide-cellulose nanocomposite for enhanced azo dye adsorption at optimized conditions. Carbohyd. Polym. **246**, 116661 (2020)
154. P. Banerjee et al., Application of graphene oxide nanoplatelets for adsorption of ibuprofen from aqueous solutions: evaluation of process kinetics and thermodynamics. Process Saf. Environ. Prot. **101**, 45–53 (2016)
155. P. Banerjee et al., Ultrasound assisted mixed azo dye adsorption by chitosan–graphene oxide nanocomposite. Chem. Eng. Res. Des. **117**, 43–56 (2017)
156. M. Adel, M.A. Ahmed, A.A. Mohamed, Effective removal of cationic dyes from aqueous solutions using reduced graphene oxide functionalized with manganese ferrite nanoparticles. Compos. Commun. **22**, 100450 (2020)

157. W. Xu et al., Novel ternary nanohybrids of tetraethylenepentamine and graphene oxide decorated with MnFe2O4 magnetic nanoparticles for the adsorption of Pb (II). J. Hazard. Mater. **358**, 337–345 (2018)
158. A. Pourjavadi, M. Nazari, S.H. Hosseini, Synthesis of magnetic graphene oxide-containing nanocomposite hydrogels for adsorption of crystal violet from aqueous solution. RSC Adv. **5**(41), 32263–32271 (2015)
159. L. Fan et al., Fabrication of novel magnetic chitosan grafted with graphene oxide to enhance adsorption properties for methyl blue. J. Hazard. Mater. **215**, 272–279 (2012)
160. S. Bhattacharya et al., Removal of aqueous carbamazepine using graphene oxide nanoplatelets: process modelling and optimization. Sustain. Environ. Res. **30**(1), 1–12 (2020)
161. R. Foroutan et al., Performance of montmorillonite/graphene oxide/CoFe2O4 as a magnetic and recyclable nanocomposite for cleaning methyl violet dye-laden wastewater. Adv. Powder Technol. **31**(9), 3993–4004 (2020)
162. M. Cao et al., Preparation of graphene oxide composite nitrogen-doped carbon (GO@NCs) by one-step carbonization with enhanced electrosorption performance for U(VI). J. Water Proc. Eng. **48**, 102930 (2022)
163. L.P. Lingamdinne et al., Preparation and characterization of porous reduced graphene oxide based inverse spinel nickel ferrite nanocomposite for adsorption removal of radionuclides. J. Hazard. Mater. **326**, 145–156 (2017)
164. K. Li et al., Design of MXene/graphene oxide nanocomposites with micro-wrinkle structure for efficient separating of uranium(VI) from wastewater. Chem. Eng. J. **433**, 134449 (2022)
165. L.P. Lingamdinne et al., Process optimization and adsorption modeling of Pb(II) on nickel ferrite-reduced graphene oxide nano-composite. J. Mol. Liq. **250**, 202–211 (2018)
166. S. Zhu et al., Adsorption of emerging contaminant metformin using graphene oxide. Chemosphere **179**, 20–28 (2017)
167. Y. Tang et al., Synthesis of reduced graphene oxide/magnetite composites and investigation of their adsorption performance of fluoroquinolone antibiotics. Colloids Surf., A **424**, 74–80 (2013)
168. S. Anuma, P. Mishra, B.R. Bhat, Polypyrrole functionalized Cobalt oxide Graphene (COPYGO) nanocomposite for the efficient removal of dyes and heavy metal pollutants from aqueous effluents. J. Hazard. Mater. **416**, 125929 (2021)
169. M. Sivakumar et al., Porous graphene nanoplatelets encompassed with nitrogen and sulfur group for heavy metal ions removal of adsorption and desorption from single or mixed aqueous solution. Sep. Purif. Technol. **288**, 120485 (2022)
170. B. Du et al., Preparation of functionalized magnetic graphene oxide/lignin composite nanoparticles for adsorption of heavy metal ions and reuse as electromagnetic wave absorbers. Sep. Purif. Technol. **297**, 121509 (2022)
171. Z.U. Khan et al., Graphene oxide/PVC composite papers functionalized with p-Phenylenediamine as high-performance sorbent for the removal of heavy metal ions. J. Environ. Chem. Eng. **9**(5), 105916 (2021)
172. M. Verma et al., Synthesis of EDTA-functionalized graphene oxide-chitosan nanocomposite for simultaneous removal of inorganic and organic pollutants from complex wastewater. Chemosphere **287**, 132385 (2022)
173. X. Hao et al., Graphene oxide/montmorillonite composite aerogel with slit-shaped pores: Selective removal of Cu2+ from wastewater. J. Alloy. Compd. **923**, 166335 (2022)
174. Y.-H. Zhu et al., The synthesis of tannin-based graphene aerogel by hydrothermal treatment for removal of heavy metal ions. Ind. Crops Prod. **176**, 114304 (2022)
175. J. Yan, R. Li, Simple and low-cost production of magnetite/graphene nanocomposites for heavy metal ions adsorption. Sci. Total Environ. **813**, 152604 (2022)
176. J. Yan, K. Li, A magnetically recyclable polyampholyte hydrogel adsorbent functionalized with β-cyclodextrin and graphene oxide for cationic/anionic dyes and heavy metal ion wastewater remediation. Sep. Purif. Technol. **277**, 119469 (2021)
177. L.A. Al-Khateeb, S. Almotiry, M.A. Salam, Adsorption of pharmaceutical pollutants onto graphene nanoplatelets. Chem. Eng. J. **248**, 191–199 (2014)

178. S. Zhou et al., Montmorillonite-reduced graphene oxide composite aerogel (M−rGO): a green adsorbent for the dynamic removal of cadmium and methylene blue from wastewater. Sep. Purif. Technol. **296**, 121416 (2022)
179. X. Shi et al., Adsorption properties of graphene materials for pesticides: structure effect. J. Mol. Liq. **364**, 119967 (2022)
180. F.A. Rosli et al., Efficient removal of pharmaceuticals from water using graphene nanoplatelets as adsorbent. R. Soc. Open Sci. **8**(1), 201076 (2021)
181. E. Çalışkan Salihi et al., Graphene oxide as a new generation adsorbent for the removal of antibiotics from waters. Sep. Sci. Technol. **56**(3), 453–461 (2021)
182. Y. Gao et al., Adsorption and removal of tetracycline antibiotics from aqueous solution by graphene oxide. J. Colloid Interface Sci. **368**(1), 540–546 (2012)
183. P. Joshi et al., Fruit waste-derived cellulose and graphene-based aerogels: plausible adsorption pathways for fast and efficient removal of organic dyes. J. Colloid Interface Sci. **608**, 2870–2883 (2022)
184. T. Liu et al., Adsorption-photocatalysis performance of polyaniline/dicarboxyl acid cellulose@graphene oxide for dye removal. Int. J. Biol. Macromol. **182**, 492–501 (2021)
185. Y. Qi et al., Natural polysaccharides-modified graphene oxide for adsorption of organic dyes from aqueous solutions. J. Colloid Interface Sci. **486**, 84–96 (2017)
186. P.M.M. da Silva et al., Instantaneous adsorption and synergic effect in simultaneous removal of complex dyes through nanocellulose/graphene oxide nanocomposites: batch, fixed-bed experiments and mechanism. Environ. Nanotechnol. Monit. Manag. **16**, 100584 (2021)
187. D.R. Rout, H.M. Jena, Removal of malachite green dye from aqueous solution using reduced graphene oxide as an adsorbent. Mater. Today Proc. **47**, 1173–1182 (2021)
188. H. Du et al., Multifunctional magnetic bio-nanoporous carbon material based on zero-valent iron, Angelicae Dahuricae Radix slag and graphene oxide: an efficient adsorbent of pesticides. Arab. J. Chem. **14**(8), 103267 (2021)
189. A. Bibi et al., New material of polyacrylic acid-modified graphene oxide composite for phenol remediation from synthetic and real wastewater. Environ. Technol. Innov. **27**, 102795 (2022)
190. M.A. Al-Ghouti et al., Effective removal of phenol from wastewater using a hybrid process of graphene oxide adsorption and UV-irradiation. Environ. Technol. Innov. **27**, 102525 (2022)
191. H. Li et al., Reduced graphene oxide based aerogels: doped with ternary Prussian blue analogs and selective removal of Cs+ from effluent. J. Water Proc. Eng. **47**, 102741 (2022)
192. X. Liu, J. Wu, J. Wang, Removal of Cs (I) from simulated radioactive wastewater by three forward osmosis membranes. Chem. Eng. J. **344**, 353–362 (2018)
193. V. Chandra et al., Water-dispersible magnetite-reduced graphene oxide composites for arsenic removal. ACS Nano **4**(7), 3979–3986 (2010)
194. P. Sun et al., Selective trans-membrane transport of alkali and alkaline earth cations through graphene oxide membranes based on cation−π interactions. ACS Nano **8**(1), 850–859 (2014)
195. M. Sharma et al., Selective removal of uranium from an aqueous solution of mixed radionuclides of uranium, cesium, and strontium via a viable recyclable GO@chitosan based magnetic nanocomposite. Mater. Today Commun. **32**, 104020 (2022)
196. S. Zhou et al., Amidoxime modified chitosan/graphene oxide composite for efficient adsorption of U(VI) from aqueous solutions. J. Environ. Chem. Eng. **9**(6), 106363 (2021)
197. X. Zhong et al., Ultra-high capacity of graphene oxide conjugated covalent organic framework nanohybrid for U(VI) and Eu(III) adsorption removal. J. Mol. Liq. **323**, 114603 (2021)
198. M. Su et al., Graphene oxide functionalized with nano hydroxyapatite for the efficient removal of U(VI) from aqueous solution. Environ. Pollut. **268**, 115786 (2021)
199. J.-B. Huo, G. Yu, J. Wang, Adsorptive removal of Sr(II) from aqueous solution by polyvinyl alcohol/graphene oxide aerogel. Chemosphere **278**, 130492 (2021)
200. C. Mu et al., Removal of bisphenol A over a separation free 3D Ag3PO4-graphene hydrogel via an adsorption-photocatalysis synergy. Appl. Catal. B **212**, 41–49 (2017)
201. W. Wang et al., Adsorption and competition investigation of phenolic compounds on the solid-liquid interface of three-dimensional foam-like graphene oxide. Chem. Eng. J. **378**, 122085 (2019)

202. T.R. Das et al., Bismuth oxide decorated graphene oxide nanocomposites synthesized via sonochemical assisted hydrothermal method for adsorption of cationic organic dyes. J. Colloid Interface Sci. **509**, 82–93 (2018)
203. M. Deng, Y. Huang, The phenomena and mechanism for the enhanced adsorption and photocatalytic decomposition of organic dyes with Ag3PO4/graphene oxide aerogel composites. Ceram. Int. **46**(2), 2565–2570 (2020)
204. N. Liu et al., Stabilized magnetic enzyme aggregates on graphene oxide for high performance phenol and bisphenol A removal. Chem. Eng. J. **306**, 1026–1034 (2016)
205. R. Hu et al., Efficient removal of phenol and aniline from aqueous solutions using graphene oxide/polypyrrole composites. J. Mol. Liq. **203**, 80–89 (2015)
206. E.R. Hugo et al., Bisphenol A at environmentally relevant doses inhibits adiponectin release from human adipose tissue explants and adipocytes. Environ. Health Perspect. **116**(12), 1642–1647 (2008)
207. F. Chen et al., Highly efficient removal of bisphenol A by a three-dimensional graphene hydrogel-AgBr@ rGO exhibiting adsorption/photocatalysis synergy. Appl. Catal. B **217**, 65–80 (2017)

Chapter 8
Fabrication of Advanced 2D Nanomaterials Membranes for Desalination and Wastewater Treatment

Koena Selatile, Suprakas Sinha Ray, Neeraj Kumar, Vincent Ojijo, and Rotimi Emmanuel Sadiku

Abstract Environmental concerns regarding water shortages due to industrialization and pollution have led to escalation in research towards efficiency in wastewater treatment and desalination. To date, nanotechnology is the most effective solution towards water shortages and is currently used for wastewater treatment and desalination. Among the emerging nanosheets, two-dimensional (2D) nanosheets have gained much attention since graphene was discovered in the fabrication of cost-effective and sustainable membranes for environmental remediation. Recently, 2D nanoengineered membrane technologies have revealed a new potential for removing hazardous compounds from our surroundings. The present chapter gives an overview of the concept of membrane technology, membrane fabrication techniques and the importance of 2D nanomaterials in the desalination and wastewater treatment membranes. Firstly, popular fabrication methods for membranes, such as electrospinning, drop-casting, spin-coating, solution casting and phase inversion, will be discussed. This will be followed by the application of 2D nanoengineered membranes incorporated with graphene, MXenes, molybdenum disulfide (MoS_2) and other nanosheets, in their 2D form, for excellent improvement in desalination and wastewater treatment. Notably, the chapter emphasizes the wide range of membrane applications as well as their potential and challenges for use in the development of nanotechnology-based environmental remediation.

K. Selatile · S. S. Ray · N. Kumar (✉) · V. Ojijo
Centre for Nanostructures and Advanced Materials, DSI-CSIR Nanotechnology Innovation Centre, Council for Scientific and Industrial Research, Pretoria 0001, South Africa
e-mail: nkumar@csir.co.za; ynk.neeraj@gmail.com

S. S. Ray · N. Kumar
Department of Chemical Sciences, University of Johannesburg, Doornfontein, Johannesburg 2028, South Africa

K. Selatile · R. E. Sadiku
Division of Polymer Technology, Department of Chemical, Metallurgical and Materials Engineering and Institute of NanoEngineering Research, Tshwane University of Technology, Pretoria, South Africa

N. Kumar et al. (eds.), *Two-Dimensional Materials for Environmental Applications*, Springer Series in Materials Science 332, https://doi.org/10.1007/978-3-031-28756-5_8

8.1 Introduction

For centuries, inadequate water supply (for industrial, domestic and agricultural water use) has been one of the greatest global challenges facing modern society [1]. Given The amount of wastewater produced industrially, most water supplies are currently heavily polluted with hazardous/toxic chemicals such as heavy metal ions, synthetic dyes, and radioactive substances. Therefore, wastewater treatment serves as a solution to alleviate these challenges. On the other hand, the divalent ions and the minute monovalent salt ions are removed by desalination of brackish and seawater. Additionally, desalination offers another alternative fresh water supply, owing to its abundance (with more than 97% of the overall water resources covering the land) and easily accessible seawater. Various contaminants in wastewater include artificial and natural components such as natural organic matter, microorganisms, heavy metals, and synthetic dyes [2–7]. Many wastewater treatment plants use various techniques for water purification of such contaminants. These techniques include flocculation/sedimentation filtration, chlorine disinfection, adsorption, ion-exchange adsorption, ion exchange, membrane filtration, adsorption, and precipitation [8–10]. Membrane-based separation technology rapidly dominates wastewater treatment and desalination, including microfiltration, ultrafiltration and nanofiltration. Membrane filtration technology has been employed in chemical, pharmaceutical and water industries, amongst others. It offers several advantages such as cost-effectiveness, energy saving, good selectivity, compactness, ease of fabrication, simple operation and environmental friendliness [11]. Selectivity and permeability are the two factors that contribute towards membrane performance in water purification. However, the compromise between membrane rejection properties and porosity has led to more interest in nanotechnology as a solution amongst the various wastewater treatment methods, such as nanomaterials for the fabrication of innovative membranes [12].

The discovery and isolation of mono-sheet graphene in 2004 unlocked more access to fresh opportunities for exploring advanced 2D nanomaterials due to their distinctive features, such as atomic thickness, large lateral sizes, and reactive sites [13–15]. Various 2D materials, namely graphene-based materials, MXenes, h-boron nitride, and others, have been widely used to fabricate advanced 2D membranes with improved filtration performance [16, 17]. The 2D nanomaterials are thus promising as potential candidates to ensure improvement in the quality and availability of water treatment while also ensuring growing global water supplies.

8.2 2D Materials: Their Exceptional Characteristics for Membrane Fabrication

Figure 8.1 shows the structure of various popular 2D materials. The 2D nanostructures have unique features, shapes and properties. They are considered ideal replacements for traditional desalination and water treatment membranes. Owing to their

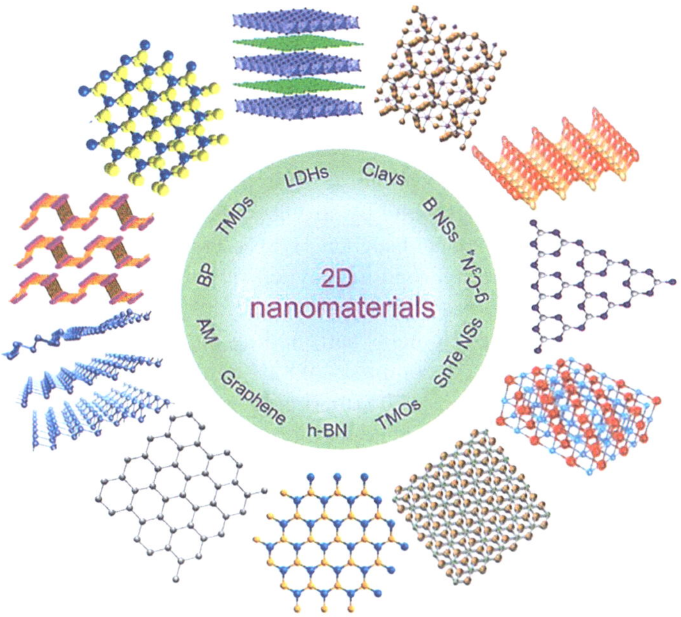

Fig. 8.1 Structure of various 2D nanomaterials. Reproduced with permission from ref. [16]. Copyright 2022, Frontiers

distinctive structural properties (atomic thickness and robustness) and precise molecular filtration capability, 2D nanomaterials are fascinating materials for outstanding nanofiltration operations such as organic solvent nanofiltration membranes [18, 19].

8.2.1 Graphene-Based Materials

Graphene, a very thin and light 2D material made of a single graphite atomic layer, is one of the strongest materials ever known due to the presence of sp^2 bonds. Due to the repulsive forces of dense and delocalized π-orbital cloud repulsive forces, graphene sheets are impermeable to even the smallest molecules, that is, monoatomic-sized particles such as helium and hydrogen, regardless of their thickness of one atom. The presence of defects in graphene sheets will impart permeability. Therefore, the introduction of controlled pores on the sheets gives them potential in membrane and separation processes (to allow water permeation while rejecting ions). The high mechanical strength, chemical resistance, and unique physiochemical properties make graphene-based nanomaterials [graphene, graphene oxide (GO) and reduced GO (rGO)] highly beneficial as a membrane material that is very suitable for water desalination. These properties include high surface area, hydrophilicity, reactivity, and chlorine resistance, in addition to the antibacterial property for biofouling control [20, 21]. Graphene-based nanomaterials, mainly GO, are the most superior carbon-based nanomaterials for ease of fabrication and economic viability [22].

GO is essentially a single-layer-thick with a hexagonal pattern lattice that contains many oxygen-rich sites on its base and edges, which lead to good hydrophilic and antifouling properties [23, 24]. GO is synthesized by various methods; these include the Hummers and Brodies methods, and their improved versions [25, 26]. GO is produced by the oxidation of graphite by strong oxidizing agents such as $KMnO_4$ and $NaNO_3$ along with a strong acid such as H_2SO_4, using a well-known improved Hummers method. With the hydrophilic oxygen-rich surface, the edges possess carboxylic groups, while the basal plane contains hydroxyl and epoxy groups. GO sheet is composed of oxidized domains surrounding the isolated pristine graphene domains, which did not oxidize. It comprises of 82% oxidized, 16% pristine graphene and 2% holes [27, 28]. Reduction of GO (thermally or chemically) produces a structure similar to that of pristine graphene. Although in its reduced form (rGO), the carbon-to-oxygen ratio is reduced from 2:1 to 3:1 [14].

Graphene is an emerging 2D material not only with a large surface area, but also with good elasticity, exceptional electrical conductivity and highly active, thus good enough characteristics for desalination [29]. Owing to its high rejecting capabilities and good permeability, GO has opened new opportunities as a membrane material. The water flow through GO membranes is through the GO galleries, gaps between the edges, wrinkles and surface openings created by the functional groups. GO membranes are chemically and thermally stable for use in processes such as membrane pervarporation [30]. As GO sheets tend to form laminates, as microcrystalites, they form interconnected layers stacked above and adjacent to each other with interlayer spacing between them. The interlayer gap is due to the oxygen functional groups attached to the nanosheets. The GO interlayer gap may require adjustment to get the desired performance regarding the rejection of solutes and permeation (water flux) [31]. The 2D interlayer gap between the GO sheets acts as the nanochannels for water transport and solute rejection, which can further be modified or adjusted to achieve desired rejection based on the size of solute particles [32].

Unlike GO, rGO is less hydrophilic with tighter nanochannels and less swelling and, therefore, more stable in water environments as it contains less oxygen functional groups [14, 29]. However, water transport through the lamellar GO or rGO sheets may be limited owing to the strong capillary force of the tight nanochannels [33]. The rGO nanochannels can thus be adjusted to control the water transport behavior and to achieve size-selective molecular filtration [32]. This can be achieved by the intercalation of materials within the nanochannels, such as soft polymers and polyelectrolytes [34]. Intercalation of more rigid materials is to avoid unstable/instability or fluctuation in the interlayer spacing over time. This led to the use or inclusion of more rigid metal oxide and metal sulfide materials. These include metal oxides such as Zinc Oxide, Copper Oxide, Magnesium Oxide, etc., and metal sulfides such as Zinc Sulphide and MoS_2 nanoparticles. The rigid spacers also enable efficient filtration by size exclusion by expanding their lateral tortuous paths for water permeability, thereby solving the permeability-selectivity trade-off. The metal oxide nanoparticles enlarge the interlayer gap of the rGO sheets, thereby, also expanding their lateral tortuous paths [29, 35].

8.2.2 *Transition Metal Dichalcogenides (MoS_2)*

MoS_2, another category of 2D nanomaterials comparable to graphene, is available as single layer or a few layers through exfoliation of the bulk MoS_2 [36, 37]. Like graphene, MoS_2 can also be produced by a bottom-up synthesis approach using Molybdenum (Mo) and sulphur (S) salts [38, 39]. MoS_2 is a graphene-like 2D transition metal dichalcogenide with formula MX_2, with "M" representing a transition metal element such as titanium, molybdenum, tungsten, while "X_2" denotes chalcogen atoms such as selenium, sulphur, telluride (i.e., MoS_2 is a negatively charged layered analogue of graphene). Compared to graphene, MoS_2 single-layer is diatomic as it contains (Mo) and sulphur (S) (i.e., a monolayer contains three atomic layers, whereas graphene is a single layer). MoS_2 consists of monolayers with strong covalent bonds between Mo and S_2 atoms and weak Van der Waals forces amongst neighboring sheets. MoS_2 single-layer membrane is 1 nm thick, with hydrophilic Mo edge and hydrophobic S edge and mechanical strength of steel (~270 GPa) [40]. The Mo atom is packed between the two S atoms, giving an interlayer gap of 0.62 nm between two Mo atoms and a free gap of about 0.30 nm as the laminar channel space, smaller than the salt ions Na^+ (0.716 nm) and Cl^- (0.664 nm), and larger than water molecules (0.276 nm), thus making them excellent candidates for desalination [37, 41]. Due to their tremendous physiochemical, biological, antifouling, and mechanical properties, MoS_2 nanosheets have been used in catalysis, sensing and optoelectronics and have recently gained considerable attention in water separation and purification [42–44]. Like graphene, it is a promising material for membrane separation due to its exceptional mechanical, thermal, and antibacterial properties. Lack of π-π bonds in MoS_2 prevents membrane scaling and biofouling. This makes it a good candidate as adsorption filtration membrane for contaminants and disinfection. Single sheet membranes of 1–10 nm sized nanopores can be sculpted by a highly focused electron beam, large enough to allow water passage while restricting ions through the membrane [45].

The single-layer nanosheets function as nanoporous membranes that can effectively filter ions while permitting fast water passage without modification when compared to other nanoporous membranes of GO and carbon nanotubes (CNT). The separation or isolation of Mo and S_2 affects membrane performance; for example, having Mo only on the nanopore induces hydrophilicity, thus increasing water transport through the nanopore. These hydrophilic sites give more permeability to the membrane as they can attract water molecules through hydrogen bonding. Molecular simulations have shown that MoS_2 membranes have higher flux than GO-based membranes [46]. MoS_2 nanopores thus have a pore chemistry good enough to control the water flux 70% higher than that of graphene nanopores without needing any further surface chemical modification, unlike the graphene nanoporous membranes that require further modification. The functionalization affects salt rejection (trade-off effect), and the rough surface caused by functional groups exposed on the carbon plane causes water flow resistance [45, 46].

Moreover, the termination of Mo leads to the creation of charged edges of MoS_2 nanopore edges. Therefore, it creates positive charges on nanopores and repels Na^+ ions, unlike graphene with neutral carbon atoms. At the same time, hydrophilic Mo at the tip of the nozzle-like nanopore draws water to the inside of the MoS_2. It was found that a nanopore area of 20–60 A^2 can easily reject 88% of salt ions. Unlike GO, the smoothness is due to the lack of functional groups. Yet, pore chemistry can control the flux and salt rejection without needing further modification/pore functionalization. Therefore, it can be said that membranes with stacked MoS_2 sheets thus have higher water permeability than GO membranes. Chemically exfoliated MoS_2 is hydrophilic, negatively charged, and smooth as the original 2D structure is maintained. On the other hand, the hydrophilicity and negative charge of GO are due to chemical oxidation, which produces oxygenated hydrophilic functional groups protruding from the carbon plane, thus, a rough surface [45, 47].

MoS_2 multi-layer membranes have successfully been applied in gas separation and water treatment for heavy metal ions and organic/inorganic contaminants removal by adsorption (i.e., intercalating contaminants into the interlayer spacing or by the S atoms that can be exposed by the exfoliation process of MoS_2) and electrostatic interaction due to the negative surface charge [48–51]. The hydrophilicity and negative charge of MoS_2 make it highly useful in modifying the physiochemical properties of the thin film nanocomposite (TFNC) membrane surface. MoS_2 are also stacked as multi-layers whereby the exfoliated monolayers are restacked by vacuum filtration for modification of physiochemical properties of the TFNC reverse osmosis desalination membranes, which include hydrophilicity, surface charge and roughness. Advantages such as thinner selective skin layer and a decrease in membrane smoothness, enhanced surface hydrophilicity, negative surface charge and interlayer spacing) contribute towards the separation efficiency of the MoS_2 multi-layer TFNC membrane. MoS_2 is a good candidate for membrane-based separation as nanoporous membranes and multi-layer (laminar) membranes [37].

8.2.3 MXenes

Since the discovery of MXenes in 2011 by Drexel University, the production of multi-layer MXenes has become quite popular for preparing robust lamellar water purification membranes [11]. MXenes are a new addition to the existing 2D materials as a new collection of 2D transition metal carbides, nitrides and carbonitrides [52]. MXenes are one of the rapidly-expanding 2D nanomaterials. The three atomic structures are shown in Fig. 8.2, M2X to M3X_2 and M_4X_3 represent Ti_2C, Ti_3C_2 and Ti_4C_3, as the three types of MXene, respectively [53]. MXenes are commonly characterized by the formula, $M_{n+1}X_nT_x$; where M is an early transition metal such as Ti, Mo, Sc, W, V, Zr, with interleaving $n + 1$ layers; X symbolizes carbon or

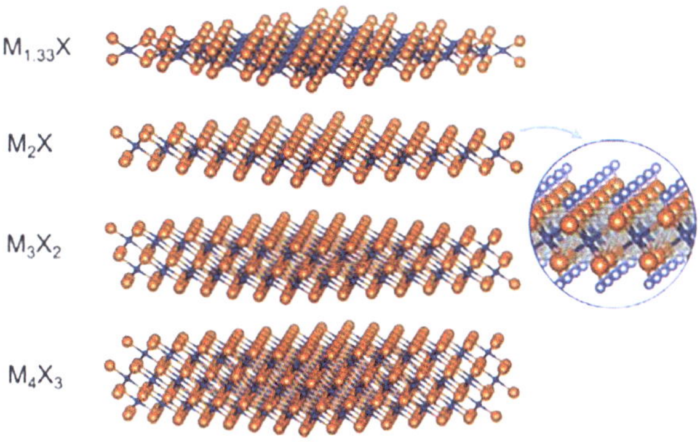

Fig. 8.2 Structure of mono-MXenes. Reproduced with permission from ref. [53]. Copyright 2020, Elsevier Science Ltd.

nitrogen with n layers; T signifies surface end groups such as −Cl, −F, −O, and −OH; the x denotes the number of surface functional groups, and n is from 1 to 4 [54]. MXenes materials have recently emerged as exceptional materials for application in various fields such as energy storage [52, 55], catalysis [56], electromagnetic interference shielding [57] and novel membranes for water desalination [58]. MXenes have exceptional properties such as hydrophilicity, large interlayer gap, high surface area and sufficient amount of very active surface sites (the surface chemistry is easily tunable (owing to the existence of the free electrons associated with transition metal atoms) and are negatively charged materials [53, 59]. The intriguing structural and chemical properties are highly influenced by their synthesis conditions: starting material for MXene, material used for etching, etching and intercalation techniques. The properties can further be tailored to suit the particular application MXene M and X arrangements and surface terminations (–F, –OH, and –O). This includes chemical stabilities, outstanding electrical conductivities, and environmentally benign nature. MXenes as their ability to host and Li^+, Na^+, Mg^{2+}, K^+, intercalate cations, such as and $NH_4{}^+$ inside the Ti_3C_2 sheet in aqueous environments. With these properties, they are proficient in removing heavy metals, dyes, and radioactive nuclides through adsorption or adsorptive, electrostatic interactions and steric exclusion [53].

The remarkable structural properties of MXenes include simplicity of functionalization and activated metallic hydroxide sites, biocompatibility, and superior adsorption-reduction capability. In addition to their antifouling properties, their electrical conductivity, mechanical, thermal, chemically stable, and adsorbent and catalytic properties make them highly suited as desalination membranes and nanofiltration (NF) membranes for removal of pollutants such as dyes and ions (metal and salt ions), oil/water emulsion/mixtures from wastewater [60].

MXene membranes have proven great potential as water treatment and desalination membranes. MXenes are materials as barrier layers of high flux filtration membranes owing to their ultrathin 2D structure and ease of surface functionalization [61]. They have shown the ability to remove pollutants from the water as pristine and modified membranes. MXenes are synthesized economically at room temperature. Multi-layer MXenes are synthesized through an exfoliation process by selective etching of group A elements from their MAX phases (Fig. 8.3). It is a wet-chemical etching process in hydrofluoric acid (HF) or HF-containing or HF-forming etching materials that will add –O, –F, or –OH functionalities (represented by T_x in the MXene formula). MAX phases are transition metal layered ternary carbides, nitrides, and carbonitrides comprising of M, A and X layers, "M" is an early transition metal, the "A" groups are mostly groups 13 and 14 elements, and "X" is C and/or N. Some of the MAX phases, amongst the 60^+ others that are known to exist, include: Ti_2AlC, Ta_4AlC_3, $(Ti_{0.5},Nb_{0.5})_2AlC$, $(V_{0.5},Cr_{0.5})_3AlC_2$, and Ti_3AlCN. These powders are treated in HF acid at room temperature and then sonicated. Thus, the "A" group layer is eliminated from the MAX phases, forming the 2-D sheets and referred to as MXenes (the term highlights their similarities to graphene, as these have the same structure as the graphene, but with hydrophilic surfaces) [62]. Titanium carbides ($Ti_{n+1}C_n$, for example, $Ti_3C_2T_x$ (Titanium (III) carbide (II)); with T_x representing surface groups such as $–OH_x$ and $–F_x$) is the most investigated class of MXenes, synthesized using ammonium bifluoride salt (NH_4HF_2) as a HF-containing etchant that replaces HF [54, 63]. The toxicity of concentrated HF has led to a less hazardous synthesis route that uses milder etchants for exfoliation and delamination. For example, a combination of less hazardous hydrochloric acid (HCl) and lithium fluoride (LiF) salts can be used to etch Al from Ti_3AlC_2 [64]. This procedure is followed by intercalating water molecules and cations to form large interlayer spacing between the MXene layers. Another etchant example is ammonium bifluoride (NH_4HF_2), followed by intercalation of molecules such as ammonium species $–NH_3$ and $–NH_4^+$, for the delamination of sheets from each other to form single sheets. Other safer etchants include fluoride salts. These include potassium, sodium, calcium tetrabutyl ammonium and cesium fluorides. These are used with either HCl or sulphuric acid (H_2SO_4) [63].

The equations below represent the wet-etching synthesis using HF at room temperature [53]:

$$Mn + 1AXn(s) + 3HF(aq) \rightarrow Mn + 1Xn(s) + AF_3(s) + 3/2H_2(g) \quad (8.1)$$

$$Mn + 1Xn(s) + 2H_2O\ (l) \rightarrow Mn + 1Xn(OH)_2(s) + H_2(g) \quad (8.2)$$

$$Mn + 1Xn(s) + 2HF\ (aq) \rightarrow Mn + 1XnF_2(s) + H_2(g) \quad (8.3)$$

8.3 Fabrication of 2D-Enabled Membranes

The fabrication methods for 2D-based membranes make use of three commonly used routes: (i) 2D nanosheets are used as functional materials for the fabrication of lamellar membranes; (ii) 2D nanosheets as additives for mixed matrix membranes;

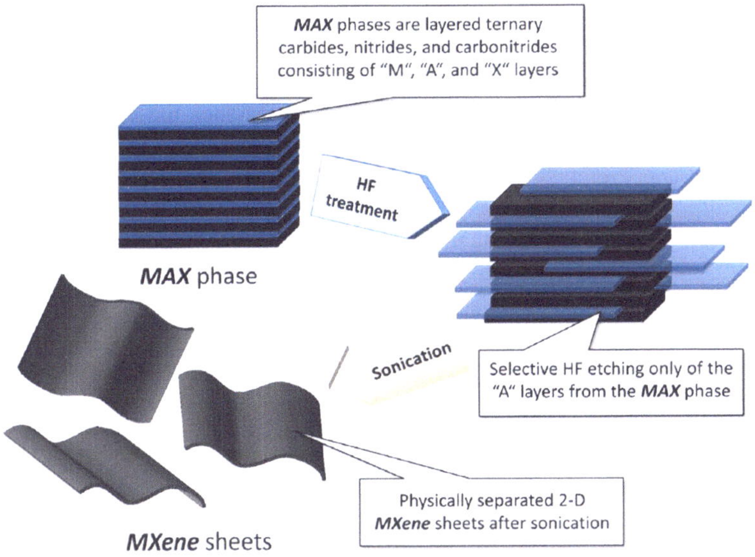

Fig. 8.3 Synthesis of MXenes using HF as the etchant. Reproduced with permission from ref. [62], Copyright 2012, American Chemical Society

and (iii) the 2D sheets used as coating layers to improve the membrane substrates [65].

8.3.1 Fabrication of Layered Membranes

Multi-layer 2D sheets of the lamellar structure can be used for layered membranes that are needed for the NF method. NF membrane filtration method can efficiently filter multi-valent salts and organic contaminants. NF membranes have nano-sized pore sizes ranging from 0.5 to 5 nm (nanoporous) with a nominal molecular weight cut-off of 200–1000 Da [29]. The filtration mechanism of NF membranes is based on size and charge exclusion, which is determined by the membrane structure and its interactions with the solute particles. On the other hand, desalination membranes have tighter pore sizes below 0.5 nm and usually use 2D single sheets as nonporous membranes that reject the monovalent ions [66].

8.3.1.1 Fabrication of Layered or Lamella Membranes by a Vacuum-Assisted Filtration

The vacuum-assisted assembly is a solution-processing membrane fabrication method, particularly for the preparation of ultrafine nanofiltration membranes

(Fig. 8.4b) [29]. In this method, the nanosheets are piled together and interlocked in a parallel fashion, leading to the construction of a robust 2D nanomaterial laminar membrane. The narrow interlayer spaces formed are highly adaptable and facilitate water permeability while blocking particles larger than the interlayer spacing. Laminar membranes can be prepared as free-standing films or supported onto porous support such as polyethersulfone, poly(vinylidene fluoride) (PVDF), polycarbonate or Nylon ultrafiltration membranes. The laminar sheets of the 2D nanomaterials (GO, rGO, MXenes, MoS_2), as free-standing membranes, are highly promising for water treatment and desalination owing to their unique structures. These sub-μm thick membranes are robust yet flexible [31].

In the case of MXenes, following the synthesis (etching) step (e.g., $Ti_3C_2T_x$ synthesized from Ti_3AlC_2 as the corresponding MAX precursor) [61], the exfoliation of the multi-layered MXene flakes into a single-layer or a few layers is due to the intercalation of guest particles. These enlarge the c-lattice factor and deteriorate the hydrogen and the electrostatic forces between the 2D MXene layers. The exfoliated MXene layers have a negative surface charge, possess a higher surface area than multi-layered MXene layers and have C atoms, –F and –OH groups exposed on their surface, which are responsible for capturing contaminants in the wastewater [68]. The exfoliated sheets are then dispersed in a suitable solvent, forming a homogenous

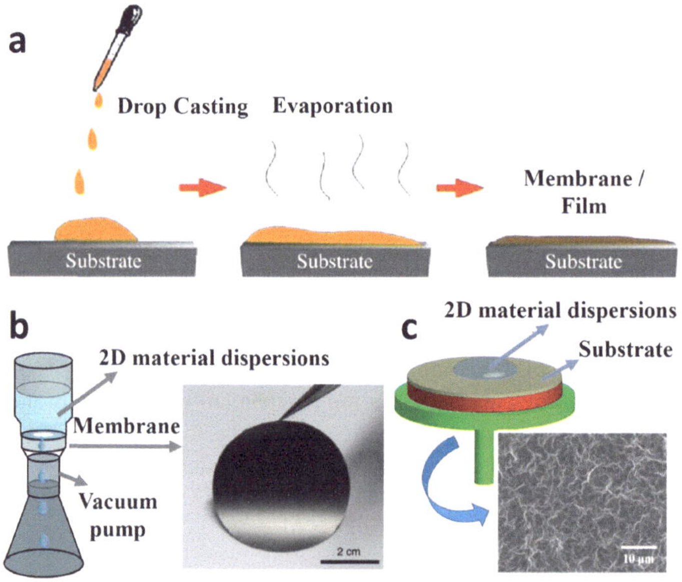

Fig. 8.4 Common membrane fabrication methods used for 2D nanomaterial-based lamellar membranes: **a** Evaporation-assisted self-assembly and drop casting methods. **b** Vacuum-assisted method and **c** spin-coating methods. Reproduced with permission from ref. [67]. Copyright 2022, Taylor & Francis

dispersion. The sheets disperse well in polar solvents, e.g., H_2O, ethanol, *N*-methyl-2-pyrrolidone, propylene carbonate, *N*, *N*′-dimethylformamide, and DMSO [69]. The dispersions are deposited onto a porous substrate using a pressure filtration device. Usually, the pressure of 1 bar is utilized as it is sufficient enough to tightly pack the nanosheets in an orderly manner [70]. The vacuum filtration deposition is accompanied by calcination or sintering at temperatures up to 500 °C. This yields a thin film MXene active layer deposited onto a porous support layer such as polymer materials of pore sizes of ± 200 nm. The interlayer spacing of MXene sheets is a result of moisture loss. The assembly of the resulting lamellar membrane thickness can be controlled by simply controlling the loading of MXene nanosheets. In addition to the vacuum- assisted filtration other methods for laminar MXene membranes include layer assembly, dip-coating, slot-die coating, and phase-inversion methods.

Lamellar membranes separate target molecules by sieving through the stacked layers with the aid of electrostatic interaction between the target species and the negatively charged layers. Unlike impermeable nanoporous GO, free-standing lamellar GO membranes permit water molecules to pass through the interlayer spaces without any hindrance. However, the water molecules can show low-friction contact with the hydrophobic area of the GO sheets, thus limiting the water passage through the nanochannels. Providentially, one advantage for the lamellar membranes is the ability to tune the interlayer spacing between the sheets, enhancing water permeability or selectivity towards the pollutants [71]. Evidently, a decrease in interlayer spacing enhances ion rejection, while an increase in interlayer spacing enhances membrane permeability.

8.3.1.2 Drop Casting Method

Drop casting (or 2D film fabrication method) is another technique to fabricate laminar membranes. In this method, the prepared nanosheet dispersion solutions are drop-cast onto a support layer (e.g., silica, polymer (e.g., polysulfone), or anodic aluminum oxide (AAO)) followed by solvent evaporation at ambient temperature. However, low vapor-pressure solvents such as H_2O may cause poor membrane uniformity due to insufficient evaporation [67, 72] (Fig. 8.4a).

8.3.1.3 Spin Coating Method

Spin Coating Method is a method whereby laminar membranes are prepared by spreading the prepared 2D nanomaterial solutions onto a substrate surface (Fig. 8.4c). The spin coating removes water deposited between the 2D nanosheets in the solution, causing strong, attractive capillary forces for additional deposition. The deposition of the nanosheets is due to the electrostatic and hydrophilic interaction with the substrate. The repulsive and attractive interactions between the sheets allow for homogeneous deposition of the assembled sheets [67].

8.3.1.4 Modification of 2D Nanosheets and Post-Treatment of Fabricated Lamellar Membranes

The fabricated $Ti_3C_2T_x$ MXene lamellar membranes can often have defects along the edges of the sheets, thereby allowing penetration of target species and resulting in poor membrane performance. The defects are usually due to the synthesis route of using harsh etchants such as HF. MXene is thus blended with other 2D materials, such as GO, to improve the rejection performance of the MXene lamellar membranes. GO combined with $Ti_3C_2T_x$ with a certain percentage loading of 10–30 wt%, followed by sonication in H_2O to form a homogenous dispersion and filtered as described earlier [71].

Thermal crosslinking is self-crosslinking by drying in an oven for 90 °C to remove free water between the sheets to control the interlayer spacing. Thermal crosslinking often requires enhancement of the MXene negative surface charge by hydroxylation. The fabricated MXene membrane is then hydroxylated after the self-curing using a potassium hydroxide mixture to substitute –F with –OH groups (improving the negative charge and wettability) [61, 73].

8.3.2 Fabrication of Mixed Matrix (Hybrid) Membranes

8.3.2.1 Solution Casting

2D nanomaterials, such as MXenes, are blended with polymers to fabricate composite polymer membranes to improve the microstructural and physicochemical properties of the membranes, such as enhanced antifouling properties. For example, when mixing MXene with polymers, the MXene oxygen-containing functional groups form chemical bonds with the polymer, thus, considerably reducing polymer chain motion and creating well-formed galleries for transporting molecules. Figure 8.5 shows the synthesis procedure of polyethersulfone (PES) and nickel (Ni)/MXene composites. MXene is blended with the Nickel nanoparticles (Ni-NPs) and incorporated into the PES polymer matrix to produce composite membranes through a phase inversion process. Fabrication of mixed matrix membranes is an easy and economical process. However, with this procedure, most of the MXene nanosheets remain in the bulk of the polymer matrix rather than being exposed to the membrane surface. Therefore, the membrane antifouling, hydrophilicity and permeability functionalities are compromised. The use of Ni-coated MXene is an attempt used to overcome this shortcoming. The magnetic NPs can thus be brought onto the membrane's outer surface using a magnet during the solution-based membrane fabrication process [74].

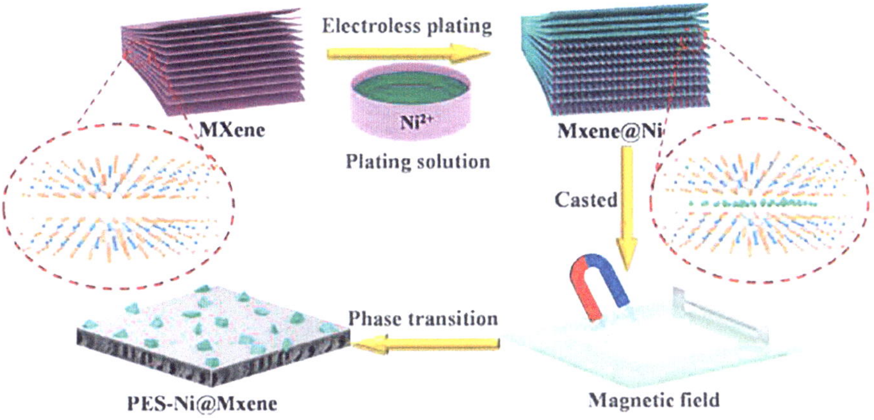

Fig. 8.5 Preparation process of PES-Ni@MXene composite membranes. Reproduced with permission from ref. [74]. Copyright 2021, Elsevier Science Ltd.

8.3.2.2 Electrospinning

Electrospinning, like the solution casting method, is a polymer solution-based membrane fabrication method used to load 2D nanosheets into the polymer matrix of the nanofibrous membranes. It is a more efficient and versatile method for fabricating mixed matrix membranes and is the preferred method over the solution casting process. It is a more suitable option for those nanosheets that tend to aggregate in solution-based processes. Electrospun membranes have recently gained attention in water treatment and are used as carrier substrates for loading MXene sheets. Although MXenes are hydrophilic, they tend to undergo oxidation in aqueous environments. MXenes are easily modified, owing to the inherent hydroxyl, fluorine and oxygen terminal groups, by using small particles, such as some organic compounds, polymers and NPs, as one of the strategies used to avoid aggregation. Therefore, electrospinning of polymer composites containing MXenes sheets in conjunction with palladium (Pd) as catalytic NPs, led to the successful fabrication of catalytic membranes for catalytic degradation of contaminants from the wastewater, e.g., PVA/PAA/MXene@PdNPs composites (Fig. 8.6) [75, 76].

8.3.2.3 The 2D Sheets as Coating Materials to Improve the Membrane Support Layer

GO sheets are coated onto existing polymer membrane surfaces by covalent bonding (e.g., TFNC desalination membranes) to enhance the membrane's antimicrobial ability, fouling resistance and hydrophilicity. In this process, the carboxylic groups of the GO nanosheets are linked with the other carboxylic groups on the surface of the polyamide (PA) film of the TFNC surface through amide coupling (Fig. 8.7) [77].

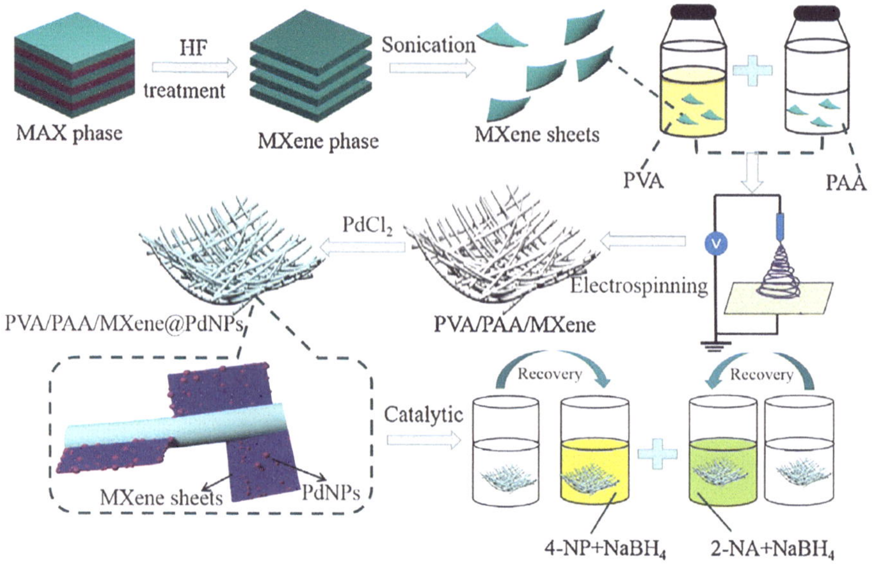

Fig. 8.6 Fabrication of mixed matrix 2D enabled membranes by electrospinning method. Reproduced with permission from ref [76]. Copyright 2019, Elsevier Science Ltd.

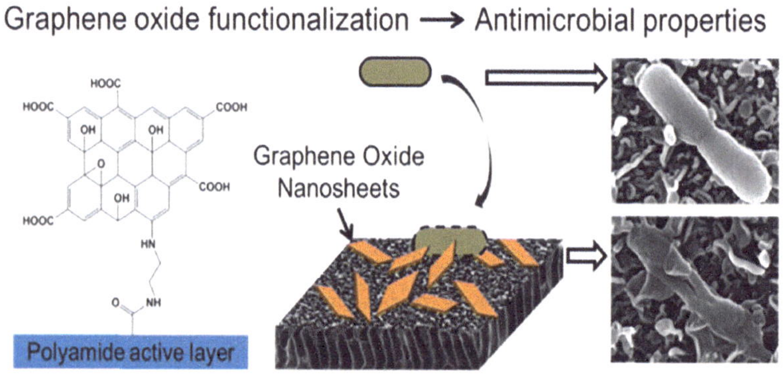

Fig. 8.7 GO surface coating onto PA-TFNC membrane. Reproduced with permission from ref. [77]. Copyright 2013, American Chemical Society

8.3.3 Layer-By-Layer Self-Assembly of 2D Nanomaterials onto Membrane Surfaces

Layer-by-layer self-assembly is the deposition of oppositely charged 2D nanomaterial sheets by the dip-coating method (Fig. 8.8). The coating is usually on membranes of negatively charged membrane surfaces such as the nanofiltration and RO desalination membranes. GO undergoes modification into positively charged sheets to form

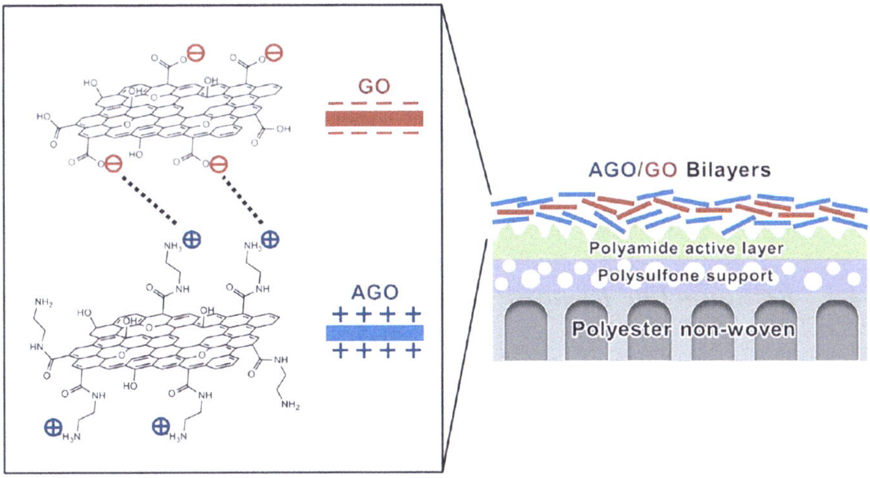

Fig. 8.8 Schematic illustration of a multi-layered graphene oxide (GO) coating on a polyamide thin-film composite membrane surface via layer-by-layer (LbL) deposition of oppositely charged GO and aminated-GO (aGO) nanosheets. Reproduced with permission from ref. [78]. Copyright 2013, American Chemical Society

aminated GO sheets (aGO) as follows: GO sheets are treated with ethylenediamine and EDC (1-(3-dimethylaminopropyl)-3-ethylcarbodiimide). EDC methiodide is a carboxyl modifying reagent (crosslinking agent for amide bond-forming reactions), which acts as a linking agent that induces the amide group between the carboxylic acid group of GO and the amine terminal group of ethylenediamine. GO sheets are coated onto the membrane surfaces via a non-covalent bonding modification process whereby the negatively charged GO is coated onto the membrane surface, followed by a layer of positively charged aminated GO (aGO). The aGO functions as a bridging nanomaterial [78]. This is followed by the preparation of uniformly dispersed aqueous suspensions of the positive aGO and the negative GO sheets by sonication. The positively charged aGO sheets are then attached to a negatively charged membrane surface by dipping it into the positive aGO suspension, followed by dipping into a negative GO suspension. This forms a bilayer of GO/aGO onto the membrane surface, then dried in the oven. The process can be repeated to get the desired bilayer thickness.

8.4 Application of 2D-Enabled Membrane Materials in Environmental Applications

Modified and unmodified 2D-based membranes have been employed to remove different contaminants from water. The functionalization of the 2D nanomaterials, such as GO and MXenes, could also enhance the membrane rejection performance and anti-fouling properties through adsorption, charges, and sieving [60]. However, the performance is highly dependent on membrane selectivity and permeability. As the stacked sheets of laminar membranes tend to have reduced permeability compared to the traditional polymer water purification membranes, the inclusion of NPs within the galleries of laminar membranes has a positive effect on membrane flux and pollutant rejection [65].

8.4.1 Radionuclide Removal

Radionuclide contamination originates from nuclear accidents during the mismanagement of radioactive liquid waste [10]. Barium isotopes (^{133}Ba and ^{140}Ba) and strontium (^{90}Sr) Ba^{2+}, Sr^{2+}, are the most toxic radionuclide substances. Radioactive barium isotopes adversely affect breathing, liver, heart, and kidneys and Sr^{2+} causes oxygen shortage in the lungs [10, 79]. Uranium (U) is the critical constituent of conventional nuclear fuel cycles, is a highly hazardous radionuclide with a long half-life and threatens the environment and ecological system. Hexavalent uranium U(VI) is exceptionally mobile, easily dissolves in water, forms the most stable uranyl ion (UO^{22+}) in solution, and has been mostly mitigated by adsorbents such as GO and activated carbon. In addition, MXenes have shown great potential as effective adsorbents for removing these pollutants [53].

8.4.2 Uranium Exclusion

Uranium is a naturally occurring radioactive substance found in soil, rocks, air and water. However, high concentration level in drinking water and groundwater used for agriculture is toxic to humans and causes kidney damage. With extensive exposure, uranium decays into other cancer-causing radioactive substances, such as radium. Uranium in the soil and human tissue may be sourced from phosphate fertilizers. Therefore, long-term usage of uranium-bearing fertilizers leads to an increase in the uranium concentrations in fertilized soil [80]. 2D materials such as MXenes are resilient to damage caused by radiation, have excellent chemical and mechanical properties, and can remove such contaminants from wastewater. They have shown the ability to function as desired adsorbents for radio-active nuclides from the environment, such as uranium (238U), thorium (232Th), cesium (137Cs), and strontium (90Sr) [53]. The mechanism/interaction pathways for MXenes for radionuclides removal include coordination, ion exchange, and reduction immobilization. Adsorption of Ba^{2+}, Sr^{2+}, Pd^{2+}, Cs^{+}, Th^{4+}, Eu^{3+}, Re^{7+}, and U^{6+} ions on MXene materials has been investiagted [53]. Wang et al. [81] developed a multi-layered $Ti_3C_2T_x$ MXene membrane to capture U(VI) in aqueous solutions. They explored the adsorptive behavior of the pristine MXene and that of the hydrated MXene with DMSO. As the hydrated MXene was intercalated and had enlarged interlayer spacing, they

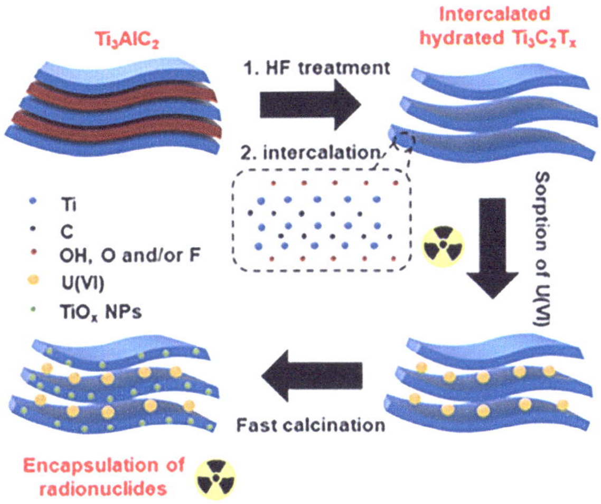

Fig. 8.9 Hydrated intercalation synthesis strategy of $Ti_3C_2T_x$ MXene for efficient U(VII) uptake and imprisonment. Reproduced with permission from ref. [82]. Copyright 2022, Elsevier Science Ltd.

could capture the U^{6+} radionuclide within the MXene galleries. This was termed a "radionuclide imprisoning" approach (Fig. 8.9), which is a fast low-temperature post-adsorption calcination intercalation strategy of increasing the adsorption capacity of MXenes [81].

8.4.3 Dyes and Other Colorants Removal

Dyes are organic pollutants mainly from wastewater discharged by textile industries, thus posing a serious threat to the ecological environment. Removing dyes and other organic contaminants is dominated by adsorption and catalytic degradation [83]. MXenes have exceptional photocatalytic performance and thus can quickly and effectively eliminate dyes from the discharged wastewater. The 2D nanomaterials as sheets stacked into multi-layered laminar membranes function as NF membranes, and the dyes are removed mainly by means of size-exclusion separation of methyl blue [29]. However, the 2D laminar sheets such as MXenes and graphene-based laminar NF membranes produced by the vacuum filtration method result in a dense film with decreased porosity, thus reducing water permeability. The removal of the large dye molecules with different hydration radii, therefore, requires MXene sheets to be used in conjunction with other materials (intercalants) to enlarge/increase the interlayer spacing for enhanced permeability. Khang et al. [71] produced GO laminar membrane intercalated MXene sheets to enhance the membrane permeability (Fig. 8.10). The improvement was mostly because of the short transport pathway together with additional nanopores and interlayer gaps on the hydrophilic MXene nanosheet area [71, 84]. As the stability of GO-based membranes is weak owing to electrostatic repulsion between the nanosheet layers, MXene is incorporated to offer more stability due to reduced electrostatic repulsion and better π-π interaction.

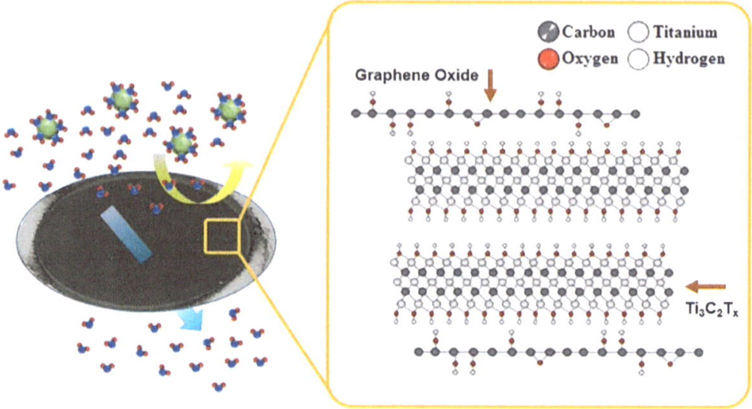

Fig. 8.10 MXene sheets incorporated/intercalated within the GO galleries (Ti_3C_2Tx/GO composite membrane). Reproduced with permission from ref. [71]. Copyright 2017, American Chemical Society

8.4.4 Removal of Heavy Metal Salts

Heavy metals such as lead (Pb), Chromium (Cr), Copper (Cu) and Cadmium (Cd) are representative heavy metal ions extensively applied in numerous industries, such as oil refineries. Pb species mainly exist in Pb^{2+}, $Pb(OH)^+$, and $Pb_2(OH)^+$ in water solution. These pose a health threat when their concentrations exceed the maximum acceptable Pb(II) concentration in drinking water. Despite their lower surface area, 2D nanomaterials such as MXenes have proven to have higher removal capacities than other adsorbents like activated carbon. Therefore, MXene has recently been employed as a new, highly efficient material for the removal of Pb(II) from the water through electrostatic interaction and ion exchange mechanisms. In the high surface negative surface charge of MXenes, the heavy metal ions are mainly removed by ion exchange, electrostatic attraction, and inner-sphere complex formation mechanisms [85]. It was noticed that MXene-based laminar nanofiltration membranes perform well as composites. This was shown by using Fe_3O_4@MXene membrane in a study by Yang et al. [30] that showed that the inclusion of NPs within the interlayer sheet spacing helps improve membrane permeability due to the construction of large permeation channels for water molecules. Fe_3O_4 contributed to MXene nanosheets adsorbing metal ions effectively.

8.4.5 2D Nanomaterial Membranes for Desalination

Nanoporous membranes are single-layer ultrathin 2D membranes with a sub-nanometer thickness (e.g., single-layer graphene with a thickness of ~ 3.4 Å). They are impermeable to liquids and gases, including helium. Using focused ion beam and transmission electron microscopy, the nanoscale pores (nanoporous) are drilled or produced by eliminating atoms in the center of the membrane until the preferred pore size is attained. The nanoscale pores may range from 50 to 80 Å^2 [46, 66, 86]. Nanoporous membranes are usually employed for effective desalination and targeted to remove monovalent salt ions (Na^+, Cl^-). Nanoporous membrane performance is affected by factors such as the size, structure of 2D materials, pore chemistry, and geometry [66].

With simulations, water flux is calculated by determining the average rate of permeate molecules over time by fitting the gradient of the permeate molecules against the simulation time curve, often operated for desalination using the RO process (Fig. 8.11). For example, the selectivity against salt ions is calculated as the ratio of ions transported during the simulation time of 10 ns. Comparison can be made to compare the performance of the various 2D nanoporous membrane materials to determine their efficiencies. The pore area of 50 Å^2 for MXene nanoporous membranes gives a desired salt rejection of 99% and maximum water flux at a pressure of 100 MPa. The rejection rate decreases with an increase in pressure and an increase in pore volume. Ultimately, the performance comparison between the different MXenes (Ti_2C, Ti_3C_2, Ti_4C_3) shows that the membrane selectivity depends more on the pore than the membrane material [87].

Permeation coefficient is used to compute pressure invariant water permeability of

$$p = \frac{J_w}{-V_w \Delta C_s + \frac{V_w}{N_A k_b T} \Delta P}$$

where, Jw is the H_2O flux (#/ns), Vm is the H_2O molar volume (18.91 mL/mol), ΔCs is the concentration gradient of the solute (2.28 M), NA is Avogadro's number, kB is the Boltzmann constant, T is the temperature (300 K), and ΔP is the applied hydrodynamic pressure (50 – 300 MPa) [66].

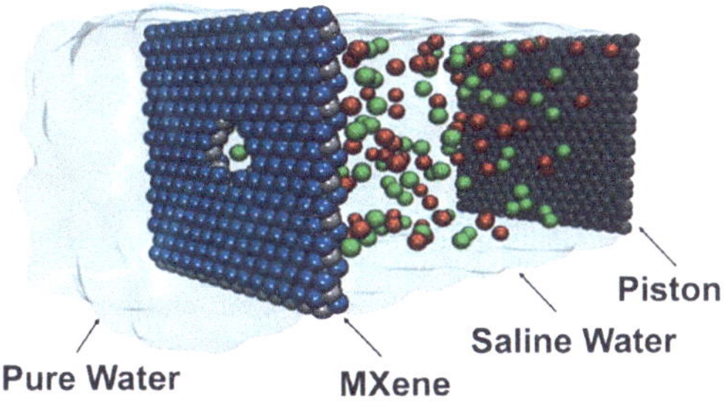

Fig. 8.11 A typical simulation box shows a graphene sheet piston, saline feed water with K^+ and Cl^- ions, a MXene membrane with drilled pores, and a fresh/filtered water section. Reproduced with permission from ref. [66]. Copyright 2021, American Chemical Society

In addition to the single-sheet nanoporous membranes, MXene sheets could be used to produce salt rejection membranes for the precise filtration of ions from saline water and used as desalination membranes. Cations of various sizes, including the monovalent Na^+, have shown the ability to intercalate between $Ti_3C_2T_x$ MXene layers, resulting in the expansion or contraction of the multi-layer laminar membrane produced by the vacuum filtration method [58].

8.5 Conclusion

This chapter covers the properties of 2D nanomaterial-based membranes in water treatment and desalination. It mainly focuses on the most commonly used emerging nanomaterials, namely graphene-based materials, MXenes and MoS_2, and their membrane fabrication methods. These membranes have shown great potential in removing toxic organic and inorganic compounds in water treatment and desalination. Therefore, using nanomaterials in wastewater treatment is promising in efficiently removing toxic contaminants from wastewater streams. Considering the membrane fouling challenge faced by the currently used polymer membranes, the use of 2D nanomaterials in either improving or replacing the wastewater and desalination membrane properties is a promising solution. In addition, using 2D nanomaterials with their exceptional properties, desalination can become more sustainable and economical. Currently, desalination using the existing polymer membranes is facing challenges of membrane permeability. Thus, the low throughput of desalination plants makes them expensive. Therefore, using a membrane with improved permeability will make desalination possible to be used globally as an everyday water treatment technique. In this regard, a sustainable water supply can be secured using membrane technologies.

Acknowledgements The authors acknowledge the funding support from the Council for Scientific and Industrial Research and the University of Johannesburg for their financial support.

References

1. M.A. Shannon, P.W. Bohn, M. Elimelech, J.G. Georgiadis, B.J. Mariñas, A.M. Mayes, Science and technology for water purification in the coming decades. Nature **452**, 301–310 (2008)
2. B. Schreiber, T. Brinkmann, V. Schmalz, E. Worch, Adsorption of dissolved organic matter onto activated carbon—the influence of temperature, absorption wavelength, and molecular size. Water Res. **39**, 3449–3456 (2005)
3. E.L. Sharp, S.A. Parsons, B. Jefferson, Seasonal variations in natural organic matter and its impact on coagulation in water treatment. Sci. Total Environ. **363**, 183–194 (2006)
4. E.P. Holinger, K.A. Ross, C.E. Robertson, M.J. Stevens, J.K. Harris, N.R. Pace, Molecular analysis of point-of-use municipal drinking water microbiology. Water Res. **49**, 225–235 (2014)
5. S.S. Ray, R. Gusain, N. Kumar, Carbon nanomaterial-based adsorbents for water purification: Fundamentals and applications. (Elsevier, 2020)

6. N. Kumar, E. Fosso-Kankeu, S.S. Ray, Achieving controllable MoS2 nanostructures with increased interlayer spacing for efficient removal of Pb(II) from aquatic systems. ACS Appl. Mater. Interfaces **11**, 19141–19155 (2019)
7. A.R. Mahdavian, M.A.-S. Mirrahimi, Efficient separation of heavy metal cations by anchoring polyacrylic acid on superparamagnetic magnetite nanoparticles through surface modification. Chem. Eng. J. 159:264–271 (2010)
8. S.S. Ray, R. Gusain, N. Kumar, Chapter three–Water purification using various technologies and their advantages and disadvantages, in *Carbon nanomaterial-based adsorbents for water purification*, ed. by Ray S.S. Gusain, R.N. Kumar (Elsevier; 2020), pp. 37–66
9. A. Ghaemi, M. Torab-Mostaedi, M. Ghannadi-Maragheh, Characterizations of strontium(II) and barium(II) adsorption from aqueous solutions using dolomite powder. J. Hazard. Mater. **190**, 916–921 (2011)
10. A.K. Fard, G. McKay, R. Chamoun, T. Rhadfi, H. Preud'Homme, M.A. Atieh, Barium removal from synthetic natural and produced water using MXene as two dimensional (2-D) nanosheet adsorbent. Chem. Eng. J. **317**, 331–342 (2017)
11. Z. Ahmed, F. Rehman, U. Ali, A. Ali, M. Iqbal, K.H. Thebo, Recent advances in MXene-based separation membranes. ChemBioEng Rev. **8**, 110–120 (2021)
12. Ihsanullah, Carbon nanotube membranes for water purification: Developments, challenges, and prospects for the future. Separat. Purificat. Technol. **209**, 307–337 (2019)
13. K.S. Novoselov, A.K. Geim, S.V. Morozov, D. Jiang, Y. Zhang, S.V. Dubonos et al., Electric field effect in atomically thin carbon films. Science **306**, 666–669 (2004)
14. N. Kumar, K. Setshedi, M. Masukume, S.S. Ray, Facile scalable synthesis of graphene oxide and reduced graphene oxide: comparative investigation of different reduction methods. Carbon Lett. **32**, 1031–1046 (2022)
15. N. Kumar, R.S. Sinha, Synthesis and functionalization of nanomaterials, in *Processing of Polymer-based Nanocomposites: Introduction*. ed. by S. Sinha Ray (Springer International Publishing, Cham, 2018), pp.15–55
16. J. Fatima, A.N. Shah, M.B. Tahir, T. Mehmood, A.A. Shah, M. Tanveer, et al., Tunable 2D nanomaterials; their key roles and mechanisms in water purification and monitoring. Front. Environ. Sci. **10** (2022)
17. L. Cheng, G. Liu, J. Zhao, W. Jin, Two-dimensional-material membranes: manipulating the transport pathway for molecular separation. Acc. Mater. Res. **2**, 114–128 (2021)
18. S. Liu, G. Liu, G. Chen, G. Liu, W. Jin, Scale-up fabrication of two-dimensional material membranes: challenges and opportunities. Curr. Opin. Chem. Eng. **39**, 100892 (2023)
19. P. Liu, J. Hou, Y. Zhang, L. Li, X. Lu, Z. Tang, Two-dimensional material membranes for critical separations. Inorganic Chem. Front. **7**, 2560–2581 (2020)
20. P. Zhang, J.-L. Gong, G.-M. Zeng, B. Song, W. Cao, H.-Y. Liu et al., Novel "loose" GO/MoS2 composites membranes with enhanced permeability for effective salts and dyes rejection at low pressure. J. Membr. Sci. **574**, 112–123 (2019)
21. D. Cohen-Tanugi, J.C. Grossman, Water desalination across nanoporous graphene. Nano Lett. **12**, 3602–3608 (2012)
22. Q. Liu, G.-R. Xu, Graphene oxide (GO) as functional material in tailoring polyamide thin film composite (PA-TFC) reverse osmosis (RO) membranes. Desalination **394**, 162–175 (2016)
23. J. Wang, X. Gao, H. Yu, Q. Wang, Z. Ma, Z. Li et al., Accessing of graphene oxide (GO) nanofiltration membranes for microbial and fouling resistance. Sep. Purif. Technol. **215**, 91–101 (2019)
24. R. Gusain, N. Kumar, S.S. Ray, Recent advances in carbon nanomaterial-based adsorbents for water purification. Coord. Chem. Rev. **405**, 213111 (2020)
25. N. Kumar, R. Salehiyan, V. Chauke, O. Joseph Botlhoko, K. Setshedi, M. Scriba et al., Top-down synthesis of graphene: a comprehensive review. FlatChem. **27**, 100224 (2021)
26. S. Verma, H.P. Mungse, N. Kumar, S. Choudhary, S.L. Jain, B. Sain et al., Graphene oxide: an efficient and reusable carbocatalyst for aza-Michael addition of amines to activated alkenes. Chem. Commun. **47**, 12673–12675 (2011)

27. D. An, L. Yang, T.-J. Wang, B. Liu, Separation performance of graphene oxide membrane in aqueous solution. Ind. Eng. Chem. Res. **55**, 4803–4810 (2016)
28. J. Chang, G. Zhou, E.R. Christensen, R. Heideman, J. Chen, Graphene-based sensors for detection of heavy metals in water: a review. Anal. Bioanal. Chem. **406**, 3957–3975 (2014)
29. W. Zhang, H. Xu, F. Xie, X. Ma, B. Niu, M. Chen et al., General synthesis of ultrafine metal oxide/reduced graphene oxide nanocomposites for ultrahigh-flux nanofiltration membrane. Nat. Commun. **13**, 471 (2022)
30. K. Almarzooqi, M. Ashrafi, T. Kanthan, A. Elkamel, M.A. Pope, Graphene oxide membranes for high salinity, produced water separation by pervaporation. Membranes **11**, 475 (2021)
31. H.M. Hegab, L. Zou, Graphene oxide-assisted membranes: fabrication and potential applications in desalination and water purification. J. Membr. Sci. **484**, 95–106 (2015)
32. S.P. Koenig, L. Wang, J. Pellegrino, J.S. Bunch, Selective molecular sieving through porous graphene. Nat. Nanotechnol. **7**, 728–732 (2012)
33. Y. Wei, Y. Zhang, X. Gao, Y. Yuan, B. Su, C. Gao, Declining flux and narrowing nanochannels under wrinkles of compacted graphene oxide nanofiltration membranes. Carbon **108**, 568–575 (2016)
34. W. Zhang, M. Shi, Z. Heng, W. Zhang, B. Pan, Soft particles enable fast and selective water transport through graphene oxide membranes. Nano Lett. **20**, 7327–7332 (2020)
35. X. Yang, Y. Liu, S. Hu, F. Yu, Z. He, G. Zeng et al., Construction of Fe3O4@MXene composite nanofiltration membrane for heavy metal ions removal from wastewater. Polym. Adv. Technol. **32**, 1000–1010 (2021)
36. M. Chhowalla, H.S. Shin, G. Eda, L.-J. Li, K.P. Loh, H. Zhang, The chemistry of two-dimensional layered transition metal dichalcogenide nanosheets. Nat. Chem. **5**, 263–275 (2013)
37. Y. Li, S. Yang, K. Zhang, B. Van der Bruggen, Thin film nanocomposite reverse osmosis membrane modified by two dimensional laminar MoS2 with improved desalination performance and fouling-resistant characteristics. Desalination **454**, 48–58 (2019)
38. N. Kumar, B.P.A. George, H. Abrahamse, V. Parashar, J.C. Ngila, Sustainable one-step synthesis of hierarchical microspheres of PEGylated MoS2 nanosheets and MoO3 nanorods: their cytotoxicity towards lung and breast cancer cells. Appl. Surf. Sci. **396**, 8–18 (2017)
39. K. Malkappa, S.S. Ray, N. Kumar, Enhanced thermo-mechanical stiffness, thermal stability, and fire retardant performance of surface-modified 2D MoS2 nanosheet-reinforced polyurethane composites. Macromol. Mater. Eng. **304**, 1800562 (2019)
40. S. Bertolazzi, J. Brivio, A. Kis, Stretching and breaking of ultrathin MoS2. ACS Nano **5**, 9703–9709 (2011)
41. K. Ai, C. Ruan, M. Shen, L. Lu, MoS2 nanosheets with widened interlayer spacing for high-efficiency removal of mercury in aquatic systems. Adv. Func. Mater. **26**, 5542–5549 (2016)
42. R. Gusain, N. Kumar, F. Opoku, P.P. Govender, S.S. Ray, MoS2 Nanosheet/ZnS composites for the visible-light-assisted photocatalytic degradation of oxytetracycline. ACS Appl. Nano Mater. **4**, 4721–4734 (2021)
43. S. Pandey, E. Fosso-Kankeu, M.J. Spiro, F. Waanders, N. Kumar, S.S. Ray et al., Equilibrium, kinetic, and thermodynamic studies of lead ion adsorption from mine wastewater onto MoS2-clinoptilolite composite. Mater. Today Chem. **18**, 100376 (2020)
44. X. Li, H. Zhu, Two-dimensional MoS2: properties, preparation, and applications. J. Materiom. **1**, 33–44 (2015)
45. Z. Wang, B. Mi, Environmental applications of 2D Molybdenum Disulfide (MoS2) Nanosheets. Environ. Sci. Technol. **51**, 8229–8244 (2017)
46. M. Heiranian, A.B. Farimani, N.R. Aluru, Water desalination with a single-layer MoS2 nanopore. Nat. Commun. **6**, 8616 (2015)
47. S. Yang, Q. Jiang, K. Zhang, Few-layers 2D O-MoS2 TFN nanofiltration membranes for future desalination. J. Membr. Sci. **604**, 118052 (2020)
48. D. Chen, W. Ying, Y. Guo, Y. Ying, X. Peng, Enhanced gas separation through nanoconfined ionic liquid in laminated MoS2 membrane. ACS Appl. Mater. Interfaces. **9**, 44251–44257 (2017)

49. W. Hirunpinyopas, E. Prestat, S.D. Worrall, S.J. Haigh, R.A.W. Dryfe, M.A. Bissett, Desalination and nanofiltration through functionalized laminar MoS2 membranes. ACS Nano **11**, 11082–11090 (2017)
50. J. Azamat, A. Khataee, Improving the performance of heavy metal separation from water using MoS2 membrane: molecular dynamics simulation. Comput. Mater. Sci. **137**, 201–207 (2017)
51. N. Kumar, S. Kumar, R. Gusain, N. Manyala, S. Eslava, S.S. Ray, Polypyrrole-promoted rGO–MoS2 nanocomposites for enhanced photocatalytic conversion of CO2 and H2O to CO, CH4, and H2 products. ACS Appl. Energy Mater. **3**, 9897–9909 (2020)
52. O. Folorunso, N. Kumar, Y. Hamam, R. Sadiku, S.S. Ray, Recent progress on 2D metal carbide/nitride (MXene) nanocomposites for lithium-based batteries. FlatChem. **29**, 100281 (2021)
53. S.K. Hwang, S.-M. Kang, M. Rethinasabapathy, C. Roh, Y.S. Huh, MXene: an emerging two-dimensional layered material for removal of radioactive pollutants. Chem. Eng. J. **397**, 125428 (2020)
54. M. Alhabeb, K. Maleski, B. Anasori, P. Lelyukh, L. Clark, S. Sin et al., Guidelines for synthesis and processing of two-dimensional titanium carbide (Ti3C2Tx MXene). Chem. Mater. **29**, 7633–7644 (2017)
55. M.R. Lukatskaya, O. Mashtalir, C.E. Ren, Y. Dall'Agnese, P. Rozier, P.L. Taberna, et al., Cation intercalation and high volumetric capacitance of two-dimensional titanium carbide. Science **341**, 1502–1505 (2013)
56. Z.W. Seh, K.D. Fredrickson, B. Anasori, J. Kibsgaard, A.L. Strickler, M.R. Lukatskaya et al., Two-dimensional molybdenum carbide (MXene) as an efficient electrocatalyst for hydrogen evolution. ACS Energy Lett. **1**, 589–594 (2016)
57. F. Shahzad, M. Alhabeb, C.B. Hatter, B. Anasori, S. Man Hong, C.M. Koo et al., Electromagnetic interference shielding with 2D transition metal carbides (MXenes). Science **353**, 1137–1140 (2016)
58. C.E. Ren, K.B. Hatzell, M. Alhabeb, Z. Ling, K.A. Mahmoud, Y. Gogotsi, Charge- and size-selective ion sieving through Ti3C2Tx MXene membranes. J. Phys. Chem. Lett. **6**, 4026–4031 (2015)
59. J. Peng, X. Chen, W.-J. Ong, X. Zhao, N. Li, Surface and heterointerface engineering of 2D MXenes and their nanocomposites: insights into electro- and photocatalysis. Chem **5**, 18–50 (2019)
60. I. Ihsanullah, M. Bilal, Potential of MXene-based membranes in water treatment and desalination: a critical review. Chemosphere **303**, 135234 (2022)
61. Y. Sun, S. Li, Y. Zhuang, G. Liu, W. Xing, W. Jing, Adjustable interlayer spacing of ultrathin MXene-derived membranes for ion rejection. J. Membr. Sci. **591**, 117350 (2019)
62. M. Naguib, O. Mashtalir, J. Carle, V. Presser, J. Lu, L. Hultman et al., Two-dimensional transition metal carbides. ACS Nano **6**, 1322–1331 (2012)
63. J. Halim, M.R. Lukatskaya, K.M. Cook, J. Lu, C.R. Smith, L.-Å. Näslund et al., Transparent conductive two-dimensional titanium carbide epitaxial thin films. Chem. Mater. **26**, 2374–2381 (2014)
64. M. Ghidiu, M.R. Lukatskaya, M.-Q. Zhao, Y. Gogotsi, M.W. Barsoum, Conductive two-dimensional titanium carbide 'clay' with high volumetric capacitance. Nature **516**, 78–81 (2014)
65. Y.A.J. Al-Hamadani, B.-M. Jun, M. Yoon, N. Taheri-Qazvini, S.A. Snyder, M. Jang et al., Applications of MXene-based membranes in water purification: a review. Chemosphere **254**, 126821 (2020)
66. K. Meidani, Z. Cao, F.A. Barati, Titanium carbide MXene for water desalination: a molecular dynamics study. ACS Appl. Nano Mater. **4**, 6145–6151 (2021)
67. F.H. Memon, F. Rehman, J. Lee, F. Soomro, M. Iqbal, S.M. Khan, et al., Transition metal dichalcogenide-based membranes for water desalination, gas separation, and energy storage. Separat. Purific. Rev. 1–15 (2022)
68. X. Wang, L. Chen, L. Wang, Q. Fan, D. Pan, J. Li et al., Synthesis of novel nanomaterials and their application in efficient removal of radionuclides. Sci. China Chem. **62**, 933–967 (2019)

69. K. Maleski, V.N. Mochalin, Y. Gogotsi, Dispersions of two-dimensional titanium carbide MXene in organic solvents. Chem. Mater. **29**, 1632–1640 (2017)
70. K.W. Putz, O.C. Compton, C. Segar, Z. An, S.T. Nguyen, L.C. Brinson, Evolution of order during vacuum-assisted self-assembly of graphene oxide paper and associated polymer nanocomposites. ACS Nano **5**, 6601–6609 (2011)
71. K.M. Kang, D.W. Kim, C.E. Ren, K.M. Cho, S.J. Kim, J.H. Choi et al., Selective molecular separation on Ti3C2Tx–graphene oxide membranes during pressure-driven filtration: comparison with graphene oxide and MXenes. ACS Appl. Mater. Interfaces **9**, 44687–44694 (2017)
72. A. Aher, Y. Cai, M. Majumder, D. Bhattacharyya, Synthesis of graphene oxide membranes and their behavior in water and isopropanol. Carbon **116**, 145–153 (2017)
73. S. Wang, F. Wang, Y. Jin, X. Meng, B. Meng, N. Yang et al., Removal of heavy metal cations and co-existing anions in simulated wastewater by two separated hydroxylated MXene membranes under an external voltage. J. Membr. Sci. **638**, 119697 (2021)
74. Z. Huang, J. Liu, Y. Liu, Y. Xu, R. Li, H. Hong et al., Enhanced permeability and antifouling performance of polyether sulfone (PES) membrane via elevating magnetic Ni@MXene nanoparticles to upper layer in phase inversion process. J. Membr. Sci. **623**, 119080 (2021)
75. B. Anasori, M.R. Lukatskaya, Y. Gogotsi, 2D metal carbides and nitrides (MXenes) for energy storage. Nat. Rev. Mater. **2**, 16098 (2017)
76. J. Yin, F. Zhan, T. Jiao, H. Deng, G. Zou, Z. Bai et al., Highly efficient catalytic performances of nitro compounds via hierarchical PdNPs-loaded MXene/polymer nanocomposites synthesized through electrospinning strategy for wastewater treatment. Chin. Chem. Lett. **31**, 992–995 (2020)
77. F. Perreault, M.E. Tousley, M. Elimelech, Thin-film composite polyamide membranes functionalized with biocidal graphene oxide nanosheets. Environ. Sci. Technol. Lett. **1**, 71–76 (2014)
78. W. Choi, J. Choi, J. Bang, J.-H. Lee, Layer-by-layer assembly of graphene oxide nanosheets on polyamide membranes for durable reverse-osmosis applications. ACS Appl. Mater. Interfaces **5**, 12510–12519 (2013)
79. B.-M. Jun, C.M. Park, J. Heo, Y. Yoon, Adsorption of Ba2+ and Sr2+ on Ti3C2Tx MXene in model fracking wastewater. J. Environ. Manage. **256**, 109940 (2020)
80. G. Bjørklund, Y. Semenova, L. Pivina, M. Dadar, M.M. Rahman, J. Aaseth et al., Uranium in drinking water: a public health threat. Arch. Toxicol. **94**, 1551–1560 (2020)
81. L. Wang, W. Tao, L. Yuan, Z. Liu, Q. Huang, Z. Chai et al., Rational control of the interlayer space inside two-dimensional titanium carbides for highly efficient uranium removal and imprisonment. Chem. Commun. **53**, 12084–12087 (2017)
82. S. Yu, H. Tang, D. Zhang, S. Wang, M. Qiu, G. Song et al., MXenes as emerging nanomaterials in water purification and environmental remediation. Sci. Total Environ. **811**, 152280 (2022)
83. M. Jeon, B.-M. Jun, S. Kim, M. Jang, C.M. Park, S.A. Snyder et al., A review on MXene-based nanomaterials as adsorbents in aqueous solution. Chemosphere **261**, 127781 (2020)
84. R.P. Pandey, K. Rasool, V.E. Madhavan, B. Aïssa, Y. Gogotsi, K.A. Mahmoud, Ultrahigh-flux and fouling-resistant membranes based on layered silver/MXene (Ti3C2Tx) nanosheets. J. Mater. Chem. A **6**, 3522–3533 (2018)
85. Q. Shafiq, S.M. Husnain, F. Shahzad, M. Taqi Mehran, S.A. Raza Kazmi, S. Mujtaba-ul-Hassan et al., Rational design of MXene coated polyurethane foam for the removal of Pb2+. Mater. Lett. **304**, 130600 (2021)
86. G.-R. Xu, J.-M. Xu, H.-C. Su, X.-Y. Liu, L. Lu, H.-L. Zhao et al., Two-dimensional (2D) nanoporous membranes with sub-nanopores in reverse osmosis desalination: latest developments and future directions. Desalination **451**, 18–34 (2019)
87. K. Qu, K. Huang, Z. Xu, Recent progress in the design and fabrication of MXene-based membranes. Front. Chem. Sci. Eng. **15**, 820–836 (2021)

Chapter 9
Development of 2D Nanomaterials-Based Sensors for Detection of Toxic Environmental Pollutants

S. Irem Kaya, Merve Yence, Goksu Ozcelikay, Ahmet Cetinkaya, Fatma Budak, and Sibel A. Ozkan

Abstract The global environmental pollution crisis is an ever-increasing issue due to human-induced factors such as urbanization and industrialization. Toxic environmental pollutants such as pesticides, dyes, polycyclic aromatic hydrocarbons, heavy metals, etc., threaten human health and the environment. Therefore, the development of rapid, affordable, selective, and sensitive sensing platforms for determining toxic environmental pollutants is a significant necessity. Today, many methods with the features to overcome the disadvantages of traditional sensors are being developed. Among these, electrochemical sensors stand out with features, such as short analysis time, high sensitivity, versatility, miniaturization, and low cost. Due to their unique chemical and physical characteristics, two-dimensional (2D) nanomaterials, such as graphene and its derivatives, transition metal oxides, graphitic carbon nitride, and metal dichalcogenides, are widely used to improve electrochemical sensors performance thanks to their high surface area, porosity, and catalytic effects. In this chapter, the determination of the most important toxic environmental pollutants in various environmental samples with 2D nanomaterials-based optical and electrochemical sensors are overviewed, covering the years between 2016 and 2022.

9.1 Introduction

Recently, environmental pollution-related issues have reached a global crisis level. The various toxic environmental pollutants seriously threaten all living organisms through environmental sources, ecosystems, and natural balance. Although the

S. I. Kaya
Gulhane Faculty of Pharmacy, Department of Analytical Chemistry, University of Health Sciences, 06018 Ankara, Türkiye

M. Yence · G. Ozcelikay · A. Cetinkaya · F. Budak · S. A. Ozkan (✉)
Faculty of Pharmacy, Department of Analytical Chemistry, Ankara University, 06560 Ankara, Türkiye
e-mail: ozkan@pharmacy.ankara.edu.tr

M. Yence · A. Cetinkaya · F. Budak
Graduate School of Health Sciences, Ankara University, Ankara, Türkiye

N. Kumar et al. (eds.), *Two-Dimensional Materials for Environmental Applications*, Springer Series in Materials Science 332, https://doi.org/10.1007/978-3-031-28756-5_9

extent of the negative effects of environmental pollution has been noticed since the mid1900s, human-induced factors such as industrialization, urbanization, uncontrolled disposal of harmful wastes, and changes in agricultural practices have accelerated the process, especially in recent years [1, 2]. Clean natural resources and air, water, and soil quality are essential for the survival of all living things. Since all ecosystem components are interrelated, in case of contamination and pollution, starting from a small area, harmful consequences arise that affect plants, humans, and animals [3].

Scientists with current studies have focused their attention on pollutants such as heavy metals, pesticides, bisphenols, and pharmaceutical residues due to their toxicity profiles, carcinogenic effects, and non-degradable and persistent properties [1, 4]. These environmental pollutants can cause toxic effects even at low concentrations and with acute or chronic exposure. Even though metal ions are vital materials for both plants and humans, heavy metals such as arsenic (As), cadmium (Cd), chromium (Cr), copper (Cu), mercury (Hg), lead (Pb), and zinc (Zn) are toxic contaminants [3]. The leading causes of heavy metal pollution are intense industrial activity and uncontrolled waste disposal. Direct exposure to heavy metals in various routes or consumption of contaminated food and plants may disrupt the functions of the heart, kidneys, liver, and pancreas [3, 5]. Additionally, the contamination of the soil affects plant growth and agricultural yield. The contamination of the aquatic environment causes serious health issues in marine animals [1].

Pesticides are one of the environmental pollutants that pose the greatest threat to the environment. Because pesticides are compounds that prevent harmful organisms, they have high toxicity and persistent profile [6, 7]. Although pesticide use is controlled and limited, incorrect and excessive use causes acute toxicity and poisoning, as well as water and soil contamination [8]. Since bisphenols, one of the most important elements of the plastics industry, is found in many products used in daily life, they are important in terms of both food contamination and waste polluting water and soil. In addition, the endocrine disruptive effects of bisphenols cause serious health issues; therefore, they are considered significant environmental pollutants [7].

Considering the ubiquity, contamination potential and possibly serious health risks of environmental pollutants, their determination and monitoring have always been critical research topics. From this point of view, electrochemical methods come to the fore as one of the most popular analytical methods to meet the constant need to develop fast, effective, low-cost, sensitive, and selective sensors [9]. Another advantage of electrochemical methods comes from their versatility, which allows their integration and modification with different materials. One of the most significant modification materials is the nanomaterials that enable enhanced sensor performance due to the catalytic and conductive effects [10]. Nanomaterials encompass a large group of which new members are still being added. 2D nanomaterials stand out with their advantages, such as high specific surface area, enabling functionalization, and unique optical, anisotropic, and mechanical properties. The most significant representatives of 2D nanomaterials are graphene and graphene derivatives, MXenes, metal–organic frameworks, and transition metal dichalcogenides (e.g., molybdenum sulfide (MoS_2)), and they find applications in various fields (biomedical applications, electrochemical sensors, biosensors, etc.) [11, 12]. The different characteristics of each of them highlight 2D nanomaterials as necessary modification materials in electrochemical sensor applications. For example, graphene and its derivatives are prominent in nanosensor applications with their high electrical conductivity, mechanical strength, and biocompatibility properties. As another example, MoS_2 has a fast electron transfer rate and provides a short analysis time with improved sensitivity [13]. MXenes, relatively new 2D nanomaterials, have catalytic, conductive, and electrochemical properties and can be used in energy storage and sensor devices [11, 12, 14]. Depending on the characteristics of the sensor to be developed, these materials can be preferred alone or in combination with other 2D nanomaterials or nanomaterials. In addition, 2D nanomaterials are also utilized in various optical sensor applications.

This chapter overviews the most important toxic pollutants in the environment in different categories. Additionally, the 2D nanomaterials-based sensor applications for their determination are evaluated in optical and electrochemical sensors. Finally, this chapter covers the most recent and prominent studies in the literature and highlights the importance of 2D nanomaterials.

9.2 Overview of Toxic Pollutants in the Environment

Environmental pollution is the intensive mixing of foreign chemicals in the air, water, and soil, which harms the living and non-living elements of the environment, causes structural damage and degrades their quality. A significant source of this environmental problem is the ever-increasing industrial activities and the regulated or uncontrolled emission of harmful substances from numerous industrial sectors [15]. Toxic pollutants include wastes from mining and petroleum production, industry, agriculture, and residential activities, as well as treated and untreated sludge. They contain compounds that are hazardous to human health and the environment. Human-made chemicals, xenoestrogens, organochlorine, and organophosphate pesticides, heavy metals, polycyclic aromatic hydrocarbons, polychlorinated biphenyls, dioxins, radioactive materials, plastics, dyes, and radioactive materials are major toxic pollutants and were presented in Fig. 9.1 [1, 16, 17].

9.2.1 Heavy Metals

The most frequent heavy metals that can poison the environment include arsenic, cadmium, chromium, copper, nickel, lead, and mercury, which have high density and atomic weight compared to water [18]. Because of their numerous industrial, residential, agricultural, medical, and technical applications, they are widely distributed in the environment and lead to raising worries about their possible effects on human health and the environment. Metal corrosion, air deposition, soil erosion of metal

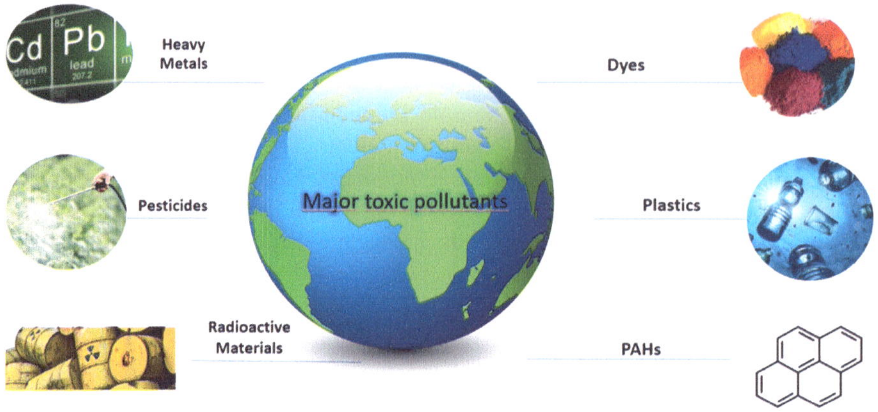

Fig. 9.1 Major toxic pollutants in the environment

ions and leaching of heavy metals, sediment re-suspension, and metal evaporation from water supplies to soil and groundwater can cause air, water, and soil pollution [18]. In addition, some of them are carcinogenic and can influence biological functioning and growth, resulting in various dangerous disorders like cancer [19]. Thus, heavy metals greatly impact the environment and human lives.

9.2.2 *Pesticides*

Pesticides protect agricultural land, stored food, and flower beds and exterminate pests that spread harmful infectious diseases [20]. With the increase in pesticide use, synthetic pesticides have been widely used. However, organochlorine (OC) pesticides were prohibited or restricted in most technologically advanced countries after the 1960s [21]. Pesticides directly impact humans and the environment, especially surface water contamination, groundwater contamination, and soil contamination. For example, pesticides can contaminate surface water due to runoff from treated plants and soil. This affects aquatic life even if drinking water samples come from groundwater. Also, pesticides interact with soil and its microbes, affecting microbial diversity, biochemical processes, and enzymatic activity [20].

9.2.3 *Radioactive Materials*

Nuclear radiations are another dangerous form of pollution because radioactive wastes cannot be eliminated; they stay in our environment for a long time, especially in the fields of radioisotopes in medicine, science, industry, or nuclear and radiological accidents [22]. Even the disposal of radioactive materials can pollute the environment and harm the environment. Radioactive waste that is not correctly disposed of could contaminate and pollute the soil. The radioactive materials in this form of trash may react with the soil components, making it very poisonous and infertile. Moreover, plants cultivated in such soils can absorb the radiation and store it in their bodies so that radiation can pass to animals and humans.

9.2.4 Dyes

Dyes are water-based soluble organic compounds that are widely used in garments, textiles, paintings, papers, and personal-care products [1]. Textile dyes hinder photosynthesis, restrict plant growth, penetrate the food chain, give recalcitrance and bioaccumulation, and may enhance toxicity, genotoxicity, and carcinogenic effects due to consisting of heavy metals and polycyclic aromatic and nitro compounds [23]. Especially, industrial dye wastes affect the aquatic environment and soils. When the dyes are released into the water, a transparent film forms on the surface of the water. It prevents the light from passing, leading to a decrease in the efficiency of photosynthesis and the amount of dissolved oxygen in the water affecting the entire aquatic biota [23]. In addition, azo dye compounds hurt soil microbial populations as well as plant germination and growth [24].

9.2.5 Plastics

Plastic pollution has been one of the most important environmental challenges because plastics take a long time to degrade so soil pollution can be seen worldwide. Floating plastics are generally attractive to aquatic animals when they are thrown into the water due to their color and odor, so marine animals consume them. Then, they affect the entire environment by participating in the food chain [1]. Chlorinated plastic can leach hazardous chemicals into the soil, which can then leak into groundwater or other nearby water sources, harming the environment and the ecosystem. This can have various negative consequences for the creatures that consume the water. Poisonous compounds emitted by plastics and some additivities like phthalates and bisphenols alter the soil structure and pollute groundwater.

9.2.6 Polycyclic Aromatic Hydrocarbons (PAHs)

Polycyclic aromatic hydrocarbons (PAHs) are organic compounds with two or more single or fused aromatic rings. They are commonly detected in air, soil, and water produced by incomplete combustion of organic materials such as coal, oil, petrol, and wood [25]. Oil leaks discharge PAHs into aqueous areas, which has an impact on the marine ecology. PAHs are predominantly released into the atmosphere due to the incomplete combustion of organic materials. PAHs are cancerous micropollutants because of their very hydrophobic structure; they are resistant to natural degradation so that they can remain on the soil surface for many years, and as a result of the plant's failure to photosynthesize, it can disrupt the food chain. [26].

9.3 An Important Tool in Sensor Applications: A General Outlook of 2D Nanomaterials

Nanomaterials are classified according to their dimensions. The same chemical compounds can exhibit different dimensional crystal structures configured as zero (0D)-, one (1D)-, two (2D)-, and three (3D) dimensions [27]. In recent years, 2D nanomaterials such as graphene/ reduced graphene oxide (rGO)/graphene oxide (GO), transition metal oxides (TMOs), silicate clays, black phosphorus (BP), layered double hydroxides (LDHs), transition metal dichalcogenides (TMDs), graphitic carbon nitride (g-C_3N_4) and hexagonal boron nitride (hBN), boron (B) and tin telluride (SnTe) nanosheets, metal dichalcogenides (MXenes) (Fig. 9.2) have used in the electronics, sensors, catalysts, energy storage facilities and solar cells because of high surface-to-volume ratios, surface charge and uniform shapes [28].

2D nanomaterials are categorized as layered and non-layered nanomaterials. Graphene-based nanomaterials can be given as an example of layered nanomaterials. Strong chemical bonds connect atoms on the plane in each layer. For non-layered nanomaterials, which are included in the other above-referred 2D nanomaterials, are formed by bulk crystals with a weak van der Waals interaction [28]. The 2D Nanomaterials were characterized using SEM, TEM, and XRD to explain their physical and chemical properties [29].

Graphene is widely used in electrochemical sensors ranging from impedance-based to voltammetric/amperometric sensors. Graphene is combined with different nanomaterials to form synergistic effects. 2D carbon-based hybrid materials combine through non-covalent interaction via π-π stacking. The synergistic effects provide excellent electrical properties, high surface areas, and facile chemical functionality to fabricate novel transducing Technologies [30]. Graphene-based sensors have great attention thanks to fast electron transfer kinetics and high sensitivities/selectivities.

Moreover, graphene and its derivatives can be integrated with molecularly imprinted polymer (MIP) based sensors to get high selectivity. Al-Ammari et al. [31] report that MIP-based zinc oxide nanoparticle/graphene nanoplatelet modified GCE was developed for the determination of 4-chlorophenol in real water samples. MIPs provide many imprinted cavities on nanomaterials modified sensors. MIPs as highly selective electrochemical sensors for chlorophenols based on nanomaterials, such as graphene, to increase the number of imprinted cavities.

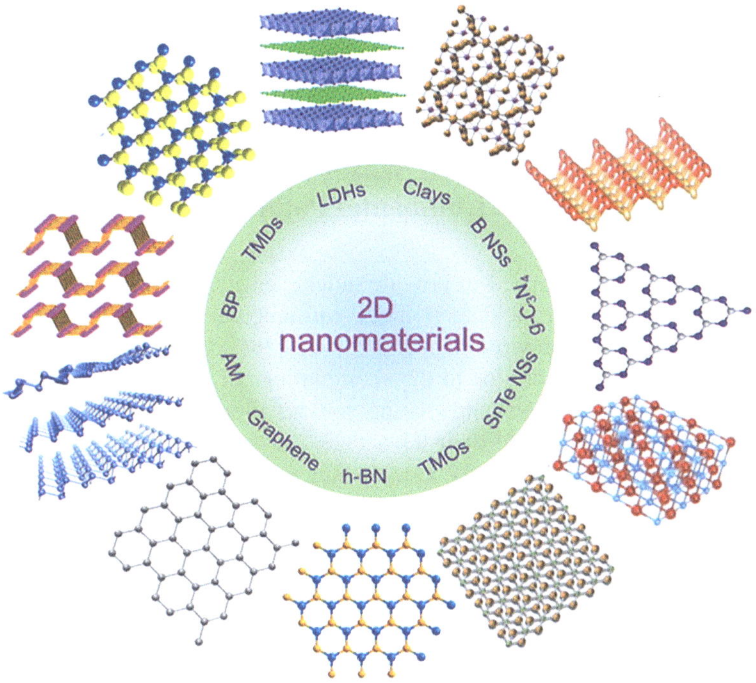

Fig. 9.2 The examples of 2D carbon nanomaterials

The graphene configurations can be changed by doping with the heteroatoms such as nitrogen, boron, and phosphorus atoms which causes the region activation in the catalytic reaction. Therefore, heteroatoms doped graphene contribute high conductivity and electrocatalytic activity in electrochemical applications [32]. Chen X. et al. report that the electrochemical immunosensor was developed for the determination of Aflatoxin B1. The dual heteroatoms -doped graphene can contribute to the interaction between metal catalyst and graphene. Gold/platinum nanorods can accelerate electron transfer between electrode surface and analytes. The linear range and LOD value were found as 0.025–60.00 ngmL^{-1} and 0.0052 ngmL^{-1}, respectively [33].

rGO and its derivatives were applied for developing smartphone-based electrochemical portable devices. The GO containing much oxygen has played a crucial role because of easy functionalization. Therefore, the point of care testing is generally preferred for its on-site investigations. Toxic environmental pollutants can be determined from the groundwater system to the environment.

Ionic liquids (ILs) are used as a stabilizer for functionalized graphene. ILs consisting of imidazolium cations can be linked with graphene via π- π bonding [34]. Therefore, it can prevent the aggregation of graphene nanosheets. Redhi et al. [35] used RGO-Fe_3O_4 nanorods-IL as a significant enhancement to fabricate the aflatoxin B1 (AFA-B1) sensor. As a result, the RGO-Fe_3O_4 nanorods-IL/GCE sensor was obtained with excellent stability and reproducibility. This nanosensor was applied to real ground paprika samples. The linear range and LOD value were found as 0.02–0.33 ngmL^{-1}, and 0.02 ngmL^{-1}, respectively.

9.4 2D Nanomaterials-Based Sensor Applications for Detection of Environmental Pollutants

There has been a pressing need to provide quick, sensitive, affordable, and multiplexed sensing systems for the detection of environmental toxins due to the rapidly growing concerns about environmental pollution [36]. Analytical techniques are widely used today. However, due to the shortcomings of traditional methods, new analytical techniques are constantly being sought. Electrochemical sensors have attracted great attention due to their advantages, such as short analysis time, low cost, high sensitivity, and selectivity. Establishing new analytical methodologies for the assessment of hazards from drugs to food contaminants in water and soil resources is critical [37, 38].

Due to their unique chemical and physical characteristics, two-dimensional (2D) nanomaterials have considerable potential and have been widely used to monitor environmental toxins in combination with various detection methods [36]. 2D nanomaterials are frequently employed to improve electrochemical sensors' porosity and surface area. For the detection of environmental pollutants, carbon nanotubes and nanosensors embedded in graphene (and its derivatives) are the most preferred materials. Pharmaceutical forms, fruits, tap water, lake water, drinks, and soil sources have been used as targets for optical and electrochemical sensors based on different improved 2D-carbon nanomaterials. This section aims to highlight the importance of 2D nanomaterials for sensors and summarize published work on environmental pollutant analysis [37, 38]. Moreover, we examine current developments in 2D nanomaterials-based sensors for detecting environmental contaminants such as heavy metal ions, organic chemicals, pesticides, antibiotics, etc., using different methods like optical and electrochemical applications, and we discuss the potential and difficulties of using 2D nanomaterials for environmental monitoring [36].

9.4.1 Optical

Nanomaterial-based optical sensors provide highly sensitive, selective, and quick monitoring of environmental contaminants without the need for substantial sample preparation. 2D nanomaterials such as graphene, MXenes, transition metal dichalcogenides, metal oxides, and black phosphorus have unique optical and electrical properties like size-tunable, high conductivity, shape-dependent, easy surface modification, and large specific surface area [39]. The combination of 2D nanomaterials and optical sensors has been used particularly for the detection of heavy metals, organic dyes, pesticides, and microorganisms.

More recently, Sridhar et al. [40] developed a paper-based Surface-enhanced Raman spectroscopy (SERS) method for the detection of rhodamine 6G (Rh6G) that food colorant classified as an organic toxin. They used the synergistic effect of gold nanoparticles/graphene oxide (AuNPs/GO) as SERS substrate using the green synthesis method as grape skin waste extract and paper forms as GE-AuNPs/GO (Fig. 9.3). Compared to normal Raman spectra, SERS signals of Rh6G give an enhancement factor of 5.8×10^4 for GE-AuNPs and 1.92×10^9 for GE-AuNPs/GO. This increase is a sign of the influence of charge transfer, electromagnetic surface plasmon, and molecular resonances that results in low detection of limit (7.33×10^{-11} M) and high recovery of tap water and many foods e.g., red chilli powder, red glutinous rice ball and tomato ketchup.

Due to graphene's large surface area and special electrocatalytic features, graphene has attracted remarkable attention for detecting pesticide residues, heavy metals, and dyes. Graphene and graphene oxide have been used to remove toxic heavy metals [41]. Yun et al. [42] developed a novel fluorescent detection method for multiple metal ions such as Cu^{2+}, Pb^{2+}, and Mg^{2+} based on the DNAzymes and graphene oxide. This mechanism is based on the on–off fluorescence system. For a generation of the DNAzyme branching structure, enzyme-strand and substrate-strand of three fluorophores labeled DNA sequences were fused. After introducing metal ions, cleavage of fluorophore-labeled substrate-strand DNA occurred, then DNA fragments were broken, which were absorbed by graphene oxide via pi stacking (Fig. 9.4). This situation ends up with the quenching of the fluorescent signal. This method was applied for the simultaneous detection of heavy metals in human serum.

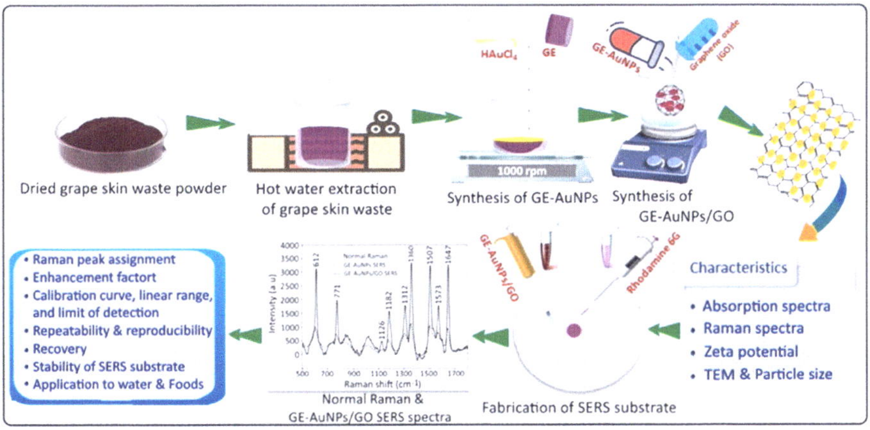

Fig. 9.3 Schematic representation of developed SERS method and its properties. Reproduced with permission from Elsevier [40]

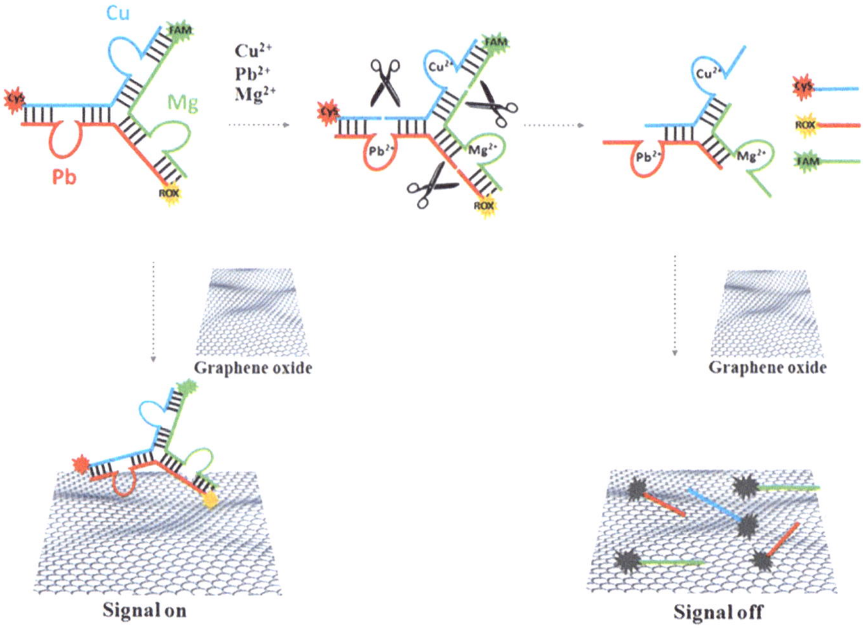

Fig. 9.4 Schematic representation of developed fluorescent detection of heavy metals based on graphene oxide structure. Reproduced with permission from Elsevier [42]

The search for new strategies to fabricate surface-enhanced Raman scattering (SERS) substrates has received a lot of scientific attention as high sensitivity, reproducibility, and cost-effectiveness are difficult to achieve. Daoudi et al. [43], fabricated silver nanoprism/graphene oxide/silicon nanowire (Ag/GO/SiNWs) arrays for effective detection of atrazine. Homogeneous and vertically aligned SiNWs were successfully synthesized using silver-assisted chemical etching. Moreover, Ag/GO/SiNWs arrays were then spin-coated of GO and then assembled by drop-casting of Ag nanoprisms. As fabricated, microstructural properties of nanoarrays were well studied by electron scanning and atomic force microscopic techniques (Fig. 9.5). Nanowire bundles were decorated with GO layers; this further facilitated the homogeneous deposition of Ag nanoprisms. The high SERS activity and remarkable reproducibility of substrates as produced were established using rhodamine 6G (R6G) as the probe molecule. The synergistic charge transfer between GO and Ag nanoprisms on large surface area Si nanowires enabled high efficiency to be obtained for the fabricated sensors. As the array is fabricated, a charge transfer mechanism was designed to contribute to the surface plasmon resonance effect. In addition, these inexpensive Ag/GO/SiNWs sensors have demonstrated ultra-sensitive detection of contaminants down to picomolar levels, including methylene blue and atrazine residues.

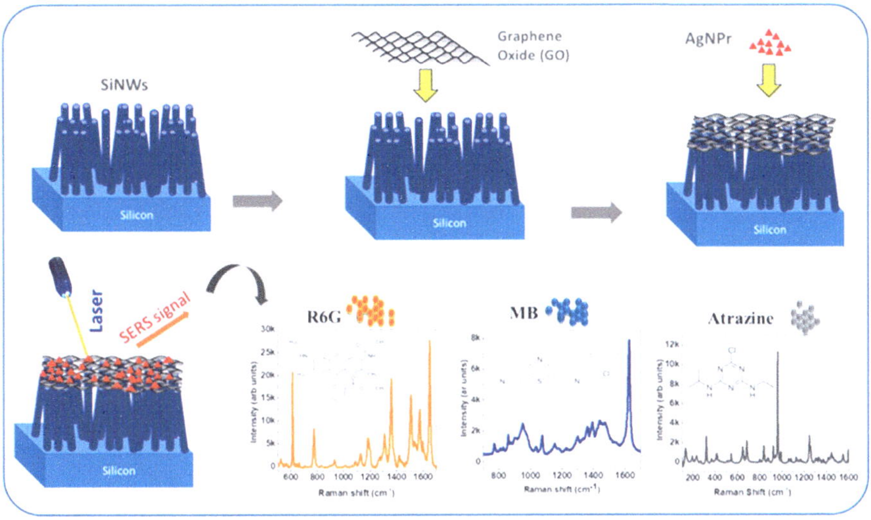

Fig. 9.5 Schematic representation of Ag/GO/SiNWs sensor for detecting organic pollutants. Reproduced with permission from Elsevier [43]

9.4.2 Electrochemical Applications

2D nanomaterial-based sensor platforms are frequently used in environmental analysis due to their fast electron transfer, high selectivity, sensitivity, and stability. Table 9.1 contains the latest applications (2016–2022) of modified electrochemical sensors with 2D nanomaterial-based sensor platforms for the detection of pollutants such as organic compounds, pesticides, and heavy metals in environmental samples. In addition, details such as recovery, sensitivity, medium, linear range, and method of sensors developed in environmental examples in the literature were given in Table 9.1.

A nanohybrid-based sensor was developed by Zhu et al. [44]. During the fabrication of this sensor, phosphorone, and Titanium carbide MXene (Ti_3C_2-MXene) nanohybrid were dropped onto the surface of laser-induced porous graphene (LIPG) in order to obtain high stability, flexibility, and large surface area. In this study, a nanohybrid-based, easily portable wireless smart nanozyme and a flexible sensing platform were developed. The BP/Ti_3C_2-MXene/GCE sensor, developed using an artificial neural network algorithm, was designed to be mounted unmanned. Ultratrace intelligent analysis of naphthalene acetic acid residues in agricultural areas through machine learning model LIPG has a wide range of environmental applicability (Fig. 9.6). BP/Ti_3C_2-MXene exhibited a wide linear range of 0.02–40 μM with a detection limit of 1.6 nM. This sensor will provide cheap, simple, fast trace detection as well as a flexible detection option for the analysis of environmental samples.

Wang et al. [55] developed an electrochemical sensor to detect trace level mercury in water samples. The working electrode consists of a chitosan/graphene oxide (GO)

Table 9.1 Selected latest applications of 2D nanomaterial-based electrochemical sensor platforms for the determination of toxic pollutants in environmental samples

Analyte	Modified electrode	Method	LOD	Linear range	Medium	Real sample	Recovery (%)	Sensitivity	Refs.
Naphthalene acetic acid (NAA)	Ti_3C_2-MXene/BP/LIPG	LSV EIS	1.6 nM	0.02–40 μM	pH 4 PBS	Soil Water Food	98.98 98.78 97.28–100.4	NA	[44]
4-Nitrophenol	Nb_2CTX/Zn-Co-NC	DPV	0.070 μM	1–500 μM	pH 7 PBS	River water Tap water Industrial water	97.2–101.7	4.65 μA μM^{-1}	[45]
Cadmium (II) Lead (II)	Naf/Bi-SPE	SWV	1.5 ppb 0.3 ppb	5–60 ppb	pH 4.6 ABS	Sea water	NA	NA	[46]
Cadmium (II) Lead (II)	GO-Fe_3O_4-PAMAM/GCE	SWASV	70 ng L^{-1} 130 ng L^{-1}	0.2–140 mg L^{-1} 0.4–120 mg L^{-1}	pH 4.6 ABS	Drinking water	96.7–104.0 96.0–103.0	0.1897 μA μg L^{-1} 0.1541 μA μg L^{-1}	[47]
Cadmium (II) Lead (II)	Fe_3O_4/MWCNTs/LSG/CS/GCE	SWASV	0.1 μg L^{-1} 0.07 μg L^{-1}	1–200 μg L^{-1}	pH 5 ABS	Drinking water Tap water	97.8–105.96	NA	[48]
Lead (II)	PVA/chitosan-TRG/GCE	SWASV	0.05 ppb	1–50 ppb	pH 5 ABS	River water Lake water	NA	NA	[49]
Mercury (II)	Pt/g-C_3N_4/PAn NCs	DPV	0.014 nM	1–500 nM	pH 7 PBS	Waste water	97.73–98.99	0. 6 μA nM^{-1}	[50]
Arsenic (III)	Gr/MOF-GCE	DPASV	0.06 ppb	0.2–25 ppb	pH 5 ABS	Groundwater Mineral water River water	96.0–104.0	0.623 μA ppb^{-1}	[51]

(continued)

Table 9.1 (continued)

Analyte	Modified electrode	Method	LOD	Linear range	Medium	Real sample	Recovery (%)	Sensitivity	Refs.
Lead (II) Mercury (II)	NMO-GR/GCE ZMO-GR/GCE	SWASV	0.040 μM (Pb^{2+}) 0.015 μM (Hg^{2+}) 0.060 μM(Pb^{2+}) 0.025 μM (Hg^{2+}) (Individual)	1.4–7.9 μM (Pb^{2+}) 0.7–7.0 μM (Hg^{2+}) 1.4–9.4 μM (Pb^{2+}) 0.7–8.0 0 μM (Hg^{2+}) (Individual)	pH 6 ABS	River water	92.93–105.70 (Pb^{2+}) 96.50–105.30 (Hg^{2+}) 92.58–98.10 (Pb^{2+}) 93.21–106.90 (Hg^{2+})	3.92 μA μM^{-1} (Pb^{2+}) 9.89 μA μM^{-1} (Hg^{2+}) 2.41 μA μM^{-1} (Pb^{2+}) 6.06 μM^{-1} (Hg^{2+}) (Individual)	[52]
Cadmium (II) Lead (II) Mercury (II)	STB/Gs/GCE	DPV	2.30 nM (Cd^{2+}) 0.78 nM (Pb^{2+}) 1.15 nM (Hg^{2+}) (Individual)	0.05–5 μM (Cd^{2+}) 0.025–8.5 μM (Pb^{2+}) 0.05–7.5 μM (Hg^{2+}) (Individual)	pH 5 ABS	River water Tap water	96–104 96–103	1013 μA μM^{-1} 21.68 μA μM^{-1} 15.71 μA μM^{-1} (Individual)	[53]
Cadmium (II)	GR/GO/GCE	DPV	0.087 μM	0–10 μM	pH 4 ABS	Tap water	91.79–102.45	2.726 μA μM^{-1}	[54]
Mercury (II)	chitosan/GO/MoS_2 /AuNPs	DPSV	2.9 nM	0.05–20 nM	pH 7.4 Tris–HCl buffer	Tap water	80.87–127.92	1.50 μA $\mu g.L^{-1}$	[55]

(continued)

Table 9.1 (continued)

Analyte	Modified electrode	Method	LOD	Linear range	Medium	Real sample	Recovery (%)	Sensitivity	Refs.
Cupper (II) Lead (II) Cadmium (II)	NH_2-MIL-88(Fe)-rGO/GCE	DPASV	3.6 nM 10 nM 4.9 nM	0.005–0.05 μM 0.01–0.3 μM 0.005–0.3 μM	pH 5.17 PBS	Lake water	97.2–97.9 99.8–104.0 98.1–100.2	0.389 μA μM^{-1} 0.425 μA μM^{-1} 0.895 μA μM^{-1}	[56]
Mercury (II)	Pt/g-C_3N_4/PTh NCs	DPV	0.009 nM	1–500 nM	pH 7 PBS	Wastewater	97.44–95.39	1.0787 μA nM^{-1}	[57]
Lead (II)	2D MOF	EIS	3.3 nM	5 nM-1 μM	PBS	Tap water	93.33–106.7	NA	[58]
Trichloroacetamide	AgNPR@MoS_2/GCE	SWV	0.17 μM	0.5–10 μM 10–80 μM	pH 6 PBS	Drinking water	93.2–112.4	0.24 μA μM^{-1} 0.06 μA μM^{-1}	[59]
Bisphenol A	AuNPs/MoS_2-NFs/IL-graphene	DPV	0.028 μM	0.05–0.8 μM 0.8–4.0 μM	pH 7 PBS	Tap water Lake water	95.7–105.4	4.576 μA μM^{-1}	[60]
Sulfamethoxazole Trimethoprim	rGNR/SPCE	DPV	0.09 μM 0.04 μM	1–10 μM	pH 6 PBS	Tap water	97.83–102.04 100.14–100.48	0.014 μA μM^{-1} 0.030 μA μM^{-1}	[61]
Mercury(II)	rGO/MoS_2/GCE	DPASV	1 nM	0.001–10.0 μM	pH 7 PBS	Tap water	NA	24.599 μA μM^{-1}	[62]
Cadmium (II) Lead (II)	Ti_3C_2@N–C/GCE	SWASV	2.55 nM 1.10 nM	0.1–4 μM 0.05–2 μM	pH 6 ABS	Tap water Sea water	88.81–102.64	49.8500 μA μM^{-1} 177.3327 μA μM^{-1}	[63]

(continued)

Table 9.1 (continued)

Analyte	Modified electrode	Method	LOD	Linear range	Medium	Real sample	Recovery (%)	Sensitivity	Refs.
Lead (II) Bizmut (III) Copper (II)	Fe_3O_4:Gr/CPE	SWASV	43 ng L^{-1} 31 ng L^{-1} 57 ng L^{-1}	0.8–200 nM 0.6–90 nM 4–300 nM	pH 3.6 ABS	River water Sea water	97.00–105.00	46.26 µA μM^{-1} 60.10 µA μM^{-1} 37.54 µA μM^{-1}	[64]
Sulfamethoxazole Trimethoprim	GR-ZnO/GCE	DPV	0.4 µM 0.3 µM	1–220 µM 1–180 µM	pH 7 PBS	Lake water Tap water	98.2–108	NA	[65]
Nitrofurantoin	GS/PMB/GCE	DPV	55 nM	5–100 µM	pH 7.2 PBS	Tap water	95.5–103.0	0.297 µA μM^{-1}	[66]
Methiocarb Diethofencarb	Ti_3C_2Tx/GCE	DPV	0.19 µg mL^{-1} 0.46 µg mL^{-1}	3–55 µg mL^{-1} 3–50 µg mL^{-1}	0.5 M H_2SO_4	River water	95.13–103.35 93.8–101.18	NA	[67]
Nitrite	TiO_2 –$Ti_3C_2T_X$ /CTAB/CS/GCE	DPV	0.85 µM	0.003–0.25 mM 0.25–1.25 mM	pH 7 PBS	Sea water Tap water Milk	94.0–100.0	70.461 µA mM^{-1} 36.704 µA mM^{-1}	[68]
Sulfamethoxazole	2D Bi_2WO_6	DPV	3.56 µM (Light) 5.44 µM (Dark)	0.25–40 µM	pH 5.8 PBS	Lake water Tap water	96.1–103.5 95.5–102.9	0.033 µA mg L^{-1} (Light) 0.026 µA mg L^{-1} (Dark)	[69]

(continued)

Table 9.1 (continued)

Analyte	Modified electrode	Method	LOD	Linear range	Medium	Real sample	Recovery (%)	Sensitivity	Refs.
Carbendazim	GO/GCE	SWV	13.8 nM	0.1–2.4 μM	pH 4.2 PBS	Tap water Lake water Reverse osmosis, dam water Pond water Soil	94.0–96.0	109.17 μA μM $^{-1}$	[70]
Methyl parathion Diazinon Chlorpyrifos	BNQDs/GO/GCE	DPV	0.31 pM 0.067 pM 0.033 pM	0.001–1 nM	pH 6 PBS	Industrial pool water	101.59–102.33 97.06–101.85 98.36–101.96	11.953 μA nM $^{-1}$ 59.765 μA nM $^{-1}$ 111.23 μA μM $^{-1}$	[71]
Carbendazim	TiO_2/GCE	SWV	171 nM	0.1–0.42 μM	pH 7 PBS	Dam water Lake water Tap water Pond water Reverse osmosis water Soil	89.0–95.0 92.5–98.5	9.414 μA μM $^{-1}$	[72]
Paraoxon	MoS_2NPs@MWCNTs/GCE	DPV	2 pM	0.10 pM-1 nM	pH 7 PBS	Tap water	100	15.911 μA nM $^{-1}$	[73]
Mercury (II)	Au@HS-rGO/GCE AuPd@UiO-67/GCE	EIS	0.16 nM	1.0 nM-1.0 mM	pH 7.4 PBS	Tap water Lake water	99.36–101.7 99.96- 103.5	NA	[74]

(continued)

Table 9.1 (continued)

Analyte	Modified electrode	Method	LOD	Linear range	Medium	Real sample	Recovery (%)	Sensitivity	Refs.
Cadmium (II) Lead (II)	pg-C_3N_4/$CoMn_2O_4$/ GCE	SWASV	0.021 μM 0.014 μM	0.5–7.0 μM 0.2–4.4 μM	pH 5 ABS	River water Lake water Tap water	93.0–106.0	5.178 μA nM^{-1} 1.660 μA nM^{-1}	[75]
Mercury (II) Chromium (VI)	T-GO-C/GCE	SWV	1 ppb 20 ppb	5–600 ppb	pH 4.5 ABS	Tap water Tannery water	93.0–98.0	6 × 10–8 μA ppb^{-1} 5 × 10–9 μA ppb^{-1}	[76]
Amaranth	SnO_2/rGO/Nafion/GCE	DPV	0.68 nM 0.0027 μM	1–800 nM 1–60 μM	pH 6 PBS	Soft drinks River water	96.4–101.3	NA	[77]
Lead (II) Cadmium (II)	PM/g-C_3N_4/ASPE	DPV	0.008 μM 0.02 μM	0.1–1 μM	pH 5 ABS	River water Lake water Tap water	92.5–106.4 90.0–102.60	NA	[78]
Catechol	GA/Cu-MOF/GCE	DPV	1.65 μM	12.5–900 μM	pH 7 PBS	River water	97.53–101.67	0.0812 μA $μM^{-1}$	[79]
Mercury (II)	rGO-ZVI-P/GCE	SWASV	1.20 nM	0.05–0.60 μM	pH 5 PBS	River water	100 ± 3	41.42 μA $μM^{-1}$	[80]
Cadmium (II) Lead (II)	Bi/LIGF	SWASV	0.4 μg/L	1.0–140.0 μg/L	pH 4.5 ABS	Drinking water Tap water	NA	0.19 μA $μg^{-1}$ L 0.20 μA $μg^{-1}$ L	[81]
Chlorpyrifos	AgBr/Ti_3C_2 MXene /ITO	EIS	0.33 pg L^{-1}	1 × 10^{-3}-1 ng L^{-1}	pH 7 PBS	River water	96.0–106.0	0.3098 μA gL^{-1}	[82]

(continued)

Table 9.1 (continued)

Analyte	Modified electrode	Method	LOD	Linear range	Medium	Real sample	Recovery (%)	Sensitivity	Refs.
Cadmium (II)	IIP-Cd/GCE	DPV	3.51 nM	0.01–0.1 μM	pH 5 ABS	Tap water Lake water	96.8–102	2.9758 μA μM $^{-1}$	[83]
Chromium (III) Copper (II)	ZnO-NGs@GO ZnO-NGs@rGO	DPV	7.05 μM 14.9 μM	10–100 μM	pH 7 PBS	NA	NA	49.28 μA μM $^{-1}$ 185.32 μA μM $^{-1}$	[84]
Cadmium (II) Lead (II)	Co:ZnO/RGO/GCE	DPV	0.94 μg/L 0.83 μg/L	10–90 μg/L	pH 4.5 ABS	Tap water	99.06–101.33	55.20 μA 54.47 μA	[85]
Cadmium (II) Lead (II) Copper (II)	rGO/Ala/ PANI/GCE	SWASV	0.03 nM 0.045 nM 0.063 nM	100–0.08 nM	pH 5 ABS	Tap water	98.0–104.0	0.082 μA nM $^{-1}$ 0.116 μA nM $^{-1}$ 0.135 μA nM $^{-1}$	[86]
Mercury (II)	RS-gRGO/GCE	DPASV	57 ppt	1–40 ppb	pH 7 PBS	Waste water Tap water Drinking water	92.0–99.5	70.22 μA ppb	[87]
Cadmium (II)	g-C_3N_4 nanosheet/GCE	SWASV	3.9 nM	0.05–0.7 μM	pH 5 ABS	Natural water	NA	22.668 μA $μM^{-1}$	[88]

(continued)

Table 9.1 (continued)

Analyte	Modified electrode	Method	LOD	Linear range	Medium	Real sample	Recovery (%)	Sensitivity	Refs.
Cadmium (II)	NSCN/GCE	SWASV	0.35 nM	0.001–0.1 μM 0.1–5 μM 5.18 μM	pH 5.5 ABS	Tap water Lake water River water	94.6–104.7	43.9 μA μM^{-1} 10.9 μA μM^{-1} 3.01 μA μM^{-1}	[89]
Carbaryl Fenobucarb Carbosulfan	MnO_2-GNPs/SPCE	DPV	0.30 μM 1.30 μM 14.90 μM (Individual)	1–40 μM 5–150 μM 50–600 μM (Individual)	pH 7 PBS	Field water	98.2–105.9 92.1–106.1 92.7–108.5	0.1380 μA μM^{-1} 0.0479 μA μM^{-1} 0.0041 μA μM^{-1}	[90]
Tetrabromobisphenol-A	g-C_3N_4-823 K/GCE	DPV	5 nM	0.02–1 μM	pH 5 PBS	Outdoor pool water	NA	− 1.12 μA μM^{-1}	[91]
Cadmium (II)	g-C_3N_4/Nafion/GCE	DPASV	0.5 nM	1 nM-1 μM 1–100 μM	pH 3.5 ABS	Drinking water	NA	0.0153 μA μM^{-1} 0.837 μA μM^{-1}	[92]

NA: Not applicable, GCE: Glass Carbon Electrode, SCE: Saturated Calomel electrode, SPCE: Screen Printed Electrode, PGE: pencil graphite electrode, SC-ISE: solid-contact ion-selective electrode, ITO: Indium tin oxide electrode, DPV: Differential Pulse Voltammetry, EIS: Electrochemical Impedance Spectroscopy, CV: cyclic voltammetry, SWV: square-wave voltammetry, LSV: Linear sweep voltammetry, CA: Chronoamperometry, AMP: Amperometry, SWASV: Square-waveanodic stripping voltammetry, DPSV: Differential pulse stripping voltammetry, DPASV:Differential pulse anodic stripping voltammetry, PBS: Phosphate Buffer Solutions, ABS: Acetate Buffer Solutions, BRB: Britton Robinson Buffer,DWCNTs: Double-walled carbon nanotubes, ChOx: Cholesterol oxidase, C: Carbon, LSG:Laser scribed Graphene, TRG: Thermally reduced Graphene, PVA: Polyvinyl alcohol, MOF:Metal–organic framework, STB: Sulfur-doped C3N4 tube bundles, GS: Graphene nanosheets, MWCNTs: multi-walled carbon nanotubes, SWCNTs: Single-walled carbon nanotubes, NMO-GR: $NiMn_2O_4$-graphene, ZMO-GR: $ZnMn_2O_4$-graphene, CMO-GR: $CuMn_2O_4$-graphene, Gr: Graphene,, GO: Graphene oxide, Au NPs: Gold Nanoparticles, MoS_2: Molybdenum disulfide, PTh NCs: Polythiophene nanocomposite, EG: Electrochemically deposited graphene, AgNPR: Silver nanoprism, MoS_2-NFs: MoS_2 nanoflowers rGNR: Reduced graphene nanoribbons, MPL: Mechanical pencil lead, GS: Graphite sheets, PMB: poly(methylene blue), BNODs: Boron nitride quantum dots, TiO_2: Titanium dioxide, Au@HS-rGO: Gold modified thiol graphene, AuPd@UiO-67: Gold–palladium modified zirconium metal–organic frameworks, pg-C_3N_4: Porous graphitic carbon nitride, T-GO-C: Thymine-GO-Carbohydrazide, PM: poly(melamine), GA: graphene oxide aerogel, ZVI-P: polyphenols mediated zero-valent iron, LIGF: Laser-induced graphene fiber, N-RGO: Chitosan-N-doped graphene oxide, IIPCd: Ion-imprinted polymer sensor, NGs: Nanoglobules,Co: Cobalt, Ala: Alanine, PANI: polyaniline, RS: Rhodamine hydrazide, NSCN: g-C_3N_4 ultrathin nanosheets, g-C_3N_4: Graphitic carbon nitride

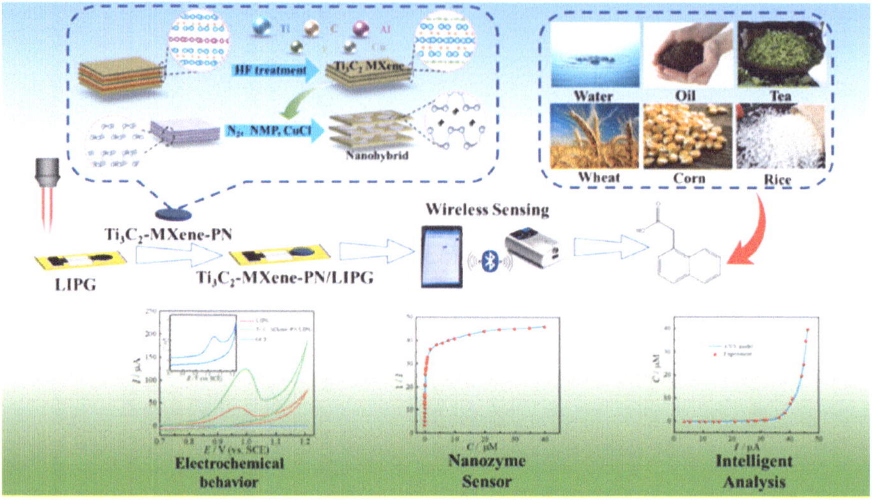

Fig. 9.6 Schematic representation of Ti_3C_2-MXene/BP nanohybrid modified LIPG for the portable wireless intelligent sensing of ultra-trace NAA in environmental samples. Reprinted from Ref. [44] with permission from Elsevier

layer, MoS_2 nanolayers, AuNPs, and a thiolated DNA probe (DNA P) (Fig. 9.7). AuNPs in this sensor was used to increase the DNA adsorption on MoS_2 by forming an Au–S bond, thus creating a synergistic catalytic effect. The sensor they developed achieved very satisfactory recovery in water samples, thus providing the potential for application for mercury detection in real samples. Under optimized conditions, this sensor showed a detection limit of 5.8 ng/L and dynamic linearity in the 0.01–4 μg/L range.

In another work, a sensor was developed by Wang et al. [74] using gold-modified thiol graphene (Au@HS-rGO) and preparing a sensing area to create an electrochemical aptasensor. Gold–palladium modified zirconium metal–organic frameworks (AuPd@UiO-67) nanozyme was developed as a signal enhancer for the determination of mercury (Fig. 9.8). The prepared electrochemical aptasensor showed a wide concentration range of 1.0 nmol/L to 1.0 mmol/L with a low detection limit of 0.16 nmol/L. Furthermore, this electrochemical aptasensor showed highly selective, stable, and repeatable with excellent performance during analysis in ambient water samples. Therefore, this recommended aptasensor may have promising applications providing a reference for environmental management and monitoring purposes. In this study, the application of AuPd@UiO-67 nanozyme as a signal enhancer was reported for the first time to label the DNA strand for sensitive detection of mercury.

In a recent work by Jayaraman et al. [76], graphene oxide was covalently functionalized with thymine and carbohydrazide (Thymine-GO-Carbohydrazide, T-GO-C) via an epoxide ring cleavage. Then a multiplex electrochemical sensor was developed with a simultaneous reduction approach. This sensor was used to detect heavy metals Hg (II) and Cr (VI) in drinking water. The prepared electrochemical sensor

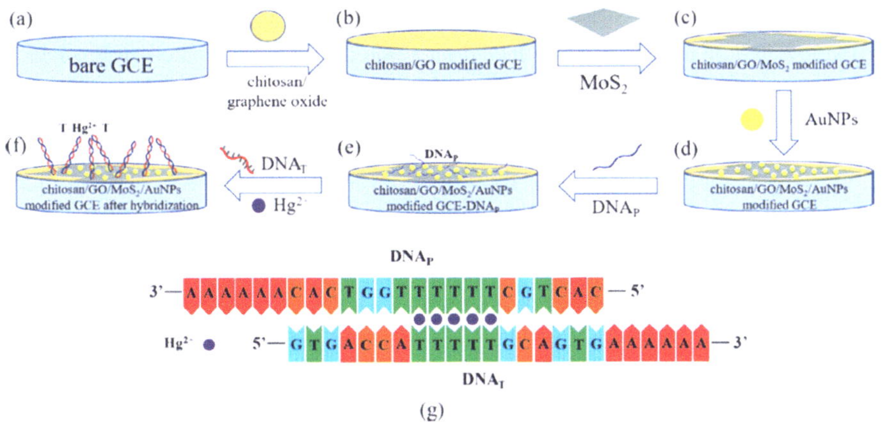

Fig. 9.7 Schematic representation of the sensor preparation and mercury detection processes of the developed sensor. Reprinted from Ref. [55] with permission from Elsevier

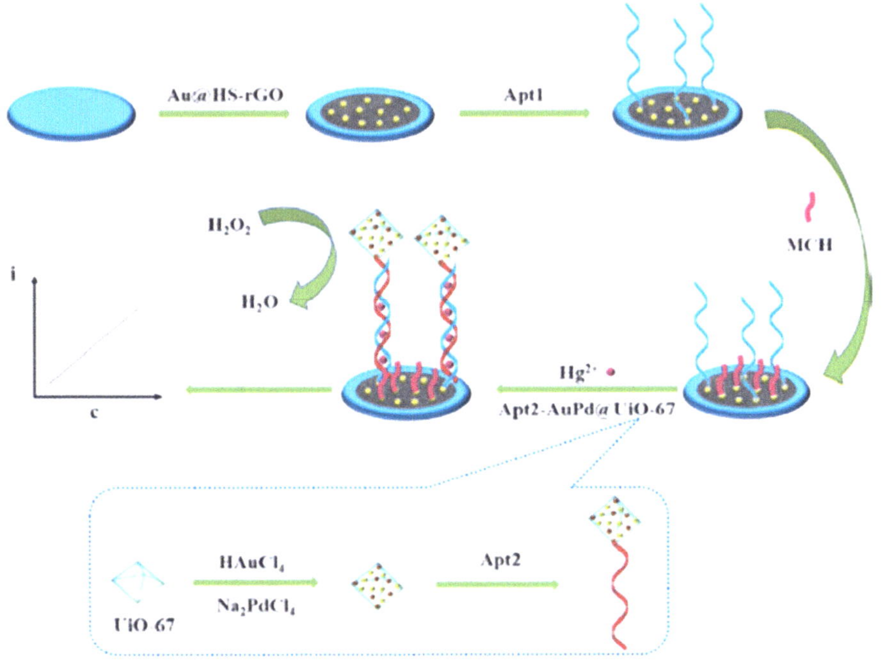

Fig. 9.8 **a** Schematic preparation of Apt2-AuPd@UiO-67; **b** Schematic diagram for the designed electrochemical aptasensor. Reprinted from Ref. [74] with permission from Elsevier

showed a wide concentration range of 5–600 ppb and a recovery of 93.0–98.0%. In this study, nanomaterials were collected on a fine drop-casting coated GCE surface to form the T-GO-C/GCE sensor, and analysis was performed by and square wave voltammetry technique (Fig. 9.9). The specificity and sensitivity of the recommended electrochemical sensor were also tested in real water samples.

In 2022, Bölükbaşı et al. [73] developed the MoS_2NPs@MWCNTs sensor to detect paraoxon in environmental samples. The prepared electrochemical sensor showed a wide concentration range of 0.10 pM to 1 nM, a limit of detection of 2 pM, and sensitivity of 15.911 μA nM $^{-1}$. In this study, measurements were taken using differential pulse voltammetry in a pH 7 PBS medium. The preparation steps of the MoS_2NPs@MWCNTs sensor are shown in Fig. 9.10. The nanocomposite produced for the purpose of complex determination in this study showed good recovery results. The development of this sensor potentially paves the way for a highly selective, stable, and sensitive electrochemical sensor for the detection of organophosphorus pesticides.

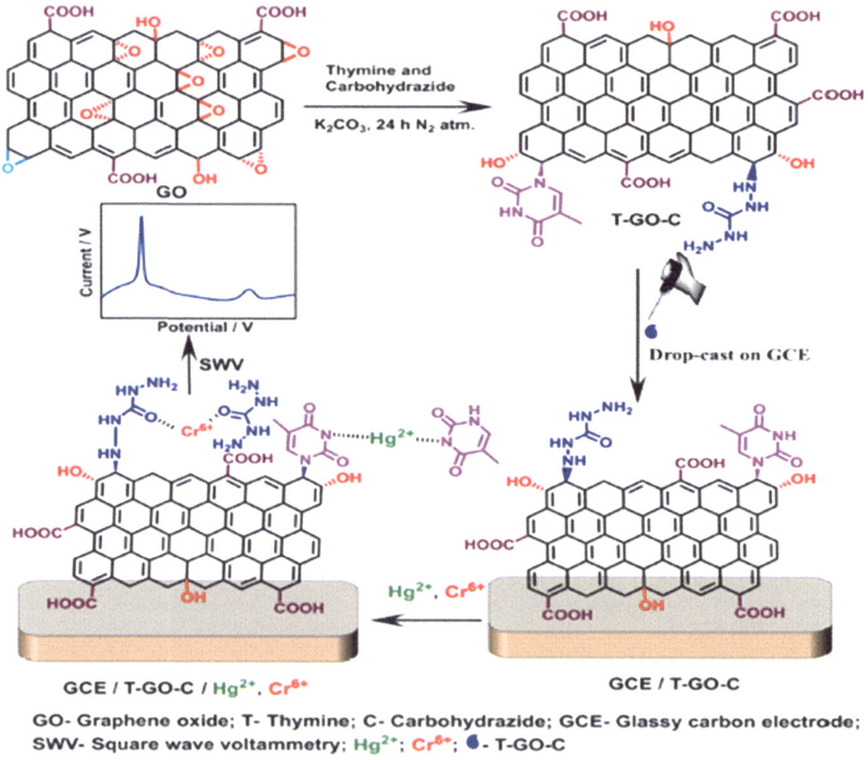

Fig. 9.9 Schematic representation of the electrochemical sensor produced with T-GO-C nanomaterials). Reprinted from ref. [76] with permission from Elsevier

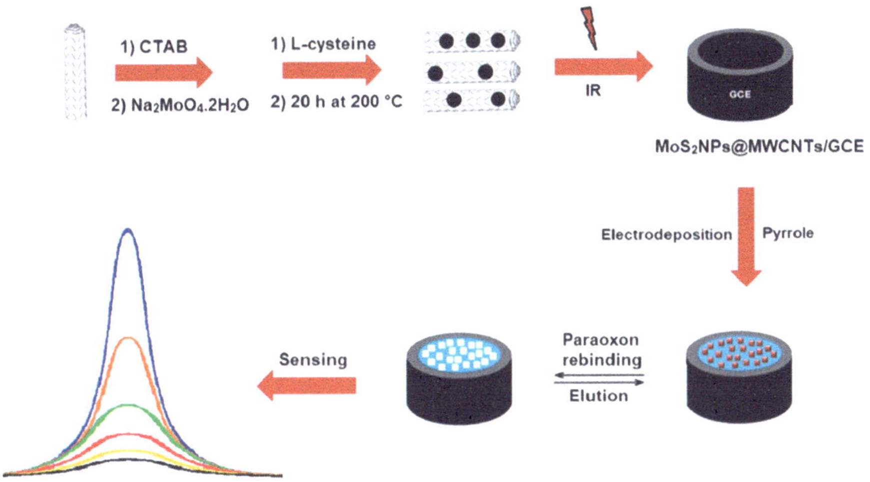

Fig. 9.10 Schematic representation of the preparation steps of MoS2NPs@MWCNTs nanocomposite and PAR imprinted electrochemical electrode. Reprinted from Ref. [73] with permission from Elsevier

9.5 Conclusion and Future Prospects

Environmental pollution is becoming more of a problem and a source of concern owing to the negative consequences it is causing all over the world. This chapter aims to explain 2D nanomaterials-based optical and electrochemical sensors for the detection of toxic environmental pollutants with the latest research studies. Additionally, it gives a general outlook of 2D nanomaterials, toxic environmental pollutants, and their effects on the environment.

When the most current and prominent studies (Table 9.1) in the literature in recent years are examined, it is seen that heavy metals and pesticides are mostly analyzed among environmental pollutants. Heavy metals such as Cd, Hg, Pb, and Cr have a high incidence in environmental sources because they are widely used in the industrial sense. In addition, it is a pollutant group that researchers give importance to due to its serious toxicity profiles. As the uncontrolled and excessive use of pesticides is quite common, it significantly reduces the water and soil quality. In addition, pesticides are persistent due to their chemical structure and may cause long-term exposure to humans. In the studies in the literature, it is seen that most of the environmental samples are analyzed from the lake and river waters. This can be explained by the fact that they are highly prone to contamination, and they are diverse systems that include many aquatic organisms. Graphene and its derivatives and MoS_2 were the most preferred 2D nanomaterials as electrode modification materials in electrochemical sensor studies. It is common for these materials and other 2D nanomaterials to be used in combination to increase the catalytic effect. The mostly used electrochemical techniques are DPV and SWV due to their short analysis time and high sensitivity.

2D Nanomaterials-based sensors are expected to become commercialized throughout time as a result of their extensive laboratory research and will improve people's quality of life in a variety of ways. The current trend is to develop sensors to detect a wide range of environmental contaminants or toxic substances in real-time, quickly, and uniquely, with high sensitivity and selectivity. Consequently, sensors are fast evolving into more automated, high-speed, and low-cost for real-time detection of various environmentally toxic pollutants in the coming years.

References

1. A. Saravanan, P.S. Kumar, D.V.N. Vo, S. Jeevanantham, S. Karishma, P.R. Yaashikaa, A review on catalytic-enzyme degradation of toxic environmental pollutants: microbial enzymes. J. Hazard. Mater. **419**, 126451 (2021)
2. A. Rhouati, M. Berkani, Y. Vasseghian, N. Golzadeh, MXene-based electrochemical sensors for detection of environmental pollutants: a comprehensive review. Chemosphere **291**, 132921 (2022)
3. R. Chormare, M. Anil, Environmental health and risk assessment metrics with special mention to biotransfer, bioaccumulation and biomagnification of environmental pollutants. Chemosphere **302**, 134836 (2022)
4. S. Wasi, S. Tabrez, M. Ahmad, Toxicological effects of major environmental pollutants: an overview. Environ. Monit. Assess. **185**, 2585–2593 (2013)
5. A. Saravanan, P.S. Kumar, R.V. Hemavathy, S. Jeevanantham, P. Harikumar, G. Priyanka, D.R.A. Devakirubai, A comprehensive review on sources, analysis and toxicity of environmental pollutants and its removal methods from water environment. Sci. Total Environ. **812**, 152456 (2022)
6. T.M. Murray, PBT (Persistent, Bioaccumulative, and Toxic) Chemicals, Third Edit. Encycl Toxicol Third Ed (2014). https://doi.org/10.1016/B978-0-12-386454-3.00641-2
7. H. Xu, Y. Jia, Z. Sun, J. Su, Q.S. Liu, Q. Zhou, G. Jiang, Environmental pollution, a hidden culprit for health issues. Eco-Environ. Heal. (2022). https://doi.org/10.1016/j.eehl.2022.04.003
8. A. Shakeel, K. Rizwan, U. Farooq, S. Iqbal, T. Iqbal, N.S. Awwad, H.A. Ibrahium, Polymer based nanocomposites: a strategic tool for detection of toxic pollutants in environmental matrices. Chemosphere (2022). https://doi.org/10.1016/j.chemosphere.2022.134923
9. X. Wu, P. Ma, Y. Sun, F. Du, D. Song, G. Xu, Application of MXene in electrochemical sensors: a review. Electroanalysis **33**, 1827–1851 (2021)
10. S. Kurbanoglu, S.A. Ozkan, Electrochemical carbon based nanosensors: a promising tool in pharmaceutical and biomedical analysis. J. Pharm. Biomed. Anal. **147**, 439–457 (2018)
11. A. Murali, G. Lokhande, K.A. Deo, A. Brokesh, A.K. Gaharwar, Emerging 2D nanomaterials for biomedical applications. Mater. Today **50**, 276–302 (2021)
12. Y. Zhou, H. Yin, S. Ai, Applications of two-dimensional layered nanomaterials in photoelectrochemical sensors: a comprehensive review. Coord. Chem. Rev. **447**, 214156 (2021)
13. V. Revuri, Y.K. Lee, 2D material-based hybrid nanostructure for diagnosis and therapy, in *Biomedical Applications of Graphene and 2D Nanomaterials* (Elsevier Inc., 2019), pp. 143–164
14. J. Singhal, S. Verma, S. Kumar, The physio-chemical properties and applications of 2D nanomaterials in agricultural and environmental sustainability. Sci. Total Environ. (2022). https://doi.org/10.1016/j.scitotenv.2022.155669
15. T. Rasheed, C. Li, M. Bilal, C. Yu, H.M.N. Iqbal, Potentially toxic elements and environmentally-related pollutants recognition using colorimetric and ratiometric fluorescent probes. Sci. Total Environ. **640–641**, 174–193 (2018)
16. A. Valavanidis, T. Vlahogianni, M. Dassenakis, M. Scoullos, Molecular biomarkers of oxidative stress in aquatic organisms in relation to toxic environmental pollutants. Ecotoxicol. Environ. Saf. **64**, 178–189 (2006)

17. K.H. Al-Gubory, Environmental pollutants and lifestyle factors induce oxidative stress and poor prenatal development. Reprod. Biomed. Online **29**, 17–31 (2014)
18. P.B. Tchounwou, C.G. Yedjou, A.K. Patlolla, D.J. Sutton, *Heavy Metal Toxicity and the Environment* (2012), pp. 133–164
19. J. Briffa, E. Sinagra, R. Blundell, Heavy metal pollution in the environment and their toxicological effects on humans. Heliyon **6**, e04691 (2020)
20. H. Kaur, H. Garg, Pesticides: Environmental Impacts and Management Strategies. Pestic. Toxic Asp. (2014). https://doi.org/10.5772/57399
21. W. Aktar, D. Sengupta, A. Chowdhury, Impact of pesticides use in agriculture: their benefits and hazards. Interdiscip. Toxicol. **2**, 1–12 (2009)
22. M. Laraia, Radioactive contamination and other environmental impacts of waste from nuclear and conventional power plants, medical and other industrial sources, in *Environmental Remediation and Restoration of Contaminated Nuclear and Norm Sites* (Elsevier, 2015), pp. 35–56
23. B. Lellis, C.Z. Fávaro-Polonio, J.A. Pamphile, J.C. Polonio, Effects of textile dyes on health and the environment and bioremediation potential of living organisms. Biotechnol. Res. Innov. **3**, 275–290 (2019)
24. K. Rehman, T. Shahzad, A. Sahar, S. Hussain, F. Mahmood, M.H. Siddique, M.A. Siddique, M.I. Rashid, Effect of reactive black 5 azo dye on soil processes related to C and N cycling. PeerJ **6**, e4802 (2018)
25. H.I. Abdel-Shafy, M.S.M. Mansour, A review on polycyclic aromatic hydrocarbons: source, environmental impact, effect on human health and remediation. Egypt J. Pet. **25**, 107–123 (2016)
26. S. Gan, E.V. Lau, H.K. Ng, Remediation of soils contaminated with polycyclic aromatic hydrocarbons (PAHs). J. Hazard Mater. **172**, 532–549 (2009)
27. Z. Rafiei-Sarmazdeh, S. Morteza Zahedi-Dizaji, A. Kafi Kang, Two-dimensional nanomaterials. Nanostructures (2020). https://doi.org/10.5772/intechopen.85263
28. T. Hu, X. Mei, Y. Wang, X. Weng, R. Liang, M. Wei, Two-dimensional nanomaterials: fascinating materials in biomedical field. Sci. Bull. **64**, 1707–1727 (2019)
29. N. Baig, M. Sajid, T.A. Saleh, Recent trends in nanomaterial-modified electrodes for electroanalytical applications. TrAC-Trends Anal. Chem. **111**, 47–61 (2019)
30. T.Y. Huang, J.H. Huang, H.Y. Wei, K.C. Ho, C.W. Chu, RGO/SWCNT composites as novel electrode materials for electrochemical biosensing. Biosens. Bioelectron. **43**, 173–179 (2013)
31. R.H. AL-Ammari, A.A. Ganash, M.A. Salam, Electrochemical molecularly imprinted polymer based on zinc oxide/graphene/poly(o-phenylenediamine) for 4-chlorophenol detection. Synth. Met. **254**, 141–152 (2019)
32. V. Kumar, M. Shorie, A.K. Ganguli, P. Sabherwal, Graphene-CNT nanohybrid aptasensor for label free detection of cardiac biomarker myoglobin. Biosens. Bioelectron. **72**, 56–60 (2015)
33. X. Chen, H. Li, G. Zhang, S. Feng, G. Yang, L. Shi, Au@Pt hybrid nanorods encapsulated in B S dual-doped graphene as highly sensitive ımmunosensing platform for electrochemical determination of aflatoxin B1, vol 15 (2020), pp. 6269–6289
34. Y. Zheng, Z. Liu, H. Zhan, J. Li, C. Zhang, Studies on electrochemical organophosphate pesticide (OP) biosensor design based on ionic liquid functionalized graphene and a Co3O4 nanoparticle modified electrode. Anal. Methods **8**, 5288–5295 (2016)
35. R. Chokkareddy, G.G. Redhi, Fe3O4 nanorods-RGO-ionic liquid nanocomposite based electrochemical sensor for aflatoxin B1 in ground paprika (2021). https://doi.org/10.1002/elan.202100377
36. S. Su, S. Chen, C. Fan, Recent advances in two-dimensional nanomaterials-based electrochemical sensors for environmental analysis. Green Energy Environ. **3**, 97–106 (2018)
37. V.T. Le, Y. Vasseghian, E.N. Dragoi, M. Moradi, A. Mousavi Khaneghah, A review on graphene-based electrochemical sensor for mycotoxins detection. Food Chem. Toxicol. **148**, 111931 (2021)
38. Y. Vasseghian, E.N. Dragoi, F. Almomani, V.T. Le, Graphene-based materials for metronidazole degradation: a comprehensive review. Chemosphere **286**, 131727 (2022)

39. C.W. Lee, J.M. Suh, H.W. Jang, Chemical sensors based on two-dimensional (2D) materials for selective detection of ions and molecules in liquid. Front. Chem. (2019). https://doi.org/10.3389/fchem.2019.00708
40. K. Sridhar, B.S. Inbaraj, B.-H. Chen, An improved surface enhanced Raman spectroscopic method using a paper-based grape skin-gold nanoparticles/graphene oxide substrate for detection of rhodamine 6G in water and food. Chemosphere **301**, 134702 (2022)
41. S.K. Krishnan, E. Singh, P. Singh, M. Meyyappan, H.S. Nalwa, A review on graphene-based nanocomposites for electrochemical and fluorescent biosensors. RSC Adv. **9**, 8778–8881 (2019)
42. W. Yun, H. Wu, X. Liu, M. Fu, J. Jiang, Y. Du, L. Yang, Y. Huang, Simultaneous fluorescent detection of multiple metal ions based on the DNAzymes and graphene oxide. Anal. Chim. Acta **986**, 115–121 (2017)
43. K. Daoudi, M. Gaidi, S. Columbus, M. Shameer, H. Alawadhi, Hierarchically assembled silver nanoprism-graphene oxide-silicon nanowire arrays for ultrasensitive surface enhanced Raman spectroscopy sensing of atrazine. Mater. Sci. Semicond. Process **138**, 106288 (2022)
44. X. Zhu, L. Lin, R. Wu, Y. Zhu, Y. Sheng, P. Nie, P. Liu, L. Xu, Y. Wen, Portable wireless intelligent sensing of ultra-trace phytoregulator α-naphthalene acetic acid using self-assembled phosphorene/Ti3C2-MXene nanohybrid with high ambient stability on laser induced porous graphene as nanozyme flexible electrode. Biosens. Bioelectron. **179**, 113062 (2021)
45. R. Huang, D. Liao, Z. Liu, J. Yu, X. Jiang, Electrostatically assembling 2D hierarchical Nb2CTx and zifs-derivatives into Zn-Co-NC nanocage for the electrochemical detection of 4-nitrophenol. Sens. Actuators, B Chem. **338**, 129828 (2021)
46. N. Colozza, M.F. Gravina, L. Amendola, M. Rosati, D.E. Akretche, D. Moscone, F. Arduini, A miniaturized bismuth-based sensor to evaluate the marine organism Styela plicata bioremediation capacity toward heavy metal polluted seawater. Sci. Total Environ. **584–585**, 692–700 (2017)
47. M. Baghayeri, H. Alinezhad, M. Fayazi, M. Tarahomi, R. Ghanei-Motlagh, B. Maleki, A novel electrochemical sensor based on a glassy carbon electrode modified with dendrimer functionalized magnetic graphene oxide for simultaneous determination of trace Pb(II)and Cd(II). Electrochim. Acta **312**, 80–88 (2019)
48. Z. Xu, X. Fan, Q. Ma, B. Tang, Z. Lu, J. Zhang, G. Mo, J. Ye, J. Ye, A sensitive electrochemical sensor for simultaneous voltammetric sensing of cadmium and lead based on Fe3O4/multiwalled carbon nanotube/laser scribed graphene composites functionalized with chitosan modified electrode. Mater. Chem. Phys. **238**, 121877 (2019)
49. L.D. Nguyen, T.C.D. Doan, T.M. Huynh, V.N.P. Nguyen, H.H. Dinh, D.M.T. Dang, C.M. Dang, An electrochemical sensor based on polyvinyl alcohol/chitosan-thermally reduced graphene composite modified glassy carbon electrode for sensitive voltammetric detection of lead. Sens. Actuators B Chem. **345**, 130443 (2021)
50. M.R. Mahmoudian, Y. Alias, P. Meng Woi, R. Yousefi, W.J. Basirun, An electrochemical sensor based on Pt/g-C3N4/polyaniline nanocomposite for detection of Hg2+. Adv. Powder Technol. **31**, 3372–3380 (2020)
51. M. Baghayeri, M. Ghanei-Motlagh, R. Tayebee, M. Fayazi, F. Narenji, Application of graphene/zinc-based metal-organic framework nanocomposite for electrochemical sensing of As(III) in water resources. Anal. Chim. Acta **1099**, 60–67 (2020)
52. P. Lei, Y. Zhou, S. Zhao, C. Dong, S. Shuang, Carbon-supported X-manganate (XNi, Zn, and Cu) nanocomposites for sensitive electrochemical detection of trace heavy metal ions. J. Hazard. Mater. **435**, 129036 (2022)
53. J. Wang, P. Yu, K. Kan, H. Lv, Z. Liu, B. Sun, X. Bai, J. Chen, Y. Zhang, K. Shi, Efficient ultra-trace electrochemical detection of Cd2+, Pb2+ and Hg2+ based on hierarchical porous S-doped C3N4 tube bundles/graphene nanosheets composite. Chem. Eng. J. (2021). https://doi.org/10.1016/j.cej.2021.130317
54. Q. Shan, J. Tian, Q. Ding, W. Wu, Electrochemical sensor based on metal-free materials composed of graphene and graphene oxide for sensitive detection of cadmium ions in water. Mater. Chem. Phys. **284**, 126064 (2022)

55. R. Wang, C.Y. Xiong, Y. Xie, M.J. Han, Y.H. Xu, C. Bian, S.H. Xia, Electrochemical sensor based on MoS2 nanosheets and DNA hybridization for trace mercury detection. Chin. J. Anal. Chem. **50**, 100066 (2022)
56. S. Duan, Y. Huang, Electrochemical sensor using NH2-MIL-88(Fe)-rGO composite for trace Cd2+, Pb2+, and Cu2+ detection. J. Electroanal. Chem. **807**, 253–260 (2017)
57. M.R. Mahmoudian, W.J. Basirun, Y. Alias, P. MengWoi, Investigating the effectiveness of g-C3N4 on Pt /g-C3N4/ polythiophene nanocomposites performance as an electrochemical sensor for Hg2+ detection. J Environ Chem Eng **8**, 104204 (2020)
58. G. Chen, W. Bai, Y. Jin, J. Zheng, Fluorescence and electrochemical assay for bimodal detection of lead ions based on Metal-Organic framework nanosheets. Talanta **232**, 122405 (2021)
59. X. Fang, Z. Zeng, Q. Li, Y. Liu, W. Chu, T. Maiyalagan, S. Mao, Ultrasensitive detection of disinfection byproduct trichloroacetamide in drinking water with Ag nanoprism@MoS2 heterostructure-based electrochemical sensor. Sens. Actuators B Chem. **332**, 129526 (2021)
60. Y. Wang, Y. Liang, S. Zhang, T. Wang, X. Zhuang, C. Tian, F. Luan, S.Q. Ni, X. Fu, Enhanced electrochemical sensor based on gold nanoparticles and MoS2 nanoflowers decorated ionic liquid-functionalized graphene for sensitive detection of bisphenol A in environmental water. Microchem. J. **161**, 105769 (2021)
61. T.S. Martins, J.L. Bott-Neto, O.N. Oliveira, S.A.S. Machado, Paper-based electrochemical sensors with reduced graphene nanoribbons for simultaneous detection of sulfamethoxazole and trimethoprim in water samples. J. Electroanal. Chem. (2021). https://doi.org/10.1016/j.jelechem.2021.114985
62. X. Chen, R. Li, Y. Li, Y. Wang, F. Zhang, M. Zhang, Research of the multifunctional rGO/MoS2 material in the sensing field: Human breathing and Hg(II) pollution detection. Mater. Sci. Semicond. Process (2022). https://doi.org/10.1016/j.mssp.2021.106268
63. X. Zhang, D. An, Z. Bi, W. Shan, B. Zhu, L. Zhou, L. Yu, H. Zhang, S. Xia, M. Qiu, Ti3C2-MXene@N-doped carbon heterostructure-based electrochemical sensor for simultaneous detection of heavy metals. J. Electroanal. Chem. **911**, 116239 (2022)
64. H.S. El-Desoky, A.M. Beltagi, M.M. Ghoneim, A.I. El-Hadad, The first utilization of graphene nano-sheets and synthesized Fe3O4 nanoparticles as a synergistic electrodeposition platform for simultaneous voltammetric determination of some toxic heavy metal ions in various real environmental water samples. Microchem J **175**, 106966 (2022)
65. X. Yue, Z. Li, S. Zhao, A new electrochemical sensor for simultaneous detection of sulfamethoxazole and trimethoprim antibiotics based on graphene and ZnO nanorods modified glassy carbon electrode. Microchem J. **159**, 105440 (2020)
66. N.S. Prado, L.A.J. Silva, R.M. Takeuchi, E.M. Richter, A.L. dos Santos, E.H.L. Falcão, Graphite sheets modified with poly(methylene blue) films: A cost-effective approach for the electrochemical sensing of the antibiotic nitrofurantoin. Microchem J (2022). https://doi.org/10.1016/j.microc.2022.107289
67. A. Sinha, K. Ma, H. Zhao, 2D Ti3C2Tx flakes prepared by in-situ HF etchant for simultaneous screening of carbamate pesticides. J. Colloid Interface Sci. **590**, 365–374 (2021)
68. X. Wang, M. Li, S. Yang, J. Shan, A novel electrochemical sensor based on TiO2–Ti3C2TX/CTAB/chitosan composite for the detection of nitrite. Electrochim. Acta **359**, 136938 (2020)
69. S. Hu, Q. Fei, Y. Li, B. Wang, Y. Yu, Br-terminated 2D Bi2WO6 nanosheets as a sensitive light-regenerated electrochemical sensor for detecting sulfamethoxazole antibiotic. Surfaces and Interfaces **25**, 101302 (2021)
70. D. Ilager, S.J. Malode, N.P. Shetti, Chemosphere Development of 2D graphene oxide sheets-based voltammetric sensor for electrochemical sensing of fungicide, carbendazim. Chemosphere **303**, 134919 (2022)
71. M.L. Yola, Electrochemical activity enhancement of monodisperse boron nitride quantum dots on graphene oxide: Its application for simultaneous detection of organophosphate pesticides in real samples. J. Mol. Liq. **277**, 50–57 (2019)
72. L. Killedar, D. Ilager, S.J. Malode, N.P. Shetti, Fast and facile electrochemical detection and determination of fungicide carbendazim at titanium dioxide designed carbon-based sensor. Mater. Chem. Phys. **285**, 126131 (2022)

73. Ö.S. Bölükbaşı, B.B. Yola, H. Boyacıoğlu, M.L. Yola, A novel paraoxon imprinted electrochemical sensor based on MoS2NPs@MWCNTs and its application to tap water samples. Food Chem. Toxicol. (2022). https://doi.org/10.1016/j.fct.2022.112994
74. Y. Wang, Y. Wang, F. Wang, H. Chi, G. Zhao, Y. Zhang, T. Li, Q. Wei, Electrochemical aptasensor based on gold modified thiol graphene as sensing platform and gold-palladium modified zirconium metal-organic frameworks nanozyme as signal enhancer for ultrasensitive detection of mercury ions. J. Colloid Interface Sci. **606**, 510–517 (2022)
75. Y. Wang, Z. Nie, X. Li, Y. Zhao, H. Wang, Highly sensitive and selective electrochemical sensor based on porous graphitic carbon nitride/CoMn2O4 nanocomposite toward heavy metal ions. Sens. Actuators B Chem. **346**, 130539 (2021)
76. N. Jayaraman, Y. Palani, R. Rao, Sensors and Actuators: B. Chemical Covalently dual functionalized graphene oxide-based multiplex electrochemical sensor for Hg (II) and Cr (VI) detection. Sens. Actuators B Chem. **367**, 132165 (2022)
77. J.A. Buledi, A.R. Solangi, A. Hyder et al., Selective oxidation of amaranth dye in soft drinks through tin oxide decorated reduced graphene oxide nanocomposite based electrochemical sensor. Food Chem. Toxicol. **165**, 113177 (2022)
78. M. Eswaran, P.C. Tsai, M.T. Wu, V.K. Ponnusamy, Novel nano-engineered environmental sensor based on polymelamine/graphitic-carbon nitride nanohybrid material for sensitive and simultaneous monitoring of toxic heavy metals. J. Hazard. Mater. **418**, 126267 (2021)
79. Y. Xu, Y. Yu, S. Xue, X. Ma, H. Tao, Innovative electrochemical sensor based on graphene oxide aerogel wrapped copper centered metal-organic framework to detect catechol. J. Electroanal. Chem. **899**, 115686 (2021)
80. Q.-X. Bao, Y. Liu, Y.-Q. Liang, R. Weerasooriya, H. Li, Y.-C. Wu, X. Chen, Tea polyphenols mediated Zero-valent Iron/Reduced graphene oxide nanocomposites for electrochemical determination of Hg2+. J. Electroanal. Chem. **917**, 116428 (2022)
81. S.E. Jeong, S. Kim, J.H. Han, J.J. Pak, Simple laser-induced graphene fiber electrode fabrication for high-performance heavy-metal sensing. Microchem. J. **172**, 106950 (2022)
82. X. Du, W. Du, J. Sun, D. Jiang, Self-powered photoelectrochemical sensor for chlorpyrifos detection in fruit and vegetables based on metal–ligand charge transfer effect by Ti3C2 based Schottky junction. Food Chem. **385**, 132731 (2022)
83. J. Chen, Y. Chen, Y. Liang, Application of chitosan-N-doped graphene oxide ion-imprinted sensor in Cd (II) ions detection. Diam. Relat. Mater. **119**, 108591 (2021)
84. E.B. Kim, M. Imran, E.H. Lee, M.S. Akhtar, S. Ameen, Multiple ions detection by field-effect transistor sensors based on ZnO@GO and ZnO@rGO nanomaterials: Application to trace detection of Cr (III) and Cu (II). Chemosphere **286**, 131695 (2022)
85. R. Karthik, S. Thambidurai, Synthesis of cobalt doped ZnO/reduced graphene oxide nanorods as active material for heavy metal ions sensor and antibacterial activity. J. Alloys Compd. **715**, 254–265 (2017)
86. M. Akhtar, A. Tahir, S. Zulfiqar, F. Hanif, M.F. Warsi, P.O. Agboola, I. Shakir, Ternary hybrid of polyaniline-alanine-reduced graphene oxide for electrochemical sensing of heavy metal ions. Synth. Met. (2020). https://doi.org/10.1016/j.synthmet.2020.116410
87. K. Pramanik, P. Sarkar, D. Bhattacharyay, Semi-quantitative colorimetric and supersensitive electrochemical sensors for mercury using rhodamine b hydrazide thio derivative. J. Mol. Liq. **276**, 141–152 (2019)
88. Y. Liu, G.L. Wen, X. Chen, R. Weerasooriya, Z.Y. Hong, L.C. Wang, Z.J. Huang, Y.C. Wu, Construction of electrochemical sensing interface towards Cd(II) based on activated g-C3N4 nanosheets: considering the effects of exfoliation and protonation treatment. Anal. Bioanal. Chem. **412**, 343–353 (2020)
89. M. Wang, R. Fu, C. Jin, Z. Li, J. Sun, P. Yu, M. You, The facile fabrication of g-C 3 N 4 ultrathin nanosheets with higher specific surface areas for highly sensitive detection of trace cadmium. Meas. J. Int. Meas. Confed. **140**, 548–556 (2019)
90. K. Kunpatee, K. Kaewdorn, J. Duangtong, S. Chaiyo, O. Chailapakul, K. Kalcher, M. Kerr, A. Samphao, A new disposable electrochemical sensor for the individual and simultaneous determination of carbamate pesticides using a nanocomposite modified screen-printed electrode. Microchem. J. **177**, 107318 (2022)

91. Q. Zhao, W. Wu, X. Wei, S. Jiang, T. Zhou, Q. Li, Q. Lu, Graphitic carbon nitride as electrode sensing material for tetrabromobisphenol-A determination. Sens. Actuators B Chem. **248**, 673–681 (2017)
92. W. Gao, X. Wang, P. Li, Q. Wu, F. Qi, S. Wu, Y. Yu, K. Ding, Highly sensitive and selective detection of cadmium with a graphite carbon nitride nanosheets/Nafion electrode. RSC Adv. **6**, 113570–113575 (2016)

Chapter 10
2D Nanomaterial Photoelectrodes for Photoelectrochemical Degradation of Pollutants and Hydrogen Generation

Pooja Singh, Sweta Sharma, and Pooja Devi

Abstract The development of sustainable solutions for meeting escalating needs, such as clean energy and safe drinking water, is of the utmost importance to the modern world. Hydrogen as a fuel can be worthiest for this purpose, and further generating it from wastewater via green routes, i.e. photo/electrocatalytic splitting, can make it a sustainable solution, overcoming challenges of wastewater treatment simultaneously. In this chapter, we have discussed different materials that can be utilized as photoelectrocatalyst focusing on 2D materials for hydrogen generation from wastewater (textile, pharmaceutical, food industry, etc.). The potential catalytic properties of transition metal dichalcogenides (TMDs), transition metal oxides (TMOs), MXenes, graphene, nitrides, carbides, and their hybrids are discussed for the same. The standard diagnostic parameter for evaluating photoelectrocatalyst is photo response, incident photon to current efficiency, faradaic efficiency, and wastewater treatment in terms of percentage degradation, COD, TOC, etc., are presented and compared for 2D materials. Further, material performance in terms of the band gap, appropriately positioned valence and conduction bands, stability, economics, etc., are also compared for the wastewater systems. Last but not the least, the future outlook of the field is also presented with respect to challenges and research directions to tap this important unused energy source, i.e. wastewater.

Singh and Sharma: Equally contributed.

P. Singh · S. Sharma · P. Devi (✉)
Academy of Scientific and Innovative Research, New Delhi, India
e-mail: poojaiitr@csio.res.in

P. Devi
Materials Science and Sensor Application, Central Scientific Instruments Organisation, Chandigarh 160030, India

N. Kumar et al. (eds.), *Two-Dimensional Materials for Environmental Applications*, Springer Series in Materials Science 332, https://doi.org/10.1007/978-3-031-28756-5_10

10.1 Introduction

To meet the global demand for clean energy and clean water, developing and progressing sustainable solutions and technologies is necessary. In current times, a decarbonized world powered by renewable energy is transitioning, and hydrogen is seen as a clean energy carrier. Thus, hydrogen as a fuel can be worthiest for this and generating it from wastewater in greenways, like the photo/electrocatalytic splitting of water, can also enable wastewater utilization as well as treatment [30]. Hydrogen as a fuel has found tremendous stationery and mobile usage, such as a propellant of non-polluting vehicles, home heating, cooking, chemical transformations, and aviation. Therefore, it is projected that hydrogen will join solar power as the primary energy carrier for a sustainable energy system [4, 27, 90]. Likewise, photocatalytic degradation of organic wastes in a photoelectrochemical (PEC) system is an appealing approach for their complete mineralization. Therefore, a dual system utilizing wastewater as feedstock can address both challenges at the same time [19]. The fundamental photo(electro) catalytic setups of both processes share the same roots, but they are rarely explored concurrently. The most well-known scientific paper that impacted this field's progress was by Fujishima and Honda [1] in 1972, which described photocatalytic water splitting in a PEC cell employing semiconductor photoelectrode.

As a feedstock, wastewater has a rich supply of nutrients like organic compounds, phosphate, nitrogen and minerals, which makes it an enormous source for energy recovery. The current conventional wastewater treatment facilities use a lot of energy, yet many organic/inorganic constituents still remain untreated. Alternatively, wastewater can also be used to generate power, heat, or fuels like hydrogen or methane. Anaerobic digestion, which produces biogas, is one of the most popular and is still the most widely adopted technology for recovering energy from wastewater around the world [5]. This process results in the production of biogas, a gas combination primarily made up of methane and CO_2. Alternatively, hydrogen generation from wastewater is a promising and less explored approach. Some efforts have been made in the direction of biological processes such as fermentation, in particular, photo-fermentation and dark fermentation, for producing biohydrogen [89]. In photo-fermentation, sunlight-powered photosynthetic bacteria convert organic molecules into hydrogen and carbon dioxide. In general, dark fermentation is frequently combined with photo-fermentation; wherein photosynthetic bacteria convert the organic acids produced as a by-product of dark fermentation into hydrogen [56]. Another way of hydrogen production from wastewater is via splitting the water under light (photocatalysis) or electricity (electrocatalysis) or the combination of both (photo (electro)catalysis). On exposure to the light to a photocatalyst system, charge carriers are generated, which in turn can be utilized to drive hydrogen evolution and organic pollutants oxidation in a wastewater environment [70]. The first such system reported by Fujishima and Honda utilized TiO_2 electrodes [1]. Since that time, interest in photocatalysts research has been continuously increasing, leading to the development of plethora of materials in this field. Neverthless, only a few findings concentrate on producing hydrogen from the wastewater. This may be due to despite of simplicity of the underlying chemistry of water splitting, effective wastewater-to-hydrogen generation is a complicated process due to the involvement of two thermodynamic processes of pollutant mineralization as well as water reduction. It requires specifically designed catalytic electrodes and a significant amount of energy to drive the process. The presence of impurities in the wastewater environment can not only damage the catalysts but also reduce the overall efficiency along with the generation of unwanted by-products. Thus the present chapter is focused on discussing and revieweing 2D materials-based photoelectrodes for a dual-system photo electrochemistry i.e. the simultaneous removal of pollutants from wastewater and hydrogen generation. It also covers the basic understanding of the type of wastewater pollutants, an overview of the PEC fundamentals, a discussion on the class of 2-D materials, and factors affecting their performance in PEC. A critical evaluation of the current state of the art and their limitations in this field is also presented with the projection of future opportunities.

10.2 Wastewater Characteristics

Wastewater is a mixture of liquid or water-carried pollutants removed from households, organizations, and institutions with groundwater, surface water, and stormwater. Wastewater generally includes suspended solids that comprise metals, organic/inorganic substances, disease-causing microorganisms, faecal matter, organic waste, and more [108]. Before entering the treatment facility, wastewater is referred to as influent; after it is released from the facility, it is referred to as effluent. Sludge is the term used to describe the leftover semi-solid waste in the working unit [77]. The following four categories could be employed to categorize it based on the source of contamination:

(A) Domestic wastewater: It is further divided into grey and black garbage. Grey water is the waste from laundry, washrooms, and kitchens, whereas black water is made up of urine and faeces.
(B) Industrial wastewater: Water released by organizations (such as pharma, paper industry, and textile mills) and livestock farming.
(C) Infiltration wastewater: The water is made up of sewerage water that gets contaminated through leaks in joints, gaps, porous roofs, and walls.
(D) Storm wastewater (runoff): It is a combination of sewerage water and flood water from rain [82].

Wastewater can further be categorized according to the physicochemical parameters depending upon the source. Based on these physical and chemical characteristics (as shown in Fig. 10.1), the pre-treatment methods and treatment technologies are selected [9, 81], wherein the water is analyzed for its temperature, pH, color, turbidity, total suspended solids (TSS), and total dissolved solids (TDS), Chemical oxygen demand (COD), biochemical oxygen demand (BOD), total organic carbon

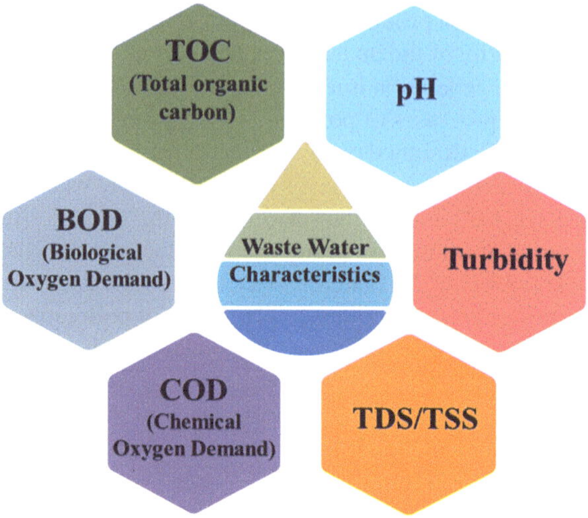

Fig. 10.1 Physical and chemical characteristics of wastewater

(TOC) before and after the treatment. The temperature of influent wastewater mainly ranges from 26.2–35.4 °C [103], while its pH varies from the source of its origin and generally covers a wide range. The broad range of pH of wastewater can change the rate of biochemical reactions in aquatic organisms [85]. The color of the water is also a qualitative factor for the representation of the storage circumstance as well as the presence of a type of organic constituents in the water [85]. Turbidity is another important physical parameter, which is a direct indication of pollutants present in water. The higher amount of turbidity signifies a high concentration of waste or vice-versa [69]. The chemical parameters, i.e. TOC, COD, and BOD, are related to organic constituents present in water [83]. COD determines the amount of oxygen needed for chemically oxidizing organic waste, whereas the overall amount of oxygen microbes required to decompose organic waste in an aerobic environment completely is referred to as BOD. The high level of BOD, COD and TOC signifies a higher level of pollution [61, 64].

10.3 Photoelectrocatalysis Mechanism

A PEC system generally contains three electrodes or a two-electrodes: a counter electrode, reference electrode, and the working electrode (photoanode & photocathode). At least one of the electrodes needs to be a semiconductor capable of absorbing light with higher energy than its bandgap to generate electron and hole pairs. Under external bias, these charge carriers can be separated to perform water reduction to generate hydrogen and pollutant oxidation for mineralization. In general, an externally generated electrical/chemical bias provides the extra voltage required for the reaction i.e. water splitting. While in PEC, the photocatalytic electrodes can collect solar energy that can help to reduce this energy or completely replace electrical energy based on material characteritsics. A PEC system also has advantages over the standalone photocatalytic system in terms of (i) self and external bias can alleviate the electron–hole recombination and enhance the charge transfer kinetics, (ii) easy separation of hydrogen and oxygen at both the electrodes, (iii) coating of the semiconductor materials onto large area substrates such as conductive and semiconductor substrate, and (iv) increased stability of the catalyst in bound form than the powder catalyst.

In general, based on the type of semiconductor at the working electrode, i.e. n-type-photoanode or p-type-photocathode, the two types of reactions, i.e. hydrogen evolution reaction (HER) and oxygen evolution reaction (OER), occur at the anode and cathode, respectively. In general, the processes in a PEC system are (i) absorption of light and generation of photo charges, (ii) charge separation & charge transfer and (iii) water splitting/redox reaction [71].

(i) **Absorption of light and generation of excitons**: The semiconductor's electrical structure interacts with the electromagnetic wave as light passes through it. This interaction excites certain electrons in the valence band (VB) to enter the conduction band (CB) and create holes in VB. The bandgap of semiconductors is thus the energy difference between the CB and VB. When the photon energy is greater than the semiconductor's band gap, the excited electrons transfer the extra energy to nearby atoms by thermal vibration, stabilizing the CB. Due to their excellent chemical resistance, metal oxide semiconductors have been examined as photoelectrodes; however, due to their large band gap energy, which accounts for 4% of the solar spectrum, their activity is primarily restricted in UV light. Thus efforts are being made to design narrow bandgap semiconductors to overcome this limitation.

(ii) **Charge separation and transfer**: To stop the recombination process, the photo-generated charge carriers at the photoelectrode surface should be separated. Recombination is the principal barrier to charge carriers' transport to the photo electrode's active sites, as it makes it challenging to extract current from charge carriers due to the high recombination degree. The crystal structure and size of the particle of the material affects the rate of recombination. The defects/imperfections can act as a trapping centre, while a high crystalline material exhibits significant photocatalytic activity due to less recombination. In the case of nanomaterials, the recombination is less due to the high surface-to-volume ratio and ease of transport of excitons to the catalyst surface. The transfer of photogenerated charge is further influenced by the morphology of the catalyst on the substrate, it's defect characteristics, and electrolyte composition. For example, charge transfer across the linear transport path can be improved by using 1D nanostructures like nanotubes. Furthermore, an exterior bias or a built-in internal potential can enhance the electron-holes separation and reduce the recombination. The inbuilt potential could be generated by doping metal/non-metal elements, integration of metals onto the semiconductor, and making heterojunctions with materials of different band gaps. Some

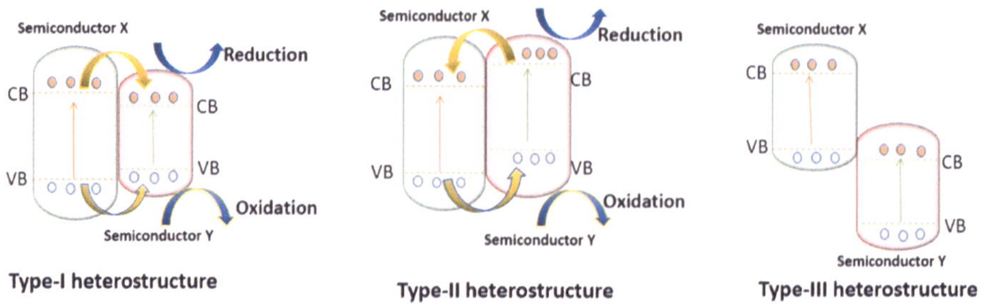

Fig. 10.2 Types of heterostructures

junctions are: (A) p-n homojunction: In this junction, a semiconductor interface of the same semiconducting material, and band gap but of different doping types is used. In most cases, n-type and p-type materials play the role of donor and acceptor, respectively. (B) p-n heterojunction: An interface between two dissimilar semiconductors with unequal band gaps. A heterojunction structure for better charge separation at the interface between two semiconductors. Semiconductor interfaces can therefore be organized into three types of heterojunction, as shown in Fig. 10.2.

(a) Straddling gap: Type I, in this case, the photogenerated electrons and holes drift from semiconductor X to Y, which causes charge recombination due to the large accumulation of the electron/holes. In this heterostructure, the charge carriers make the quantum wells in both bands.
(b) Staggered gap: Type II in which holes photogenerated on semiconductor X are supplied to semiconductor Y, and photogenerated electrons in semiconductor Y are transferred to semiconductor X for charge separation. In this heterostructure formation of a quantum well in one band and an energy barrier in the other band. Generally, the type-II heterojunction includes type-II, Z-Scheme, and S-Scheme. In the Z-scheme heterojunction, the electron transfer occurs from the higher CB of one semiconductor to the lower VB potential of the other semiconductor due to the large driving force [54]. S-Scheme heterojunction is similar to the type-II heterojunction. But both are completely different due to the charge-transfer path. In the case of type II, photoelectrons and holes are accumulated on the CB of the oxidation photocatalyst, and VB of the reduction photocatalyst, respectively, in the weak redox ability. But for the S-Scheme, it is vice-versa [101].
(c) Broken gap: Type III, in this heterojunction, there is no directional transport of photogenerated carriers between the semiconductor X and Y. The type-III heterojunction is identical to that of the type-II heterojunction, with the exception that the severe staggered gap prevents the band gaps from overlapping. Therefore, the type-III heterojunction is not ideal for boosting the separation of electron–hole pairs because electron–hole

migration and separation between the two semiconductors cannot occur [111].

(iii) **Water splitting and redox reaction**: The electric field that is already present can separate the photogenerated charge. At the photoanode, the photo-excited electrons are collected via the back contact and channelized to the counter cathode, wherein the protons are reduced to H_2. At the same time, holes migrate to the surface of the photoelectrode for an OER. In the case of a photocathode, the HER is carried out by driving the electrons to the surface of the electrode. The splitting of water is a non-spontaneous reaction and requires a Gibbs' free energy of 237.2 kJ/mol under the ideal conditions of 1 mol ion concentration, 1 atm gas pressure, and 25 °C temperature. This value corresponds to a theoretical thermodynamic reversible potential of 1.23 V. The Nernst equation states that the value of the potential changes with the pH of the electrolyte according to the relation: $E_{RHE} = E_{NHE} + 0.059pH$, where E_{RHE} is the potential w.r.t reversible hydrogen electrode, and E_{NHE} is the potential in relation to a normal hydrogen electrode.

10.4 Nanomaterials as Photoelectrocatalyst

At the nanoscale dimension, the scattering rate is greatly reduced, and carrier collecting efficiency is increased as their diameter is equivalent to carrier scattering lengths. Thus, nanomaterials have substantial absorption coefficients due to the rise in oscillator strength, enabling great conversion efficiency. The nanomaterial's band gap (QDs and semiconductors) can cover the entire solar spectrum and alter the absorb specific wavelengths by changing their size. Further, with the help of doping the electronic band structure and heterostructure can be modified. Moreover, it enables rapid charge transfer and mass, enhanced light absorption, and reduced dispersion of light. The photocatalytic activity can be increased by coating the electrode with nanoparticles or dispersing them in the water used as the reaction medium. In the case of PEC, the basic idea is to convert solar energy into hydrogen by applying an exterior bias to a photovoltaic material dipped in an electrolyte. When using the reference electrode to do measurements, one vital understanding in a PEC system is knowing the Fermi Energy (E_f), which compares the E_f of the semiconductor with its own constant Fermi level. The probability of discovering an electron at this energy level is cut in half. In an intrinsic semiconductor, E_f lies exactly between E_c and E_v or at the center of the band gap. Depending on the kind of dopant, E_f moves closer to or farther away from Ec. Equilibration occurs at the interface by adjusting the semiconductor's Fermi Level to match the redox couple of the electrolyte. Depending on the type of semiconductor (n-type/p-type), this creates a small region of space charge layer near the semiconductor surface that bends the band up or down [75]. To put it simply, increasing the effectiveness of water splitting is highly dependent on the photoelectrodes' electrical characteristics. It has been demonstrated that it is difficult to optimize every process in a single component. Numerous efforts have been made to increase effectiveness and absorb a broad spectrum of light including the construction of photocatalysts with heterostructures (n–n/n-p/p-p transitions) [60]. In this direction, several classes of materials have been investigated as photoelectrodes, including metal oxide semiconductors, metal-doped semiconductors, nitrides and carbides. The 2D materials are usually composed of sturdy covalent bonds and weak Van der Waals bonds, which lead to stability of in-plane and stacked layer structure formation, respectively.Trailing the breakthrough of graphene (2004), a new perspective opened for investigating other 2D materials, for example, transition metal oxides (TMOs), transition metal dichalchalcogens (TMDs), graphene, and MXenes [6].

10.4.1 Factors Affecting the PEC Efficiency of 2D Nanomaterials

The orderliness, homogeneity, and morphology of the catalysts heavily influence their performance and efficiency in water splitting. Controlling the shape and the structure, in addition to shrinking the size, is crucial for an effective mechanism for splitting water, and the same are discussed below.

10.4.1.1 Crystallanity

When compared to amorphous materials, highly organized crystalline materials perform substantially better. Such as, the photocurrent characteristics of crystalline/annealed TiO_2 nanotubes are greater than the amorphous TiO_2 nanotubes. At a high temperature of around 300 °C, amorphous TiO_2 nanotubes are capable of crystallization. It is evident that a structural attribute can affect the photocurrent efficiency as the number of defects and the site for e/h + recombination decreases as crystallinity increases. The e/h + transfer and OER/HER reactions of semiconductors depend heavily on the crystallinity and surface-active sites, respectively. High crystallinity materials with few flaws can reduce the recombination of e/h^+. OER/HER reactions should proceed more quickly than backward reactions, which mostly depend on the quantity of active sites [28]. Shi et al. showed that high-temperature annealing (540 °C) improved the crystallinity of the $BiVO_4$ film, which reduced the electron–hole recombination and boosted PEC water splitting when exposed to light [87]. By Kim et al. showed the efficiency of the PEC water splitting was also discovered to be directly enhanced by enhancing the crystallinity of the produced photoanodes [48].

10.4.1.2 Dimensionality

Based on their dimensions, nanomaterials can be divided into 0D, 1D, 2D, and 3D structures. In which 2D nanostructures such as TiO_2 thin films [40], MoS_2 nanosheets

[105], hematite films, and g-C_3N_4 [62] nanosheets are utilized for the oxidation in PEC water splitting processes. It can effectively harvest a significant amount of UV radiation because of its thin profile and substantial surface area, which in turn can increase hydrogen generation efficiency by enabling charge transfer onto the surface. Compared to thin films, nanorods and nanowires transfer charge carriers more effectively and are more photoactive. Nanotubes have a higher surface area than nanorods and nanowires, allowing for a faster redox reaction rate even when only a small portion of them are exposed to light absorption [45].

10.4.1.3 Band Gap

The ability to precisely tune the band gap is nanomaterials' main benefit. The band gap's narrowing is thought to improve the PEC's characteristics. This aids in the vast range of solar spectrum absorption. Large band gaps cannot absorb the requisite intensities of light energy to divide the water effectively. For effective PEC water splitting, the band gap should be between 1.6 and 2.2 eV. The band edge position is exact, and the mobility of the photogenerated charge carrier is enhanced in this band gap range. Donor–acceptor can be included in the semiconductor to reduce the band gap. Band edge position and carrier diffusion length are important factors when choosing a suitable material. In contrast to the H_2 formation potential, CB should be more negative, and VB should be more favorable than his O_2 formation potential [98].

10.4.1.4 pH Dependency

PEC cells work in a variety of pH environments. The pH of the electrolyte measures whether the net charge adsorbed on the surface is positive, negative, zero, or and has a large impact on the equilibrium of the water-splitting reaction [45]. To ensure efficiency, photo corrosion of the electrode should be minimized even under hard and corrosive electrolytic conditions. The electrode surface may become less durable due to ion migration occurring during reactions. Photoelectrodes with nanomaterial integration display stability under various pH settings. Buffering the electrolyte solution increased the photocatalytic stability and property [21].

10.4.1.5 Size

The performance of PEC can be improved with the help of co-catalysts' size effects. One method to increase PEC efficiency is to create a good catalyst. The electro kinetics, which results in increased electron–hole recombination, dominates in smaller particles. Greater charge extraction at the electrode and electrolyte interface is possible due to the band-bending features of larger particles. It follows that larger cocatalysts are suitable for greater PEC performance [76].

10.4.1.6 Temperature and Pressure

Temperature is a key factor in improving PEC effectiveness. Experiments are frequently carried out in hot environments. Low temperature treatments exhibit increased PEC efficacy in a study. When the temperature was low, the incident photon to current efficiency (IPCE) rose by up to 95%. Researchers found that increasing working pressure improves the band gap [2].

10.4.1.7 Type of Pollutants

The co-existence of organic substances and heavy metals complicates the situation due to their complex interactions as compared to a single pollutant [104]. Numerous studies have shown that PEC performs better treating individual pollutants that can be organic and heavy metals. Chen et al. showed the removal of Cu-EDTA mixture (pH 3.18) by photolysis and electrocatalysis techniques. These methods demonstrate a low removal rate of pollutants at a current density of 0.5 mA/cm^2 while PEC achieved approx. 74% removal of Cu-complex and 75.54% recovery of Cu(II) recapture, indicating its potential over PC and EC [16]. A similar behavior of PEC treatment is reported by Wang et al. [97] for the treatment of mixed (organic-metal). Operation variables such as the applied bias potential or current, the solution pH, the concentration of organic-heavy metal mixed pollutants, the type of electrolyte and the electrolyte concentration reveal notable influence in the PEC removal of these pollutants [38]. The authors presented evidence of effect of pH; an increasing trend in pollutant removal was observed when pH was varied from 10.0 to 3.0. Similarly, Zhao et al. [113] showed the influence of the initial concentration of pollutants on PEC activity. It happen due to a limited amount of photogenerated holes and OH oxidants.

10.4.2 Synthesis Methods for 2D Photoelectrodes

2D materials can be synthesized via the hydrothermal route, sol–gel method, co-precipitation, and impregnation methods, as shown in Fig. 10.3. Table 10.1 also lists the synthesis approach of various 2D photoelectrodes. A brief discussion about each method is discussed below:

10.4.2.1 Hydrothermal Synthesis

Hydrothermal synthesis is a straightforward and practical method for fabricating semiconductor photocatalysts with high purity [51, 78, 100, 102]. Substances from the precursors are first added to distilled water, agitated for a while, placed in the stainless steel autoclave (Teflon-lined) and heated up to ~220 °C for several hours

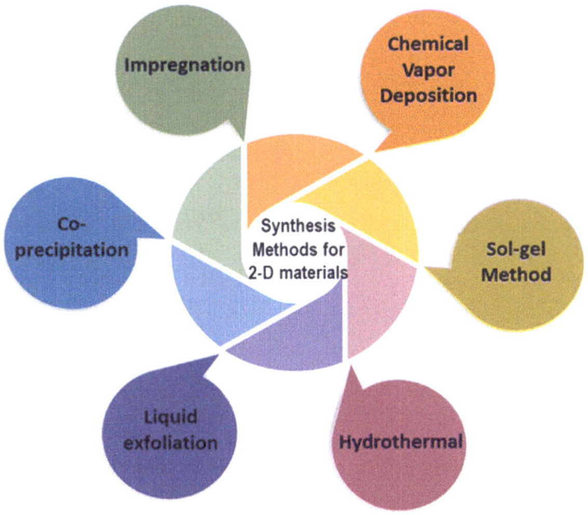

Fig. 10.3 Various methods for the synthesis of 2D materials

[8]. If non-aqueous chemicals are used as a solvent in place of water, the technique is known as solvothermal synthesis [59]. This is a typical technique that has been frequently utilized to synthesize 2D nanomaterials [100, 102]. A generalized solvothermal approach is shown by Dou et al. [20] for the assembly of several metal oxide nanosheets, including ZnO, TiO_2, MnO_2, Co_3O_4, Fe_3O_4, and WO_3. These metal oxide nanosheets ranged in thickness from 1.6 to 5.2 nm, or layers of 2- 5 stack forming the monolayer. Additionally, ZnS, ZnSe, CeO_2, and In_2O_3 were among the non-layer structured 2D nanosheets that Xie and co-worker [50, 93, 94] developed using the hydro/solvothermal approaches.

10.4.2.2 Sol–gel Synthesis

The sol–gel method is a widely used technique for the fabrication of photoelectrodes. This procedure has the potential to take care of the surface and textural parameters of the catalysts [74]. This technique is among the easy-to-process and most efficient techniques for engineering different nanomaterials [53]. In order to obtain the appropriate chemical composition, nanostructure, and surface characteristics, Parashar et al. [73] investigated stable and effective sol–gel synthesis protocols of TiO_2, SnO_2, WO_3, and ZnO nanomaterials. Along with that, Panimalar et al. [72] reported the successful fabrication of MnO_2/g-C_3N_4 hybrid nanocomposites via the facile sol–gel method with a combination of annealing. Compared to physical approaches, the sol–gel pathway is more cost-effective since it uses low operating temperatures and less energy [80].

Table 10.1 Various material syntheses methods for 2D photocatalyst

S/n	Photocatalyst material	Precursors	Synthesis process	References
1	NiMnLDHs/g-C_3N_4	$Ni(NO_3)_2.6H_2O$, $MnCl_2.4H_2O$ and dicyandiamide	Hydrothermal method	[84]
2	Nb_2O_5/ZnAl LDHs	$Zn(NO_3)_2.6H_2O$, Al $(NO_3)_3.9H_2O$	Hydrothermal method	[41]
3	Bi_2O_3/Bi_2WO_6/MgAl	$MgSO_4{\cdot}7H_2O$, Al $(NO_3)_3{\cdot}9H_2O$ and $Bi(NO_3)_3{\cdot}5H_2O$	Hydrothermal method	[7]
4	MnO_2/g-C_3N_4	Urea, Potassium permanganate ($KMnO_4$), Manganese (II) sulfate ($MnSO_4.H_2O$)	Sol–gel method	[72]
5	Mo_2C/graphene	Gold foil, Mo foil, CH_4	CVD	[92]
6	Graphene/Cu foil	Copper foils, a mixture of methane and hydrogen	CVD	[52]
7	MoS_2/SiO_2/Si	MoO_3 and S powders	CVD	[49]
8	WS_2	Tungsten (IV) disulfide, poly(vinylidene fluoride) (PVDF), copper foil, acetone and ethanol	Liquid exfoliation	[18]
9	ZnAl/g-C_3N_4/CuO	$Al(NO_3)_3{\cdot}9H_2O$, $Zn(NO_3)_2{\cdot}6H_2O$, melamine and $Cu(O)_2{\cdot}H_2O$	Co-precipitation method	[63]
10	g-$C_3N_4/SnIn_4S_8$	Tin (IV) chloride pentahydrate ($SnCl_4{\cdot}5H_2O$), indium (III) chloride tetrahydrate ($InCl_3{\cdot}4H_2O$), Dicyandiamide and urea	Co-precipitation method	[47]
11	Bi_2WO_6	Bi_2WO_6, 8.3416 g of $Bi(NO_3)_3{\cdot}5H_2O$, $(NH_4)_{10}(W_{12}O_{41}){\cdot}H_2O$ and HNO_3	Co-precipitation method	[3]
12	MoS_2/CdS	$(NH_4)_2MoS_4$, Commercial CdS powder	Impregnation method	[120]
13	WS_2/CdS	Commercial CdS powder, $(NH_4)_2WS_4$	Impregnation method	[121]

10.4.2.3 Chemical Vapour Deposition (CVD)

Another intriguing way to grow high-quality single crystalline 2D sheets on substrates is the CVD technique [42, 88, 91]. It allows adjustable thickness, scalable size, and outstanding electrical characteristics of grown material. At temperatures as high as

1000 °C, Li et al. [52] used a variety of methane and hydrogen to grow graphene using carbon by CVD on copper foils. Additionally, Shi et al. utilize MoO_3 and S powders as the reactants to fabricate the MoS_2 layer through CVD directly on SiO_2/Si substrates [88]. However, the downside of the CVD process is that it requires particular substrates, high temperatures, and high vacuums.

10.4.2.4 Liquid Exfoliation

The process of liquid exfoliation consists of the sonication of multilayer bulk crystals in surfactant/polymer solutions that result in high yield and industrial scale production of ultrathin 2D nanomaterials [36, 65, 117]. Though, this approach produces a low yield of single-layer sheets along with the challenge of dissociating the monolayer sheets from the mixture. As a substitute, the ion-intercalation and exfoliation approach was established to fabricate monolayer nanosheets at high yield and on a large scale, including graphene, MoS_2, WS_2, TiS_2, ZrS_2, and others [18, 109, 110], Castellanos-[10]. The 2D Nanosheets produced by this method have several flaws and limited lateral size [114]. Additionally, this technology is extremely susceptible to oxidation.

10.4.2.5 Co-precipitation Technique

The co-precipitation technique is a frequently used method to fabricate various nanomaterials, including semiconductor-based composites [58]. It is a very facile method to synthesize iron oxide nanoparticles (Fe_3O_4 or γ-Fe_2O_3) specifically [26]. This technique is mostly utilized for developing mixed oxides [43]. García-Pérez et al. [25]utilized the co-precipitation process for the formation transition metal tungstates having the typical formula $M^{2+}WO_4$ (M^{2+} = Co, Cu, Mn, and Ni) and studied their photocatalytic activity [25]. In another work, Kianpour et al. claimed the first low-temperature co-precipitation approach for the development of g-C_3N_4/$SnIn_4S_8$ material in their work [47]. However, this approach does not perform effectively if the reactants have significantly varied precipitation rates [79].

10.4.2.6 Impregnation Method

This method allows the final characteristic and structure to be controlled in advance as well as efficient and stable. The amount and heating of the precursor are two factors that have an impact on the impregnation technique-synthesized composites [31]. In regard to this, Zong et al. developed TMD-based photocatalysts, including MoS_2/CdS [119] and WS_2/CdS [121] that showed strong photocatalytic activity by impregnating host materials with TMD precursors and high-temperature calcination. However, obtaining precise shapes and layer-number controlled with this method is very challenging. Additionally, the material aggregation brought on by high temperatures may result in a reduction in the specific surface area and active catalytic sites [120].

10.5 2D Nanomaterials in a Dual PEC System

10.5.1 Metal Oxide Semiconductors

Single TMOs are highly widely employed as photocatalytic semiconductors due to their exceptional chemical stability and efficiency. TiO_2 and ZnO are prominent catalysts for waste degradation and hydrogen production [22, 35, 106]. The conduction and valance band edge potentials of TMOs (as described in Fig. 10.4) are beneficial for a diversity of applications involving water splitting and pollutant treatment by thermodynamical aspect [33]. Though, due to their extensive optical bandgaps of 3.2–3.4 eV only the UV region of the solar spectrum is available for light absorption [23]. Thus, TMOs, which comprise significantly lesser bandgap and absorb visible light, are highly needed. TMO like WO_3, MoO_3, and V_2O_5 consist of a bandgap between 2.6 and 3.0 eV that can absorb light in a visible area. Along with that, they show better electron transport traits, so there will be a decrease in the fast charge recombination, which is otherwise a frequent demerit of TMOs [15, 44, 67]. A unique PEC reactor was designed by utilizing TiO_2 by Ghassan Chehade et al. to produce hydrogen and treat the waste product of galvanizing industry [11]. This unique reactor design enables chlorine gas evolution at the surface of the photoanode without dissolution and hydrogen gas to flow out of the reactor through the cathode compartment having aqueous 5 M HCl as an electrolyte. An excellent proton and H^+ transfer rate are also enhanced by this innovative design of the corrosion-resistance cathode, while the photoanode improves the charge transfer process. The reactor is used for the PEC studies with and without sunlight. It showed the generation rate of hydrogen was 3 ml/min [11]. Similarly, T.T. Guaraldo et al. studied $Ti/TiO_2/WO_3$ electrodes as photoanode fabricated via electrochemical anodization and cathodic electrodeposition for simultaneous hydrogen production and dye degradation [29]. The photocurrent density of 11 mA/cm^2 was achieved with a hybrid electrode, while it was 8 mA/cm^2 in the case of pristine TiO_2. The hydrogen production efficiency attained was 46%, while an 85% reduction in TOC was observed [29]. In another work by Ouyang et al., $BiVO_4/WO_3$ is utilized as a photoanode, and MnO_2/graphene oxide (GO) hybrid was used as a cathode [68]. The inverse opal-shaped photoanode ($BiVO_4/WO_3$) attained a photocurrent density of ~5.04 mA/cm^2, which was greater than bare material at 1.2 V versus Ag/AgCl. The inverse opal $BiVO_4/WO_3$ photoanode was merged with the improved cathode (MnO_2/GO) to design a visible-light accessible PEC system. The modified system indicated better photocurrent generation of 593.5 mA/cm^2 compared to only

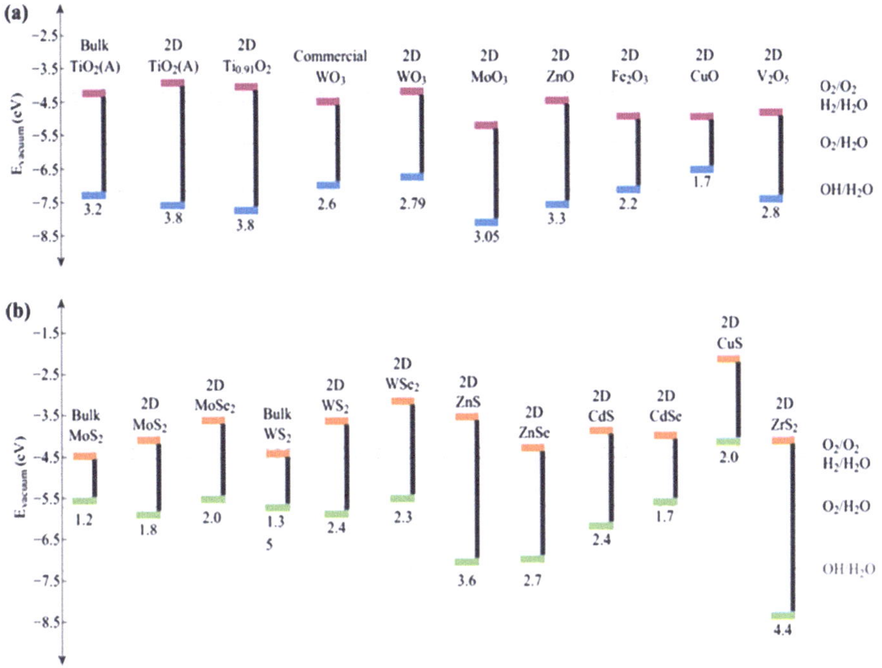

Fig. 10.4 Band structure of **a** TMOs and **b** TMDs (Reproduced with permission from [33])

MnO_2 cathode, i.e. 255.9 mA/cm^2. Additionally, Rhodamine B degradation was also confirmed for successfully establishing pollutant degradation application [68].

10.5.2 Transitional Metal Dichalcogenides

Transition metal dichalcogenides are also studied due to their earth abundance [39]. The TMDs like MoS_2, $MoSe_2$, WS_2, and $TiSe_2$ shown in Fig. 10.4 are a group of layered substances with typical formula of MX_2, where M is an element from transition metal group 4–10, e.g. Ti, Zr, V, Hf, Nb, Ta, W, Mo, and X is the from chalcogens like S, Se, and Te [17, 20]. For PEC applications, TMD nanosheets can be utilized by various means i.e. by boosting light absorption in the visible part of the solar spectrum, as a photosensitizer, as a charge separator via band alignment, and charge transporter [13, 37, 96, 112, 115].

Chen et al. studied a sequence of direct Z-scheme by synthesis of $MnIn_2S_4$/g-C_3N_4 photocatalysts via hydrothermal route as a result of in-situ stocking of $MnIn_2S_4$ flakes on the top of g-C_3N_4 nanosheets [12]. To achieve high efficiency for the simultaneous degradation of pharmaceutical wastewater and hydrogen generation, the weight ratio between $MnIn_2S_4$ nano-flakes and mesoporous g-C_3N_4 nanosheets was optimized. The development of Z-scheme $MnIn_2S_4/g\text{-}C_3N_4$ structures efficiently enhanced the transfer and estrangement of photogenerated charge carriers via close interface connections established between these two materials. The good stability of $MnIn_2S_4/g\text{-}C_3N_4$ nanocomposites was demonstrated by re-utilization of it for pharmaceutical wastewater treatment [99].

Zheng et al. proposed a multifunctional PEC procedure for the instantaneous degradation of pharmaceuticals and personal care products (PPCPs), generation of H_2 and disinfection of *E. coli* based upon MoS_2@BL-$BiVO_4$ photoanode [116]. An improved MoS_2@BL-$BiVO_4$ photoanode (Fig. 10.5) with decreased recombination of surface charge was fabricated that showed enhanced photocurrent density (2.21 mA/cm^2) in real sewage water, which was 8.1 times higher of pristine $BiVO_4$. The hybrid photoanode attained absolute removal of 2 ppm benzophenone-3, along with 89.32 μmol of hydrogen production and decontamination of *E. coli* bacteria in a time limit of 30 min at a bias of 1.0 V versus Ag/AgCl. The PEC system exhibits outstanding reusability and stability while degrading PPCPs with reasonable efficiency [116].

10.5.3 Graphene and Graphene like Materials

Graphene is a monolayer of sp_2-bonded carbon atoms, wherein each carbon atom is a surface atom, giving the material a large exposed area and p-conjugated structure. Graphene frequently exhibits remarkable electrical and thermal conductivities due to its honeycomb-like organizational structure and one layer of carbon atoms arranged in a 2D hexagonal lattice. These properties enable graphene sheets to accept and transport excited semiconductor electrons efficiently, decrease charge recombination, improve charge transfer at interfaces and photocatalytic reaction centres and provide extra active adsorption sites. and enhance the production of photocatalytic H_2 activity [45].

Graphene oxide (GO) is an oxidized form of graphene that has oxygen-containing functional groups that are advantageous. Examples include carbonyl, hydroxyl, carboxyl, and epoxy groups. Graphene is easier to functionalize with other user groups even though it is more conductive than reduced GO (rGO). Natural remediation is one of the many domains where graphene-based materials have inspired research [66].

Graphene-based materials are beneficial due to their high bonding density and oxygen-containing functional groups. Every graphene-based substance has a significant surface area. Graphene-based nanomaterials may act as adsorbents via hydrogen bonds, pie-bonds, hydrophobic, electrostatic, and covalent interactions, among other conceivable interactions. Thus, materials made of graphene are effective at absorbing many types of pollutants from water. A nanostructured composite $FeMoO_4$-GO catalyst synthesized via a simple one-pot hydrothermal technique is reported as a superior electrode with high degrading capacity and quick hydrogen generation rates. The electrode could perform 90% of berberine elimination within 30 min. At a

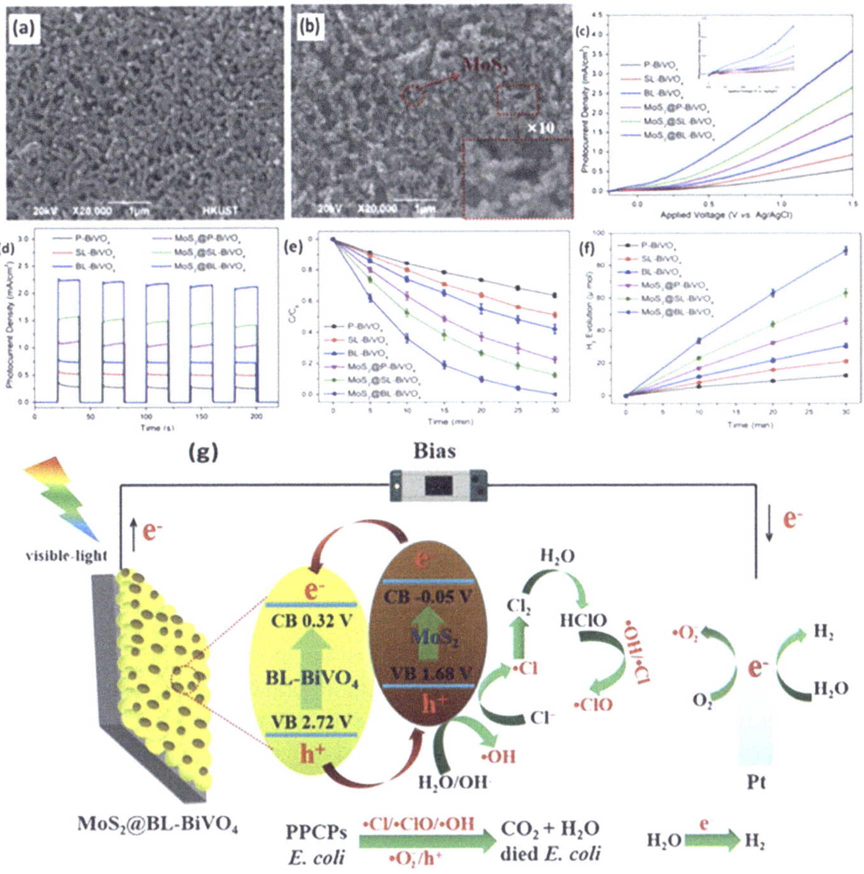

Fig. 10.5 **a** SEM image of BL-$BiVO_4$, **b** hybrid MoS_2@BL-$BiVO_4$, **c** Linear sweep voltammetry (LSV) graph and **d** transient photoresponse of the photoanodes in the sewage at 1.0 V versus Ag/AgCl applied, **e** degradation of BZP, **f** H_2 generation using various photoanodes at 1.0 V versus Ag/AgCl in real sewage water. **g** schematic diagram showing the process of multifunctional system for removal of PPCPs, disinfection of *E. coli* and generation of H_2 in sewage by MoS_2@BL-$BiVO_4$ hybrid electrode. (Reproduced with permission from [116])

current density of 10 mA cm^{-2}, the fabricated system generated hydrogen at a rate of 3.1 mmol cm^{-2} min^{-1} [95]. In another work by Hongqiang Shen et al., a two-step solvothermal fabrication process was used to create a 2D/2D/3D Z-scheme system of Cu_2O/RGO/$BiVO_4$, which could not only produce H_2 but also degrade TC simultaneously [86]. An attempt is also made by fusing the optoelectric properties of AgNPs with those of MoS_2/RGO/$NiWO_4$ (Ag-MRGON) heterostructures to design a Z-scheme-based photosystem with enhanced light absorption capacity. The applied bias to photocurrent efficiency (ABPE) is reported as 0.52% and 17.3 times, and 4.3 times better than the pure MoS_2 and MoS_2/$NiWO_4$ photoanode, respectively. The photocurrent density of Ag-MRGON was increased to 3.5 mA/cm^2 and the charge recombination was controlled using AgNPs as photo enhancers and RGO as electron mediators to reach photo stability of about 2 h. Furthermore, the photocurrent onset potential of the Ag-MRGON heterojunction (i.e., 0.61 V_{RHE}) cathodic was reduced in comparison to $NiWO_4$ (0.83 V_{RHE}), MoS_2 (0.80 V_{RHE}), and $MoS_2/NiWO_4$ heterojunction (0.73 V_{RHE}). In similar lines, a hybrid nanostructures of $BiVO_4$/Ag/rGO is shown for improved PEC-WS performance over $BiVO_4$. The system exhibited an IPCE of 3.5% and solar to hydrogen (STH) of 0.9% and was successful in degrading MB and phenol [34]. Another hydrothermally synthesized Z-scheme system $ZnIn_2S_4$/RGO/$BiVO_4$ was shown hydrogen production and simultaneous organic pollutant degradation. The organic contaminants can encourage the production of hydrogen from the breakdown of water. The hydrogen production rate, formaldehyde degradation rate, apparent quantum yield (A.Q.Y.) and the n_c are calculated to be 1687 mol g^{-1} h^{-1}, 22.2 mmol g^{-1} h^{-1}, 22.9%, and 13.3%, respectively, at a formaldehyde concentration of 5 mol l^{-1}, catalyst concentration of 1.5 g l^{-1} and pH 13 [118]. The system can also have good activity for hydrogen production and organic pollutant breakdown when mixed organic pollutants are supplied.

Similarly, Wei et al. worked on the synthesis of bismuth-doped g-C_3N_4 nanohybrid through a two-stage solvothermal technique. In accordance with the concept of turning waste into energy, efficiencies were quantified by measuring the evolution of hydrogen from antibiotic wastewater. The optimized sample (Bi/g-C_3N_4) showed 35.9 μmol/h of H_2 evolution rate under visible light that is about three times of bare g-C_3N_4. Results showed that it is possible to combine hydrogen evolution with concurrent antimicrobial wastewater degradation. Additionally, the samples displayed excellent stability throughout cycling tests after being zanalyzed by XRD and XPS descriptions [99].

The mixture of graphene and these hybrids can enhance the unique qualities of graphene and produce new qualities, such as effective charge separation and transfer properties, in addition to excellent stability, which can increase the overall photocatalytic efficiency in water splitting for the production of O_2 and H_2 [100, 102]. However, several obstacles still exist, even though significant advancement has been made recently. First off, the microstructure of the graphene and TMDCs composites plays a significant role in determining their properties and uses. Therefore, the composites must be designed with great caution to achieve great quality, high equivalent morphology on the nanoscale and better photocatalytic capabilities. Second, the absence of hydrophilic functional groups on pure graphene sheets presents significant hurdles for the fabrication of graphene-based nanomaterials or high-quality graphene. As a result, it is important to improve suitable reduction processes to lessen flaws in graphene sheets while maintaining their conductivity. Thirdly, in TMDCs materials with low coordination, the atoms near the grain boundaries or edges are usually very active for electrochemical interaction. However, the density of active sites and inadequate electronic transport severely restrict their electrochemical activity. To get around these restrictions, TMDCs with extremely active sites must be zsynthesized and zhybridized with extremely conductive materials like graphene. Last but not least, it is not entirely clear how adding graphene, and TMDCs improves photocatalytic performance. Therefore, effective fusion of theoretical calculations and actual results will necessitate several investigations to grasp the enhancing mechanism of graphene-based composite photocatalysts in wastewater treatment and hydrogen generation [46].

10.5.4 MXenes

With the discovery of a new class of 2D materials, MXenes, several reports have been obtained on the increased PEC efficiency of the photocatalysts containing them. This is because the 2D layered structure of MXenes can expose multiple unsaturated metal sites (for example, Ti, Nb, or V), leading to higher redox reactivity than single elemental carbon materials [107]. Moreover, the hydrophilic functional groups (such as - ^{-}OH, F and O) present on the surface of MXenes are advantageous for binding to various device components such as semiconductors and the long-term stabilization of 2D MXenes in aqueous solution [24]. MXene also possesses strong metallic conductivity, which, like graphene, also contributes to efficient charge transfer. Due to the MXene family's exceptional features, 2D MXenes are prospective building blocks for effective photoelectrodes. Monolayer MXene has a hexagonal lattice with a rhombohedral unit cell, similar to graphene in the top view. However, MXene materials that function as HER electrodes need a large overpotential to split water.

Consequently, a workable solution to address this problem is to integrate MXene photoelectrode with other functional components [46]. MXene hybrids are highly suitable for PC/PEC water splitting to capture sunlight due to their effective H^+ adsorption, suitable EC water splitting capacity, and close matching of Fermi energy (E_F) positions with H^+/H_2 reduction potentials. PC processes, in contrast to EC, demand semiconductor photo absorbers, whereelectron–hole pairs are produced. Therefore, other photoactive materials must be combined with MXenes since most MXenes are metallic and not semiconducting.

Lately, nano carbons have been incorporated into MXene-based systems to enhance interfacial bonding, increase photogenerated charge carrier separation and improve injection efficiency. These results indicate that 2D-MXene/Nano carbons are one of the most promising materials for photoelectrodes in PEC systems for effective renewable energy conversion. But, 2D-MXenes still lack the expected photo response to generate photo-induced electrons and holes, which is important for photovoltaic PEC-HERs, and thus nowadays researchers are finding for 2D-MXene/nanocarbon photocathodes. Further progress in PEC-HER activities is needed.

Due to the excellent metallic conductivity of 2D MXenes/nanocarbons, it is necessary to add photoactive components to create triad photocathodes by interacting with them as cocatalysts. MXene hybrids have been used as versatile PEC-HER, OER, and water-splitting catalysts in several publications. Lin et al. reported a $Ti_3C_2T_x$/InGaN photoanode with a tenfold increase in HER photocurrent density with the addition of MXene, confirming the significant HER activity on the MXene surface [57].

The TTL($Ti_3C_2T_x$/TiO_2/NiFeCo-LDH)/$BiVO_4$ photoanode presented a much better photocurrent density of 2.25 mA /cm^2 at 1.23 V (vs. RHE) in comparison to the pristine $BiVO_4$ (0.39 mA/cm^{-2}) in the potential range of 0.2 to 1.4 V (vs. RHE), which is about five times more than the pristine $BiVO_4$ [32]. A $Ti_3C_2T_x$ system is shown in combination with TiO_2 as an efficient photoanode over bare TiO_2. The $Ti_3C_2T_x$/TiO_2 as an OER photoanode exhibited a sixfold increase in the ABPE. The photoanode Ti_3C_2/TiO_2 showed a super high photocurrent density of 1.94 mA/cm^2 at 1.23 V versus RHE, which is six times higher than bare TiO_2 [14]. The MXene surface functioned as a hole scavenger and OER active site in this instance, where PEC OER took place [55]. MXenes are primarily used as conductive supports and electronic modulators since, as previously mentioned, they typically only have minimal OER activity. As a result, they must be zhybridized with OER-active catalysts (such TMDC and LDHs) to produce OER. To benefit from MXenes' HER activity, most PC/PEC applications for MXenes and their hybrids are focused on HER rather than OER. However, MXene systems are not yet studied in wastewater environments for simultaneous treatment and hydrogen generation, which requires detailed study to establish their potential in this domain.

10.6 Conclusions and Future Prospective

In the last decade, the utilization of 2D materials have been established for the PEC treatment of wastewater and the production of hydrogen simultaneously. These PEC studies showed the effects of pH, kind of electrolyte, applied bias, and pollutant content-like variables at the rate of hydrogen generation and removal efficiency of pollutants which confirmed the requirement of optimization in the fabrication of 2D photocatalysts. Due to solar energy's renewable significance, research on light sources has steadily shifted from ultraviolet (UV) to visible light and sunshine. The fabrication of several nanostructured, doped and composite semiconductor electrodes like TMOs, TMDs, graphene and MXenes with improved ability made it possible to absorb visible light and utilize solar energy. However, irrespective of the good performances of laboratory-level experimental outcomes, further research is certainly required to improve PEC technology to remove pollutants and produce hydrogen at an industrial scale. The development of extremely effective large-scale photocatalysts and fabricating adaptable PEC reactors according to real samples is still a challenge that needs the focus of researchers. Also, there is a need to develop hybrid 2D materials which can absorb near-infrared light that cover the largest area of the sun-spectrum. There is no question that PEC's forthcoming studies on wastewater management will be crucial, but there is still a long way to go. It is expected that the PEC technique can successfully use solar power to simultaneously treat pollutant wastewater and generate hydrogen on industrial scale.

Acknowledgements Authors are thankful to Director, CSIO for his kind permission. The funding support under SERB WEA Project WEA/2020/000022 is highly acknowledged. PS and SS are thankful to GATE-JRF and DST INSPIRE, respectively for doctoral study fellowship.

References

1. A. Fujishima, K.Honda, Electrochemical photolysis of water at a semiconductor electrode. Nature **238**, 37–38 (1972)
2. F.A. Al-Agel, J. Suleiman, S.A. Khan, Studies on silicon quantum dots prepared at different working pressure. Results Phys. **7**, 1128–1134 (2017). https://doi.org/10.1016/j.rinp.2017.03.002
3. S.O. Alfaro, A. Martínez-De La Cruz, Synthesis, characterization and visible-light photocatalytic properties of Bi2WO6 and Bi2W2O9 obtained by co-precipitation method. Appl. Catal. A Gen **383**(1–2), (2010). https://doi.org/10.1016/j.apcata.2010.05.034
4. M.Y. Azwar, M.A. Hussain, A.K. Abdul-Wahab, Development of biohydrogen production by photobiological, fermentation and electrochemical processes: a review. Renew. Sustain. Energy Rev. **31**, 158–173 (2014). https://doi.org/10.1016/J.RSER.2013.11.022
5. D.S. Babu et al., Detoxification of water and wastewater by advanced oxidation processes. Sci. Total Environ. (2019). https://doi.org/10.1016/j.scitotenv.2019.133961
6. G.R. Bhimanapati et al., Recent advances in two-dimensional materials beyond graphene. ACS Nano (2015). https://doi.org/10.1021/acsnano.5b05556
7. X. Bing et al., Biomimetic synthesis of Bi2O3/Bi2WO6/MgAl-CLDH hybrids from lotus pollen and their enhanced adsorption and photocatalysis performance J. Photochem. Photobiol. A Chem. **364** (2018). https://doi.org/10.1016/j.jphotochem.2018.06.030
8. P. Blake et al., Making graphene visible. Appl. Phys. Lett. **91**(6) (2007). https://doi.org/10.1063/1.2768624.
9. E. Brown, M.W. Skougstad, M.J. Fishman, Methods for collection and analysis of water samples for dissolved minerals and gases, in *Techniques of Water-Resources Investigations 05-A1* [Preprint] (1970)
10. A. Castellanos-Gomez et al., Single-layer MoS2 mechanical resonators. Adv. Mater. **25**(46) (2013). https://doi.org/10.1002/adma.201303569
11. G. Chehade et al., A photoelectrochemical system for hydrogen and chlorine production from industrial waste acids. Sci. Total Environ. **712** (2020). https://doi.org/10.1016/j.scitotenv.2019.136358
12. W. Chen et al., Direct Z-scheme 2D/2D MnIn2S4/g-C3N4 architectures with highly efficient photocatalytic activities towards treatment of pharmaceutical wastewater and hydrogen evolution. Chem. Eng. J. **359** (2019). https://doi.org/10.1016/j.cej.2018.11.141
13. Chen, X. *et al.* (2018) 'RuII Photosensitizer-Functionalized Two-Dimensional MoS2 for Light-Driven Hydrogen Evolution', *Chemistry - A European Journal*, 24(2). doi:https://doi.org/10.1002/chem.201705203.
14. X. Chen et al., Ti3C2 MXene quantum dots/TiO2 inverse opal heterojunction electrode platform for superior photoelectrochemical biosensing. Sens. Actuators B Chem. **289**(March), 131–137 (2019). https://doi.org/10.1016/j.snb.2019.03.052
15. Y. Chen et al., Single-crystalline orthorhombic molybdenum oxide nanobelts: Synthesis and photocatalytic properties. CrystEngComm **12**(11) (2010). https://doi.org/10.1039/c000744g
16. Y. Chen et al., Photoelectrocatalytic oxidation of metal-EDTA and recovery of metals by electrodeposition with a rotating cathode. Chem. Eng. J. **324** (2017). https://doi.org/10.1016/j.cej.2017.05.031
17. M. Chhowalla et al., The chemistry of two-dimensional layered transition metal dichalcogenide nanosheets. Nat. Chem. (2013). https://doi.org/10.1038/nchem.1589

18. M.B. Dines, Lithium intercalation via n-Butyllithium of the layered transition metal dichalcogenides. Mater. Res. Bull. **10**(4) (1975). https://doi.org/10.1016/0025-5408(75)90115-4
19. M.H. Do et al., Challenges in the application of microbial fuel cells to wastewater treatment and energy production: a mini review. Sci. Total Environ. **639**, 910–920 (2018). https://doi.org/10.1016/J.SCITOTENV.2018.05.136
20. X. Duan et al., Two-dimensional transition metal dichalcogenides as atomically thin semiconductors: opportunities and challenges. Chem. Soc. Rev. (2015). https://doi.org/10.1039/c5cs00507h
21. M. Fekete et al., Photoelectrochemical water oxidation by screen printed ZnO nanoparticle films: effect of pH on catalytic activity and stability. Nanoscale **6**(13), 7585–7593 (2014). https://doi.org/10.1039/c4nr01935k
22. A. Fujishima, T.N. Rao, D.A. Tryk, Titanium dioxide photocatalysis. J. Photochem. Photobiol. C (2000). https://doi.org/10.1016/S1389-5567(00)00002-2
23. A. Fujishima, X. Zhang, D.A. Tryk, TiO2 photocatalysis and related surface phenomena. Surf. Sci. Rep. (2008). https://doi.org/10.1016/j.surfrep.2008.10.001
24. G. Gao, A.P. O'Mullane, A. Du, 2D MXenes: a new family of promising catalysts for the hydrogen evolution reaction. ACS Catal. **7**(1), 494–500 (2017). https://doi.org/10.1021/acscatal.6b02754
25. U.M. García-Pérez, A. Martínez-De La Cruz, J. Peral, Transition metal tungstates synthesized by co-precipitation method: Basic photocatalytic properties Electrochim. Acta **81** (2012). https://doi.org/10.1016/j.electacta.2012.07.045
26. R.K. Gautam, M.C. Chattopadhyaya, Functionalized magnetic nanoparticles: adsorbents and applications BT-nanomaterials for wastewater remediation, in *Nanomaterials for Wastewater Remediation* (2016)
27. D. Gielen, W. Bank, *Hydrogen: A Renewable Energy Perspective Some of the authors of this publication are also working on these related projects: A Global Renewable Energy Roadmap: Comparing Energy Systems Models with IRENA's REmap 2030 Project View project Innovation for the* (no date)
28. J. Gong, Y. Lai, C. Lin, Electrochemically multi-anodized TiO2 nanotube arrays for enhancing hydrogen generation by photoelectrocatalytic water splitting. Electrochim. Acta **55**(16), 4776–4782 (2010). https://doi.org/10.1016/j.electacta.2010.03.055
29. T.T. Guaraldo et al., On the application of nanostructured electrodes prepared by Ti/TiO2/WO3 "template": a case study of removing toxicity of indigo using visible irradiation. Chemosphere **91**(5) (2013). https://doi.org/10.1016/j.chemosphere.2012.12.027
30. J.S. Guest et al., A new planning and design paradigm to achieve sustainable resource recovery from wastewater. Environ. Sci. Technol. **43**(16), 6126–6130 (2009). https://doi.org/10.1021/es9010515
31. M. Hadnadjev-Kostic, T. Vulic, R. Marinkovic-Neducin, Solar light induced rhodamine B degradation assisted by TiO2-Zn-Al LDH based photocatalysts. Adv Powder Technol. **25**(5) (2014). https://doi.org/10.1016/j.apt.2014.05.015
32. N. Hao et al., In situ hybridization of an MXene/TiO2/NiFeCo-layered double hydroxide composite for electrochemical and photoelectrochemical oxygen evolution. RSC Adv. **8**(37), 20576–20584 (2018). https://doi.org/10.1039/c8ra02349b
33. F. Haque et al., Two-dimensional transition metal oxide and chalcogenide-based photocatalysts. Nano-Micro Lett. (2018). https://doi.org/10.1007/s40820-017-0176-y
34. A.H. Hendi et al., Visible light-driven photoelectrocatalytic water splitting using Z-scheme Ag-decorated MoS2/RGO/NiWO4 heterostructure. ACS Omega **5**(49), 31644–31656 (2020). https://doi.org/10.1021/acsomega.0c03985
35. M.D. Hernández-Alonso et al., Development of alternative photocatalysts to TiO2: challenges and opportunities. Energy Environ. Sci. (2009). https://doi.org/10.1039/b907933e
36. Y. Hernandez et al., High-yield production of graphene by liquid-phase exfoliation of graphite. Nat. Nanotechnol. **3**(9) (2008). https://doi.org/10.1038/nnano.2008.215

37. M.C. Hsiao et al., Ultrathin 1T-phase MoS2 nanosheets decorated hollow carbon microspheres as highly efficient catalysts for solar energy harvesting and storage J. Power Sources **345** (2017). https://doi.org/10.1016/j.jpowsour.2017.01.132
38. H.T. Hsu et al., Enhanced photocatalytic activity of chromium(VI) reduction and EDTA oxidization by photoelectrocatalysis combining cationic exchange membrane processes. J. Hazard. Mater. **248–249**(1) (2013). https://doi.org/10.1016/j.jhazmat.2012.12.058
39. J.S. Hu et al., Mass production and high photocatalytic activity of ZnS nanoporous nanoparticles. Angew. Chemie-Int. Edition **44**(8) (2005). https://doi.org/10.1002/anie.200462057
40. C.W. Huang et al., Hydrogen generation from photocatalytic water splitting over TiO 2 thin film prepared by electron beam-induced deposition. Int. J. Hydrogen Energy **35**(21), 12005–12010 (2010). https://doi.org/10.1016/j.ijhydene.2010.08.113
41. G. Huang et al., Nb2O5/ZnAl-LDH composites and its calcined products for photocatalytic degradation of congo red under visible light irradiation Mater. Lett. **173** (2016). https://doi.org/10.1016/j.matlet.2016.03.073
42. Q. Ji et al., Chemical vapour deposition of group-VIB metal dichalcogenide monolayers: engineered substrates from amorphous to single crystalline. Chem. Soc. Rev. (2015). https://doi.org/10.1039/c4cs00258j
43. B. Jia et al., Study of NO removal and resistance to SO2 and H2O of MnOx/TiO2, MnOx/ZrO2 and MnOx/ZrO2–TiO2. Appl. Catal. A **553** (2018). https://doi.org/10.1016/j.apcata.2017.12.016
44. H. Jiang, M. Nagai, K. Kobayashi, Enhanced photocatalytic activity for degradation of methylene blue over V2O5/BiVO4 composite. J. Alloys Compd. **479**(1–2) (2009). https://doi.org/10.1016/j.jallcom.2009.01.051
45. J. Joy, J. Mathew, S.C. George, Nanomaterials for photoelectrochemical water splitting–review. Int. J. Hydrogen Energy **43**(10), 4804–4817 (2018). https://doi.org/10.1016/j.ijhydene.2018.01.099
46. Z.B. Khalid et al., Application of 2D graphene-based nanomaterials for pollutant removal from advanced water and wastewater treatment processes. ACS Symp. Ser. **1353**(August), 191–217 (2020). https://doi.org/10.1021/bk-2020-1353.ch009
47. G. Kianpour et al., Facile synthesis and characterization of nickel molybdate nanorods as an effective photocatalyst by co-precipitation method. J. Mater. Sci. Mater. Electron. **27**(10) (2016). https://doi.org/10.1007/s10854-016-5103-3
48. J.Y. Kim et al., Charge transfer in iron oxide photoanode modified with carbon nanotubes for photoelectrochemical water oxidation: an electrochemical impedance study. Int. J. Hydrogen Energy **36**(16), 9462–9468 (2011). https://doi.org/10.1016/j.ijhydene.2011.05.046
49. N. Kumar et al., Growth of highly crystalline and large scale monolayer MoS2 by CVD: the role of substrate position. Cryst. Res. Technol. **53**(6) (2018). https://doi.org/10.1002/crat.201800002
50. F. Lei et al. Oxygen vacancies confined in ultrathin indium oxide porous sheets for promoted visible-light water splitting. J. Am. Chem. Soc. **136**(19) (2014). https://doi.org/10.1021/ja501866r
51. H. Li et al., Preparation and applications of mechanically exfoliated single-layer and multilayer MoS2 and WSe2 nanosheets. Accounts Chem. Res. **47**(4) (2014). https://doi.org/10.1021/ar4002312
52. X. Li et al., Large-area synthesis of high-quality and uniform graphene films on copper foils. Science **324**(5932) (2009). https://doi.org/10.1126/science.1171245
53. Y. Li, T. White, S.H. Lim, Structure control and its influence on photoactivity and phase transformation of TiO2 nano-particles. Rev. Adva. Mater. Sci. **5**(3) (2003)
54. G. Liao et al., Emerging polymeric carbon nitride Z-scheme systems for photocatalysis. Cell Rep. Phys. Sci. **2**(3), 100355 (2021). https://doi.org/10.1016/j.xcrp.2021.100355
55. K.R.G. Lim et al., Rational Design of Two-Dimensional Transition Metal Carbide/Nitride (MXene) Hybrids and Nanocomposites for Catalytic Energy Storage and Conversion. ACS Nano **14**(9), 10834–10864 (2020). https://doi.org/10.1021/acsnano.0c05482

56. C.Y. Lin et al., Fermentative hydrogen production from wastewaters: a review and prognosis. Int. J. Hydrog. Energy (2012). https://doi.org/10.1016/j.ijhydene.2012.02.072
57. J. Lin et al., A novel approach for achieving high-efficiency photoelectrochemical water oxidation in InGaN nanorods grown on Si system: MXene nanosheets as multifunctional interfacial modifier. Adv. Func. Mater. **30**(13), 1–11 (2020). https://doi.org/10.1002/adfm.201910479
58. J. Liu et al., ZnCr-LDH/N-doped graphitic carbon-incorporated g-C3N4 2D/2D nanosheet heterojunction with enhanced charge transfer for photocatalysis. Mater. Res. Bull. **102** (2018). https://doi.org/10.1016/j.materresbull.2018.03.010
59. Q. Liu et al., BiOCl and TiO2 deposited on exfoliated ZnCr-LDH to enhance visible-light photocatalytic decolorization of Rhodamine B. Ceram. Int. **43**(7) (2017). https://doi.org/10.1016/j.ceramint.2017.01.119
60. J. Low et al., Heterojunction photocatalysts. Adv. Mater. **29**(20), 1–20 (2017). https://doi.org/10.1002/adma.201601694
61. S.D. Melvin, F.D.L. Leusch, Removal of trace organic contaminants from domestic wastewater: a meta-analysis comparison of sewage treatment technologies. Environ. Int. (2016). https://doi.org/10.1016/j.envint.2016.03.031
62. H. Miao et al., A novel strategy to prepare 2D g-C3N4nanosheets and their photoelectrochemical properties. J. Alloy. Compd. **690**, 669–676 (2017). https://doi.org/10.1016/j.jallcom.2016.08.184
63. M. Mureseanu et al., Green synthesis of g-C3N4/CuONP/LDH composites and derived g-C3N4/MMO and their photocatalytic performance for phenol reduction from aqueous solutions. Appl. Clay Sci. **141** (2017). https://doi.org/10.1016/j.clay.2017.02.012
64. F. Murphy et al., Wastewater treatment works (WwTW) as a source of microplastics in the aquatic environment. Environ. Sci. Technol. **50**(11), 5800–5808 (2016). https://doi.org/10.1021/acs.est.5b05416
65. V. Nicolosi, (Invited) Liquid-phase exfoliated two-dimensional nanosheets: from large scale production to energy storage applcations, in *ECS Meeting Abstracts*, MA2016–01(26) (2016). https://doi.org/10.1149/ma2016-01/26/1328
66. F. Ning et al., TiO2/graphene/NiFe-layered double hydroxide nanorod array photoanodes for efficient photoelectrochemical water splitting. Energy Environ. Sci. **9**(8), 2633–2643 (2016). https://doi.org/10.1039/c6ee01092j
67. J.Z. Ou et al., Anodic formation of a thick three-dimensional nanoporous WO3 film and its photocatalytic property. Electrochem. Commun. **27** (2013). https://doi.org/10.1016/j.elecom.2012.11.009
68. K. Ouyang et al., A novel visible-light responsive photocatalytic fuel cell with a highly efficient BiVO4/WO3 inverse opal photoanode and a MnO2/graphene oxide nanocomposite modified cathode. Int. J. Hydrogen Energy **44**(14), 7288–7299 (2019). https://doi.org/10.1016/J.IJHYDENE.2019.01.241
69. M.P. Shah, A.K. Pate, Optimization of environmental parameters on microbial degradation of reactive black dye. J. Bioremediation Biodegrad. **4**(3) (2013). https://doi.org/10.4172/2155-6199.1000183
70. S. Pahra et al., A noble metal-free candle soot derived carbon electrocatalyst for simultaneous H2 generation and wastewater treatment. J. Phys. Chem. Solids **173**, 111106 (2023). https://doi.org/10.1016/J.JPCS.2022.111106
71. S. Pahra, S. Sharma, P. Devi, Fundamental understanding and figure of merits for electrocatalytic and photoelectrocatalytic H_2 production, in *Green Energy Harvesting* (2022), pp. 46–74. https://doi.org/10.1002/9781119776086.CH3
72. S. Panimalar et al., Studies of MnO2/g-C3N4 hetrostructure efficient of visible light photocatalyst for pollutants degradation by sol-gel technique. Surf. Interfaces **20** (2020). https://doi.org/10.1016/j.surfin.2020.100512
73. M. Parashar, V.K. Shukla, R. Singh, Metal oxides nanoparticles via sol–gel method: a review on synthesis, characterization and applications. J. Mater. Sci. Mater. Electron. (2020). https://doi.org/10.1007/s10854-020-02994-8

74. S.P. Paredes et al., Sol-gel synthesis of hydrotalcite-like compounds. J. Mater. Sci. **41**(11) (2006). https://doi.org/10.1007/s10853-005-5347-4
75. F. Petronella, R. Comparelli, Nanomaterials in photo (Electro) catalysis. Catalysts **11**(2), 1–4 (2021). https://doi.org/10.3390/catal11020149
76. S. Pokrant et al., Size effects of cocatalysts in photoelectrochemical and photocatalytic water splitting. Mater. Today Energy **5**, 158–163 (2017). https://doi.org/10.1016/j.mtener.2017.06.005
77. S. Popli, U.D. Patel, Destruction of azo dyes by anaerobic–aerobic sequential biological treatment: a review. Int. J. Environ. Sci. Technol. (2015). https://doi.org/10.1007/s13762-014-0499-x
78. C. Prasad et al., An overview of semiconductors/layered double hydroxides composites: properties, synthesis, photocatalytic and photoelectrochemical applications. J. Mol. Liq. (2019). https://doi.org/10.1016/j.molliq.2019.111114
79. A.V. Rane et al., Methods for synthesis of nanoparticles and fabrication of nanocomposites, in *Synthesis of Inorganic Nanomaterials: Advances and Key Technologies* (2018). https://doi.org/10.1016/B978-0-08-101975-7.00005-1
80. B.G. Rao, D. Mukherjee, B.M. Reddy, Novel approaches for preparation of nanoparticles, in *Nanostructures for Novel Therapy: Synthesis Characterization and Applications* (2017). https://doi.org/10.1016/B978-0-323-46142-9.00001-3
81. R.S. Lokhande, U.P. Singare, S.D. Pimple, Study on physico-chemical parameters of waste water effluents from Taloja industrial area of Mumbai, India. Int. J. Ecosyst. **1**(1) (2012). https://doi.org/10.5923/j.ije.20110101.01
82. S. Samsami et al., Recent advances in the treatment of dye-containing wastewater from textile industries: overview and perspectives. Process Saf. Environ. Prot. (2020). https://doi.org/10.1016/j.psep.2020.05.034
83. A.B. dos Santos, F.J. Cervantes, J.B. van Lier, Review paper on current technologies for decolourisation of textile wastewaters: perspectives for anaerobic biotechnology. Biores. Technol. (2007). https://doi.org/10.1016/j.biortech.2006.11.013
84. M. Shakeel et al., Layered by layered Ni-Mn-LDH/g-C3N4 nanohybrid for multi-purpose photo/electrocatalysis: morphology controlled strategy for effective charge carriers separation. Appl. Catal. B **242** (2019). https://doi.org/10.1016/j.apcatb.2018.10.005
85. K.P. Sharma et al., A comparative study on characterization of textile wastewaters (untreated and treated) toxicity by chemical and biological tests. Chemosphere **69**(1) (2007). https://doi.org/10.1016/j.chemosphere.2007.04.086
86. H. Shen et al., 2D/2D/3D architecture Z-scheme system for simultaneous H2 generation and antibiotic degradation. Fuel **280**(July), 1–9 (2020). https://doi.org/10.1016/j.fuel.2020.118618
87. L. Shi et al., Annealing temperature effects on photoelectrochemical performance of bismuth vanadate thin film photoelectrodes. RSC Adv. **8**(51), 29179–29188 (2018). https://doi.org/10.1039/c8ra04887h
88. Y. Shi, H. Li, L.J. Li, Recent advances in controlled synthesis of two-dimensional transition metal dichalcogenides via vapour deposition techniques. Chem. Soc. Rev. (2015). https://doi.org/10.1039/c4cs00256c
89. P. Singh et al., Recent advances in bacteriorhodopsin-based energy harvesters and sensing devices. Nano Energy (2021). https://doi.org/10.1016/j.nanoen.2020.105482
90. S. Singla et al., BiVO4/MoSe2 Photocatalyst for the photocatalytic Abatement of tetracycline and photoelectrocatalytic water splitting. Mater. Chem. Phys. **295**, 127111 (2023). https://doi.org/10.1016/J.MATCHEMPHYS.2022.127111
91. L. Song et al., Large scale growth and characterization of atomic hexagonal boron nitride layers. Nano Lett. **10**(8) (2010). https://doi.org/10.1021/nl1022139
92. W. Sun et al. Controlled synthesis of 2D Mo2C/graphene heterostructure on liquid Au substrates as enhanced electrocatalytic electrodes. Nanotechnology **30**(38) (2019). https://doi.org/10.1088/1361-6528/ab2c0d

93. Y. Sun et al., Fabrication of flexible and freestanding zinc chalcogenide single layers. Nat. Commun. **3** (2012). https://doi.org/10.1038/ncomms2066
94. Y. Sun et al., Pits confined in ultrathin cerium(IV) oxide for studying catalytic centers in carbon monoxide oxidation. Nat. Commun. **4** (2013). https://doi.org/10.1038/ncomms3899
95. L. Tang, L. Liu, F. Yang, FeMoO4-graphene oxide photo-electro-catalyst for berberine removal and hydrogen evolution. Int. J. Hydrogen Energy **44**(36), 19755–19761 (2019). https://doi.org/10.1016/j.ijhydene.2019.03.078
96. T.N. Trung et al., Enhanced photoelectrochemical activity in the heterostructure of vertically aligned few-layer MoS2 flakes on ZnO Electrochim. Acta **260** (2018). https://doi.org/10.1016/j.electacta.2017.11.089
97. H. Wang et al., Simultaneous removal of phenol and Cr(VI) by TiO2 nanotube array photoelectrocatalysis. Cuihua Xuebao/Chin. J. Catal. 32(4) (2011). https://doi.org/10.3724/SP.J.1088.2011.01246
98. J. Wang et al., Band structure tuning of TiO2 for enhanced photoelectrochemical water splitting. J. Phys. Chem. C **118**(14), 7451–7457 (2014). https://doi.org/10.1021/jp5004775
99. Z. Wei et al., Photocatalytic hydrogen evolution with simultaneous antibiotic wastewater degradation via the visible-light-responsive bismuth spheres-g-C3N4 nanohybrid: waste to energy insight. Chem. Eng. J. **358** (2019). https://doi.org/10.1016/j.cej.2018.10.096
100. M. Xu et al., Graphene-like two-dimensional materials. Chem. Rev. **113**(5), 3766–3798 (2013). https://doi.org/10.1021/cr300263a
101. Q. Xu et al., S-scheme heterojunction photocatalyst. Chem **6**(7), 1543–1559 (2020). https://doi.org/10.1016/j.chempr.2020.06.010
102. Y. Xu et al., Synthesis of ultrathin CdS nanosheets as efficient visible-light-driven water splitting photocatalysts for hydrogen evolution. Chem. Commun. **49**(84) (2013). https://doi.org/10.1039/c3cc46342g
103. D.A. Yaseen, M. Scholz, Shallow pond systems planted with Lemna minor treating azo dyes. Ecol. Eng. **94** (2016). https://doi.org/10.1016/j.ecoleng.2016.05.081
104. S. Ye et al., Simultaneous removal of organic pollutants and heavy metals in wastewater by photoelectrocatalysis: a review. Chemosphere **273**(xxxx), 128503 (2021). https://doi.org/10.1016/j.chemosphere.2020.128503
105. Z. Yin et al., Au nanoparticle-modified MoS2 nanosheet-based photoelectrochemical cells for water splitting. Small **10**(17), 3537–3543 (2014). https://doi.org/10.1002/smll.201400124
106. J. Yu, X. Yu, Hydrothermal synthesis and photocatalytic activity of zinc oxide hollow spheres. Environ. Sci. Technol. **42**(13) (2008). https://doi.org/10.1021/es800036n
107. X. Yu et al., Ti3C2 MXene nanoparticles modified metal oxide composites for enhanced photoelectrochemical water splitting. Int. J. Hydrogen Energy **44**(5), 2704–2710 (2019). https://doi.org/10.1016/j.ijhydene.2018.11.221
108. C. Zaharia, D. Suteu, *Characteristics, Polluting Effects and Separation/Elimination Procedures from Industrial Effluents – A Critical Overview, Organic Pollutants Ten Years After the Stockholm Convention - Environmental and Analytical, Organic Pollutants Ten Years After the Stockholm Convention - Environmental and Analytical Update* (2012)
109. Z. Zeng et al., Single-layer semiconducting nanosheets: High-yield preparation and device fabrication. Angew. Chemie-Int. Edition **50**(47) (2011). https://doi.org/10.1002/anie.201106004
110. Z. Zeng et al., An effective method for the fabrication of few-layer-thick inorganic nanosheets. Angew. Chemie-Int. Edition **51**(36) (2012). https://doi.org/10.1002/anie.201204208
111. G. Zhang, Z. Wang, J. Wu, Construction of a Z-scheme heterojunction for high-efficiency visible-light-driven photocatalytic CO2reduction. Nanoscale **13**(8), 4359–4389 (2021). https://doi.org/10.1039/d0nr08442e
112. W. Zhao et al., Fabrication of a novel p-n heterojunction photocatalyst n-BiVO4@p-MoS2 with core-shell structure and its excellent visible-light photocatalytic reduction and oxidation activities. Appl. Catal. B **185** (2016). https://doi.org/10.1016/j.apcatb.2015.12.023
113. X. Zhao et al., Photoelectrocatalytic oxidation of CuII-EDTA at the TiO 2 electrode and simultaneous recovery of CuII by electrodeposition. Environ. Sci. Technol. **47**(9) (2013). https://doi.org/10.1021/es3046982

114. J. Zheng et al., High yield exfoliation of two-dimensional chalcogenides using sodium naphthalenide. Nat. Commun. **5** (2014). https://doi.org/10.1038/ncomms3995
115. L. Zheng et al., Hierarchical MoS2 nanosheet@TiO2 nanotube array composites with enhanced photocatalytic and photocurrent performances. Small **12**(11) (2016). https://doi.org/10.1002/smll.201503441
116. Z. Zheng, I.M.C. Lo., Fabrication of MoS2@BL-BiVO4 photoanode with promoted charge separation for photoelectrochemical sewage treatment to simultaneously degrade PPCPs, disinfect E. coli, and produce H2: Performance, mechanisms, and influence factors. Appl. Catal. B **299** (2021). https://doi.org/10.1016/j.apcatb.2021.120636
117. C. Zhi et al., Large-scale fabrication of boron nitride nanosheets and their utilization in polymeric composites with improved thermal and mechanical properties. Adv. Mater. **21**(28) (2009). https://doi.org/10.1002/adma.200900323
118. R. Zhu et al., Z scheme system ZnIn2S4/RGO/BiVO4 for hydrogen generation from water splitting and simultaneous degradation of organic pollutants under visible light. Renew. Energy **139**, 22–27 (2019). https://doi.org/10.1016/j.renene.2019.02.049
119. X. Zong et al., Enhancement of photocatalytic H2 evolution on CdS by loading MoS2 as cocatalyst under visible light irradiation. J. Am. Chem. Soc. **130**(23) (2008). https://doi.org/10.1021/ja8007825
120. X. Zong et al., Photocatalytic H2 evolution on MoS2/CdS catalysts under visible light irradiation. J. Phys. Chem. C 114(4) (2010). https://doi.org/10.1021/jp904350e
121. X. Zong et al. Photocatalytic H2 evolution on CdS loaded with WS2 as cocatalyst under visible light irradiation. J. Phys. Chem. C 115(24) (2011). https://doi.org/10.1021/jp2006777

Chapter 11
Advances in 2D MOFs for Environmental Applications

Goksu Ozcelikay, Ahmet Cetinkaya, Merve Yence, Fatma Budak, S. Irem Kaya, and Sibel A. Ozkan

Abstract Toxic environmental pollutants include chemical contaminants such as heavy metals and pesticides. Moreover, pesticides are used for modern extensive agricultural practices, affecting human health badly. Meanwhile, pharmaceutical and industrial wastes, personal care products, and endocrine disruptors can be cited as chemical pollutants. Environmental monitoring has been important in providing better safety measures in various sectors of life. Fabricating a new sensing platform with high selectivity and sensitivity is the most needed environmental detection tool. Electrochemical sensors draw attention to on-site environmental analysis. They are selective and sensitive toward electroactive compounds, fast, moveable, and cheap. Due to its important properties, many researchers use two-dimensional metal–organic frameworks (2D MOFs). The applications of 2D MOFs are broad ranges, including sensors, catalysis, gas adsorption, gas separation, supercapacitor, and so on. 2D MOFs are based on the coordination of unique metal ions and organic ligands, including two-dimensional layered structures. The typical 2D materials consist of transition metal carbonitrides (Mxenes), graphene, transition metal disulfide compounds (TMDC), silene and carbon nitride, and hexagonal boron nitride. The chapter focused on 2D MOFs materials, sensor design strategies, and environmental application. Moreover, this chapter summarizes and evaluates recent advances in the development and application of 2D MOFs-based sensors for determining toxic environmental pollutants.

G. Ozcelikay · A. Cetinkaya · M. Yence · F. Budak · S. A. Ozkan (✉)
Faculty of Pharmacy, Department of Analytical Chemistry Ankara, Ankara University, Ankara, Türkiye
e-mail: ozkan@pharmacy.ankara.edu.tr

S. I. Kaya
Gulhane Faculty of Pharmacy, Department of Analytical Chemistry, University of Health Sciences, Ankara, Türkiye

N. Kumar et al. (eds.), *Two-Dimensional Materials for Environmental Applications*, Springer Series in Materials Science 332, https://doi.org/10.1007/978-3-031-28756-5_11

11.1 Introduction

A two-dimensional (2D) material is a type of substance in which electrons may only travel unrestrictedly (planar motion) in two dimensions at the nanoscale (1–100 nm). Numerous two-dimensional compounds have continuously been found since the successful separation of graphene from graphite in 2004. Reduced graphene oxide (rGO) is an example of a typical 2D material. It is a possible candidate material for electron carriers and features 2D honeycomb carbon lamellae. In the process of breaking down water, it may encourage the movement of photoexcited electrons [1]. The upcoming wave of nano-electrical technology is seen as a promising material. Because of its various benefits, graphite carbon nitride (g-C_3N_4) has emerged in this industry as a new type of visible light catalyst. However, the first g-C_3N_4 generally exhibits limited activity. The ability of the doped photocatalyst to generate photocatalytic hydrogen with visible light has been significantly improved with increased specific surface area, a smaller band gap, and more efficient carrier transfer [2]. The past few years have witnessed the rapid development of MXenes, a new class of 2D metal carbide consisting of transition metal nitrides, carbides, and carbonitrides. Another 2D material, an analog of graphene known as white graphite, hexagonal boron nitride (h-BN), has a hexagonal honeycomb structure and a smooth surface. The lattice structure is similar to that of graphene [3].

When metal ions or metal clusters are coordinated with organic ligands and have unlimited network architecture in 1D, 2D, and 3D, a type of coordination compound known as a metal–organic framework (MOF) is formed. They are also a group of coordination polymerized species, and their structures are often porous. Due to the pores in their structure, MOFs that can hold various molecules and have large voids can create different effects [4]. Catalysis, gas storage, gas adsorption/separation, chiral solubility, drug administration, bioimaging, sensing, etc., are a few application areas for MOF. In addition, MOFs are a novel porous material with a large specific surface area, which can be synthesized green and easily modified. It also has many potential applications [5]. 2D MOFs are an important subset of MOF materials but have been less studied. The main goal in 2D MOF design and synthesis is to prevent the growth of the third dimension without affecting its two-dimensional scale. The two main categories of 2D MOF synthesis strategies are top-down and bottom-up. The rational design and synthesis of 2D MOFs are extremely important as it determines the structure properties. Moreover, 2D MOFs have excellent application potential and properties comparable to those of other 2D materials, such as light transmittance and electrical conductivity [6, 7].

Toxic environmental pollutants include chemical pollutants such as heavy metals and pesticides. Moreover, pesticides are used for modern extensive agricultural practices, affecting human health badly. Meanwhile, pharmaceutical and industrial wastes, personal care products, and endocrine disruptors can be cited as chemical pollutants.

Environmental monitoring has been important in providing better safety measures in various sectors of life. Fabricating a new sensing platform with high selectivity and sensitivity is the most needed environmental detection tool. Electrochemical sensors draw attention to on-site environmental analysis. They are selective and sensitive toward electroactive compounds, fast, moveable, and cheap.

11.2 Overview of 2D MOF Materials

Porous MOF structures formed after the coordination of metal ions and organic binders are generally synthesized by interacting with metal salts and ligands in a solvent under a certain pH, temperature, and pressure. 2D MOFs were applied in different applications such as biosensors, separation, molecular imaging, biocatalysis, storage, cancer and other disease treatments, and energy conversion because 2D MOFs have multifunctional properties. Moreover, 2D-MOF was used for biomedical applications because of its tailored structure, biodegradation, and biocompatible [8].

The combination of the 2D-MOF structure with the 2D-layered structure (graphene, etc.) would create excellent opportunities and obtain maximum functionality. The 2D MOFs possess exposed active sites, various topologies, flexible porosity, and large surface area, creating great potential for various technological applications. Although 2D-MOF nanostructures have some advantages, several problems should be handled, such as low conductivity and fast aggregation. For this reason, 2D-MOF is used as singular materials and combined with hybrid structures. The carbonaceous nanomaterials such as CNTs, GO, and rGO can use with 2D-MOF. The 2D-MOF/hybrid materials contribute to the increase in conductivity and obstacle to aggregation [9].

M-TCPP nanofilm (M = Metal, TCPP = Tetrakis(4-carboxyphenyl)porphyrin)) was based on the interaction between metal nodes and TCPP nanofilm by van der Waals (vdW) force. M can be Co, Zn, Ni, and Cu. Therefore, the M-TCPP nanofilm can be characterized by TEM, SEM, EDS, and AFM (Fig. 11.1) [10].

The metal oxides/sulfides linked with MOF consisting of carbonaceous nanomaterials (Mn-MOF NSs) can be synthesized to obtain stable and porous 2D-MOF compounds in supercapacitor applications.

The metal–ligand ratio, metal ions/clusters, coordination number and geometry, structural configuration, electronic properties, and structure of ligands are parameters required for designing the 2D-MOF materials. The synthesized 2D-MOF structures can be evaluated in terms of their optical, electrical, chiral, catalytic, porosity, and magnetic properties [10].

Synthesis methods of MOF structures can be grouped under eight headings: solvothermal, hydrothermal, microwave, ultrasonic waves, electrochemical, mechanical, ionic liquid, microfluidic, and dry gel. MOF structures are commonly synthesized by solvothermal and hydrothermal methods. The starting materials and the solvent are heated at the specified temperature and time while preparing the framework. In this method, if water is used as a solvent, it is called hydrothermal; if alcohol or another organic solvent is used, it is called the solvothermal method [11].

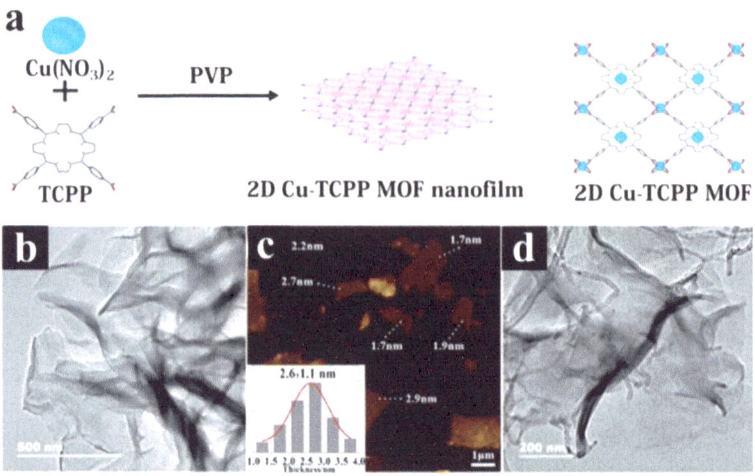

Fig. 11.1 The synthesis mechanism of M-TCPP MOF (**a**), TEM (**b**, **d**), and AFM (**c**) Reproduced with permission from Elsevier [10]

The preparation of MOF structures by microwave and ultrasonic synthesis methods is newer than other methods. In these methods, microwave and ultrasonic waves are used as energy sources. In the microwave method, the reactions can take place in a shorter time compared to other methods. It was first used to synthesize MOF-5 in 2006. The synthesis solution prepared in this method was placed in a microwave oven, and the desired temperature was reached. Crystal formation was observed for a short time, and the crystals were collected. In addition to these methods, the electrochemical method uses conductive salts such as nitrate, perchlorate, and chloride to synthesize MOF structure. HKUST-1 (known as MOF-199) was synthesized with this method and used copper as a metal center electrode [12].

MOFs are also synthesized by the mechanochemical method. For this, the starting materials are mechanically ground (in micro size) into powder, and the reactions are carried out by preparing mixtures of these powders [13].

11.3 Fabrication and Electrocatalytic Properties of 2D MOFs

2D MOFs nanosheets and their composite materials consisting of metal ions and organic linkers are broadly studied because of their numerous active sites, adjustable ion dispersion, and morphologies [14]. To improve the ion diffusion capabilities of MOFs and MOF derivatives, researchers have worked hard to create low-dimensional MOFs with a controllable shape and dimension with a greater specific surface area.

Over the last few years, remarkable progress has been achieved in the fabrications and applications of 2D MOFs and their derivatives.

Various synthesis procedures have been used to fabricate 2D MOFs so far. Generally, the fabrication of 2D MOFs can be grouped into two parts top-down and bottom-up.

11.3.1 Top-Down Approach

The top-down approach means that bulk materials are transformed into nanoparticles using physical methods such as ball milling, thermal evaporation, sputtering, and laser ablation [15]. Layered 2D MOFs have intramolecular and intermolecular forces that can be broken with physical and chemical exfoliation. Physical exfoliation is a simple and efficient way of breaking bonds by applying external forces like ultrasonication, wet ball milling, shaking, and freeze–thaw [14]. Chemical exfoliation uses in-situ chemical reactions to destroy bonds and weaken interactions by using intercalators and introducing a non-planar structure to achieve exfoliation [16].

11.3.2 Bottom-Up Approach

The bottom-up approach forms nanomaterials or MOFs by self-assembling atoms, molecules, and organic linkers with coordination interactions [14]. This coordination is achieved by template, surfactant, and molecule-assisted methods. Also, metal surfaces with chemical vapor deposition and liquid/gas or liquid/liquid surfaces can be used to grow MOFs [17]. This strategy is thought to be more efficient in fabricating 2D MOFs with better uniformity than the top-down approach due to easy formation control and certain directions [14].

2D MOFs have a promising future in various fields, such as energy storage, sensing, catalysis, gas adsorption, and separation [18–20]. Therefore, more and more synthesis methodologies, different organic ligands or metals, and composite materials have been searched for in developing new 2D MOFs. Researchers have aimed to synthesize 2D MOFs and find a suitable application. One of the most important applications is the usage of 2D MOFs in environmental applications. For instance, Lou and co-workers have designed a facile synthesis of BiOBr/Mg metal–organic frameworks and used it to degrade traditional dye rhodamine B (RhB) under visible light irradiation [21]. A simple hydrothermal method was used to create binary composites, which were then coated onto BiOBr substrates. The removal rate of this dye can get nearly 98% in almost one hour. This study shows that easy synthetic modified 2D MOFs are used for environmental cleaning applications.

Another elaborate synthesis design for double-ligand MOFs was introduced by Li et al. [22] in 2022 to detect metal ions and nitroaromatic compounds (NACs). Fe^{3+}, $Cr_2O_7^{2-}$, and picric acid (TNP) are common toxic heavy pollutants, so their detection

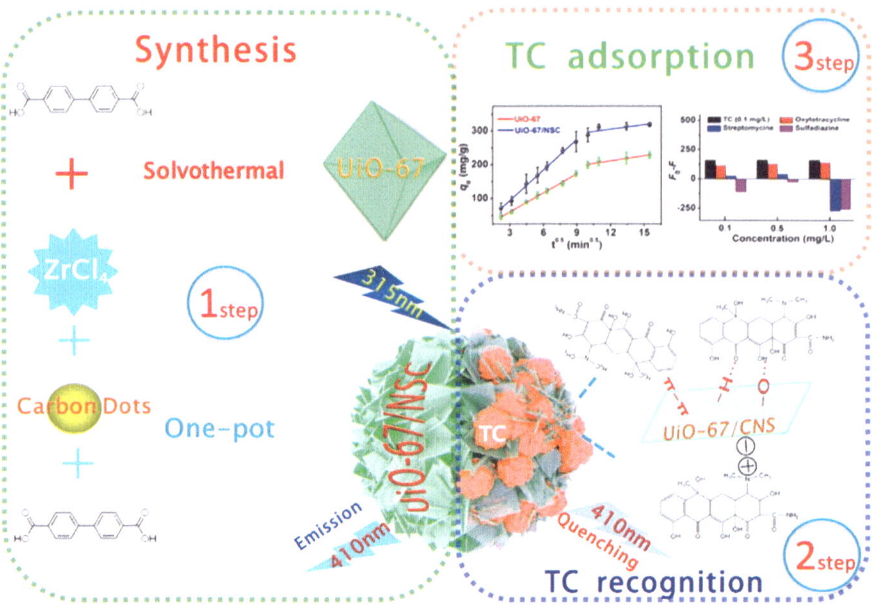

Fig. 11.2 The schematic representation of the synthesis of carbon dots hybrid Zr-based MOFs for removal of tetracycline. Reproduced with permission from Elsevier [22]

is critical for environmental protection. Hence, this study explains the synthesis of five novel luminescent Cd(II)-MOFs based on anthracene ligands with solvothermal reactions. These MOFs-based fluorescent sensors effectively detect Fe^{3+}, $Cr_2O_7{}^{2-}$, and picric acid (TNP) and provide a practical and feasible technique with a lot of potential in environmental sensing.

For the first time, Yang and co-workers [22] fabricated a 3D hierarchical Zr-MOF-based smart adsorbent with carbon dots and 2D MOF sheets, with an easy solvothermal technique for detection of one of the major antibiotic residues (tetracycline) in contaminated water (Fig. 11.2). This hybrid smart MOF material has a high surface area, huge pores, specific adsorption sites, and a special functional group for the recognition and removal of tetracycline. Therefore, this rational synthesis of MOF material enables the purification of wastewater.

11.3.3 Electrocatalytic Properties

Easily accessible active sites, exchangeability of the coordination positions, and easy formation of MOF-based films on the electrode surface make 2D MOFs valuable materials for electrochemical applications. Thanks to their versatility and tunability, a large surface area, porous structure, and catalytic effects can be obtained in the synthesis and preparation processes of 2D MOFs. By this means, they are widely utilized in catalytic and electrocatalytic applications. Furthermore, the exchangeable coordination sites on the large surface area are available for interactions with the analytes offering an effective catalytic activity compared to other MOF materials. This can be obtained thanks to the higher charge transfer rate and capacity on the specific surface area [23, 24].

Especially for electrochemical applications, 2D MOFs provide high electrocatalytic activity due to the fact that they can easily form a thin film layer on the electrode surface. Hence, they are used as modification materials in electrochemical sensors [25]. In addition to their use in electrochemical sensor applications, 2D MOFs also find a place in various electrocatalytic reduction reactions (e.g., O_2, CO_2, N_2) because of the slow kinetics of these kinds of reactions [24, 26]. As an example of this approach, Chi et al. [27] reported the synthesis of the catalytic composite combining the 2D Ni-based MOF with carbon dots. The resulting catalyst has a higher conductivity, better stability, and larger surface area thanks to the unique features of both nanomaterials and 2D MOFs. Sustainable and green energy production is one of the most important issues in today's environmental crisis. Fuel cells are powerful energy tools of recent times with their high efficiency and environmental friendliness to meet this need. The catalysts used to improve fuel cell performance must be highly efficient and stable. To achieve this, 2D MOF-based catalysts have an important place. The catalyst developed in this study is a highly effective 2D MOF material that can be used in fuel cells based on ethanol oxidation reaction. To obtain the required catalytic effects, it is important to design the synthesis and preparation processes carefully. For the improvement and enhancement of the electrocatalytic activity of MOFs, it is possible to incorporate them with different metal nanoparticles [28].

The most significant drawback of 1D and 3D MOFs is poor electrical conductivity and low charge transfer rate; hence their electrocatalytic applications are limited. On the other hand, 2D MOFs are good alternatives with great conductive features [24]. In addition to the existing properties of 2D MOFs, selecting suitable secondary functional groups such as NH_2, Br, OH, and Cl for organic ligands of MOFs during synthesis improves the desired catalytic effect. Another option to acquire desired features for MOFs is the modifications after synthesis [29]. The expected electrocatalytic activity is to promote and then maintain the catalysis. In addition, the presence of the metal ions in 2D MOF structure can also provide a catalytic activity [11].

11.4 Effective Strategies for Modification of MOFs

An important step in improving the sensitivity and selectivity of target analytes is the appropriate and correct selection of the modification to be applied to the electrode surface. In studies featured in the literature, researchers used materials such as quantum dots (QDs), carbon-based nanomaterials, metal oxides, conductive polymers, and MOFs to improve the performance of electrodes. The developed sensor platforms provided fast response time, easy fabrication, and more sensitive and selective sensing performances. In addition, these modifications offer advantages such as high signal amplification and remarkable thermal stability at the electrode surface. In this context, MOFs are crystalline materials that exhibit properties such as large surface area, high porosity, adjustable pore size and absorbability due to their structural diversity [30].

Furthermore, MOFs-based derivatives exhibit exceptional conductivity and high biocompatibility. MOF-derived nanomaterials are frequently produced by subjecting MOFs to various thermal treatments. In such cases, MOFs are often preferred to form metal compounds, carbon materials, and composites. Thanks to the careful selection of MOFs, metal oxides derived from MOFs and their composites can be selected according to the material used, thereby improving electrochemical properties and biocompatibility. As a result, carbon materials from MOFs and related metal oxide composites in electrocatalysis, environmental cleanup, and energy conversion are increasing daily [30, 31].

The electrochemical activity of materials based on MOFs can also be increased by adding metal nanoparticles (NPs) or carbon compounds, but their preparation is rather difficult. The specific surface areas, nanostructures, and stability of MOFs can vary significantly with the addition of different components. To increase the sensing capacity of MOFs, combining several metal components into a single MOF or to produce separate MOFs are viable ways to improve electron transport [30]. A new way to exploit the synergistic effects brought about by incorporating a second metal ion into the inorganic MOF nodes is provided by synthesizing heterometallic centers in bimetallic MOFs. Different transition metal ions in the bimetallic MOF improve the electrochemical properties of the MOF due to induced lattice distortions, improved porosity, and combinational effects between various transition metal ions. In addition, Bimetallic MOFs have very stable exposed active sites that are advantageous for their use in catalyzing processes based on electron transfer. Some of the most recent findings regarding the use of bimetallic MOFs for electrochemical sensing are Ni/Cu bimetallic MOF for glucose, Fe/Ni MOF for dopamine, Ag/Zn MOF for HCV-RNA, Co/Ni MOF 126 for miRNA, and Cu/Ce MOF derivatives for Malathion. However, even against non-electroactive objects, these studies have low detection limits [32]. Moreover, bimetallic MOFs may be made using several different techniques, including the one-pot method [33], the step-by-step method [34], the epitaxial growth approach [35], and the MOF-on-MOF method [36], and the template-oriented approach [35].

Carbon materials have important properties such as large surface area, adjustable pore size, high electrical conductivity, and chemical stability. These materials can be produced in substantial quantities and can be synthesized using various techniques. MOFs are considered promising templates for the synthesis of carbon compounds due to their ordered structure, homogeneous composition, and high carbon content. Carbons derived from MOF offer significant benefits over traditional synthesizing ways of carbon-based compounds during the preparation step. The distinctive variety of MOFs provides the basis for precise control of a material's physical and chemical properties [37]. The benefits of carbon-based composite materials made from MOFs include: (1) because different MOFs have different morphologies and structures, they can be transformed into porous carbon materials with various compositions and functions and (2) to prevent agglomeration of metal or metal oxide nanoparticles during the pyrolysis process and to ensure uniform distribution. For this reason, it is advantageous to regularly cascade metal ions or metal ion clusters and organic ligands in the structure of MOFs; (3) The MOF-making technique is generally simple [31]. Carbon-based materials, including graphene, carbon nanotubes, and carbon dots (CDs), are frequently used to build electrochemical sensors due to their high electrochemical activity [38]. Applications for electrochemical sensors have been found when these carbon-based conductive materials are combined with MOFs [39].

11.5 2D MOFS-Based Sensor for Environmental Applications

When we look at the literature in recent years, MOFs were attracted great attention because they act as supports for semiconductor materials such as TiO_2, CuO, and ZnO. Due to its highly acidic properties, adsorption, and surface area properties, it was advantageous, such as very good electron abstraction property and a small scattering of UV light. Table 11.1 contains the latest applications (2018–2022) of modified electrochemical sensors with 2D MOFs-based sensor platforms for detecting actual samples such as organic compounds, antibiotics, pesticides, hydroquinone, and heavy metals. Details such as recovery, linear range, the limit of detection (LOD), and the method of sensors developed in environmental examples in the literature were given in Table 11.1.

The new magnet electrochemical method was developed by Zhai et al. [41]. Magnetic Fe_3O_4-loaded silver metal–organic framework (Fe_3O_4@Ag-MOF) was synthesized by a hydrothermal method (Fig. 11.3). The magnetic-MOF nanomaterials were applied on the glassy carbon electrode surface. Hydrogen sulfide (H_2S) as a non-color gas was determined in aquaculture water (fish, shrimp, and crab). Using the linear sweep voltammetry (LSV) technique, the linear range and LOD were found as 4–1400 nM and < 2.0 nM for H_2S, respectively.

Xiao J. et al. report that the electrochemical immunosensors were fabricated to determine sulfonamides (SAs) (Fig. 11.4). The synthesized nanozyme was formed peroxidase-like 2D Cu-TCPP(Fe). Polyethyleneimine (PEI) was applied to form a better affinity between PEI and Cu^{2+}. The 2D-MOF/hybrid materials were coated to the surface of bare GCE. The immonosensor was followed by amperometric technique. The developed immunosensor was applied to different kinds of water. This immunosensor was compared with a commercial ELISA kit. The standard curve was established after the measurement of a serial concentration of SMM (Fig. 11.4). Good performance was obtained with a low limit of detection (LOD: 0.395 ng/mL) and a wide range of SAs (1.186–28.051 ng/mL) [43].

Table 11.1 Latest applications of electrochemical sensors for 2D MOFS sensor for environmental samples

AAnalyte	Modified electrode	Method	LOD	Linear range	Real sample	Recovery (%)	References
Tetracycline	MCH/aptamer/AuInCN/ITO electrodes	EIS	3.3 pM	0.01–500 nM	Lake water	NA	[40]
Hydrogen Sulfide	Fe_3O_4@Ag-MOF/MGCE	LSV	2.0 nM	4.0–1400 nM	Freshwater	NA	[41]
Hydroxylamine Chlorogenic acid	CuTCPP/pOMC-3/GCE	AMP	0.8 μM 0.019 μM	5.8–733.8 μM 733.8–2933.8 μM 0.1–2 μM 2–15 μM	Tap water Lake water	99.03–102.7 99.32–102.52	[42]
Sulfonamides	2D Cu-TCPP(Fe)/MWCNTs/ GCE	CV	0.395 ng mL^{-1}	1.186–28.051 ng mL^{-1}	Pure water Pond water Tap water River water	64–118	[43]
Cadmium (II)	CNT-Cu-MOF/GCE	DPV	0.275 nM	0.2–10 μM	Tap water	99.4–100.4	[44]
Bisphenol A	$Mo_2Ti_2AlC_3$/MWCNT/GCE	DPV	2.7 nM	0.01–8.50 μM	Milk pack plastic bottle cans	95.67–100.60	[45]
Arsenic (III)	Fe-MOF/MXene/GCE	SWASV	0.58 ng L^{-1}	1–10 ng L^{-1} 10–100 ng L^{-1}	Tap water Lake water River water	97.4–98.4	[46]
Chlorpyrifos	MnFe-MOF/SPCE	SWV	0.85 nM	1–0.01 nM	Tap water Garden soil	98.80–106.81	[32]
Enrofloxacin	Au@ZnNi-MOF/AuE	SWV	1.02×10^{-7} ng mL^{-1}	5.0×10^{-6}-1.0×10^{-2} ng mL^{-1}	Water Milk	97.9–116.8	[47]
Trace As(III)	Au_{nano}/Fe-MOF/GCE	SWASV	0.0085 ppb	0.02–2 ppb 2–30 ppb	Tap water Lake water River water	95.6–104.6	[48]

(continued)

Table 11.1 (continued)

AAnalyte	Modified electrode	Method	LOD	Linear range	Real sample	Recovery (%)	References
Enrofloxacin	AuTNSs-cDNA-Thi/ BSA/Apt/AuNPs/Ni-MOF/AuE	DPV	0.056 pg·mL^{-1}	0.01–1000 pg·mL^{-1}	Honey Milk	94.6–107.2	[49]
Fenamiphos	Co_3O_4@MOF-74/cycle4/GCE	EIS	3 pM	0.01–1.0 nM	Orange juice	99.67–100.45	[50]
Lead (II)	PtNPs@Cu-MOF-HP/Pb^{2+}/MCH/DNA walker-SS/AuE	DPV	0.2 pM	3 pM-5 μM	Orange juice Tap water	89.7–110.5	[51]
Hydroquinone	GCE/CAU-1 MOF/GCE	DPV	0.067 μM	0.5–1500 μM	Water	99.34–103.93	[52]
Salmonella typhimurium	Bimetallic Fe/Mn MOF/MβCD/AuNPs stabilized on MWCNTs/AuE	DPV	0.07 pM	1 pM-1 μM	Milk	95.6–104	[53]
Streptomycin	STR/MB-aptamer/ MCH/cDNA/PEIU–PANITi/GCE	DPV	0.0033 nM	0.01–200 nM	Milk	94.0–105.0	[54]
Catechol Hydroquinone	Fe.N-CNs based on MOF/GCE	DPV	0.17 μM	0.50–80.00 μM	Tap water	95.65–100.9 98.7–103.2	[55]
Norfloxacin	MIP/CoFe-MOFs/AuNPs/GCE	DPV	0.131 pM	5–100 pM 100–1000 pM 1000–6000 pM	Milk	97.04–110.2	[56]
Metol Hydroquinone Catechol	Co_3O_4@C/GNPs/GCE	DPV	5.1 nM 14.7 nM 169 nM	0.01–80 μM 0. 04–30 μM 0. 5–30 μM	Lake water	93.4–101.4 94.0–104 97.3–103.5	[57]
Hydrazine	Au_{90}NPs@NiFe2O4/NFE	EIS	0.05 μM	0.3–150 μM	Tap water	97.2–104.2	[58]
Malathion	CP-MOF-Fc/MCH/Apt/AuNPs/GCE	DPV	17.18 ng L^{-1}	25–850 ng L^{-1}	Cucumber Long bean	95.0–103.16	[59]
Oxytetracycline	mMOFs@MIPs/MGCE	DPV	4.1 × 10^{-10} g mL^{-1}	1 × 10^{-9}-2 × 10^{-4} g mL^{-1}	Milk	89.0–103.1	[60]

(continued)

Table 11.1 (continued)

AAnalyte	Modified electrode	Method	LOD	Linear range	Real sample	Recovery (%)	References
Phthalates	UiO-66-$(NH2)_{0.58}(4F)_{0.42}$/GCE	EIS	0.0039–0.016 ng L^{-1}	0.03–1000 ng L^{-1}	Milk Liquor	83.7–118.1	[61]
Cadmium (II) Lead (II)	AgNP@Bt/TC/GCE	DPASV	0.79 μg/L 0.88 μg/L	5.0–60.0 μg/L	Spring water	93–108 99–113	[62]
Cyanide	Cu-MOF/CPE	DPV	0.60 μM	1.87–25 μM	River water Mineral water Well water	96–105	[63]
Nitrate	AgNPs/Cu-BDC MOF/SPCE	APM	0.24 μM	0.5–1000 μM	Drinking water	97.8–102	[64]
Resorcinol	Cu-MOF/rGO/CuO/α-MnO_2/GCE	DPV	0.2 μM	0.2–22 μM	Tap water River water	96–106.2	[65]
Acetamiprid residual	Co-MOF/GCE	CV	0.42 nM	1–200 nM	Agricultural soil Tap water River water Agricultural runoff water	82.6–110.8	[66]
Nitrite Nitric oxide	CNFs-Bi-MOF/GCE	DPV	0.184 nM 3.463 nM	2 nM-2 mM 10 nM-1 mM	Tap water	NA	[67]
Copper (II) Lead (II) Cadmium (II)	Fc-NH2-Ni-MOF/GCE	DPASV	6.3 nM 0.2 nM 7.1 nM	0.01–2.0 μM 0.001–2.0 μM 0.001–2.0 μM	Tap water River water	97.0–110.0 96.0–100.0 95.0–110.0	[68]
Copper (II) Lead (II) Cadmium (II) Mercury (II)	UiO-66-NH2/GaOOH/GCE	DPV	0.019 μM 0.028 μM 0.016 μM 0.006 μM	0.30–1.40 μM 0.55–2.50 μM 0.35–1.60 μM 0.10–0.45 μM	Water	94.5–105.1 96.3–105.4 92.1–100.4 96.9–104.3	[69]

(continued)

Table 11.1 (continued)

AAnalyte	Modified electrode	Method	LOD	Linear range	Real sample	Recovery (%)	References
Estrone	US AuNPs@AuZn-MOF/CPE	DPV	12.3 nM	0.05–5 μM	Lake water	93.0–103.5	[70]
Copper (II) Lead (II) Cadmium (II) Mercury (II)	Co-TMC4R-BDC/GCE	SWASV	13 nM 11 nM 26 nM 18 nM	0.05–12.0 μM 0.05–13.0 μM 0.1–17.0 μM 0.75–18.0 μM	Tap water Mineral water Lake water	98.80–102.92 99.80–103.70 98.70–101.28 99.80–104.39	[71]
Methylmercury	Au/ZIF67/GCE	DPSV	0.05 μg/L	1–25 μg/L	Lake water	93.0–105.0	[72]
Lead (II) Cadmium (II)	ZJU-27/GCE	SWASV	0.228 ppb 1.66 nM	0.1–1.0 μM	Lake water Drinking water	97.3–101.8 94.47–103.2	[73]
Lead (II)	Au@MoS2/rGO/Apt1/MCH/Apt2-AuPd@Fe-MOFs/GCE	AMP	0.07 pM	5.0 pM-2.0 μM	Tap water Lake water	96.0–103.0	[74]
Nitrobenzene	Ni-MOF-PVP/GCE	DPV	97 nM	0.2 μM-1 mM	Tap water Lake water	NA	[75]
4-aminophenol	ZnO/NPC/GCE	DPV	0.014 μM	5–120 μM	Wastewater	94.02–107.7	[76]
Catechol Hydroquinone	Alk-Ti_3C_2/N-PC/GCE	DPV	3.1 nM 4.8 nM	0.5–150 μM	Wastewater	97.7–106.8 98.2–106.4	[77]
Hydrazine	Zn-MOF@rGO/AuE	CV	0.087 μM	0.001–100 μM	Water	105.0–110.0	[78]
Glyphosate	Cu-BTC/ITO	DPV	0.14 pM	1 pM-1 nM 1 nM-10 μM	Soybean	98.0–105.0	[79]
Tetracycline	NH2-MIL-101(Fe)/CNF@AuNPs/aptamer electrode	EIS	0.01 nM	0.1–10^5 nM	Tap water River water	89.7–102.8	[80]
Organophosphates	BSA/Chi-AuNP-rIgG-BSA/MOF/ITO	CV	4.0 ng L^{-1}	4–100 ng L^{-1}	Okra Brinjal Capsicum	NA	[81]
Streptomycin	MCH/dsDNA/Au/SPCE	DPV	2.6 pg mL^{-1}	0.005–150 ng mL^{-1}	Milk	99.4–106.0	[82]

(continued)

Table 11.1 (continued)

AAnalyte	Modified electrode	Method	LOD	Linear range	Real sample	Recovery (%)	References
Oxytetracycline	Ce-MOF@ MCA_{500}/AuE	EIS	17.4 fg mL^{-1}	0.1–0.5 ng mL^{-1}	River water Milk	94.0–106.0	[83]
Epinephrine L-tyrosine	SnS@Co,NC-30%/GCE	DPV	3.3 nM 3.1 nM	0.01–300.0 μM 0.01–250.0 μM	Whole milk	95.2–107.8 98.0–106.8	[84]
Chloramphenicol	Apt/Co–Ni/MOF/BP/PTC-NH2/GCE	EIS	0.029 pM	0.1 pM-1 μM	Tap water	97.0–103.6	[85]
p-Chloronitrobenzene	Ni-MOF/GO/GCE	DPV	8.0 nM	0.10–300.0 μM	Tap water	99.6–101.4	[86]
17β-estradiol	MIP-PB-MIL/CNTs-GCE	DPV	6.9 fM	0.01 pM-1 nM	Pond water	96.9–103.9	[87]
Hydrogen peroxide Hydrazine	MWCNTs-$Co(OH)_2$/GCE	CV	1.36 μM 0.084 μM	5–550 μM 0.2–1106.0 μM	Rain water Tap water	98.52–99.4 96.33–100.0	[88]
Lead (II)	Fe-MOFs/PdPt NPs/GCE	AMP	2 pM	0.005–1000 nM	Sewage water Drinking water	96.0–102.31	[89]
4-nonylphenol	MIL-101(Cr)@rGO/GCE	DPV	33 nM	0.1–12.5 μM	River water Lake water	95.8–105.7	[90]
Mercury (II)	Cu-MOF/GCE	DPV	0.0633 nM	0.1–50 nM	Tap water	101.0–103.0	[91]
Chromium (III) Chromium (IV)	Fe_3O_4@SDX G-150/SPCE	LSV	0.1 μM 0.01 μM	0.5–10 μM	Mineral water	99.0–101.0	[92]
Paraoxon	Naf/AChE/NF-Fe_2O_3@C/CPE	DPV	0.12 pM	10^{-14}-10^{-8} g mL^{-1}	Tap water	98.8–104.2	[93]
Lead (II)	TCPP-MWCNTs@Fe_3O_4/MGCE	FSV	6.7×10^{-5} μg L^{-1}	2.0×10^{-4}–2.0×10^{-3} μg L^{-1}	Seawater	NA	[94]

NA: Not applicable, GCE: Glass Carbon Electrode, MGCE: Magnetic glass carbon electrode, NFE: Nickel foam electrode, AuE: Gold electrode, SPCE: Screen Printed Electrode, MGCE: Magnetic glassy carbon electrode, CPE: Carbon paste electrod, ITO: Indium tin oxide electrode, DPV: Differential Pulse Voltammetry, EIS: Electrochemical Impedance Spectroscopy, CV: cyclic voltammetry, SWV: square-wave voltammetry, LSV: Linear sweep voltammetry, AMP: Amperometry, SWASV: Square-wave anodic stripping voltammetry, DPSV: Differential pulse stripping voltammetry, DPASV:Differential pulse anodic stripping voltammetry, FSV: Fast scan voltammetry, Au NPs: Gold Nanoparticles, AuInCN: MOF-derived In2O3@g-C3N4 nanoarchitectures, MWCNTs: multi-walled carbon nanotubes, Cu-MOF: Copper metal–organic framework, Au_{nano}: Gold Nanoparticles, Ni-MOF: Nickel metal–organic framework, AuTNSs: Gold triangular nanosheets, Thi: Thionine, Apt: Aptamer, PMS: Peroxymonosulfate, SS: Substrate Strand, PtNPs: Platinum Nanoparticles, MβCD: Methyl-β-cyclodextrin, C: Carbon, NF: Nickel foam, CP: Complementary probe, Fc-COOH: Carboxy-ferrocene, mMOFs@MIPs: Molecularly imprinted magnetic metal–organic frameworks, AgNP: Silver Nanoparticles, Cu-BDC MOFs: Copper (II)-terephthalate metal–organic frameworks, Fc: Ferrocene, US AuNPs: Ultrasmall gold nanoparticles, ZIF67: Zeolitic imidazolate framework-67, PVP: Polyvinylpyrrolidone, rIgG: Polyclonal antibody, MCH: 6-mercapto-1-hexanol, AChE: Acetylcholinesterase.

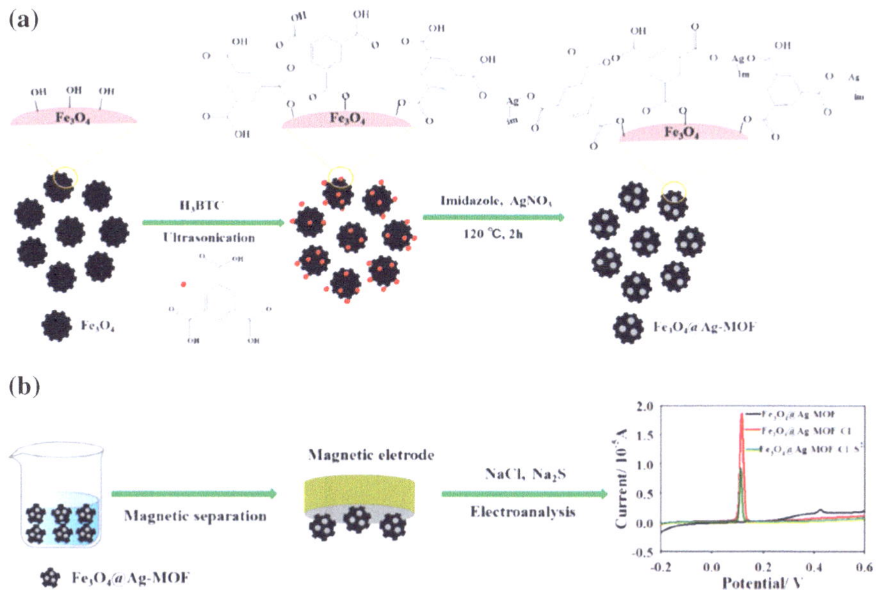

Fig. 11.3 **a** Schematic illustration of the synthetic procedure of magnetic Fe3O4@Ag-MOF. **b** Main procedure and sensing mechanism of the Fe_3O_4@Ag-MOF-based electroanalysis of H2S through the solid-state Ag/AgCl signals. Reproduced with permission from Elsevier [41]

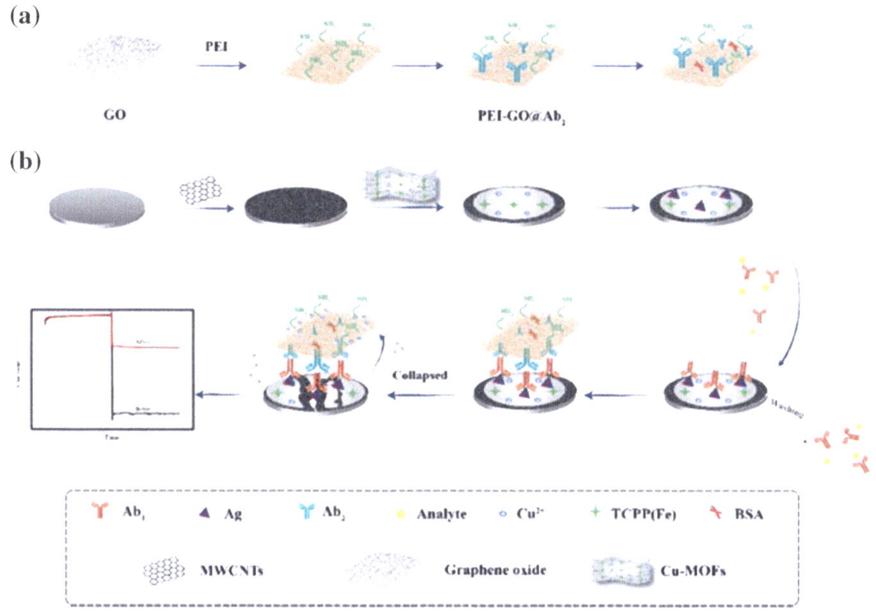

Fig. 11.4 Schematic illustration of the electrochemical immunosensor for the detection of Sulfonamides. Reproduced with permission from Elsevier [43]

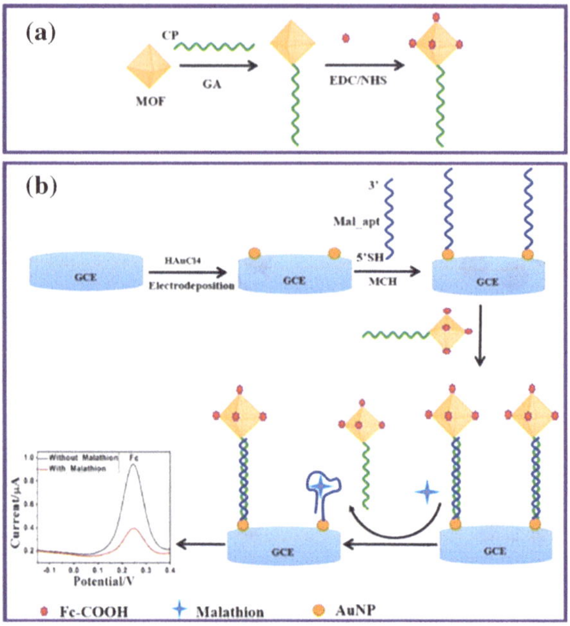

Fig. 11.5 Schematic illustration of the ultrasensitive aptamer sensor for malathion assay based on UiO-66-NH2 and Fc. Reproduced with permission from Elsevier [59]

The zirconium-based MOF was synthesized by the hydrothermal method, and complementary probe (CP) and carboxy-ferrocene (Fc-COOH) were linked by covalent bonds to form CP-MOF-Fc nanocomposite probe (Fig. 11.5). To increase the electrochemical signal intensity and improve the sensitivity, the CP-MOF-Fc complex (UiO-66-NH_2 and Fc) was prepared by hydrothermal process. A linear range of Malathion was found as 25–850 ng L^{-1} with LOD of 17.18 ng L^{-1} by DPV technique. The recovery results were obtained from 95.00 to 103.16% in cucumber and long bean [59].

The tetracycline antibiotics were detected with an electrochemical aptasensor based on electrospinning technology by Song et al. (Fig. 11.6). The electrical signals of tetracycline obtained with the interaction between antibiotics and aptamers were enhanced with NH_2-MIL-101(Fe)/CNF@AuNPs synthesized by a combination of electrodeposition, hydrothermal, pyrolysis, and electrospinning methods. Under the optimized conditions, the linearity of tetracycline was obtained in the range of 0.1–105 nM with LOD of 0.01 nM by EIS. In addition, the new analytical strategy exhibited high selectivity and stability in real water samples [80].

Wang et al. report that a sensitive electrochemical aptasensor using AuPd@Fe-MOFs was developed to determine Pb^{2+} (Fig. 11.7). For the fabrication of aptasensor, firstly, Au@MoS2/rGO nanocomposite was coated on bare GCE. The mercapto-group-labeled substrate DNA strand (Apt1) was immobilized through Au–S interaction. Then mercaptohexanol (MCH) was incubated to block the surface against the nonspecific adsorption. Then, catalytic strand Apt2 linked with AuPd@Fe-MOFs solution was immobilized on the electrode surface. For the measurement of Pb^{2+},

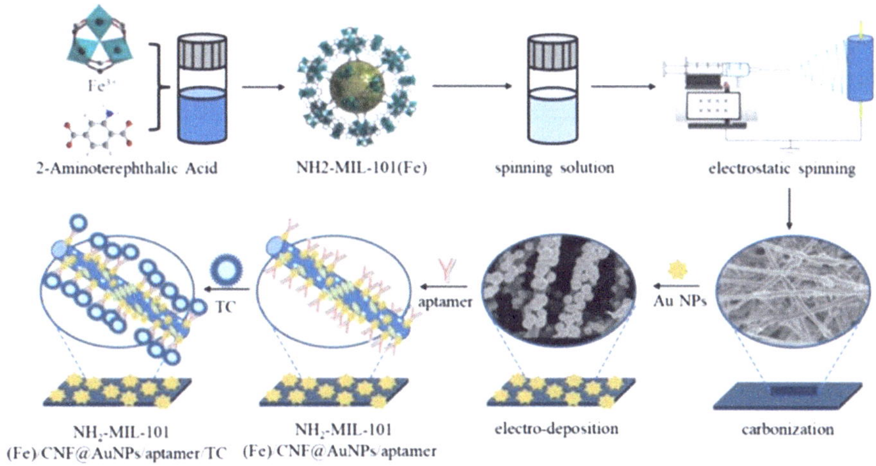

Fig. 11.6 Schematic representation of the NH2-MIL-101(Fe)/CNF aptasensor preparation. Reproduced with permission from Elsevier [80]

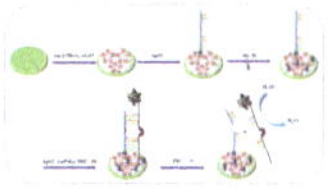

Fig. 11.7 Schematic illustration of the fabrication process of the electrochemical aptasensor. Reproduced with permission from Elsevier [74]

the amperometric technique was performed in PBS (pH 7.4) to the excellent current response of Pb^{2+}. The fabricated aptasensor has a highly linear response from 5–2000 pmol/L with a LOD of 0.07 pmol/L for Pb^{2+}. Moreover, the proposed aptasensor for Pb^{2+} detection demonstrated good performance in environmental monitoring [74].

11.6 Conclusion and Perspectives

This chapter summarizes the properties of 2D MOFs, their usage applications, synthesis approaches, modifications, and the most recent and prominent environmental applications of sensors. Metal–organic frameworks have piqued interest in materials science, nanotechnology, and chemical science as a new family of 2D materials due to their large specific surface areas, high porous structures, tunable morphologies, rich electron environment, and availability of functional groups. As a result, several methodologies have been developed for the fabrication of 2D MOFs, especially for catalysis, energy storage/conversion, and adsorption.

2D MOFs with excellent electrocatalytic properties, which are not available in 1D and 3D MOFs and can be improved by integrating with various materials, have also started to occupy a large place in sensor applications. In particular, their excellent conductivity properties provide a great advantage for electrochemical sensor studies. The novel works in the literature were evaluated and compared in terms of the analyte, used technique, 2D MOF material, linear range, LOD, and applied environmental sample. It is aimed to shed light on future studies by presenting a comprehensive perspective on the studies conducted in this field.

As it is known that the environmental crisis will become a growing problem in the coming years, it is possible to state that the importance of utilizing the advantageous properties of 2D MOFs in environmental applications will increase. In addition, it can be predicted that the tendency to develop sensor devices for rapid and on-site determination in the analysis of environmental samples will increase.

References

1. Y. Lin, Y. Li, Y. Cao, X. Wang, Two-dimensional MOFs: Design & synthesis and applications. Chem-An Asian J **16**, 3281–3298 (2021)
2. F. Wei, Y. Liu, H. Zhao, X. Ren, J. Liu, T. Hasan, L. Chen, Y. Li, B.L. Su, Oxygen self-doped g-C3N4 with tunable electronic band structure for unprecedentedly enhanced photocatalytic performance. Nanoscale **10**, 4515–4522 (2018)
3. J. Wang, F. Ma, M. Sun, Graphene, hexagonal boron nitride, and their heterostructures: properties and applications. RSC Adv **7**, 16801–16822 (2017)
4. W.P. Lustig, S. Mukherjee, N.D. Rudd, A.V. Desai, J. Li, S.K. Ghosh, Metal-organic frameworks: Functional luminescent and photonic materials for sensing applications. Chem Soc Rev **46**, 3242–3285 (2017)
5. W. Zhu, C. Zhang, Q. Li, L. Xiong, R. Chen, X. Wan, Z. Wang, W. Chen, Z. Deng, Y. Peng, Selective reduction of CO2 by conductive MOF nanosheets as an efficient co-catalyst under visible light illumination. Appl Catal B Environ **238**, 339–345 (2018)
6. W. Yang, Y. Peng, Y. Li, Y. Ban, H. Jin, W. Jiao, X. Liu, Metal-organic framework nanosheets as building blocks for molecular sieving membranes. Science. (80-) **346**:1352–1356
7. S. Sakaida, K. Otsubo, O. Sakata, C. Song, A. Fujiwara, M. Takata, H. Kitagawa, Crystalline coordination framework endowed with dynamic gate-opening behaviour by being downsized to a thin film. Nat Chem **8**, 377–383 (2016)
8. M. Hui, K. Loon, G. Zheng, Synthesis and applications of MOF-derived porous nanostructures. Green Energy Environ **2**, 218–245 (2017)
9. G. Chakraborty, I. Park, R. Medishetty, J.J. Vittal, Two-Dimensional Metal-Organic Framework Materials : Synthesis. Structures, Properties and Applications. (2021). https://doi.org/10.1021/acs.chemrev.0c01049
10. X. Tang, L. Zhao, W. Sun, Y. Wang, Two-dimensional metal-organic framework materials for energy conversion and storage. J Power Sources **477**, 228919 (2020)
11. J. Liu, H. Yu, L. Wang, Z. Deng, K. Naveed K ur R, A. Nazir, F. Haq Two-dimensional metal-organic frameworks nanosheets: Synthesis strategies and applications. Inorganica Chim Acta **483**, 550–564(2018)
12. W. Wang, Y. Yu, Y. Jin, X. Liu, M. Shang, X. Zheng, T. Liu *Two-dimensional metal-organic frameworks: from synthesis to bioapplications* (2022), pp 1–17
13. W. Bai, S. Li, J. Ma, W. Cao, J. Zheng, *Ultrathin 2D metal–organic framework (nanosheets and nanofilms)-based xD–2D hybrid nanostructures as biomimetic enzymes and supercapacitors* (2019), pp. 9086–9098
14. K. Zhao, W. Zhu, S. Liu, X. Wei, G. Ye, Y. Su, Z. He, Two-dimensional metal-organic frameworks and their derivatives for electrochemical energy storage and electrocatalysis (2020). https://doi.org/10.1039/c9na00719a

15. N. Abid, A.M. Khan, S. Shujait, K. Chaudhary, M. Ikram, M. Imran, J. Haider, M. Khan, Q. Khan, M. Maqbool, Synthesis of nanomaterials using various top-down and bottom-up approaches, influencing factors, advantages, and disadvantages: A review. Adv. Colloid. Interface. Sci. **300**, 102597 (2022)
16. W. Wang, W. Zhao, H. Xu, S. Liu, W. Huang, Q. Zhao, Fabrication of ultra-thin 2D covalent organic framework nanosheets and their application in functional electronic devices. Coord. Chem. Rev. **429**, 213616 (2021)
17. J. Duan, Y. Li, Y. Pan, N. Behera, W. Jin, Metal-organic framework nanosheets: An emerging family of multifunctional 2D materials. Coord. Chem. Rev. **395**, 25–45 (2019)
18. J.N. Joshi, G. Zhu, J.J. Lee, E.A. Carter, C.W. Jones, R.P. Lively, K.S. Walton, Probing metal-organic framework design for adsorptive natural gas purification. Langmuir **34**, 8443–8450 (2018)
19. X. Chen, S. Shen, L. Guo, S.S. Mao, Semiconductor-based photocatalytic hydrogen generation. Chem. Rev. **110**, 6503–6570 (2010)
20. M. Sohail, M. Altaf, N. Baig, R. Jamil, M. Sher, A. Fazal, A new water stable zinc metal organic framework as an electrode material for hydrazine sensing. New. J. Chem. **42**, 12486–12491 (2018)
21. W. Lou, L. Wang, Y. Zhang, Y. Xing, Synthesis of BiOBr/Mg metal organic frameworks catalyst application for degrade organic dyes rhodamine B under the visible light. Appl Organomet Chem. (2021). https://doi.org/10.1002/aoc.6324
22. J. Li, J. Tian, H. Yu, M. Fan, X. Li, F. Liu, J. Sun, Z. Su, Controllable synthesis of metal-organic frameworks based on anthracene ligands for high-sensitivity fluorescence sensing of Fe3+, Cr2O72–, and TNP. Cryst Growth Des **22**, 2954–2963 (2022)
23. A. Dhakshinamoorthy, A.M. Asiri, H. Garcia, 2D Metal-organic frameworks as multifunctional materials in heterogeneous catalysis and electro/photocatalysis. Adv. Mater. (2019). https://doi.org/10.1002/adma.201900617
24. M.K. Lee, M. Shokouhimehr, S.Y. Kim, H.W. Jang, Two-dimensional metal-organic frameworks and covalent-organic frameworks for electrocatalysis: Distinct merits by the reduced dimension. Adv. Energy Mater **12**, 1–32 (2022)
25. S.K. Das, G. Kumar, M. Das, R.S. Dey, A 2D covalent organic framework as a metal-free electrode towards electrochemical oxygen reduction reaction. Mater. Today Proc. **57**, 228–233 (2022)
26. J. Nicks, K. Sasitharan, R.R.R. Prasad, D.J. Ashworth, J.A. Foster, Metal-organic framework Nanosheets: Programmable 2D materials for catalysis, sensing, electronics, and separation applications. Adv. Funct. Mater (2021). https://doi.org/10.1002/adfm.202103723
27. X. Chi, L. Gao, W. Zhou, Y. Zhang, T. Hu, Enhanced electrocatalytic performance of 2D Ni-MOF for ethanol oxidation reaction by loading carbon dots. J. Solid. State. Chem. **311**, 123094 (2022)
28. M.E. Çorman, G. Ozcelikay, A. Cetinkaya, S.I. Kaya, C. Armutcu, E. Özgür, L. Uzun, S.A. Ozkan, Metal-organic frameworks as an alternative smart sensing platform for designing molecularly imprinted electrochemical sensors. TrAC-Trends Anal Chem. (2022). https://doi.org/10.1016/j.trac.2022.116573
29. X. Liu, H.L. Valentine, W.P. Pan, Y. Cao, B. Yan, 2D metal-organic frameworks: Syntheses, structures, and electrochemical properties. Inorganica Chim. Acta **447**, 162–167 (2016)
30. S. Zhang, F. Rong, C. Guo, F. Duan, L. He, M. Wang, Z. Zhang, M. Kang, M. Du, Metal–organic frameworks (MOFs) based electrochemical biosensors for early cancer diagnosis in vitro. Coord Chem. Rev. **439**, 213948 (2021)
31. D. Liu, W. Gu, L. Zhou, L. Wang, J. Zhang, Y. Liu, J. Lei, Recent advances in MOF-derived carbon-based nanomaterials for environmental applications in adsorption and catalytic degradation. Chem. Eng. J. **427**, 131503 (2022)
32. P. Janjani, U. Bhardwaj, R. Gupta, H. Singh Kushwaha, Bimetallic Mn/Fe MOF modified screen-printed electrodes for non-enzymatic electrochemical sensing of organophosphate. Anal. Chim. Acta. **1202**, 339676 (2022)

33. L. He, Z. Li, C. Guo, B. Hu, M. Wang, Z. Zhang, M. Du, Bifunctional bioplatform based on NiCo Prussian blue analogue: Label-free impedimetric aptasensor for the early detection of carcino-embryonic antigen and living cancer cells. Sens. Actuators, B Chem **298**, 126852 (2019)
34. N. Zhou, F. Su, Z. Li, X. Yan, C. Zhang, B. Hu, L. He, M. Wang, Z. Zhang, Gold nanoparticles conjugated to bimetallic manganese(II) and iron(II) Prussian Blue analogues for aptamer-based impedimetric determination of the human epidermal growth factor receptor-2 and living MCF-7 cells. Microchim Acta. **186**, 3–12 (2019)
35. D. Kim, A. Coskun, Template-directed approach towards the realization of ordered heterogeneity in bimetallic metal-organic frameworks. Angew Chemie–Int Ed **56**, 5071–5076 (2017)
36. N. Zhou, F. Su, C. Guo, L. He, Z. Jia, M. Wang, Q. Jia, Z. Zhang, S. Lu, Two-dimensional oriented growth of Zn-MOF-on-Zr-MOF architecture: A highly sensitive and selective platform for detecting cancer markers. Biosens Bioelectron **123**, 51–58 (2019)
37. M. Hao, M. Qiu, H. Yang, B. Hu, X. Wang, Recent advances on preparation and environmental applications of MOF-derived carbons in catalysis. Sci. Total. Environ. **760**, 143333 (2021)
38. S.J. Le, H.F. Wang, X. Zhu, C.M. Chen, X. Huang, X.D. Zhang, B.Q. Li, C. Tang, Q. Zhang, The nanostructure preservation of 3D porous graphene: New insights into the graphitization and surface chemistry of non-stacked double-layer templated graphene after high-temperature treatment. Carbon N Y **103**, 36–44 (2016)
39. Z. Tang, J. He, J. Chen, Y. Niu, Y. Zhao, Y. Zhang, C. Yu, A sensitive sandwich-type immunosensor for the detection of galectin-3 based on N-GNRs-Fe-MOFs@AuNPs nanocomposites and a novel AuPt-methylene blue nanorod. Biosens Bioelectron **101**, 253–259 (2018)
40. Y. Feng, T. Yan, T. Wu, N. Zhang, Q. Yang, M. Sun, L. Yan, B. Du, Q. Wei, A label-free photoelectrochemical aptasensing platform base on plasmon Au coupling with MOF-derived In2O3@g-C3N4 nanoarchitectures for tetracycline detection. Sens. Actuators, B Chem. **298**, 126817 (2019)
41. X. Zhai, S. Li, Y. Wang, S. Cao, W. Sun, M. Liu, G. Mao, B. Cao, H. Wang, A magnet-renewable electroanalysis strategy for hydrogen sulfide in aquaculture freshwater using magnetic silver metal-organic frameworks. Anal Chim Acta. **1195**, 339450 (2022)
42. X. Zhao, J. Bai, X. Bo, L. Guo, A novel electrochemical sensor based on 2D CuTCPP nanosheets and platelet ordered mesoporous carbon composites for hydroxylamine and chlorogenic acid. Anal Chim Acta. **1075**, 71–80 (2019)
43. J. Xiao, X. Hu, K. Wang, Y. Zou, E. Gyimah, S. Yakubu, Z. Zhang, A novel signal amplification strategy based on the competitive reaction between 2D Cu-TCPP(Fe) and polyethyleneimine (PEI) in the application of an enzyme-free and ultrasensitive electrochemical immunosensor for sulfonamide detection. Biosens. Bioelectron **150**, 111883 (2020)
44. S. Singh, A. Numan, H.H. Somaily, M.M.A. Dawsari, M.H.S. Alqarni, A. Alam, P. Kumar, A novel, eco-friendly multi-walled carbon nanotubes functionalized copper metal-organic framework for ultrasensitive potentiometric detection of cadmium ions. J. Environ. Chem. Eng. **9**, 106534 (2021)
45. V. Sanko, A. Şenocak, S.O. Tümay, Y. Orooji, E. Demirbas, A. Khataee, An electrochemical sensor for detection of trace-level endocrine disruptor bisphenol A using Mo2Ti2AlC3 MAX phase/MWCNT composite modified electrode. Environ. Res. **212**, 1–9 (2022)
46. P. Xiao, G. Zhu, X. Shang, B. Hu, B. Zhang, Z. Tang, J. Yang, J. Liu, An Fe-MOF/MXene-based ultra-sensitive electrochemical sensor for arsenic(III) measurement. J Electroanal Chem. **916**, 116382 (2022)
47. B. Zhang, L. Lv, X. Ma, L. Xie, M. Lin, H. Chen, B. He, Au@ZnNi-MOF labeled electrochemical aptasensor for detection of enrofloxacin based on AuPt@h-CeO2/MoS2 and DNAzyme-driven DNA walker triple amplification signal strategy. Biosens Bioelectron. **210**, 114296 (2022)
48. X. Zhu, G. Zhu, Y. Ge, B. Zhang, J. Yang, B. Hu, J. Liu, Aunano/Fe-MOF hybrid electrode for highly sensitive determination of trace As(III). J. Electroanal. Chem. **899**, 115642 (2021)

49. L. Lv, B. Zhang, P. Tian, L. Xie, W. Wei, J. He, M. Lin, H. Zhu, H. Chen, B. He, A "signal off" aptasensor based on AuNPs/Ni-MOF substrate-free catalyzed for detection Enrofloxacin. J. Electroanal. Chem (2022). https://doi.org/10.1016/j.jelechem.2022.116251
50. H. Karimi-Maleh, M.L. Yola, N. Atar, Y. Orooji, F. Karimi, P. Senthil Kumar, J. Rouhi, M. Baghayeri, A novel detection method for organophosphorus insecticide fenamiphos: Molecularly imprinted electrochemical sensor based on core-shell Co3O4@MOF-74 nanocomposite. J. Colloid Interface. Sci **592**, 174–185 (2021)
51. J. Dong, D. Zhang, C. Li, T. Bai, H. Jin, Z. Suo, A sensitive electrochemical sensor based on PtNPs@Cu-MOF signal probe and DNA walker signal amplification for Pb2+ detection. Bioelectrochemistry **146**, 108134 (2022)
52. S.S. Ming, N.S.K. Gowthaman, H.N. Lim, P. Arul, E. Narayanamoorthi, I. Ibrahim, H. Jaafar, S.A. John, Aluminium MOF fabricated electrochemical sensor for the ultra-sensitive detection of hydroquinone in water samples. J. Electroanal. Chem **883**, 115067 (2021)
53. H. Sohrabi, M.R. Majidi, K. Asadpour-Zeynali, A. Khataee, A. Mokhtarzadeh, Bimetallic Fe/Mn MOFs/MβCD/AuNPs stabilized on MWCNTs for developing a label-free DNA-based genosensing bio-assay applied in the determination of Salmonella typhimurium in milk samples. Chemosphere **287**, 132373 (2022)
54. H. Peng, Y. Hui, L. Zhang, F. Zhang, Y. Liu, J. Zheng, R. Jia, Y. Song, B. Wang, A novel electrochemical aptasensor based on Ti3C2-MOFs nanocomposites for rapid streptomycin detection in milk samples. Sens. Actuators B Chem. **368**, 132119 (2022)
55. Q. Wang, J. Zhang, N. Dou, J. Qu, Sensitive simultaneous determination of catechol and hydroquinone based on iron and nitrogen doped carbon nanonets derived from MOFs. J. Electroanal. Chem. **913**, 116290 (2022)
56. C. Ye, X. Chen, D. Zhang, J. Xu, H. Xi, T. Wu, D. Deng, C. Xiong, J. Zhang, G. Huang, Study on the properties and reaction mechanism of polypyrrole@norfloxacin molecularly imprinted electrochemical sensor based on three-dimensional CoFe-MOFs/AuNPs. Electrochim. Acta. **379**, 138174 (2021)
57. M. Cao, Y. Zou, Y. Zhang, T. Zeng, Q. Wan, G. Lai, N. Yang, Robust and selective electrochemical sensing of hazardous photographic developing agents using a MOF-derived 3D porous flower-like Co3O4@C/graphene nanoplate composite. Electrochim. Acta. **409**, 139967 (2022)
58. W. Wang, Z. Zhao, Q. Lei, Y. Sun, W. Zhang, S. Zhuiykov, W. Zhang, J. Hu, Constructing and electrochemical performance of AuNPs decorated MIL-53 (Fe, Ni) MOFs–derived nanostructures for highly sensitive hydrazine detection. Appl. Surf. Sci. **596**, 153573 (2022)
59. G. Xu, D. Huo, J. Hou, C. Zhang, Y. Zhao, C. Hou, J. Bao, X. Yao, M. Yang, An electrochemical aptasensor of malathion based on ferrocene/DNA-hybridized MOF, DNA coupling-gold nanoparticles and competitive DNA strand reaction. Microchem. J. **162**, 105829 (2021)
60. Y. Yang, Z. Shi, Y. Chang, X. Wang, L. Yu, C. Guo, J. Zhang, B. Bai, D. Sun, S. Fan, Surface molecularly imprinted magnetic MOFs: A novel platform coupled with magneto electrode for high throughput electrochemical sensing analysis of oxytetracycline in foods. Food. Chem. **363**, 130337 (2021)
61. N. Zhang, D. Li, M. Mu, M. Lu, Partially Fluorinated Uio-66-Nh2(Zr): Positive Effect of the Fluorine Moiety on the Adsorption Capacity for Environmental Pollutants of Metal-Organic Frameworks. SSRN Electron. J. **2**, 137467 (2022)
62. J. Lalmalsawmi, T.D. Sarikokba, D.-J. Kim, Simultaneous detection of Cd2+ and Pb2+ by differential pulse anodic stripping voltammetry: Use of highly efficient novel Ag0(NPs) decorated silane grafted bentonite material. J. Electroanal. Chem. **918**, 116490 (2022)
63. A.M. Mahmoud, M.H. Mahnashi, M.M. El-Wekil, Indirect differential pulse voltammetric analysis of cyanide at porous copper based metal organic framework modified carbon paste electrode: Application to different water samples. Talanta. **221**, 121562 (2021)
64. R.K.A. Amali, H.N. Lim, I. Ibrahim, Z. Zainal, S.A.A. Ahmad, Silver nanoparticles-loaded copper (II)-terephthalate framework nanocomposite as a screen-printed carbon electrode modifier for amperometric nitrate detection. J. Electroanal. Chem. **918**, 116440 (2022)

65. T. Iftikhar, Y. Xu, A. Aziz, G. Ashraf, G. Li, M. Asif, F. Xiao, H. Liu, Tuning electrocatalytic aptitude by incorporating α-MnO2Nanorods in Cu-MOF/rGO/CuO hybrids: Electrochemical sensing of resorcinol for practical applications. ACS Appl Mater Interfaces. **13**, 31462–31473 (2021)
66. X. Zhang, X. Huang, Z. Wang, Y. Zhang, X. Huang, Z. Li, M. Daglia, J. Xiao, J. Shi, X. Zou, Bioinspired nanozyme enabling glucometer readout for portable monitoring of pesticide under resource-scarce environments. Chem. Eng. J. **429**, 132243 (2022)
67. P. Arul, S.T. Huang, V. Mani, Y.C. Hu, Ultrasonic synthesis of bismuth-organic framework intercalated carbon nanofibers: A dual electrocatalyst for trace-level monitoring of nitro hazards. Electrochim. Acta. **381**, 138280 (2021)
68. J. Wan, Y. Shen, L. Xu, R. Xu, J. Zhang, H. Sun, C. Zhang, C. Yin, X. Wang, Ferrocene-functionalized Ni(II)-based metal-organic framework as electrochemical sensing interface for ratiometric analysis of Cu2+, Pb2+ and Cd2+. J. Electroanal. Chem. **895**, 115374 (2021)
69. J. Ru, X. Wang, X. Cui, F. Wang, H. Ji, X. Du, X. Lu, GaOOH-modified metal-organic frameworks UiO-66-NH2: Selective and sensitive sensing four heavy-metal ions in real wastewater by electrochemical method. Talanta. **234**, 122679 (2021)
70. C. Chai, J. Gao, G. Zhao, L. Li, Y. Tang, C. Wu, C. Wan, In-situ synthesis of ultrasmall Au nanoparticles on bimetallic metal-organic framework with enhanced electrochemical activity for estrone sensing. Anal. Chim Acta. **1152**, 338242 (2021)
71. F.F. Wang, C. Liu, J. Yang, H.L. Xu, W.Y. Pei, J.F. Ma, A sulfur-containing capsule-based metal-organic electrochemical sensor for Super-Sensitive capture and detection of multiple Heavy-Metal ions. Chem. Eng. J. **438**, 135639 (2022)
72. Y. Liu, R. Weerasooriya, X. Chen, The metal-organic framework supported gold nanoparticles as a highly sensitive platform for electrochemical detection of methyl mercury species in the aqueous environment. J. Hazard. Mater. **431**, 128608 (2022)
73. W. Ye, Y. Li, J. Wang, B. Li, Y. Cui, Y. Yang, G. Qian, Electrochemical detection of trace heavy metal ions using a Ln-MOF modified glass carbon electrode. J. Solid State Chem. **281**, 121032 (2020)
74. Y. Wang, G. Zhao, G. Zhang, Y. Zhang, H. Wang, W. Cao, T. Li, Q. Wei, An electrochemical aptasensor based on gold-modified MoS2/rGO nanocomposite and gold-palladium-modified Fe-MOFs for sensitive detection of lead ions. Sens. Actuators, B Chem. (2020). https://doi.org/10.1016/j.snb.2020.128313
75. P. Arul, S.A. John, Size controlled synthesis of Ni-MOF using polyvinylpyrrolidone: New electrode material for the trace level determination of nitrobenzene. J. Electroanal .Chem. **829**, 168–176 (2018)
76. S. Chen, R. Huang, J. Zou, D. Liao, J. Yu, X. Jiang, A sensitive sensor based on MOFs derived nanoporous carbons for electrochemical detection of 4-aminophenol. Ecotoxicol. Environ. Saf. **191**, 110194 (2020)
77. R. Huang, D. Liao, S. Chen, J. Yu, X. Jiang, A strategy for effective electrochemical detection of hydroquinone and catechol: Decoration of alkalization-intercalated Ti3C2 with MOF-derived N-doped porous carbon. Sens. Actuators, B Chem **320**, 128386 (2020)
78. S. Rani, S. Kapoor, B. Sharma, S. Kumar, R. Malhotra, N. Dilbaghi, Fabrication of Zn-MOF@rGO based sensitive nanosensor for the real time monitoring of hydrazine. J. Alloys. Compd. **816**, 152509 (2020)
79. Y. Cao, L. Wang, C. Shen, C. Wang, X. Hu, G. Wang, An electrochemical sensor on the hierarchically porous Cu-BTC MOF platform for glyphosate determination. Sens. Actuators, B Chem **283**, 487–494 (2019)
80. J. Song, M. Huang, X. Lin, S.F.Y. Li, N. Jiang, Y. Liu, H. Guo, Y. Li, Novel Fe-based metal–organic framework (MOF) modified carbon nanofiber as a highly selective and sensitive electrochemical sensor for tetracycline detection. Chem. Eng. J. (2021). https://doi.org/10.1016/j.cej.2021.130913
81. B.R. Chansi, R.P. Rao, I. Mukherjee, P.K. Agrawal, T. Basu, L.M. Bharadwaj, Layered construction of nano immuno-hybrid embedded MOF as an electrochemical sensor for rapid quantification of total pesticides load in vegetable extract. J. Electroanal. Chem. **873**, 114386 (2020)

82. X.Z. Meng, H.W. Gu, H.C. Yi, Y.Q. He, Y. Chen, W.Y. Sun, Sensitive detection of streptomycin in milk using a hybrid signal enhancement strategy of MOF-based bio-bar code and target recycling. Anal. Chim. Acta. **1125**, 1–7 (2020)
83. N. Zhou, Y. Ma, B. Hu, L. He, S. Wang, Z. Zhang, S. Lu, Construction of Ce-MOF@COF hybrid nanostructure: Label-free aptasensor for the ultrasensitive detection of oxytetracycline residues in aqueous solution environments. Biosens. Bioelectron. **127**, 92–100 (2019)
84. Y. Dong, J. Li, L. Zhang, 3D hierarchical hollow microrod via in-situ growth 2D SnS nanoplates on MOF derived Co, N co-doped carbon rod for electrochemical sensing. Sens. Actuators, B Chem **303**, 127208 (2020)
85. J. Wen, D. Jiang, X. Shan, W. Wang, F. Xu, H. Shiigi, Z. Chen, Ternary electrochemiluminescence biosensor based on black phosphorus quantum dots doped perylene derivative and metal organic frameworks as a coreaction accelerator for the detection of chloramphenicol. Microchem. J **172**, 106927 (2022)
86. J. Gao, P. He, T. Yang et al., Short rod-like Ni-MOF anchored on graphene oxide nanosheets: A promising voltammetric platform for highly sensitive determination of p-chloronitrobenzene. J. Electroanal. Chem. **861**, 113954 (2020)
87. D. Duan, X. Si, Y. Ding, L. Li, G. Ma, L. Zhang, B. Jian, A novel molecularly imprinted electrochemical sensor based on double sensitization by MOF/CNTs and Prussian blue for detection of 17β-estradiol. Bioelectrochemistry **129**, 211–217 (2019)
88. N.M. Umesh, rani KK, Devasenathipathy R, Sriram B, Liu YX, Wang SF, Preparation of Co-MOF derived Co(OH)2/multiwalled carbon nanotubes as an efficient bifunctional electro catalyst for hydrazine and hydrogen peroxide detections. J. Taiwan. Inst. Chem. Eng. **93**, 79–86 (2018)
89. Y. Yu, C. Yu, Y. Niu, J. Chen, Y. Zhao, Y. Zhang, R. Gao, J. He, Target triggered cleavage effect of DNAzyme: Relying on Pd-Pt alloys functionalized Fe-MOFs for amplified detection of Pb2+. Biosens. Bioelectron. **101**, 297–303 (2018)
90. Y. Zhang, P. Yan, Q. Wan, N. Yang, Integration of chromium terephthalate metal-organic frameworks with reduced graphene oxide for voltammetry of 4-nonylphenol. Carbon N Y **134**, 540–547 (2018)
91. S. Singh, A. Numan, Y. Zhan, V. Singh, T. Van Hung, N.D. Nam, A novel highly efficient and ultrasensitive electrochemical detection of toxic mercury (II) ions in canned tuna fish and tap water based on a copper metal-organic framework. J. Hazard. Mater. **399**, 123042 (2020)
92. H. Filik, A.A. Avan, Dextran modified magnetic nanoparticles based solid phase extraction coupled with linear sweep voltammetry for the speciation of Cr(VI) and Cr(III) in tea, coffee, and mineral water samples. Food Chem. **292**, 151–159 (2019)
93. W. Wei, S. Dong, G. Huang, Q. Xie, T. Huang, MOF-derived Fe2O3 nanoparticle embedded in porous carbon as electrode materials for two enzyme-based biosensors. Sens. Actuators, B Chem. **260**, 189–197 (2018)
94. C. Zhang, T. Hao, H. Lin, Q. Wang, Y. Wu, K. Kang, X. Ji, Z. Guo, One-step electrochemical sensor based on an integrated probe toward sub-ppt level Pb2+ detection by fast scan voltammetry. Anal Chim Acta **1128**, 174–183 (2020)

Chapter 12
Applications of MoS_2 Nanostructures in Wastewater Treatment

Rashi Gusain, Neeraj Kumar, and Suprakas Sinha Ray

Abstract The fascinating properties of two-dimensional (2D) nanomaterial, such as excellent mechanical strength, a high portion of active sites, ease of functionalization and tuning the physical and chemical characteristics, are attracting researchers to host their applications in various fields, including wastewater treatment. Among various 2D nanomaterials, 2D MoS_2 has stand out as a promising alternative inorganic analogue of most explored 2D graphene due to its unique characteristics such as high active surface area, low cost, excellent mechanical strength, small band gap and the possibility of surface functionalization. The excellent water remediation characteristics are attributed to the controlled morphology, specific nano-sized properties, abundant availability, and variable surface chemistry of MoS_2 nanomaterials. Additionally, the selectivity of MoS_2 towards water contaminants promotes its application in water purification. This chapter presents the recent progress, future prospects and challenges of 2D MoS_2-based nanomaterials in water remediation techniques such as adsorbent, photocatalyst, membrane and antibacterial agent. The mechanism behind the water treatment process using 2D MoS_2 is also explained. This chapter will provide a platform to the researchers, who are focused on exploring the application of MoS_2-based materials in water purification. The research demands for future water applications of 2D MoS_2 nanomaterials are also identified.

R. Gusain (✉) · N. Kumar · S. S. Ray
Department of Chemical Sciences, University of Johannesburg, Doornfontein 2028, South Africa
e-mail: dr.rashi20@gmail.com

S. S. Ray
e-mail: rsuprakas@csir.co.za

Centre for Nanostructures and Advanced Materials, DSI-CSIR Nanotechnology Innovation Centre, Council for Scientific and Industrial Research, Pretoria 0001, South Africa

N. Kumar et al. (eds.), *Two-Dimensional Materials for Environmental Applications*, Springer Series in Materials Science 332, https://doi.org/10.1007/978-3-031-28756-5_12

12.1 Introduction

To improve the healthy and ever-lasting qualitative life on the planet, the topmost priority is balancing environmental sustainability. The word environmental sustainability signifies the healthy equilibrium between the consumerist living creatures, especially humans and the living world or the resources of the living world. However, the ever-increasing population, industrialization and urbanization demand more resources and manufacturing for consumption, leading to deforestation, greenhouse gases emission and more energy utilization. This has now become an immense challenge threatening the sustainability of our global society, mainly in the form of the scarcity of freshwater availability, energy supply and climate change. The inadequate supply of clean, fresh water is one of the global challenges which should be concerned for survival and gain the attention of social workers and researchers. Although 70% of the Earth is occupied with water in the form of glaciers, oceans, icecaps, sea, rivers, and lakes, out of all that, only 3% is available for consumption as fresh, clean water, and the rest 97% is salty water, which needs to be treated before any use [1]. Also, untreated effluents from various sources (e.g. domestic, mining, industry, agriculture, pharmaceutical (Fig. 12.1), enter into the freshwater reservoirs and groundwater to take participate in water pollution, which directly/indirectly is dangerous for the human, animals and marine creatures and disturb the sustainability of the life on the planet.

Africa and Asia are the two most affected continents, which will soon run out of clean water for consumption. Therefore, significant efforts are made in water conservation and removing the toxic contaminants from wastewater effluents before discharging them into the water reservoirs or for consumption. The effluents of wastewater contain a wide range of toxic organic (dyes, pharmaceutical by-products and ingredients, pesticides, surfactants, polyaromatic hydrocarbons,

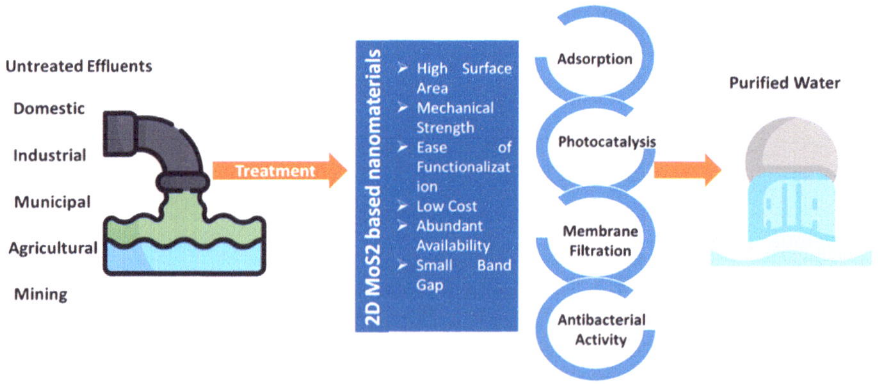

Fig. 12.1 Sources of Wastewater Effluents, properties of MoS_2-based nanomaterials and several wastewater treatment technologies using MoS_2-based nanomaterials

fertilizers, phenols, etc.) and inorganic (mining waste, heavy metal ions, radioactive substances, salts, metal oxides and metal complexes) water contaminants which are a major threat for the eco-system [2]. These toxic water contaminants can be carcinogenic and cause dysfunctional reproductive and immune systems, congenital disabilities, and risk to the physical and mental growth of infants/children [3]. The adverse impact of the toxic chemicals in the contaminated water has now become an irrefutable global issue. Therefore, to avoid water pollution, there is an immense need to raise awareness of water management and improve wastewater treatment technologies. To maintain potable water quality, the World health organization (WHO) has set a standard for the permissible limit of various elements in the water for consumption [4, 5].

Water purification techniques, such as adsorption [6], advanced oxidation process (e.g. photocatalysis, Fenton's oxidation) [7], electrocatalysis [8], photoelectrocatalysis [9], membrane filtration [10], biological precipitation [11], flocculation [12], and reserve osmosis [13], have been employed to remove or minimize the water contaminant level in wastewater. A variety of active nanomaterials with a high surface area have been investigated in the water treatment process in various methods. Among several classes of nanomaterials, 2D materials have been appreciated as the most fascinating class of nanomaterials, which can be an ideal candidate for various applications, including wastewater treatment [14]. Graphene is one of the most popular examples of 2D nanomaterials, and since its discovery in 2004 it became the popular choice to host applications [15]. With the continuous research on graphene-based materials, other 2D materials or inorganic analogues of graphene, such as MXene [16], layered double hydroxides [17], Metal–organic frameworks [18], transition metal oxides [19], and transition metal chalcogenides [20], have also become the source of attention to investigate in several fields. Among all, particularly MoS_2-based nanomaterials have also gained significant interest in several applications such as lubrication [21], energy storage [22], catalysis [23, 24], sensors [25] and water treatment [26], which is attributed to its outstanding properties, including excellent mechanical strength, high surface area, low dimension, quantum confinement, and surface defects (Fig. 12.1). Bulk MoS_2 is abundantly available as mineral molybdenite and has been used as catalysts and adsorbents for long [27–29]. However after the development of processes for isolating monolayer or few-layered MoS_2 nanosheets from bulk, with exclusive properties that are precise for the nano-sized material, the study of 2D MoS_2 nanomaterials has gained attention. Since then, the research on synthesis processes, functionalization, tuning and properties of 2D MoS_2 nanomaterials have been came into limelight and became a promising candidate in wastewater treatment [30–33].

Herein, we propose to emphasize the application of 2D MoS_2-based nanomaterials or nanocomposites to remove water contaminants from wastewater using various wastewater treatment techniques. There are several reviews published on water remediation using carbon [34], graphene or graphene-based materials [35, 36], and other 2D materials [37]. Also, the review on the MoS_2 synthesis, and properties, and several applications, especially energy applications, are published [22, 38]. However, a review or chapter focusing on the candidature of 2D MoS_2 in water remediation with a recent update is missing. The detailed mechanism of the MoS_2 materials in water treatment is also discussed. This chapter is dedicated to the application of MoS_2 nanomaterials in cleaning the wastewater through adsorption, photocatalysis, membrane filtration and antibacterial activity (Fig. 12.1) and a perspective on future work for MoS_2 nanomaterials.

12.2 Application of MoS_2 in Wastewater Treatment

Water contamination is a global issue responsible for clean water scarcity and deteriorating human, animal and marine creatures' health. Several techniques have been proposed to clean the wastewater before discharging it into the water bodies or before consumption. With the growing interest of the scientific community towards wastewater treatment to save life on Earth, the application of several nanomaterials has been investigated to clean the water. The approach of nanostructured materials in wastewater remediation offers new dimensions to evaluate, analyse and solve water pollution. Recently, MoS_2 has gained much attention in water remediation due to its unique properties, such as high active surface area, low cost, small band gap and the possibility of surface functionalization. In several studies, MoS_2 and MoS_2-based nanocomposites have been proven as excellent adsorbents and photocatalysts to remove water contaminants. The following sub-section briefly describes the application of MoS_2 and MoS_2-based nanocomposites in wastewater remediation as adsorbent, photocatalyst, membrane, and antibacterial agents.

12.2.1 Adsorption

Adsorption is one of the most explored wastewater treatment techniques, which is widely accepted and lowcost. The presence of plenty of exposed sulphur atoms on the MoS_2 surface provides the platform for the adsorption of cationic water contaminants through strong Lewis acid and base soft–soft interactions [33]. Geng et al. prepared the flower-like nanostructures of MoS_2 nanosheets for the adsorption of cationic (Rhodamine B, RhB) dye, and the adsorption capacity was noticed to be 49.2 $mg.g^{-1}$ [39]. Electrostatic interaction between the cationic dye and negative MoS_2 was found to be the major driven force for the adsorption. To confirm the adsorption behaviour of RhB on MoS_2, FTIR analysis of MoS_2 before and after the adsorption of RhB was performed, which indicates that the intensity of Mo-S peak was reduced and some new vibrational signatures assigned to aromatic rings of RhB has been introduced. This suggests the strong interaction of RhB to the MoS_2 backbone via electrostatic interaction. Further, the application of hierarchical microspheres of MoS_2 nanosheets was also compared for the various cationic (e.g. methylene blue (MB), Rhodamine (Rh) and malachite green (MG)) and anionic dyes (e.g. fuschin acid (FA) and congo red (CR)) adsorption from aqueous medium (Fig. 12.2) [26]. MoS_2 was found to exhibit excellent adsorption capacity for MB followed by others in the following order MB (297 $mg.g^{-1}$) > Rh (216 $mg.g^{-1}$) > MG (204 $mg.g^{-1}$) > FA (183 $mg.g^{-1}$) > CR (146 $mg.g^{-1}$). The high absorption capacity towards cationic dyes was also believed to be the effect of the van der Waals forces and the electrostatic interactions. However, the adsorption of anionic contaminates was only found to be compelled by van der Waals interaction. The adsorption of cationic MB dye on MoS_2 surface was further analysed FTIR spectroscopy. Figure 12.2a shows a FTIR spectra of MB dye, and MoS_2 before and after MB adsorption. The new vibrational signatures on the MB adsorbed MoS_2 are characteristic peaks of MB dye, which confirms the adsorption of MB on MoS_2. Additionally, the Mo-S vibrational peaks remains unchanged in the recovered MoS_2 after MB dye adsorption. Therefore, MoS_2 was also examined for the recyclability and showed an excellent adsorption capacity for 5 adsorption cycles (Fig. 12.2b). The adsorption was followed by Freundlich isotherm model and pseudo second order kinetics. The adsorption capacity of MoS2 for cationic and anionc dyes has shown in Fig. 12.2c.

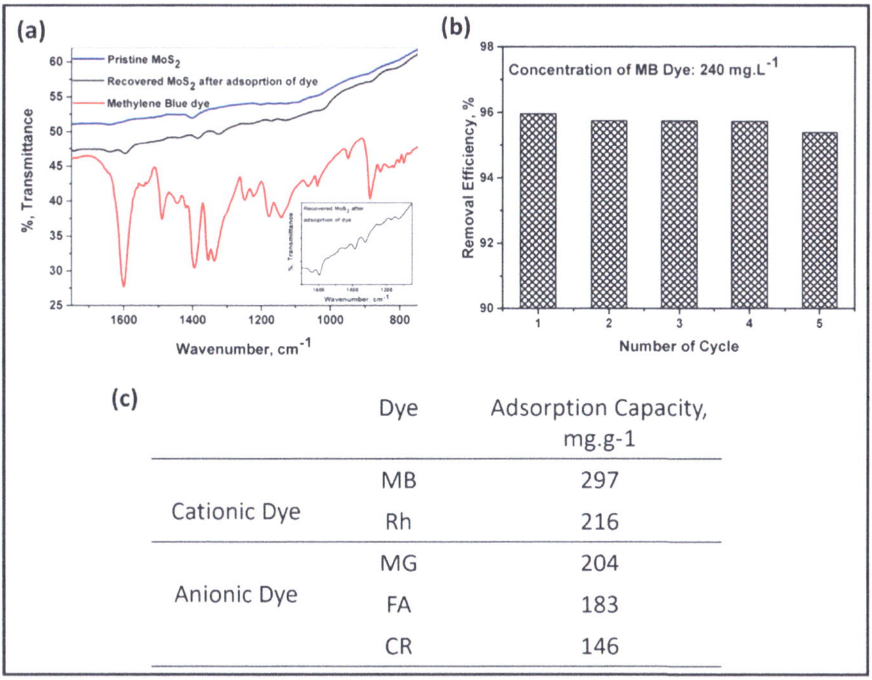

	Dye	Adsorption Capacity, mg.g-1
Cationic Dye	MB	297
	Rh	216
Anionic Dye	MG	204
	FA	183
	CR	146

Fig. 12.2 **a** FTIR analysis of MB dye, pristine MoS_2 and MoS_2 after MB dye adsorption, **b** recyclability of MoS_2 for MB dye adsorption and **c** adsorption capacity of MoS_2 nanosheets towards cationic and anionic dyes. Reproduced with permission from ref. [26]. Copyright 2016, American Chemical Society

Similar to cationic organic contaminants, MoS_2 is also valuable for removing inorganic heavy metal ions. The adsorption mechanism and all the possible interactions between the heavy metal ions and MoS_2 are nicely explained in parts (a) and (b) of Fig. 12.3 [33, 40]. Several interactions, such as electrostatic interaction, complexation formation, and ion-exchange, help in heavy metal ion adsorption (Fig. 12.3a). Ion-exchange is considered the primary adsorption mechanism for metal adsorption on the MoS_2 surface. Generally, the MoS_2 surface exhibits a negative charge with positive counter ions [40–42], which allows the metal-sulphur bond and results in complexation. Another mechanism is multilayer adsorption which involves the inner layer complex formation (metal sulphur complex formation) and outer layer complex formation (e.g. electrostatic interaction). Another possible adsorption mechanism is the intercalation of metal ions into the MoS_2 nanosheets. Several synthesis routes of MoS_2 nanosheets can introduce defects or widen the interlayer spacing. These spacing are enough to expose the interior sulphur atoms and helps the adsorption of metal ions. For example, Lu et al. followed the one-step hydrothermal route for the MoS_2 nanosheets preparation and achieved to widened the interlayer spacing to 0.94 nm from 0.62 nm (Fig. 12.3b) [40]. This helps in extremely high and fast Hg(I) adsorption (2506 $mg.g^{-1}$).

However, the adsorption of anionic contaminant on MoS_2 can be improved by the fabrication of MoS_2 with other nanomaterial/polymers as nanocomposite. For

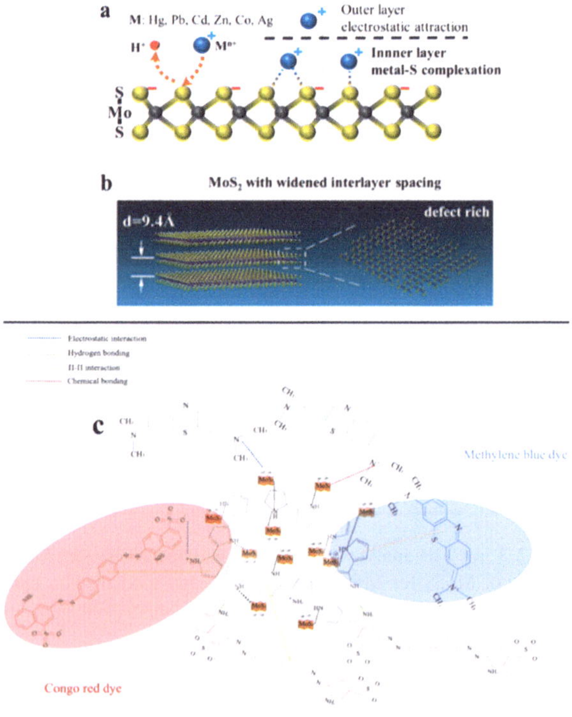

Fig. 12.3 **a** Several plausible adsorption mechanisms for heavy metal ion adsorption on MoS2 nanosheets. Reproduced with permission from ref. [33]. Copyright 2017, American Chemical Society. **b** Schematic representation of widened interlayer spacing in MoS_2 nanosheets. Reproduced with permission from ref. [40]. Copyright 2016, John Wiley and Sons, Inc. **c** Possible adsorption mechanism for the adsorption of CR and MB dye on Ppy@MoS_2. Reproduced with permission from ref. [43]. Copyright 2022, Elsevier Science Ltd.

example, Zhang et al. prepared the polypyrrole functionalized MoS_2 (Ppy@MoS_2) microtubes for MB, RhB, methyl orange (MO) and CR adsorption from wastewater [43]. The resultant composite exhibits a higher specific surface area than pristine MoS_2, which helps in the adsorption of dyes. The composite showed the best adsorption capacity for the anionic dye (CR, 598.79 mg.g^{-1}) than the cationic dye. The high surface area and synergistic effect between MoS_2 and Ppy microtubes is the key to excellent adsorption capacity. The adsorption of CR on Ppy@MOS_2 was driven by several possible interactions, such as electrostatic interaction between the anionic dye and Ppy, π-π interactions between the aromatic rings of the adsorbate and adsorbent, π-π stacking interactions, and hydrogen bonding (Fig. 12.3c). MoS_2 has also been used for the adsorption of oil and organic solvents from the water [44, 45]. Hydrophobic interactions are the major forces for the adsorption of oil and organic solvents on the MoS_2 surface. MoS_2 surface can be engineered into the superhydrophobic (water contact angle from 85° to ~150°) and used as an adsorbent for a wide range of hydrophobic oils and organic solvents [46, 47].

In summary, MoS_2 can be efficiently used as an adsorbent to remove a wide range of water pollutants, preferably cationic contaminants, from wastewater. Electrostatic interaction is considered the major driving force for the adsorption of cationic pollutants on the MoS_2 surface. The Sulphur atom on the MoS_2 surface acts as a Lewis base and exhibits high affinity toward the cationic contaminants. Therefore, the selectivity of MoS_2 is much higher than other adsorbents. However, the functionalization of MoS_2 turned the properties and provided an excellent surface area to adsorb the broad spectrum of all kinds of water contaminants. Other than electrostatic interactions, π–π interaction, hydrogen bonding, metal complexation and van der Waals interaction also participate in the adsorption of water pollutants. Table 12.1 lists examples of various MoS_2-based adsorbent materials for the adsorption of water contaminants with adsorption capacity.

12.2.2 Photocatalysis

Photocatalytic degradation of water contaminant molecules is another promising way to clean wastewater without producing secondary waste. Photocatalytic wastewater treatment is a well-known advanced oxidation process. It exhibits several advantages over other treatment techniques, such as cost-effectiveness, complete degradation or mineralization, simple practice, and mild reaction conditions [70]. 2D MoS_2 nanosheets are one of the exciting photocatalyst candidates due to their excellent charge mobility and high optical absorption characteristics [71–73]. In a typical photocatalysis reaction, a semiconductor material absorbs the photon of energy equal to or more than its band gap energy. It is excited to jump the electrons from the valence band (VB) to the conduction band (CB). This creates holes in the VB, which helps in the oxidation and the excited electrons in the CB help in the reduction to carry out the photocatalytic redox reaction and produce reactive oxygen species (ROS). These ROS species are the active generations that either mineralizes the organic

Table 12.1 Few examples of various MoS_2-based nanoadsorbents with their adsorption capacity for targeted water contaminants

MoS_2 based nanoadsorbent	Targeted water contaminant	Adsorption capacity, $mg.g^{-1}$	References
Magnetic Fe_3O_4/MoS_2	CR	71	[48]
MoS_2/Fe_3O_4	Pb(II) Hg(II)	263.6 428.9	[49]
MoS_2/CeO_2	Pb(II)	333	[50]
MoS_2	RhB	136.99	[51]
Fungus-like MoS_2	CR	285.7	[52]
Hollow MoS_2	MO	41.52	[53]
MoS_2	RhB	365	[54]
C/MoS_2	MO	450	[55]
MoS_2	MB MG RhB FA CR	297 204 216 183 146	[26]
$MoO_3@MoS_2$	RhB	326.8	[56]
$FeOCl\text{-}MoS_2$	MO	1615.11	[57]
MoS2	MB	146.43	[58]
CeO2–MoS_2	Pb(II) Humate	263 218	[59]
MoS_2–rGO	Pb(II) Ni(II)	322 294	[60]
MoS_2@Zeolite-5	Tetracycline (TC)	396.7	[61]
MoS_2	MB MO Co(II) Ni(II)	181.8 102.1 61.7 51.8	[62]
MoS_2@bentonite	Crystal violet (CV)	384.61	[63]
1 T MoS_2 2H-MoS_2	Pb(II) Cu(II) Pb(II) Cu(II)	147.09 82.13 64.16 50.74	[64]
MoS_2 with increased interlayer spacing	Pb(II)	303.04	[65]
Graphene-like layered MoS_2 (g-MoS_2)	Doxycycline (DC)	310	[66]
g-MoS_2 decorated biochar	TC	249.45	[67]
MoS_2/SH-MWCNT	Pb(II) Cd(II)	90 66.6	[68]
MoS_2/Graphene	RhB	285	[69]

molecules or degrade completely. However, one major challenge in photocatalysis is the mobility of charge carriers. The too-small band gap of the semiconductor material is responsible for the quick recombination of the generated electron and hole pairs. Hence, they won't be able to participate in the photocatalytic reaction.

Conversely, too broad-band gap does not absorb the broad spectrum of solar light, thus not appropriate for the photocatalytic response. Bulk MoS_2 exhibit a narrow band gap of ~1.3 eV, allowing the adsorption of most of the solar spectrum. Still, the fast recombination of charge carriers makes a negative impact on its photocatalytic reaction. On exfoliating the bulk MoS_2 to few-layered 2D MoS_2 nanosheets or single-layered MoS_2, the band gap increased enough to improve the life span of charge carriers, adsorption of UV–visible, visible and near-infrared light and consequently improve the photocatalytic reaction [71].

Besides band energy, band edges' position also significantly affects the photochemical reaction. The redox potential of the charge carriers strongly depends on the position of the band edges. For example, the smaller the CB potential, the stronger the reduction capability of photo-generated electrons or the larger the VB potential of semiconductor material, the stronger the oxidative capability of the holes. The band edge positions in the few-layered MoS_2 are more favourable to producing ROS generations than bulk MoS_2. Generally, MoS_2 is explored for the oxidation of water pollutants. In the photo-oxidation of water pollutants, the holes in VB play a vital role. Although the band edge potential of VB for monolayer MoS_2 (1.78 eV) is much higher than the bulk MoS_2 (1.40 eV), but is not sufficient to directly mineralise or decompose the pollutant molecules [74]. Therefore, MoS_2 is majorly used as a co-catalyst in a photocatalyst system to improve the photocatalytic degradation efficiency by increasing the charge mobility and suppressing the charge pair's recombination.

For example, MoS_2 was incorporated with Ag_3PO_4 and TiO_2 to prepare a ternary photocatalyst to degrade the MO, MB dye, and antibiotic Oxytetracycline (OTC) [75]. The ternary photocatalyst ($Ag_3PO_4/TiO_2@MoS_2$) was also compared with the $TiO_2@MoS_2$ for photocatalytic efficiency. Figure 12.4a shows the high-resolution SEM images of the ternary photocatalyst, in which TiO_2 nanofibers (180 nm diameter) are prepared by the electrospinning method and further implemented with a few layers of MoS_2 via a hydrothermal route. Ag_3PO_4 was deposited on the $TiO_2@MoS_2$ surface via chemical deposition to get $Ag_3PO_4/TiO_2@MoS_2$. Additionally, the presence of Ti, Mo, S, Ag and O energy dispersive X-ray spectrometry (EDX) analysis approves the preparation of $Ag_3PO_4/TiO_2@MoS_2$ composite. The prepared materials' optical characteristics and band gap were examined using the UV/vis diffuse reflectance spectrum. The band gap of the ternary photocatalyst was calculated to be 1.85 eV (Fig. 12.4b), which is sufficient to absorb the visible light photon.

Moreover, the ternary composite's light adsorption intensity was higher than binary Ag_3PO_4/MoS_2 and single Ag_3PO_4. The appropriate band gap and excellent light adsorption intensity significantly help the $Ag_3PO_4/TiO_2@MoS_2$ for photocatalytic application. The mechanism of the prepared nanocomposite is also explained in Fig. 12.4c. Under light irradiation, the electrons from Ag_3PO_4 VB jumped to the CB and then transferred to the MoS_2 VB. Meanwhile, the MoS_2 electrons jumped

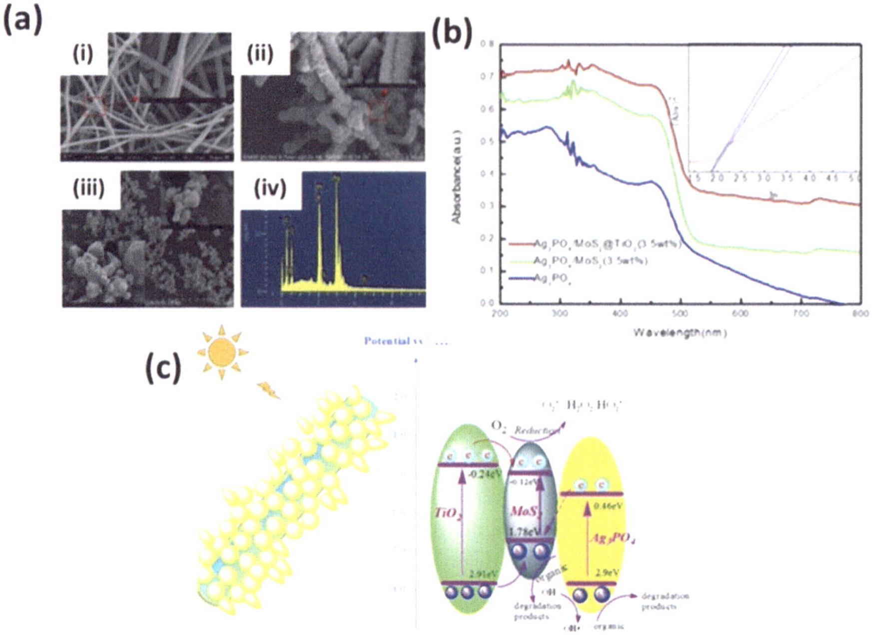

Fig. 12.4 **a** FE-SEM image of $Ag_3PO_4/TiO_2@MoS_2$ photocatalyst ((i) TiO_2 nanofibers, (ii) implantation of MoS_2 on TiO_2 nanofibers ($TiO_2@MoS_2$), (iii) ternary $Ag_3PO_4/TiO_2@MoS_2$ photocatalyst, (iv) EDS of $Ag_3PO_4/TiO_2@MoS_2$ composite), **b** UV–vis absorption graph of ternary $Ag_3PO_4/TiO_2@MoS_2$, binary Ag_3PO_4/MoS_2 and single Ag_3PO_4, and inset graph represents the kubelka–Munk transformed reflectance spectra, to calculate the bandgap values for the binary Ag_3PO4/MoS_2 (3.5 wt%) and ternary $Ag_3PO_4/TiO_2@MoS_2$, and **c** Schematic representation proposed charge mechanism with energy band structure of the ternary $Ag_3PO_4/TiO_2@MoS_2$ composites. Reproduced with permission from ref. [75]. Copyright 2017, Elsevier Science Ltd.

from VB to CB and inhibited the recombination of charge carriers. TiO_2 also trivially participates due to the small amount of UV light adsorption, and TiO_2 VB electrons also migrate to the MoS_2 CB. Subsequently, the TiO_2 fibres act as wire and help transfer the MoS_2 captured electrons in the solution for ROS generations. Also, the VB and CB positions of MoS_2 have higher potential than TiO_2, which helps for holes transfer from TiO_2 VB to MoS_2 VB. These holes can significantly oxidize the organic pollutants. The excellent charge mobility of MoS_2 not only improves the material's photocatalytic efficiency but also inhibits the photocorrosion rate of Ag(I). This improves the cyclic stability of the ternary composite, and only 10% reduction in photocatalytic efficiency was noticed after 10 cycling runs. Therefore, the excellent charge mobility and anti-photocorrosion attitude make MoS_2-based materials encouraging candidates for photocatalysis applications.

Like organic water contaminants, inorganic contaminants were also degraded using MoS_2 photocatalyst. Gao et al., employed the polyaniline (PANI) functionalized MoS_2 for the removal of Cr(VI) as an adsorbent and photocatalyst [76]. Under light irradiation, the PANI-MoS_2 nanocomposite could successfully remove

600 mg.g^{-1} Cr(VI) and photo-catalytically reduce it to Cr(III). The Cr(VI) removal was found to be highly pH dependent. The low pH conditions of the solution favour the reduction of Cr(VI).

In contrast, the reduced $Cr(OH)_3$ can quickly be precipitated on the PANI@MoS_2 surface at high pH conditions and inhibit the active sites. PANI@MoS_2 also perform excellent cyclic recyclability for several cycling runs. Cr(VI) was also photo-catalytically reduced using Fe(0) decorated g-C_3N_4-MoS_2 (GCNFM) nanocomposite [77]. The suitable band edges potential of the photocatalyst favours the photo-reduction of Cr(VI) into Cr(III). Under visible light irradiation, the electrons jumped to the Fe(0) and MoS_2 CB due to appropriate band edge alignments. This enhances the life span of charge carriers and improves the Cr(VI) reduction compared to g-C_3N_4, MoS_2 modified g-C_3N_4 and Fe(0) doped g-C_3N_4. The same photocatalyst also performs effectively for the mineralization of RhB organic dye. Several other MoS_2-modified semiconductor materials have also been examined for the photocatalytic degradation of water contaminants (Table 12.2). In summary, the advantage of photo-stability of MoS_2 than other chalcogenides against oxidation supports the application of MoS_2 in photocatalytic reactions [78]. Additionally, the band structure of a few layered or monolayer MoS_2 helps in the absorption of a broad spectrum of solar light, which is one of the essential demands for the semiconductor material. In the MoS_2-based semiconductor nanocomposite material, MoS_2 actively improve the photo-absorption response, suppress the recombination of charge carriers, providing a platform for the adsorption of contaminant and enhancing the photocatalytic reaction.

12.2.3 Membrane Filtration

The high mechanical strength, excellent thermal stability and antibacterial properties of MoS_2 make it a potent candidate for fabricating membranes with outstanding separation performance. Additionally, MoS_2-based membranes exhibit simultaneous permeability and high selectivity [99, 100], antifouling properties [101, 102] and facilitate multifunctional features [103]. In membrane filtration, the wastewater or contaminated water is passed through an appropriate membrane with specific pore sizes, allowing water to be passed and act as a barrier for contaminants. The filtration can be pressure, thermal, osmosis, and electrical driven. MoS_2 can be used to make nanoporous membranes and layer stacked membrane for wastewater treatment.

Nanoporous membranes are made using a few layers or single-layer MoS_2 nanosheets. The appropriate sizes of nanopores in the membranes are designed to block the passage of unwanted species. Pore characteristics, water filtering species characteristics and external pressure are majorly responsible for the membrane functioning in water purification [104]. The eco-friendly nature with flexible design and high quality of cleaned water through membrane filtration makes the membrane filtration technique popular to clear wastewater. Also, during water transport, the water molecules are connected inside and outside the nanopores by forming single-chain hydrogen bonding, which improves water filtration [105]. Most of the studies

Table 12.2 Few examples of various MoS_2-based nanoadsorbents with their photocatalytic efficiency in degrading the targeted water contaminant

MoS_2 based photocatalyst	Targeted water contaminant	Contaminant		Light source	Degradation efficiency	References
		Concentration $mg.L^{-1}$	Volume mL			
$MoS_2@Fe_3O_4$	RhB MB	10 30	50 50	Visible light	96% 98%	[79]
MoS_2	MB	10	200	Visible light	95.6%	[80]
$BiPO_4$–MoS_2/GR	RhB	5	100	Visible light	93%	[81]
CoS2/MoS_2@Zeolite	TC	200	50	300W Xe lamp, Visible light	96.71	[82]
BiOBr/MoS_2/GO	OTC TC Chlorotetracycline DC	10	25	300W Xe lamp, Visible light	98% 98% 98% 98%	[83]
rGO/MoS_2	MB	12.5	8.5	Simulated solar light	95%	[84]
CeO_2-MoS_2	Cr(VI)	5	20	500W Halogen lamp	99.6%	[85]
CeO_2–ZrO_2@MoS_2	Naproxen (NPX)	10	50	250 W LED lamp, Visible light	96%	[86]
MoS_2/Fe_3O_4	MO	10	50	300W Xe lamp, Visible light	80%	[87]
Fe_7S_8@MoS_2–O	MG Levofloxacin (LVX)	10	50	Xe lamp, Visible light	97.5% 92.7%	[88]
g-C_3N_4/Ag/ MoS_2	RhB	20	100	300W Xe lamp, Visible light	66.47	[89]
$ZnIn_2S_4$/MoS_2	MO	20	10	300W Xe lamp, Visible light	84	[90]
MoS_2/Ag_2CO_3	Lanasol Red 5B RhB Ciprofloxacin Metronidazole	80 40 20 20	40	500W Xe lamp, Visible light	98% 90% 80% 72%	[91]

(continued)

Table 12.2 (continued)

MoS_2 based photocatalyst	Targeted water contaminant	Contaminant		Light source	Degradation efficiency	References
		Concentration $mg.L^{-1}$	Volume mL			
CNT@MoS_2/SnS_2	Cr(VI)	50	50	300W Xe lamp, Visible light	100%	[92]
$SrFe_{12}O_{19}/MoS_2$	MB	20	80	Visible light	97%	[93]
MoS_2/Ag_3PO_4	Ofloxacin (OFL) RhB	10 10	100 100	Visible light Sunlight	79% 100%	[94]
MoS_2/ZnS	OTC	20	100	250 W Hg lamp, UV–visible light	81%	[95]
MoS_2/SnS_2	Cr(VI) MB	120 30	50 50	500W Xe lamp, Visible light	99.9% 96.5%	[96]
$MoS_2 - ZnS$	RhB	5	100	300 W Xe lamp, Visible light	100%	[97]
MoS_2/Bi_5O_7I	Bisphenol A TC Ciprofloxacin	10 20 10	100 100 100	300W Xe lamp, Visible light	99% 78% 40%	[98]

using nanoporous MoS_2-based membranes are used for desalination purposes. The performance of nanoporous MoS_2 membrane is majorly evaluated by theoretical simulation studies, not experimentally.

Heiranian et al. performed the theoretical calculations for the use of nanoporous MoS_2 membranes in water purification, and they concluded that > 88% of ions could easily be filtered using a monolayer MoS_2 membrane of a specific pore area (20 to 60 A^2) [106]. Figure 12.5a shows the typical simulation box consisting of a single-layer MoS_2 nanoporous membrane, a rigid piston (graphene sheet), water and ions. In this simulation box, three pores have been identified: Mo pores, S pores and mixed pores (consisting of Mo and S) (Fig. 12.5b). On comparing monolayer MoS_2 membrane with nanoporous graphene membrane, the water flux was found to be 70% improved with MoS_2 membrane. Figure 12.5c shows the water fluxes through various Mo pores and S pores, mixed pores and graphene sheets with respect to the applied pressure gradient. Among all types of pores in the simulation box, Mo-only pores show the highest water permeation. The mixed pore shows higher water fluxes compared to the graphene nanopores. Figure 12.5d shows the ions rejection through the pores of MoS_2 nanoporous membrane and graphene pores as a function of applied pressure. Ion rejection is lower at a higher pressure and with larger pores. The ion rejection capacity is found to be quite similar for the equivalent areas irrespective of the pore type. This suggested that ion rejection depends on the pore area only and not on the pore type. Figure 12.5e shows that the water filtration rate is also intensely increasing the pore from ~20 to ~50 $Å^2$.

This study suggests that while designing MoS_2 nanoporous membranes, the size and areas of pores are the essential characteristics to be kept in mind. The 0.44 nm or larger (diameter) pore sizes of nanoporous MoS_2 are favourable in wastewater treatment and display a minor energy barrier for water molecules and at the same time, reject the passage of salts [106, 107]. But the much larger diameter of nanopores (>1.05 nm) is not good for separating salts from water as the salt molecules can easily pass through the pores. Therefore, the identical size for the MoS_2 nano-membranes for water purification is suggested as 0.44 to 1.05 nm [33].

However, in MoS_2-based layer stacked membranes, exfoliated MoS_2 membranes are stacked to each other through the vacuum-filtration technique [108]. Efficient sieving of water contaminants (molecules, salt and ions) can be performed at the free layer spacing of capillary width between the stacked 2D MoS_2 nanosheets. The free layer spacing between the nanosheets can be controlled for the high selectivity of water filtration. The advantage of layer-stacked MoS_2 membranes is their high stability in the aqueous medium compared to other layer-stacked membranes, e.g., GO-based membranes [99]. Layer stacked MoS_2 membrane was immersed in water for 3 days to check the water stability, but it didn't swell and maintained its interlayer spacing [99]. This might be due to the absence of any hydrophilic functional group on the MoS_2 membrane surface and the van der Waals force between the stacked nanosheets, which provide the required stability against the redispersion of stacked nanosheets in water.

Additionally, the physical characteristic of 2D MoS_2-based membrane can be tuned or functionalized to boost the water flux and improve the barrier for unwanted

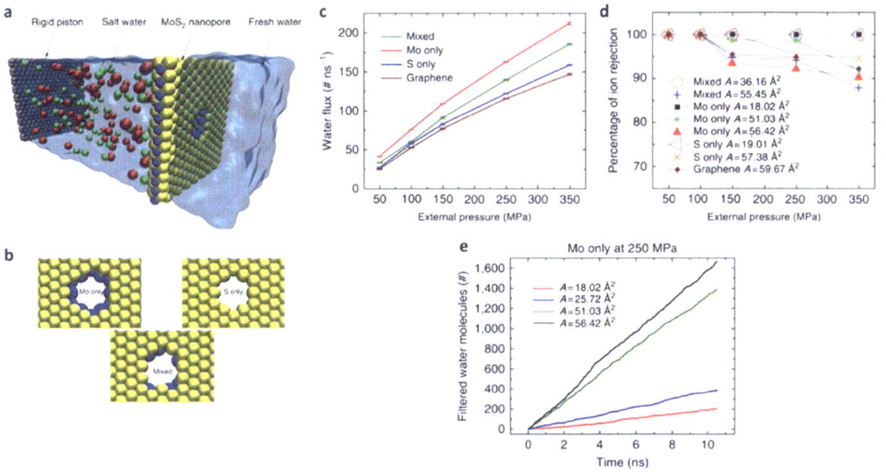

Fig. 12.5 **a** Schematic illustration of simulation box containing a MoS_2 nanosheet (blue colour represents Mo and yellow colour represents S), salt ions (red and green), graphene sheet (dark grey) and water (transparent blue). **b** Left: Pore with Mo only. Right: pore with S only. Bottom: mixed pore consisting of Mo and S. **c** Comparison of Mo-only nanopores, S-only nanopores, mixed nanopores and graphene nanopores for water flux with respect to the applied pressure with similar pore areas. **d** Percentage of ion rejection through several pores of different edge chemistries and pore ares, as a function of the applied pressure. **e** Counts of filtered water molecules through Mo only pores of different pore areas as a function of simulation time at a fixed pressure of 250 MPa. Reproduced with permission from ref. [106]. Copyright 2015, Nature Publications

contaminates [109]. For example, the interlayer spacing of the nanosheets can be modified for selective separation via the intercalation of some species. Lu et al. intercalated the amphiphilic ligand as a cross-linker in layer-stacked MoS_2 to tune the interlayer spacing for water filtration [101]. On the other hand, layer-stacked MoS_2 membranes can also be functionalized by immersing in a dye solution for desalination and nanofiltration [100]. The surface chemistry of the membrane was changed to the functionalized of dye on the surface, which contributed to minimal effect on the interlayer spacing. This helps in the rejection of the ions and salts with high selectivity. Overall 2D MoS_2-based membranes have shown a significant potential for the water treatment. However, the thorough study of MoS_2-based membranes in water purification is still underway as most studies have been performed by theoretical simulations and modellings.

12.2.4 Antibacterial Activity

Wastewater discharge, usually municipal effluents, is the perfect environment for the growth of deadly bacteria, which is a threat to the life of public health [110]. Several bacteria have shown antibiotic resistance, leading scientist to look for other effective alternative antibacterial proxies. Recently, MoS_2 has gained significant attention as an antibacterial agent due to its excellent biocompatibility, large surface area, ease of functionalization, high catalytic activity, cost-effectiveness, and chemical stability [111]. Regardless of its various outstanding characteristics, the application of pristine MoS_2 nanomaterials exhibit some limitations in the biomedical field. Therefore, MoS_2 is functionalized to improve its application as an antibacterial agent. The antibacterial activities of MoS_2 nanosheets have been found to be better than bulk MoS_2 due to the developed photo-response properties of MoS_2-based nanomaterials [112]. MoS_2 nanosheets prevent bacterial multiplication via physical contact, and the sharp ends of nano-architectures may penetrate into the cell wall of bacteria to kill it [113]. However, under the illumination of visible light, the generation of ROS on the MoS_2 surface causes bacterial inactivation, and hence photo-response of MoS_2 nanostructures helps in antibacterial activity. 1 T MoS_2 is considered a better option than 2H MoS_2, as 1 T MoS_2 exhibit higher electrical conductivity and is less resistant to electron migration.

1 T phase of chemically exfoliated MoS_2 (ce-MoS_2) nanosheets was evaluated for antibacterial purpose and compared with raw MoS_2 [115]. The antibacterial properties of ce-MoS_2 were followed by a three-step mechanism: (a) direct physical contact of bacterium-MoS_2, (b) sharp edges of MoS_2 damage the membrane, and (c) MoS_2 creates a disturbance in microbial redox reaction processes. Additionally, ce-MoS_2 is plausible to generate the ROS species under light irradiation, which raw MoS_2 cannot produce and help in high antibacterial activity. ce-MoS_2 also exhibits higher oxidation strength towards the thiols than raw MoS_2, which helps in the death of bacteria. Roy et al. proposed the one-step method to prepare the MoS_2 nanosheets using chitosan (CS-MoS_2). They found that it showed great potential for bactericidal action against both Gram-positive and Gram-negative bacteria [114]. Figure 12.6 shows the different mechanisms for the antibacterial activity of CS-MoS_2 for the bacterial cell death. The antibacterial activity of CS-MoS_2 is a collective action of oxidative stress, membrane damage by penetration and metabolic inactivation.

It is evident that the size and shape considerably affect the properties of the nanomaterials and hence the application efficiency. Xu et al. compared the two morphologies of MoS_2, i.e. nanosheets and nanoflowers, for the antibacterial activity [116]. They concluded that the nanoflowers possessed better antibacterial activity than MoS_2 nanosheets. This might be due to the nanoflower morphology's higher surface area, which provides more space for physical contact and oxidative stress to the bacterial cells. Also, nanoflower morphology shows higher oxidation strength towards the GSH, which helped in bacterial cell death.

12.3 Outlook and Future Perspectives

Among several other 2D materials, the unique properties of MoS_2 have granted its promising candidature in wastewater remediation. MoS_2 and MoS_2-based nanocomposites have been considerably studied as adsorbents, photocatalysts, membranes

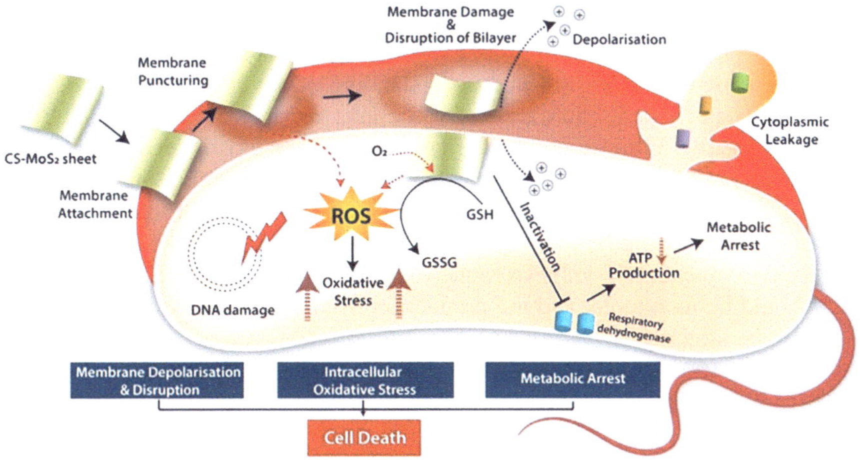

Fig. 12.6 Schematic representation of all the collective proposed mechanisms of antibacterial activity of CS-MoS_2 nanosheets. Reproduced with permission from ref. [114]. Copyright 2019, American Chemical Society

and antibacterial agents to clean wastewater. The application of MoS_2 is extensively explored as an adsorbent and photocatalyst. Despite the fact that an extensive amount of work has been done using MoS_2 in wastewater treatment, there are still a few limitations and daunting challenges for the researchers. MoS_2 surface exhibits a neutral, negative charge, and as an adsorbent material, it is majorly used to remove cationic pollutants from wastewater, which limits its broad application. To remove anionic contaminants, the modification in the MoS_2 as fabrication with other nanoparticles or functionalization with other groups is necessary. This practice usually increases the cost of the water remediation process. Additionally, it is difficult to separate the exfoliated MoS_2 nanosheets from the aqueous medium after the water-contaminant adsorption exercise. This restricts the adsorbent's reusability, increasing costs and producing secondary waste.

MoS_2 and MoS_2-based nanocomposites have also been studied as photocatalysts for the complete degradation or mineralization of water contaminants. The narrow band gap of the MoS_2 allows the absorption of broad solar spectrum photons, which makes it a popular choice. However, simultaneously, the narrow band gap of MoS_2 promotes the rapid recombination of the charge carriers, which is one of the main obstacles to the practical application of MoS_2 as a photocatalyst. To avoid this, researchers use MoS_2 functionalized nanomaterial as a photocatalyst for water decontamination. But the thorough study of band gap structural knowledge, products of water mineralization and photo-stability of the nanocomposite is still lacking. For the advancement in this field, in-depth knowledge and understanding of the photocatalytic mechanism using MoS_2-based nanocomposite is a must. MoS_2-based membranes have shown rapid water transport and a broad range of salt rejection, which is interesting in water treatment. Layer-stacked membranes also demonstrate high water stability, and the filtration process's selectivity can be enhanced by tuning the interlayer spacing of membranes. Therefore, more theoretical and simulation studies are needed for a fundamental understanding of the MoS_2 membrane interlayer and free spacing and the transport of the ions and molecules through the 2D channels [117, 118]. At the same time, the antifouling characteristics of the membrane should be maintained. Another main challenge is the preparation of uniformly and compactly distributed nanopores on the MoS_2 membrane. And at last, but not least, the main concern should be the impact of MoS_2 nanomaterials on the fabrication cost and environment. The synthesis routes of MoS_2 for various water cleaning techniques should be monitored, and the production from lab to pilot scale production should also be targeted.

In order to avoid all these limitations, there are a few suggestions to take into account while designing the MoS_2-based materials for wastewater remediation:

(a) There is a need to improve the adsorption efficiency and selectivity of MoS_2 material for a broad range of water contaminants. To improve the MoS_2 adsorbance competence towards anionic contaminants, it should be smartly functionalized with cost-effective materials. The synthesis route should also avoid the use of any hazardous chemicals.
(b) Dispersion of exfoliated MoS_2 nanostructured material is essential for the enhanced adsorption rate of water contaminant molecules. Homogenously dispersed MoS_2 performs faster adsorption, but at the same time, it faces separation challenges from the treated water. Functionalization of MoS_2 with magnetic materials can avoid this issue. Magnetic MoS_2 material can easily be separated from the treated water using the external magnetic field.
(c) Designing the 3D architecture of 2D MoS_2 nanomaterials is a current research trend to avoid agglomeration and provide more surface area and a porous network for the adsorption of water contaminants.
(d) The study on the band gap structures of MoS_2-based nanocomposites and the types of generated ROS should be performed for a better understanding of the mechanism behind MoS_2-based photocatalytic water decontamination.
(e) The development of multifunctional MoS_2-based membranes can deliver a potential direction for the researcher in future water remediation techniques. Combining the membrane filtration with other wastewater treatment techniques, such as photocatalytic degradation and electrochemical decontamination, using the same MoS_2-based membrane can reduce the water treatment cost and improve the treated water quality.

To conclude, MoS_2-based nano-structural materials nanosheets have shown immense potential for wastewater remediation. However, research using MoS_2 and MoS_2-based nanocomposites in water treatment is still underway. Further research can bring many exciting opportunities and outcomes in wastewater treatment applications.

Acknowledgements The authors would like to acknowledge the funding support from the Department of Science and Innovation (C6A0058), the Council for Scientific and Industrial Research (C6ACH20), and the University of Johannesburg (086310), for their financial support.

References

1. M. Ahmad, M. Yousaf, A. Nasir, I.A. Bhatti, A. Mahmood, X. Fang et al., Porous eleocharis@ MnPE layered hybrid for synergistic adsorption and catalytic biodegradation of toxic Azo dyes from industrial wastewater. Environ. Sci. Technol. **53**, 2161–2170 (2019)
2. S.S. Ray, R. Gusain, N. Kumar, Chapter two - Classification of water contaminants, in *Carbon Nanomaterial-Based Adsorbents for Water Purification*, ed. By S.S Ray, R. Gusain, N. Kumar (Elsevier; 2020), pp. 11–36.
3. R. Gusain, K. Gupta, P. Joshi, O.P. Khatri, Adsorptive removal and photocatalytic degradation of organic pollutants using metal oxides and their composites: A comprehensive review. Adv. Coll. Interface. Sci. **272**, 102009 (2019)
4. M. Kumar, A. Puri, A review of permissible limits of drinking water. Indian J. Occupat. Environment. Med. **16**, 40 (2012)
5. S.K. Sharma, Heavy metals in water: presence, removal and safety: R. Soc. Chem. (2014)
6. S.S. Ray, R. Gusain, N. Kumar, *Carbon Nanomaterial-Based Adsorbents for Water Purification: Fundamentals and Applications* (Elsevier; 2020)
7. D. Kanakaraju, B.D. Glass, M. Oelgemöller, Advanced oxidation process-mediated removal of pharmaceuticals from water: a review. J. Environ. Manage. **219**, 189–207 (2018)
8. H. Rabiee, L. Ge, S. Hu , H. Wang, Z. Yuan, Microtubular electrodes: an emerging electrode configuration for electrocatalysis, bioelectrochemical and water treatment applications. Chem. Eng. J. 138476 (2022)
9. B.O. Orimolade, O.A. Arotiba, Bismuth vanadate in photoelectrocatalytic water treatment systems for the degradation of organics: a review on recent trends. J. Electroanal. Chem. **878**, 114724 (2020)
10. A. Zularisam, A. Ismail, R. Salim, Behaviours of natural organic matter in membrane filtration for surface water treatment—a review. Desalination **194**, 211–231 (2006)
11. A. Magrí, M. Carreras-Sempere, C. Biel, J. Colprim, Recovery of phosphorus from waste water profiting from biological nitrogen treatment: Upstream, concomitant or downstream precipitation alternatives. Agronomy **10**, 1039 (2020)
12. R. Yang, H. Li, M. Huang, H. Yang, A. Li, A review on chitosan-based flocculants and their applications in water treatment. Water Res. **95**, 59–89 (2016)
13. R.H. Hailemariam, Y.C. Woo, M.M. Damtie, B.C. Kim, K.-D. Park, J.-S. Choi, Reverse osmosis membrane fabrication and modification technologies and future trends: A review. Adv. Coll. Interface. Sci. **276**, 102100 (2020)
14. G.H. Jeong, S.P. Sasikala, T. Yun, G.Y. Lee, W.J. Lee, S.O. Kim, Nanoscale assembly of 2D materials for energy and environmental applications. Adv. Mater. **32**, 1907006 (2020)
15. W. Choi, I. Lahiri, R. Seelaboyina, Y.S. Kang, Synthesis of graphene and its applications: a review. Crit. Rev. Solid State Mater. Sci. **35**, 52–71 (2010)
16. Y. Zhang, Y. Yan, H. Qiu, Z. Ma, K. Ruan, J. Gu, A mini-review of MXene porous films: preparation, mechanism and application. J. Mater. Sci. Technol. **103**, 42–49 (2022)
17. P.S. Jijoe, S.R. Yashas, H.P. Shivaraju, Fundamentals, synthesis, characterization and environmental applications of layered double hydroxides: a review. Environ. Chem. Lett. **19**, 2643–2661 (2021)
18. H. Furukawa, K.E. Cordova, M. O'Keeffe, O.M. Yaghi, The chemistry and applications of metal-organic frameworks. Science **341**, 1230444 (2013)

19. K. Gupta, P. Joshi, R. Gusain, O.P. Khatri, Recent advances in adsorptive removal of heavy metal and metalloid ions by metal oxide-based nanomaterials. Coord. Chem. Rev. **445**, 214100 (2021)
20. M. Dai, R. Wang, Synthesis and applications of nanostructured hollow transition metal chalcogenides. Small **17**, 2006813 (2021)
21. S. Kumari, R. Gusain, N. Kumar, O.P. Khatri, PEG-mediated hydrothermal synthesis of hierarchical microspheres of MoS2 nanosheets and their potential for lubrication application. J. Ind. Eng. Chem. **42**, 87–94 (2016)
22. O. Samy, A. El Moutaouakil, A review on MoS2 energy applications: recent developments and challenges. Energies **14**, 4586 (2021)
23. J. Mao, Y. Wang, Z. Zheng, D. Deng, The rise of two-dimensional MoS2 for catalysis. Front. Phys. **13**, 138118 (2018)
24. N. Kumar, S. Kumar, R. Gusain, N. Manyala, S. Eslava, S.S. Ray, Polypyrrole-promoted rGO–MoS2 nanocomposites for enhanced photocatalytic conversion of CO2 and H2O to CO, CH4, and H2 products. ACS Appl. Energy Mater. **3**, 9897–9909 (2020)
25. O. Samy, S. Zeng, M.D. Birowosuto, A. El Moutaouakil, A Review on MoS2 properties, synthesis. Sens. Appl. Chall. Cryst. **11**, 355 (2021)
26. A.T. Massey, R. Gusain, S. Kumari, O.P. Khatri, Hierarchical microspheres of MoS2 nanosheets: efficient and regenerative adsorbent for removal of water-soluble dyes. Ind. Eng. Chem. Res. **55**, 7124–7131 (2016)
27. P. Raybaud, J. Hafner, G. Kresse, S. Kasztelan, H. Toulhoat, Structure, energetics, and electronic properties of the surface of a promoted MoS2 catalyst: an ab initio local density functional study. J. Catal. **190**, 128–143 (2000)
28. L.S. Byskov, J.K. Nørskov, B.S. Clausen, H. Topsøe, DFT calculations of unpromoted and promoted MoS2-based hydrodesulfurization catalysts. J. Catal. **187**, 109–122 (1999)
29. M. Takaoka, N. Takeda, Y. Shimaoka, T. Fujiwara, Removal of mercury in flue gas by the reaction with sulfide compounds. Toxicol. Environ. Chem. **73**, 1–16 (1999)
30. C. Tan, H. Zhang, Two-dimensional transition metal dichalcogenide nanosheet-based composites. Chem. Soc. Rev. **44**, 2713–2731 (2015)
31. H.J. Conley, B. Wang, J.I. Ziegler, R.F. Haglund, S.T. Pantelides, K.I. Bolotin, Bandgap engineering of strained monolayer and bilayer MoS2. Nano Lett. **13**, 3626–3630 (2013)
32. D. Voiry, A. Goswami, R. Kappera, C.D.C.C.E. Silva, D. Kaplan, T. Fujita et al., Covalent functionalization of monolayered transition metal dichalcogenides by phase engineering. Nat. Chem. **7**, 45–49 (2015)
33. Z. Wang, B. Mi, Environmental applications of 2D molybdenum disulfide (MoS2) nanosheets. Environ. Sci. Technol. **51**, 8229–8244 (2017)
34. R. Gusain, N. Kumar, S.S. Ray, Recent advances in carbon nanomaterial-based adsorbents for water purification. Coord. Chem. Rev. **405**, 213111 (2020)
35. G.Z. Kyzas, E.A. Deliyanni, K.A. Matis, Graphene oxide and its application as an adsorbent for wastewater treatment. J. Chem. Technol. Biotechnol. **89**, 196–205 (2014)
36. A. Jilani, M.H.D. Othman, M.O. Ansari, S.Z. Hussain, A.F. Ismail, I.U. Khan, Graphene and its derivatives: synthesis, modifications, and applications in wastewater treatment. Environ. Chem. Lett. **16**, 1301–1323 (2018)
37. S. Dervin, D.D. Dionysiou, S.C. Pillai, 2D nanostructures for water purification: graphene and beyond. Nanoscale **8**, 15115–15131 (2016)
38. H. Wang, C. Li, P. Fang, Z. Zhang, J.Z. Zhang, Synthesis, properties, and optoelectronic applications of two-dimensional MoS 2 and MoS 2-based heterostructures. Chem. Soc. Rev. **47**, 6101–6127 (2018)
39. X. Wang, J. Ding, S. Yao, X. Wu, Q. Feng, Z. Wang et al., High supercapacitor and adsorption behaviors of flower-like MoS 2 nanostructures. J. Mater. Chem. A. **2**, 15958–15963 (2014)
40. K. Ai, C. Ruan, M. Shen, L. Lu, MoS2 nanosheets with widened interlayer spacing for high-efficiency removal of mercury in aquatic systems. Adv. Func. Mater. **26**, 5542–5549 (2016)

41. S.S. Chou, M. De, J. Kim, S. Byun, C. Dykstra, J. Yu et al., Ligand conjugation of chemically exfoliated MoS2. J. Am. Chem. Soc. **135**, 4584–4587 (2013)
42. A.E. Gash, A.L. Spain, L.M. Dysleski, C.J. Flaschenriem, A. Kalaveshi, P.K. Dorhout et al., Efficient recovery of elemental mercury from Hg (II)-contaminated aqueous media using a redox-recyclable ion-exchange material. Environ. Sci. Technol. **32**, 1007–1012 (1998)
43. Y. Zhang, S. He, Y. Zhang, Y. Feng, Z. Pan, M. Zhang, Facile synthesis of PPy@ MoS2 hollow microtubes for removal of cationic and anionic dyes in water treatment. Colloids Surf., A **632**, 127765 (2022)
44. T. Krasian, W. Punyodom, P. Worajittiphon, A hybrid of 2D materials (MoS2 and WS2) as an effective performance enhancer for poly (lactic acid) fibrous mats in oil adsorption and oil/water separation. Chem. Eng. J. **369**, 563–575 (2019)
45. J. Joy, J. Abraham, J. Sunny, J. Mathew, S.C. George, Hydrophobic, superabsorbing materials from reduced graphene oxide/MoS2 polyurethane foam as a promising sorbent for oil and organic solvents. Polym. Testing **87**, 106429 (2020)
46. X. Gao, X. Wang, X. Ouyang, C. Wen, Flexible superhydrophobic and superoleophilic MoS2 sponge for highly efficient oil-water separation. Sci. Rep. **6**, 1–8 (2016)
47. P.K. Chow, E. Singh, B.C. Viana, J. Gao, J. Luo, J. Li et al., Wetting of mono and few-layered WS2 and MoS2 films supported on Si/SiO2 substrates. ACS Nano **9**, 3023–3031 (2015)
48. H.J. Song, S. You, X.H. Jia, J. Yang, MoS2 nanosheets decorated with magnetic Fe3O4 nanoparticles and their ultrafast adsorption for wastewater treatment. Ceram. Int. **41**, 13896–13902 (2015)
49. Z. Wang, J. Zhang, T. Wen, X. Liu, Y. Wang, H. Yang et al., Highly effective remediation of Pb(II) and Hg(II) contaminated wastewater and soil by flower-like magnetic MoS2 nanohybrid. Sci. Total Environ. **699**, 134341 (2020)
50. S. Tong, H. Deng, L. Wang, T. Huang, S. Liu, J. Wang, Multi-functional nanohybrid of ultrathin molybdenum disulfide nanosheets decorated with cerium oxide nanoparticles for preferential uptake of lead (II) ions. Chem. Eng. J. **335**, 22–31 (2018)
51. Z. Li, X. Meng, Z. Zhang, Equilibrium and kinetic modelling of adsorption of Rhodamine B on MoS2. Mater. Res. Bull. **111**, 238–244 (2019)
52. H. Song, S. You, X. Jia, Synthesis of fungus-like MoS2 nanosheets with ultrafast adsorption capacities toward organic dyes. Appl. Phys. A **121**, 541–548 (2015)
53. Y. Wu, M. Su, J. Chen, Z. Xu, J. Tang, X. Chang et al., Superior adsorption of methyl orange by h-MoS2 microspheres: Isotherm, kinetics, and thermodynamic studies. Dyes Pigm. **170**, 107591 (2019)
54. X. Xiao, Y. Wang, B. Cui, X. Zhang, D. Zhang, X. Xu, Preparation of MoS2 nanoflowers with rich active sites as an efficient adsorbent for aqueous organic dyes. New J. Chem. **44**, 4558–4567 (2020)
55. H. Yang, H. Yuan, Q. Hu, W. Liu, D. Zhang, Synthesis of mesoporous C/MoS2 for adsorption of methyl orange and photocatalytic sterilization. Appl. Surf. Sci. **504**, 144445 (2020)
56. J. Chen, Y. Liao, X. Wan, S. Tie, B. Zhang, S. Lan et al., A high performance MoO3@MoS2 porous nanorods for adsorption and photodegradation of dye. J. Solid State Chem. **291**, 121652 (2020)
57. X. Liu, W. Zhang, L. Mao, Y. Yin, L. Hu, Synthesis of FeOCl-MoS2 with excellent adsorption performance for methyl orange. J. Mater. Sci. **56**, 6704–6718 (2021)
58. X.-Q. Qiao, F.-C. Hu, F.-Y. Tian, D.-F. Hou, D.-S. Li, Equilibrium and kinetic studies on MB adsorption by ultrathin 2D MoS2 nanosheets. RSC Adv. **6**, 11631–11636 (2016)
59. R. Li, H. Deng, X. Zhang, J.J. Wang, M.K. Awasthi, Q. Wang et al., High-efficiency removal of Pb(II) and humate by a CeO2–MoS2 hybrid magnetic biochar. Biores. Technol. **273**, 335–340 (2019)
60. M.J. Aghagoli, F. Shemirani, Hybrid nanosheets composed of molybdenum disulfide and reduced graphene oxide for enhanced solid phase extraction of Pb(II) and Ni(II). Microchim. Acta **184**, 237–244 (2017)
61. J. Liu, H. Lin, Y. Dong, Y. He, W. Liu, Y. Shi, The effective adsorption of tetracycline onto MoS2@Zeolite-5: Adsorption behavior and interfacial mechanism. J. Environ. Chem. Eng. **9**, 105912 (2021)

62. Y. Li, C. Tang, L. Yang, Y. Sun, W. Ju, Integrated roles of MoS2 nanosheets for water treatment and polymer flame retardant. Arab. J. Sci. Eng. **46**, 6753–6763 (2021)
63. M.K. Uddin, F. Mashkoor, I.M. AlArifi, A. Nasar, Simple one-step synthesis process of novel MoS2@bentonite magnetic nanocomposite for efficient adsorption of crystal violet from aqueous solution. Mater. Res. Bull. **139**, 111279 (2021)
64. J. Luo, K. Fu, M. Sun, K. Yin, D. Wang, X. Liu et al., Phase-mediated heavy metal adsorption from aqueous solutions using two-dimensional layered MoS2. ACS Appl. Mater. Interfaces. **11**, 38789–38797 (2019)
65. N. Kumar, E. Fosso-Kankeu, S.S. Ray, Achieving controllable MoS2 nanostructures with increased interlayer spacing for efficient removal of Pb(II) from aquatic systems. ACS Appl. Mater. Interfaces. **11**, 19141–19155 (2019)
66. Y. Chao, W. Zhu, X. Wu, F. Hou, S. Xun, P. Wu et al., Application of graphene-like layered molybdenum disulfide and its excellent adsorption behavior for doxycycline antibiotic. Chem. Eng. J. **243**, 60–67 (2014)
67. Z. Zeng, S. Ye, H. Wu, R. Xiao, G. Zeng, J. Liang et al., Research on the sustainable efficacy of g-MoS2 decorated biochar nanocomposites for removing tetracycline hydrochloride from antibiotic-polluted aqueous solution. Sci. Total Environ. **648**, 206–217 (2019)
68. R. Gusain, N. Kumar, E. Fosso-Kankeu, S.S. Ray, Efficient removal of Pb(II) and Cd(II) from industrial mine water by a hierarchical MoS2/SH-MWCNT nanocomposite. ACS Omega **4**, 13922–13935 (2019)
69. C. Wang, J. Jin, Y. Sun, J. Yao, G. Zhao, Y. Liu, In-situ synthesis and ultrasound enhanced adsorption properties of MoS2/graphene quantum dot nanocomposite. Chem. Eng. J. **327**, 774–782 (2017)
70. M.R. Hoffmann, S.T. Martin, W. Choi, D.W. Bahnemann, Environmental applications of semiconductor photocatalysis. Chem. Rev. **95**, 69–96 (1995)
71. Y.J. Yuan, H.W. Lu, Z.T. Yu, Z.G. Zou, Noble-metal-free molybdenum disulfide cocatalyst for photocatalytic hydrogen production. Chemsuschem **8**, 4113–4127 (2015)
72. X. Jin, X. Fan, J. Tian, R. Cheng, M. Li, L. Zhang, MoS 2 quantum dot decorated gC 3 N 4 composite photocatalyst with enhanced hydrogen evolution performance. RSC Adv. **6**, 52611–52619 (2016)
73. M. Ntakadzeni, W.W. Anku, N. Kumar, P.P. Govender, L. Reddy, PEGylated MoS2 Nanosheets: a dual functional photocatalyst for photodegradation of organic dyes and photoreduction of chromium from aqueous solution **2019**, 11 (2019)
74. N. Singh, G. Jabbour, U. Schwingenschlögl, Optical and photocatalytic properties of two-dimensional MoS2. Europ. Phys. J. B. **85**, 1–4 (2012)
75. N. Shao, J. Wang, D. Wang, P. Corvini, Preparation of three-dimensional Ag3PO4/TiO2@ MoS2 for enhanced visible-light photocatalytic activity and anti-photocorrosion. Appl. Catal. B **203**, 964–978 (2017)
76. Y. Gao, C. Chen, X. Tan, H. Xu, K. Zhu, Polyaniline-modified 3D-flower-like molybdenum disulfide composite for efficient adsorption/photocatalytic reduction of Cr (VI). J. Colloid Interface Sci. **476**, 62–70 (2016)
77. X. Wang, M. Hong, F. Zhang, Z. Zhuang, Y. Yu, Recyclable nanoscale zero valent iron doped g-C3N4/MoS2 for efficient photocatalysis of RhB and Cr(VI) driven by visible light. ACS Sustain. Chem. Eng. **4**, 4055–4063 (2016)
78. T. Thurston, J. Wilcoxon, Photooxidation of organic chemicals catalyzed by nanoscale MoS2. J. Phys. Chem. B **103**, 11–17 (1999)
79. Q. Wang, S. Dong, D. Zhang, C. Yu, J. Lu, D. Wang et al., Magnetically recyclable visible-light-responsive MoS2@Fe3O4 photocatalysts targeting efficient wastewater treatment. J. Mater. Sci. **53**, 1135–1147 (2018)
80. B. Sheng, J. Liu, Z. Li, M. Wang, K. Zhu, J. Qiu et al., Effects of excess sulfur source on the formation and photocatalytic properties of flower-like MoS2 spheres by hydrothermal synthesis. Mater. Lett. **144**, 153–156 (2015)
81. H. Lv, Y. Liu, H. Tang, P. Zhang, J. Wang, Synergetic effect of MoS2 and graphene as cocatalysts for enhanced photocatalytic activity of BiPO4 nanoparticles. Appl. Surf. Sci. **425**, 100–106 (2017)

82. J. Liu, H. Lin, Y. He, Y. Dong, Gueret yadiberet menzembere ER. Novel CoS2/MoS2@Zeolite with excellent adsorption and photocatalytic performance for tetracycline removal in simulated wastewater. J. Cleaner Product. **260**, 121047 (2020)
83. Y. Li, Z. Lai, Z. Huang, H. Wang, C. Zhao, G. Ruan et al., Fabrication of BiOBr/MoS2/graphene oxide composites for efficient adsorption and photocatalytic removal of tetracycline antibiotics. Appl. Surf. Sci. **550**, 149342 (2021)
84. S. Cravanzola, F. Cesano, G. Magnacca, A. Zecchina, D. Scarano, Designing rGO/MoS2 hybrid nanostructures for photocatalytic applications. RSC Adv. **6**, 59001–59008 (2016)
85. H. Wang, F. Wen, X. Li, X. Gan, Y. Yang, P. Chen et al., Cerium-doped MoS2 nanostructures: efficient visible photocatalysis for Cr(VI) removal. Sep. Purif. Technol. **170**, 190–198 (2016)
86. K. Talukdar, K. Saravanakumar, Y. Kim, A. Fayyaz, G. Kim, Y. Yoon et al., Rational construction of CeO2–ZrO2@MoS2 hybrid nanoflowers for enhanced sonophotocatalytic degradation of naproxen: mechanisms and degradation pathways. Compos. B Eng. **215**, 108780 (2021)
87. X. Lin, X. Wang, Q. Zhou, C. Wen, S. Su, J. Xiang et al., Magnetically recyclable MoS2/Fe3O4 hybrid composite as visible light responsive photocatalyst with enhanced photocatalytic performance. ACS Sustain. Chem. Eng. **7**, 1673–1682 (2019)
88. S. Wang, M.-R. Chen, S.-B. Shen, C.-H. Cheng, A.-J. Cai, A.-J. Song et al., Bifunctionalized Fe7S8@MoS2–O core-shell with efficient photocatalytic activity based on internal electric field. J. Cleaner Product. **335**, 130375 (2022)
89. D. Lu, H. Wang, X. Zhao, K.K. Kondamareddy, J. Ding, C. Li et al., Highly efficient visible-light-induced photoactivity of Z-scheme g-C3N4/Ag/MoS2 ternary photocatalysts for organic pollutant degradation and production of hydrogen. ACS Sustain. Chem. Eng. **5**, 1436–1445 (2017)
90. W.Y. Lim, M. Hong, G.W. Ho, In situ photo-assisted deposition and photocatalysis of ZnIn2S4/transition metal chalcogenides for enhanced degradation and hydrogen evolution under visible light. Dalton Trans. **45**, 552–560 (2016)
91. S. Fu, W. Yuan, Y. Yan, H. Liu, X. Shi, F. Zhao et al., Highly efficient visible-light photoactivity of Z-scheme MoS2/Ag2CO3 photocatalysts for organic pollutants degradation and bacterial inactivation. J. Environ. Manage. **252**, 109654 (2019)
92. R. Dong, Y. Zhong, D. Chen, N. Li, Q. Xu, H. Li et al., Morphology-controlled fabrication of CNT@MoS2/SnS2 nanotubes for promoting photocatalytic reduction of aqueous Cr(VI) under visible light. J. Alloy. Compd. **784**, 282–292 (2019)
93. S. Chen, Y. Di, H. Li, M. Wang, B. Jia, R. Xu et al., Efficient photocatalytic dye degradation by flowerlike MoS2/SrFe12O19 heterojunction under visible light. Appl. Surf. Sci. **559**, 149855 (2021)
94. J. Piriyanon, T. Chankhanittha, S. Youngme, K. Hemavibool, S. Nijpanich, S. Juabrum et al., Fabrication of MoS2/Ag3PO4 S-scheme photocatalyst for visible-light-active degradation of organic dye and antibiotic in wastewater. J. Mater. Sci.: Mater. Electron. **32**, 19798–19819 (2021)
95. R. Gusain, N. Kumar, F. Opoku, P.P. Govender, S.S. Ray, MoS2 nanosheet/ZnS composites for the visible-light-assisted photocatalytic degradation of oxytetracycline. ACS Appl. Nano Mater. **4**, 4721–4734 (2021)
96. T. Qiang, L. Chen, Y. Xia, X. Qin, Dual modified MoS2/SnS2 photocatalyst with Z-scheme heterojunction and vacancies defects to achieve a superior performance in Cr (VI) reduction and dyes degradation. J. Clean. Prod. **291**, 125213 (2021)
97. S.V.P. Vattikuti, C. Byon, S. Jeon, Enhanced photocatalytic activity of ZnS nanoparticles loaded with MoS2 nanoflakes by self-assembly approach. Phys. B **502**, 103–112 (2016)
98. S. Yin, R. Chen, M. Ji, Q. Jiang, K. Li, Z. Chen et al., Construction of ultrathin MoS2/Bi5O7I composites: Effective charge separation and increased photocatalytic activity. J. Colloid Interface Sci. **560**, 475–484 (2020)
99. Z. Wang, Q. Tu, S. Zheng, J.J. Urban, S. Li, B. Mi, Understanding the aqueous stability and filtration capability of MoS2 membranes. Nano Lett. **17**, 7289–7298 (2017)
100. W. Hirunpinyopas, E. Prestat, S.D. Worrall, S.J. Haigh, R.A. Dryfe, M.A. Bissett, Desalination and nanofiltration through functionalized laminar MoS2 membranes. ACS Nano **11**, 11082–11090 (2017)

101. X. Lu, U.R. Gabinet, C.L. Ritt, X. Feng, A. Deshmukh, K. Kawabata et al., Relating selectivity and separation performance of lamellar two-dimensional molybdenum disulfide (MoS2) membranes to nanosheet stacking behavior. Environ. Sci. Technol. **54**, 9640–9651 (2020)
102. M.S. Sri Abirami Saraswathi, D. Rana, P. Vijayakumar, S. Alwarappan, A. Nagendran, Tailored PVDF nanocomposite membranes using exfoliated MoS2 nanosheets for improved permeation and antifouling performance. New J. Chem. **41**, 14315–14324 (2017)
103. L. Zhang, X. He, Q. Zhou, X. Hu, Fabrication of 1T-MoS2 nanosheets and the high-efficiency removal of toxic metals in aquatic systems: performance and mechanisms. Chem. Eng. J. **386**, 123996 (2020)
104. I. Singha, P.K. Mishrab, Nano-membrane filtration a novel application of nanotechnology for waste water treatment. Mater. Today: Proc. **29**, 327–332 (2020)
105. J. Kou, J. Yao, L. Wu, X. Zhou, H. Lu, F. Wu et al., Nanoporous two-dimensional MoS2 membranes for fast saline solution purification. Phys. Chem. Chem. Phys. **18**, 22210–22216 (2016)
106. M. Heiranian, A.B. Farimani, N.R. Aluru, Water desalination with a single-layer MoS2 nanopore. Nat. Commun. **6**, 8616 (2015)
107. J. Kou, J. Yao, L. Wu, X. Zhou, H. Lu, F. Wu et al., Nanoporous two-dimensional MoS 2 membranes for fast saline solution purification. Phys. Chem. Chem. Phys. **18**, 22210–22216 (2016)
108. L. Sun, H. Huang, X. Peng, Laminar MoS 2 membranes for molecule separation. Chem. Commun. **49**, 10718–10720 (2013)
109. J. Sun, Y. Chen, C. Hu, H. Liu, J. Qu, Modulation of cation trans-membrane transport in GO-MoS2 membranes through simultaneous control of interlayer spacing and ion-nanochannel interactions. Chemosphere **222**, 156–164 (2019)
110. S.A. Wakelin, M.J. Colloff, R.S. Kookana, Effect of wastewater treatment plant effluent on microbial function and community structure in the sediment of a freshwater stream with variable seasonal flow. Appl. Environ. Microbiol. **74**, 2659–2668 (2008)
111. A. Sethulekshmi, A. Saritha, K. Joseph, A.S. Aprem, S.B. Sisupal, MoS2 based nanomaterials: Advanced antibacterial agents for future. J. Control. Release **348**, 158–185 (2022)
112. X. Tian, Y. Sun, S. Fan, M.D. Boudreau, C. Chen, C. Ge et al., Photogenerated charge carriers in molybdenum disulfide quantum dots with enhanced antibacterial activity. ACS Appl. Mater. Interfaces. **11**, 4858–4866 (2019)
113. M. Zhang, K. Wang, S. Zeng, Y. Xu, W. Nie, P. Chen et al., Visible light-induced antibacterial effect of MoS2: effect of the synthesis methods. Chem. Eng. J. **411**, 128517 (2021)
114. S. Roy, A. Mondal, V. Yadav, A. Sarkar, R. Banerjee, P. Sanpui et al., Mechanistic insight into the antibacterial activity of chitosan exfoliated MoS2 nanosheets: membrane damage, metabolic inactivation, and oxidative stress. ACS Appl. Bio Mater. **2**, 2738–2755 (2019)
115. X. Yang, J. Li, T. Liang, C. Ma, Y. Zhang, H. Chen et al., Antibacterial activity of two-dimensional MoS2 sheets. Nanoscale **6**, 10126–10133 (2014)
116. Q. Xu, P. Zhu, J. Zhang, Y. Liu, L. Cai, H. Jiang et al., Electrochemical formation of distinct nanostructured MoS2 with altered antibacterial activity. Mater. Lett. **271**, 127809 (2020)
117. Y. Xiang, Y. Liu, B. Mi, Y. Leng, Hydrated polyamide membrane and its interaction with alginate: a molecular dynamics study. Langmuir **29**, 11600–11608 (2013)
118. Y. Xiang, Y. Liu, B. Mi, Y. Leng, Molecular dynamics simulations of polyamide membrane, calcium alginate gel, and their interactions in aqueous solution. Langmuir **30**, 9098–9106 (2014)

Chapter 13
Two-Dimensional All-Metal/Metal Oxide Based Photocatalysts for Solar CO_2 Conversion

Peter Ramashadi Makgwane

Abstract Two-dimensional (2D) metal oxides consisting of stacked layers possess exceptional structure properties such as highly exposed surface-active sites, large interlayer spacing, tunability for exfoliation into single or few layers with improved mass transport and high specific surface area. The combination of these properties makes 2D materials applications very fascinating in photocatalysis because of the enhanced photoactivity performance. In this chapter, significant advancements in the design, synthesis, structure properties and photoactivity performance of 2D metal oxides in CO_2 reduction reaction are discussed. Specifically, a comprehensive insight into understanding the inter-play between strategic design of the 2D metal oxides and their multidimensional heterostructures (0D/2D, 1D/2D, 2D/2D and 2D/3D) with controlled nanoscale interface structure properties correlated to photocatalytic activity performance in CO_2 reduction reaction are highlighted. The construction strategies of 2D metal oxides heterojunctions with other materials such as non-metals, metals dopants, and metal oxides providing high interfacial intimate contacts for improved charge carriers' separations and transfer for improved high-performance CO_2 photocatalytic conversion are critically discussed.

13.1 Introduction

Layered two-dimensional (2D) nanomaterials with thin atomic thickness have enticed vast research attention in recent times due to their exceptional physicochemical and opto-electronic structure properties, which are usually very distinct from those of their bulk equivalents [1–5]. They consist of stacked layers that offers its uniqueness as class of materials from diverse building-blocks components such as 2D metal oxides [6], 2D chalcogenides [7], 2D carbon {e.g. graphene, graphitic carbon nitrides,

P. R. Makgwane (✉)
Centre for Nanostructures and Advanced Materials (CeNAM), Council for Scientific and Industrial Research, Pretoria, South Africa
e-mail: pmakgwane@csir.co.za; makgwane.peter@gmail.com

Department of Chemistry, University of the Western Cape, Western Cape, Bellville, South Africa

N. Kumar et al. (eds.), *Two-Dimensional Materials for Environmental Applications*, Springer Series in Materials Science 332, https://doi.org/10.1007/978-3-031-28756-5_13

etc.} [8], 2D black phosphorous [9], layered double hydroxides (LDHs) [10], and 2D Mxenes (metal carbides) [11] with various potential applications in environmental remediation [12, 13], sensing [14], water-splitting for hydrogen energy [15], energy storage [16, 17], photochemical reaction of CO_2 and N_2 conversions [18, 19] due to their versatile structural properties. The extraordinarily stacked layer structures of 2D nanomaterials induce flexible properties that can be constructed synthetically following the various effect such as twisted multilayered 2D [20], hybrid 2D nanostructures [21], vertical van der Waals 2D heterojunctions [22]. The tunability to fabricate the stacked layers compared to bulk offers effective strategy to attain controlled and high-quality 2D nanostructures with accurate atomic patterns for enhanced charge carrier dynamics in photocatalysis and electronic devices applications [23]. These properties of 2D nanomaterials include highly exposed active sites [24, 25], large interlayer spacing [26, 27], tuneability for exfoliation to single or few layers with improved mass transport dynamics [28, 29], and efficient charge carriers' mobility with better chemical stability [30].

The construction of highly photoactive heterojunctions interfaces based on ultrathin 2D nanomaterials integrating multidimensional such as zero dimension (0D), one-dimension (1D), and three-dimension (3D) have increased significantly because of the advantages of 2D versatility with respect to enhanced spatial charge carriers separation of electron/holes (e^-/h^+) and transfer of electron charges, which lead to improved redox potentials for high photocatalytic activity performance applications. By using 2D nanomaterials as the basis, the construction of the following heterojunction interfaces such as 0D/2D [31]; 1D/2D [32]; 2D/2D [33], and 2D/3D [34] are possible with the view to enhance on photocatalytic activity properties. In the 2D/2D heterojunction the possible fabrication of face-to-face interfacial contact present exceptional large interface area for the improved carrier separation and transfer efficiency [35], while 1D/2D due to the direction charge flow of 1D, its interface with 2D offers effective charge separations with enhanced redox capability for improvement of the photocatalytic performance [36]. The 0D/2D interface constituting decoration of nanoparticles (NPs) or quantum dots (QDs) on the surface of 2D nanosheets or nanoplates offers well distributed charge carrier separation. The decorated 0D acts as short distance active sites for electrons trapping/sink and transfer between the photoinduced carriers' separation site to the junction interface, which lead to high photoactivity performances [37]. The 3D/2D interfaces offers unique structural confinements effects due to the well-developed porous structure activity as cavities for enhanced photocatalytic mass transport efficiencies of photoexcited carriers. The assembly structure properties of these 2D layered nanostructures interfaces depend strongly on the controlled synthesis factors such as surface quality, contact angle control, shape, and size [38–40].

In this chapter, significant developments in the design, synthesis, structure properties and photoactivity performance of 2D metal oxides in CO_2 photoreduction reaction are discussed. Specifically, a comprehensive insight into understanding the interplay between strategic design of the 2D metal oxides and their heterostructures with controlled nanoscale structure properties correlated to photocatalytic activity performance in CO_2 reduction conversion are highlighted. The fabrication of 2D metal oxides heterojunctions with other materials such as non-metals, metals dopants, and metal oxides provides high interfacial intimate contacts for improved charge carriers' separations and transfer for improved high-performance CO_2 photocatalytic conversion. The discussion on the update developments of heterojunctions interfaced photocatalysts for CO_2 reduction are restricted to metal oxides semiconductors interfaced with metals (i.e. dopants), metal sulfides, perovskites, and layered double hydroxides (LDHs) semiconductors while carbon based (i.e. graphene, activated carbon, graphitic carbon nitride, polymers) are not covered.

13.2 Multidimensional Interfaces of 2D Metal Oxides for Enhanced CO_2 Conversion

Multidimension interfaces combining two or more materials with various structural properties has been an effective strategy to fabricate high performance materials with number of applications in electronic devices, catalysis, photocatalysis and energy [41–44]. Such interfaces can also include the same material reconfigured to design homojunction interfaces with effects from phases compositions and multidimensions heterointerface. In photocatalysis the heterointerface of 2D materials with other dimensions such as 0D, 1D and 3 D could induce the efficient charge carrier dynamics driven by heterojunctions such as S-scheme [45], Z-scheme [46] and the traditional type-II heterojunction [47]. The combinations of controlled synthesis procedures to yield structural characteristics such as defects and oxygen vacancies, crystal facets or planes, morphologies (1D, 0D, 3D) heterointerfaced with unique 2D planar nanostructures could modify the enhancement of the photocatalytic properties driven by redox capability, extended range for light absorption photoactivity due to aligned band energy edges, which can improve charge separation and transfer for enhanced CO_2 photoreduction activity [48–50]. The different heterojunctions interfaces of 2D materials with other materials dimensions can form 0D/2D (nanoparticles, quantum dots = 0D), 1D/2D (nanorods, nanotubes, nanowires, nanofibers = 1D), 2D/2D (nanosheets, nanoplates = 2D), and 3D/2D (microspheres, nanocubes, cuboids = 3D).

13.2.1 Pure and Heteroatom Doped 2D Metal Oxides Photocatalysts

Amongst the traditional photoactive semiconductors of SnO_2, ZnO and WO_3, TiO_2 has been the most extensively studied in various photocatalysis applications including the investigation of its 2D nanostructures photocatalytic properties [51, 52]. TiO_2 is a wide band gap material with band energy of 3.2 eV, which is mostly photoactive in the UV region of the sunlight spectrum. However, since the discovery of its excellent photocatalytic properties such as stable and resistance to photocorroision, research attention has been focused on developing various strategies to enable its visible light response photoactivity by employing the modulations of anatase/rutile compositions, crystal plane designed nanostructures and homojunction self-interfaced as pure TiO_2 [53–56]. Specific amplifications of defects and oxygen vacancies together with opto-electronics modifications in layered 2D materials have proven a success in enhancing some of the descriptors of high photocatalytic performance such effective photoexcitation of charge carrier electron and holes, their separations for prolonged life span and efficient mobility for high redox capability utilization purpose [56, 57]. The enhanced photoactivity for CO_2 reduction of 2D nanosheets of pure TiO_2 compared to other morphologies such as 1D nanotubes (NTs) and 0D NPs [58]. The high photoactivity of 2D TiO_2 nanosheets was due to carriers preferential directional flow in the specific crystal plane. The {001} crystal plane of the 2D TiO_2 nanosheets with high-energy exposure promoted efficient H_2O oxidative dissolution by photoinduced holes, which led to abundant accessible electrons for the enhanced CO_2 photoreduction on {101} plane. The surface fluorinated elemental substitutional sites in TiO_2 facilitated the creation of oxygen deficient Ti^{3+} sites, which facilitated fast photoreduction of CO_2 to CO_2^- by prolonging the life span of photoexcited e^-/h^+ pairs [58]. Similar enhanced photoactivity of TiO_2 modulated by facets was demonstrated in the fabricated 2D black TiO_2 nanosheets array films with dominant {001} plane boosting high content of oxygen vacancies that extended its visible light harvesting capacity and efficiency to separate photoinduced e^-/h^+ pairs [59]. The formed oxygen vacancies on {001} than {101} improved the photoactivity for reduction of CO_2 to CO with evolution rate of 128.5 $\mu mol \cdot g^{-1} \cdot h^{-1}$ [59]. Apart from 2D modulated oxygen vacancies, the controlled synthesis of the 2D TiO_2 nanosheets multiphases {i.e. TiO_2(A = anatase) and TiO_2(B = brookite)} can allow the formation of interface junction known as homojunction for efficient improved photocatalytic properties. This was demonstrated in the design of 2D TiO_2 nanosheets with TiO_2 (A) and TiO_2(B) phases for the enhanced CO_2 photoreduction activity into CH_3OH [60]. The enhanced performance for CO_2 photoreduction was attributed to the formed junction between the two TiO_2 crystal phases, which are capable to facilitate fast separation of photoinduced e^-/h^+ pair and prolonged lifespan of electrons for effective redox utilization [60]. By engineering 2D ultrathin TiO_2 nanosheets of 2 nm in thickness, up to 95% of dominant {100} crystal plane was achieved, which significantly improved the photoactivity for CO_2 reduction compared to bulk TiO_2 (Fig. 13.1a) [61]. The 95% composition amount of the exposed {100} facet is the highest compared to amongst the reported data for TiO_2 anatase. As illustrated in Fig. 13.1b, the 2D TiO_2 anatase (95%) exhibited remarked photoreduction activity of CO_2 that is fivefold greater than that of cuboids TiO_2 only with 53% of dominant {100} facet due to its larger surface area for abundant surface-active sites and extended photoactive visible light response range resulting from electronic band structure modifications [61]. Whilst the respective bandgap energy of 3.12 eV and 3.17 eV for TiO_2 naoncuboids and TiO_2 nanosheets are very close, hence the bandgap edges alignments (Fig. 13.1c), which enunciate that their displayed significant differences in photocatalytic CO_2 reduction performance is mainly driven by morphology effect. As a results, the TiO_2 nanosheets showed to provide efficient charge carrier separations and migrations with enhanced superior CO_2 phoreduction activity compared to TiO_2 nanocuboids. Similar enhancement in CO_2 reduction photoactivity trend influenced by the highly defective 2D nanosheets TiO_2 was demonstrated by Wang et al. [56] in the design of enriched Ti^{3+} sites ultrathin TiO_2 2D nanosheets, which exhibited high evolution rate of CH_4 (147.2 μmol g^{-1} h^{-1}) at 96.8% selectivity. The significant effect of defective sites of the exposed facets was confirmed using in-situ FTIR and in-situ EPR coupled to DFT. It showed that CO_2 is selectively activated on TiO_2 nanosheets defects surface by creating the bridging Ti···CO_2^-···Ti coordination, which promoted lower activation energy barrier and directing high selectivity towards CH_4. This was achieved due to better separation of photoinduced e^-/h^+ carriers with accelerated electrons transfer to CO_2 reduction with high photoactivity to produce CH_4 at the Ti-C coordination interface bond [56].

Both In_2O_3 and WO_3 are versatile transition metals for reductive reactions in catalysis and photocatalysis because of their capability to switch easily the redox

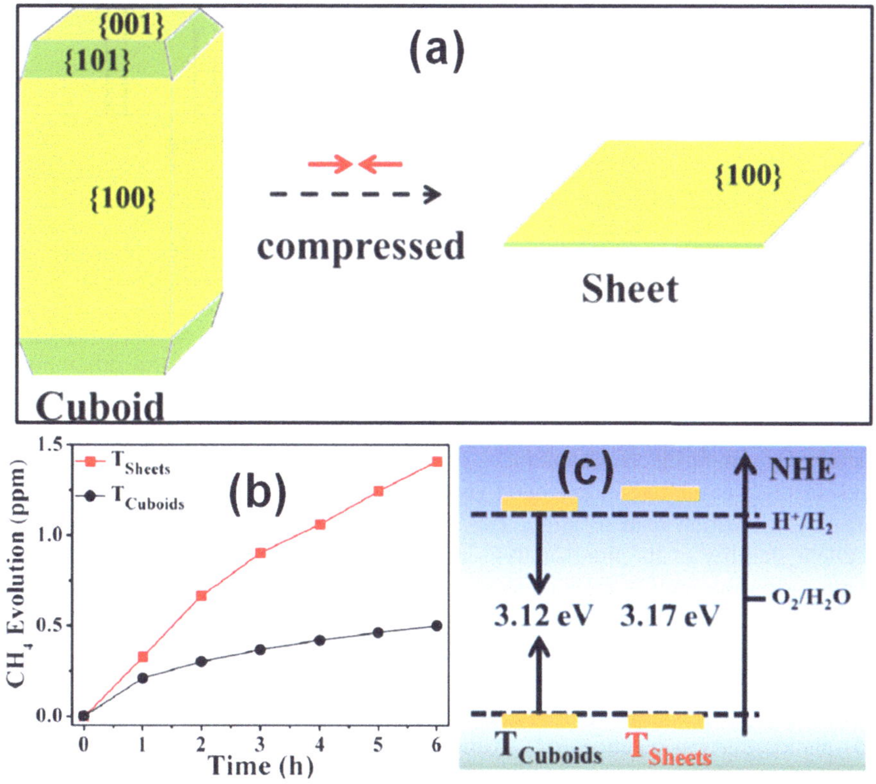

Fig. 13.1 **a** Comparison differences of the TiO_2 exposed (100) facet in 2D (nanosheets) and 3D (cuboids). **b** Photocatalytic CH_4 evolution performance of TiO_2 nanosheets and cuboids in CO_2 reduction. **c** Band energy edges of TiO_2 nanosheets and cuboids. Reproduced with permission from Ref. [61]. Copyright 2013, American Chemical Society

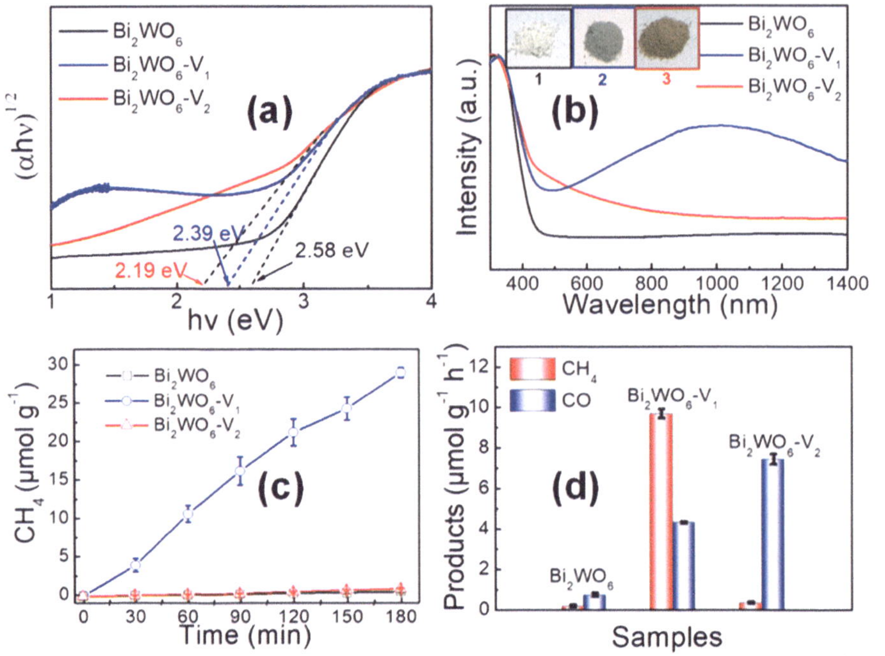

Fig. 13.2 **a** Band gap energy values of Bi_2WO_6 photocatalysts. **b** UV–vis light absorption profiles of Bi_2WO_6 photocatalysts. **c** Photocatalytic CH_4 evolution rates of various Bi_2WO_6 based photocatalysts. **d** Products selectivities formations in CO_2 reduction over Bi_2WO_6 photocatalysts. Reproduced with permission from Ref. [70]. Copyright 2021, American Chemical Society

reactivities promoted by defects oxygen vacancies compared to the other non-reducible traditional metals such as SnO_2 and ZnO [62, 63]. A 2D oxygen deficient In_2O_{3-x} was fabricated via the activation of the inert pre-synthesized $In(OH)_3$ into black In_2O_{3-x} nanosheets with induced defective active sites [63]. The defective 2D black In_2O_{3-x} exhibited an extended visible light absorption and effective adsorption of CO_2, which lead to attain 103.21 mmol g_{cat}^{-1} h^{-1} CO evolution rate at excellent stability of its photocatalytic activity [63]. In another study, an ultrathin single-crystal pure WO_3 2D nanosheet with thickness of ~4–5 nm demonstrated improved H_2O assisted CO_2 photoreduction activity in conditions into hydrocarbon fuels [64]. The effect of size-quantization in 2D nanosheets modified the band gap of WO_3 with improved visible light photoactivity capacity, and surface distributed efficient carriers separation and migration for their effective utilization in the redox CO_2 photoreduction reaction [64]. Although the photocatalytic properties of pure WO_3 have been improved using strategies such as defects structure and crystal plane exposure that modified its bandgap for better visible light absorption capabilities, it however still suffers from rapid recombination rates, thus low photocatalytic efficiencies. To improve photoactivity properties of WO_3, photocatalysts based on binary compounds of tungsten bismuth has emerged as highly photoactive materials. Bi_2WO_6, belongs to the Aurivillius family of perovskite forms with excellent visible light absorption response photocatalytic properties due to suitable band structure, high chemical stability, and non-toxic. A 2D ultrathin Bi_2WO_6 with square nanoplates morphology of ~9.5 nm thickness was synthesized using oleylamine as structure directing agent under hydrothermal route [65]. The geometry of 2D Bi_2WO_6 nanoplates facilitated improved charge carriers separation efficiency with suppressed e^-/h^+ carrier recombination due to its well-defined {001} crystal plane exposed surfaces leading to enhanced photocatalytic efficiency for CO_2 reduction into hydrocarbon fuel [65]. Further studies showed that surface modifications by chloride-doping of Bi_2WO_6 2D nanosheets provides improved photoactivity for the reduction of CO_2 to CH_4 with evolution rate of up to tenfold compared to that with no Cl^- elemental doping under H_2O conditions [66]. By varying the loading amount of elemental chloride dopant on Bi_2WO_6, the selectivity of CH_4 was increased from 51.29% to up to 94.98%, which indicated that the Bi_2WO_6 nanosheets doped with Cl^- ions not only facilitate the oxidation of H_2O to reduce the activation energy barrier for intermediate *CHO formation, but also promote the efficient generation of CH_4 [66]. In another study demonstrating elemental dopant effect coupled with 2D nanostructure effects, Bi_2WO_6 2D nanosheets doped with sulfur element were designed with high CO_2 adsorption and redox reduction ability [67]. The wider S-doping extended the visible light response activity range and improved the lifetime decay of charge carriers that lead to enhanced photoactivity for CO_2 reduction efficiency that was threefold better than the pristine 2D nanosheets [67]. Localized surface plasmon resonance (LSPR) designed photocatalytic materials offers effective strategies for extending on visible light photoactivity response capacity range for the efficient separation of photogeneration e^-/h^+ pairs, thus improved photocatalytic performances [66, 70]. A plasmonic enhanced Bi_2WO_6 with strong LSPR around the 500–1400 nm region was fabricated by electron doping to form bonding nature of W–O–W (V1) to obtain Bi_2WO_6-V1 catalyst with LSPR while creation of oxygen vacancies by the Bi–O–Bi (V2) centers to obtain Bi_2WO_6-V2 catalyst with defective absorption sites [70]. The Density functional theory (DFT) and photoluminescence (PL) confirmed that the V1 catalyst with suitable energy state facilitated electrons trapping sites which prolonged their lifespan decay that resulted in LSPR of Bi_2WO_6 (Fig. 13.2a, b). Because of the strong LSPR effect, the plasmonic V1 catalyst afforded high photoactivity and selectivity of CH_4 evolution rate of 9.95 μmol g^{-1} h^{-1} in the CO_2 photoreduction, which was 26-fold greater than the 0.37 μmol g^{-1} h^{-1} of V2 catalyst (Fig. 13.2c, d). The strong dependent of LSPR effect dependent of methane production activity was confirmed by different obtained photocatalytic results of plasmonic Bi_2WO_6 at various set light excitations [70].

Bismuth molybdate is a visible light photoactive binary semiconductor coupling bismuth and molybdenum metal with extensive interest in photocatalysis [71]. By using CTAB as structure directing agent under hydrothermal conditions, 2D Bi_2MoO_6 nanosheets were fabricated with improved photocatalytic CO_2 reduction activity to generate CO at rate of 14.38 μmol g^{-1} h^{-1} that is over tenfold higher compared to that of bulk Bi_2MoO_6. The designed 2D Bi_2MoO_6 facilitated the separation efficiency of e^-/h^+ pairs and reactive sites for adsorption of CO_2, which accounted for the enhanced photoactivity for CO_2 reduction performance [71]. The electronic and charge distribution properties of 2D metal oxide semiconductors can be relatively modified by the creation of surface defects and oxygen vacancies effect. This was demonstrated in the designed surface defective 2D Bi_2MoO_6 nanosheets with enhanced photoactivity for CO_2 reduction under H_2O without using co-catalyst and sacrificial agent [72]. The fabricated "Bi-O″" vacancy pairs of Bi_2MoO_6 nanosheets modified the local atomic structure to enable its electronic structure to promote efficient charge separation active sites. With the improved adsorption and activation of CO_2 promoted by the defective 2D surface structures of Bi_2MoO_6 nanosheets, high CO evolution rate that is 2.55-fold higher than their bulk counterpart was achieved [72]. In the family of bismuth-based semiconductors, oxyhalides BiOX (X = Cl, Br, I) are excellent monolayered 2D nanomaterials with high photocatalytic activity facilitated by their large surface area and ample active sites accompanied by high stability [73–75]. A large-scale produced monolayered 2D BiOBr nanosheets with ~0.85 nm thickness was fabricated for CO_2 photoreaction reaction [76]. The monolayered BiOBr nanosheets promoted efficient charge separation because of the large surface area and active sites with high adsorption of CO_2 for improved photoreduction into CO and CH_4 [76]. Another example of oxygen vacancies engineered ultrathin 2D nanosheet BiOBr afforded an enhanced CO_2 photoreduction activity to produce CO at 20.1 μmol g^{-1} h^{-1} under water conditions compared to the bulk BiOBr [77]. The improved CO_2 reduction photoactivity was attributed to the modified electronic surface structure, which optimized the adsorption and activation sites for CO_2 leading to formation of CO over oxygen deficient BiOBr nanosheets [77].

In the family of 2D nanomaterials, monolayer LDHs have attracted a wide interest due to the tunable defects and electronic properties amplified by the possibility for layers exfoliations [78–80]. The exfoliated LDHs nanosheets allows the confinement effect within the inter-monolayers that can improve on adsorption and desorption properties of the reacting species. The 2D monolayered NiFe-LHDs nanosheets with ~ 1 nm thickness boosting defective structure demonstrated high photoactivity for reduction of CO_2 to CH_4 with up to 81.75% selectivity under visible light irradiation [81]. Various layered 2D series of ZnM-LDHs (M = Ti^{4+}, Fe^{3+}, Co^{3+}, Ga^{3+}, Al^{3+}) photocatalysts were fabricated for CO_2 photoreduction [82]. The variation in the trivalent or tetravalent metal cations in the 2D ZnM-LDH showed to project various redox photoactivity that modified for improved products selectivity in CO_2 reduction reaction. The ZnTi-LDH achieved the best selectivity to CH_4 generation from CO_2 photoreduction as compared to others while ZnGa-LDH and ZnAl-LDH exclusively produced CO. By using diffuse refractive infrared transformation spectroscopy (DRIFTS), X-ray photoelectron spectroscopy (XPS), and DFT it was rationalized that the products selectivity of the various ZnM-LDH photocatalysts is influenced by the *d*-band center (ε_d) position of the M^{3+} or M^{4+} cations that account for the adsorption activity and the selectivity to formed products [82]. The metals cations with the *d*-band centers (i.e. Ti^{4+}, Ga^{3+} and Al^{3+}) showed relatively closer

Fermi level for enhanced strong adsorption of CO_2, and better yields of CH_4 or CO, whereas in the case of metal cations with *d*-band centers (i.e. Fe^{3+} and Co^{3+}) far away from the Fermi level demonstrated poor CO_2 adsorption and photoactivity. This observation enunciates the essence of the high redox centers nature of the LDHs characterized by abundant metal cations sites for enhanced photocatalytic properties [82]. Wang et al. [83] also demonstrated similar structural effects of metal cations in the photoactivity of LDHs for CO_2 photoreduction reaction by changing divalent-metal cations ($M = Mg^{2+}, Ni^{2+}, Cu^{2+}, Zn^{2+}$) with various *d*-bands classification in the construction of MAl-LDHs based photocatalysts. The evolution of CO and CH_4 over the 2D AL-LDHs based photocatalysts followed an increasing order of: NiAl-LDHs > CuAl-LDHs > ZnAl-LDHs > MgAl-LDHs. The DRIFTS results of both CuAl-LDHs and NiAl-LDHs showed more photoactivity for CO_2 reduction reaction which was attributed to the upward Fermi levels shifts of their d-band center as estimated by the valence-band XPS (Fig. 13.3a) [83]. The shift in the d-band center modified the visible light absorption capacity due to improved carrier separation and transfer because of the high suppressed recombination rates (Fig. 13.3b).

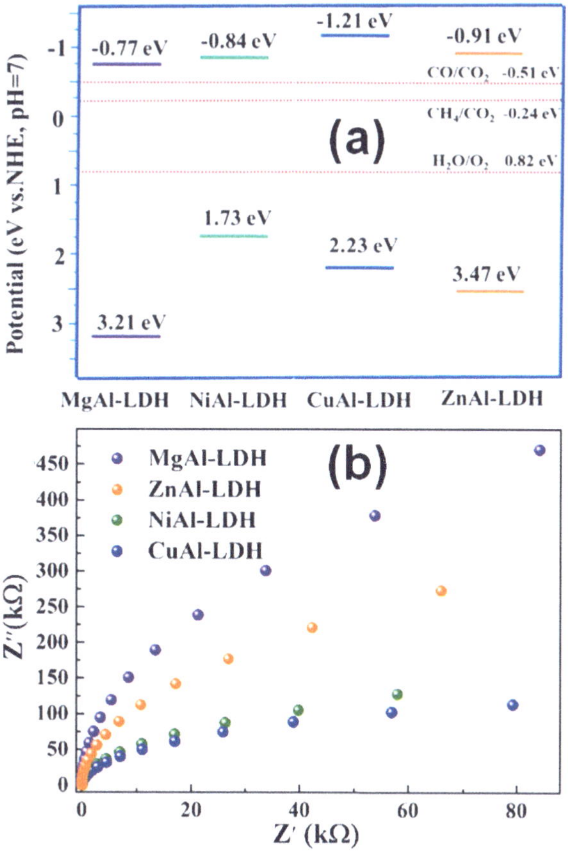

Fig. 13.3 **a** Band edges alignment of the different 2D MAl-LDHs photocatalysts. **b** Photocurrent resistance curves of 2D MAl-LDHs photocatalysts. Reproduced with permission from Ref. [83]. Copyright 2022, Elsevier

13.2.2 Heterojunction 0D-2D Metal Oxides Photocatalysts

The dispersion of 0D NPs and QDs onto layered surface 2D planar materials is effective to enhance photocatalytic properties of heterojunctions of two or more interfaced individual materials with improved efficiency to separate and transfer e^-/h^+ charge carrier for their high redox photocatalytic utilization performance [84–87]. The bimetal Ag-Pd NPs alloys decorated on TiO_2 nitrogen-doped nanosheets with 0D/2D heterojunction interface exhibited improved CO_2 phoreduction efficiency. The photocatalytic activity improvement of Ag-Pd/TiO_2 heterojunctions originated from the abundant Ti^{3+} ions that created oxygen vacancies sites on 2D TiO_2 nanosheets because of the substitutional effect of nitrogen which favored excellent adsorption of CO_2 reduction. Both substitutional N-doping of 2D TiO_2 and introduction of 0D Ag/Pd promoted an extended visible light photoactivity with improved high CO_2 phoreduction efficiency to produce CH_4 at rate of up to 79.0 μmol g^{-1} h^{-1} [88]. Wang et al. [89] constructed Au/TiO_2 plasmon Ohmic-junction 0D/2D heterointerface with enhanced CO_2 photoreduction efficiency to CH_4 and CO generated at the respective rates of 70.34 and 19.75 μmol·g^{-1}·h^{-1} with 80% selectivity of CH_4 formation [89]. The effect of synergy between Ohmic-junction and plasmonic Au NPs of Au and TiO_2 over extended visible light with efficient charge separation promoted the multi-electron photoreduction activity of CO_2 and H_2O conversions, which led to high CH_4 formation selectivity [89]. Cupric oxide (Cu_2O) is a visible light responsive semiconductor with a narrow band gap of 2.17 eV. The decoration of Cu_2O nanoclusters on dominant exposed {001} TiO_2 nanosheets allowed the construction of highly active 0D/2D hetero-interfaced junction photocatalyst with induced abundant oxygen vacancies (V_o) sites [90]. The formed interfacial contact of Cu_2O and TiO_2 in the 0D/2D heterostructure promoted the acceleration of charge carriers separation and migration for the efficient high CO_2 photocatalytic reduction assisted by the Cu(I) active species to produce CH_4, selectively at the rate that was 30-fold higher compared to pristine TiO_2 [90]. Tungsten trioxide (WO_3) is one of the traditional semiconductors with tunable visible light photoactive band gap in a range of 2.4–3.0 eV [91]. The catalytic application of WO_3 is widely explored in catalysis, energy and photocatalysis areas, including the CO_2 photoreduction [92–94]. However, the photoactivity of WO_3 suffers from high charge carrier recombination, thus low efficiencies for CO_2 photoreduction activity. As one of the strategies, designing 2D WO_3 interfaced with another 0D metal doping has demonstrated to circumvent the recombination rates in WO_3. The advantages of 2D planar surfaces and second metal active sites interfacing offers efficient carrier separation and transfer for the enhanced photoactivity [91, 96]. Xie et al. [96] by interfacing CuO QDs with WO_3 nanosheets constructed a photocatalyst with 0D/2D charge carrier driven by Z-scheme mechanism with improved CO_2 photoreduction efficiency to CO under H_2O compared to those of pure CuO and WO_3 [96]. Similar enhancement in CO_2 photoreduction activity was observed in the heterostructure of 0D Cu_2O NPs decorated on hexagonal 2D WO_3 nanosheets with predominant exposed (001) crystal plane (Cu_2O/WO_3-001) and without (Cu_2O/WO_3) in the presence of H_2O [97]. Significant improvement in

CO_2 reduction efficiency was achieved with facets exposed Cu_2O/WO_3-001 photocatalyst compared to those of WO_3-001, Cu_2O, and bulk Cu_2O/WO_3. The facet Cu_2O/WO_3-001 heterojunction produced the respective maximum yield for CO of 11.7 μmol due to Z-scheme regulated channels that promoted efficient charge separation with the enhanced redox photoactivity for CO_2 reduction [97]. CeO_2 with its highly redox exchangeable oxidation rates (Ce^{4+}/Ce^{3+}) is extensively studied in catalysis and photocatalysis [98–102]. A surface enriched active sites heterojunctions photocatalyst from interfaced 0D Pd NPs with 2D CeO_2 nanosheets exhibited selective photocatalytic CO_2 reduction activity to CH_4 formation [103]. The improvement in photocatalytic reduction efficiency for CO_2 over heterointerfaced Pd/CeO_2 was ascribed due to the formed active coordinated sites of $Pd^{\delta+}$ with oxygen vacancies Ce^{3+}-O_v confirmed by in-situ electron paramagnetic resonance (EPR) and quasi in-situ XPS results. The $Pd^{\delta+}$ sites were found to be key in H_2O oxidation activation to induce H* species for promoting the protonation of the intermediates. The doped Pd controlled the energy barrier responsible for the formation of the critical intermediates CO* and CH_3O*, which promotes the photoreduction of CO_2 to CH_4 become more easily favorable process. The heterojunction $Pd^{\delta+}/Ce^{3+}$-O_v achieved an efficient CO_2 photoreduction rate of 41.6 μmol g^{-1} h^{-1} to produce CH_4 at 100% selectivity [103]. An efficient CO_2 photocatalytic reduction was demonstrated over deigned heterojunction Ru/Bi_2MoO_6 with 0D/2D interface, which was better than the pure Bi_2MoO_6 to achieve the CO generation rates of 142.77 μmol^{-1} g^{-1} at 100% selectivity [104]. This enhanced photoactivity of Ru/Bi_2MoO_6 was attributed to the excellent electron-acceptor role promoted by metallic Ru^0. The Ru^0 was the active site for absorption and activation of CO_2 while Ru^{4+} facilitated efficiency in light absorption capacity due to modified energy levels within the bandgap of Bi_2MoO_6 2D layers favoring accelerated carrier separation and transfer [104].

The 2D bismuth compounds such as metal oxides (Bi_2WO_6, Bi_2MoO_6 and $BiVO_4$) and oxyhalides (BiOI, BiOCl, and BiOBr) have been explored extensively in photocatalysis for various applications such energy, chemical conversions, environmental remediation, and sensor devices [105–107, 107, 108, 108]. Moreover, bismuth-based oxides and oxyhalides are easily tunable layered 2D planar nanostructures with enriched surface-active sites [109]. A 2D Bi_2WO_6 nanosheets with 4.8 nm thickness synthesized via hydrothermal afforded CO_2 photoreduction rate of 19 ppm g^{-1} h^{-1} to CH_4 at low of 400 ppm CO_2 concentration under H_2O conditions [110]. The decoration of PtO_x NPs co-catalysts with average size of 2 nm on Bi_2WO_6 nanosheets improved the CH_4 production yield to up to 108.8 ppm g^{-1} h^{-1}, which was 5.7-fold greater compared to pure Bi_2WO_6 nanosheets. The heterojunction 0D/2D interface of PtO_x/Bi_2WO_6 boosted efficient charge separation and migration with accelerated H_2O oxidation by PtO_x NP to enhance the overall CO_2 photoreduction efficiency [110]. The interface of $CsPbBr_3$ QDs perovskites with Bi_2WO_6 nanosheet resulted into a highly photoactive direct Z-scheme 0D/2D heterojunctions boosting efficient charge separation and migration to achieve a total CH_4/CO production rate of 503 μmol g^{-1} that was 9.5-fold higher compared to single Bi_2WO_6 and $CsPbBr_3$ [111]. The strong intimate contact of well distributed $CsPbBr_3$ QDs on surface of 2D Bi_2WO_6 nanosheets facilitated efficient spatial separation of photoinduced carriers which preserved the effective electrons utilization with high redox photoactivity influenced by the Z-scheme heterojunction mechanism band edges and electronic structure modifications as depicted in Fig. 13.4a, b [111]. The efficient operation of Z-scheme heterojunction for transfer of separated charge carrier was confirmed by the suppressed recombination rates results of the electrochemical analysis depicted in Fig. 13.4c, d of the electrochemical impendence spectroscopy (EIS) and photocurrent resistance, respectively.

LDHs are 2D layered materials bearing highly redox surfaces from the multicomponent metal cations oxidation states endowed with reactive OH groups boosting the active basic sites. The interplanar spacing and the controlled exfoliations of the 2D layers expedite for inter-confinement effect of 0D with improved mass transfer dynamics enhancements. The layered 2D nanostructures of CoAl-LDHs heterointerfaced with Degussa TiO_2 P25 NPs were produced via hydrothermal route to construct 0D/2D heterojunctions [113]. The 0D-TiO_2/2D-CoAL-LDH was photoactive for reduction of CO_2 in the visible light range. The aligned edges of the two semiconductors boosted effective promotion of donor–acceptor carrier separation and transfer which enhanced the redox activity for the improved CO_2 photoreduction with high activity to produce CO with selectivity of over 90% [112]. The tunable bandgap depending on the types of metals compositions promoting the construction of responsive visible light photocatalysts. In the interface of Cu_2O NPs on layered ZnCr-LDH a visible light active 0D/2D heterojunction was formed with improved photoactivity for CO_2 reduction in H_2O to produce CO, which was better compared to pristine 2D ZnCr-LDHs [113]. The plasmonic Pt NPs were decorated as cocatalyst in the exfoliated MgAl-LDHs to design a heterojunction 0D/2D photocatalyst with improved photoactivity for CO_2 reduction. The decorated Pt NPs in the inter-spaced exfoliated MgAl-LDHs was achieved by electrostatic interaction. The heterojunction Pt/MgAl-LDHs resulted in improved CO_2 photoreduction activity to CO at the evolution rate of 2.64 μmol g^{-1} h^{-1}, which is 8.52-fold greater compared to pristine MgAl-LDHs due to enhanced carrier's separation efficiency [114].

13.2.3 Heterojunction 1D-2D Metal Oxides Photocatalysts

Heterojunction interface of 1D and 2D nanomaterials have been explored applied extensively in environmental remediation, energy, CO_2 and N_2 solar fuels [115–117]. The exceptional structure properties of 1D nanomaterials in photocatalysis lies in their capability to effect spatial directional flow charge carriers separation efficiency compared to random in which separation efficiency is difficult to achieve because the fast recombination of the separated hole and electron coincident with the pathway of random separation [36]. The unique spatial distribution of photoinduced electrons and holes in 1D are effective in the hybrid 1D heterostructures manipulated by the controlled morphology and multidimensions heterojunctions interfaces. The hierarchical core–shell 1D@2D of ZnO@Cu-ZnAl LDHs wrapped heterostructures was fabricated via a deposition–precipitation method combining 1D ZnO and 2D

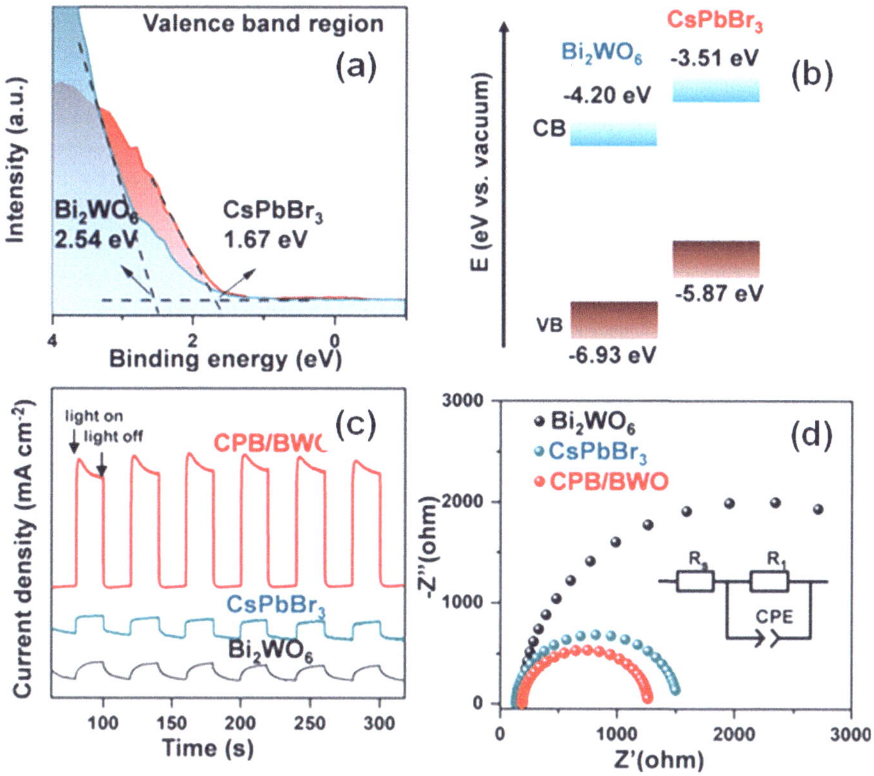

Fig. 13.4 **a** Electronic band energy values of $CsPbBr_3/Bi_2WO_6$ photocatalysts. **b** Z-scheme band edge alignment mechanism of $CsPbBr_3/Bi_2WO_6$ photocatalyst. **c** EIS profiles of $CsPbBr_3/Bi_2WO_6$ photocatalysts. **d** Photocurrent resistance profiles of $CsPbBr_3/Bi_2WO_6$ photocatalysts. Reproduced with permission from Ref. [111]. Copyright 2020, American Chemical Society

CuZnAl-LDHs for the efficient CO_2 photocatalytic reduction reaction [118]. The 1D ZnO possessed the two different morphologies of rod-like ZnO (~ 600 nm diameter and ~ 3.5 μm length) and belt-like ZnO (50 nm to 200 nm width and ~ 10 μm length). The surface coating of Cu-ZnAl-LDHs by ZnO afforded the construction of core–shell wrapped ZnO@Cu-ZnAl-LDHs heterojunction 1D/2D composite with high surface areas, and abundant active sites for CO_2 adsorption. The inhibited recombination of photoexcited e^-/h^+ pairs due to interfacial contact improvement of the separation and transfer efficiency that led to enhanced CO_2 photoreduction compared to the pristine counterparts [118]. A hierarchical $Bi_2MoO_6@In_2S_3$ composite of nanotubes/nanosheets morphology enriched with surface oxygen vacancies were produced via a multi-step synthesis strategy to form S-scheme heterojunction mechanism that enhanced fast carriers' separation and transfer of electrons [119]. The heterojunction hierarchical architectures effectively facilitated remarked photoactivity in CO_2 reduction to CO at generation rate of 28.54 μmol g^{-1} h^{-1} and 94.1%

selectivity [119]. Crystal facet or plane nano-engineering of materials with specific morphology-controlled dimensions provide excellent preferment enhancement in directional charge carrier separation and migration that is eminent for improved photocatalytic activity properties. This was demonstrated by Zhang et al. [120] in the construction of heterostructure interfaced 1D WO_3 nanowires with preferential exposed {110} plane (WO_3-110) to 2D WO_3 nanosheets predominated by {001} exposed plane (WO_3-001) doped with CuO NPs for efficient CO_2 photoreduction. This ternary multidimensional 0D/1D/2D heterojunction provided a unique interfacial contact between the WO_3 and CuO via p–n junction that promoted efficient e^-/h^+ carrier dynamics for effective redox utilization high-performance in CO_2 photocatalytic reduction to CH_4. The high CO_2 photoreduction efficiency was achieved with the {110} exposed plane of WO_3-CuO heterojunction compared to the {001} WO_3-CuO interface. This was due to the contribution effect of {110} facet WO_3 nanowires having adequate negative conduction band edge with high reduction ability facilitated accelerated carrier migration efficiency that enabled abundant photoexcited electrons to participate in the redox CO_2 reduction [120]. TiO_2 is the most established semiconductor material in photocatalysis, but its main drawback is the restricted photoactivity in UV light region. However, several approaches have been studied to improve WO_3 photoactivity to the near edges of visible light by morphology and facet engineering, phases compositions (i.e. ratio of rutile/anatase), including heterostructures [121, 122, 124, 125]. Amongst the morphologies or facets, the 1D TiO_2 based nanomaterials have shown much enhanced photocatalytic activity properties [54, 123, 124]. Kim [125] et al. demonstrated high photocatalytic performance of heterointerfaced 1D TiO_2 nanotubes (TNT) with 2D nanosheets of hexagonal $ZnIn_2S_4$ (ZIS) synthesized via hydrothermal route in CO_2 reduction as illustrated in Fig. 13.5a [125]. The improved CO_2 reduction photoactivity of TNT/ZIS was influenced by the optimal loading of TNT, with 10 wt.% achieving the best CO generation rate of 4.41 mmol g^{-1} h^{-1}, which was 1.5-fold better compared to pure TNT or ZIS [125]. The improved CO_2 photoreaction activity was ascribed because of interfacial multidimension contact of 1D and 2D with enhanced directional flow of efficiently separated charge carriers and extended visible light absorption capacity of TiO_2 nanotubes modified by $ZnIn_2S_4$ nanosheets at suppressed e-/h + recombination (Fig. 13.5b, c). Another example of improved overall photoactivity of 1D TiO_2 was demonstrated in the 1D/2D TiO_2/MoS_2 heterostructure comprising TiO_2 fibers (1D) wrapped by 2D MoS_2 nanosheets. The arrays of MoS_2 nanosheets possessed the size of $\approx$ 80 nm surface with 2 nm thickness standing vertically and uniformly upon the TiO_2 nanofibers. This heterointerface induced strong intimate chemical interaction bonding between MoS_2 and TiO_2, which was effective to promote fast e^-/h^+ carrier separation and transfer due to inhibited recombination and extended visible light which led to high redox photoactivity in the reduction reaction of CO_2 to CH_4 and CH_3OH compared to bare TiO_2 [126].

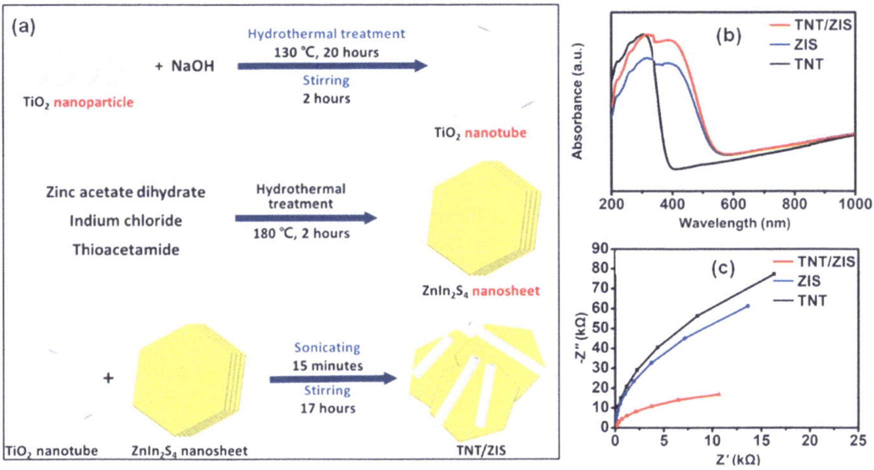

Fig. 13.5 **a** Synthesis procedures to obtain 1D/2D heterojunction TiO_2 nanotubes interfaced $ZnIn_2S_4$ nanosheets. **b** UV–vis light absorption curves of the 1D/2D heterojunction $TiO_2/ZnIn_2S_4$ photocatalysts. **c** EIS curves of the 1D/2D heterojunction $TiO_2/ZnIn_2S_4$ photocatalysts. Reproduced with permission from Ref. [125]. Copyright 2022, Elsevier

13.2.4 Heterojunction 2D-2D Metal Oxides Photocatalysts

The larger specific surface areas, of the 2D layer planar promotes short charge carrier mobility distances, which enhances the highly active participation in the photocatalytic redox reactions due to improved charge separations and migrations dynamics [127, 128]. Amongst the structure advantages of 2D materials is their flexibility to fabricate unique heterointerfaces of combining multi-components of various kinds, which enables the enhancing in photo-light absorption range and accelerated migration of efficient separated electrons with minimized recombination rates. The construction of novel heterojunctions 2D/2D with high spatial charge carrier separation and transfer efficiency, thus making available abundant redox reactive species have recently emerged as an attractive strategy to design high performance photoactive CO_2 reduction catalysts [129–132]. Xie et al. [133] designed interfaced 2D/2D heterojunction of Bi_2O_3 2D nanosheets with Bi_2WO_6 nanosheets achieved by covalently bonded epitaxial growth facilitated by the co-sharing of the coordinated Bi–O tetrahedron bond between Bi_2O_3 and Bi_2WO_6. This 2D/2D interface was characterized by excellent channels for the effective transfer boosted by efficient separation. Further, the Bi_2O_3/Bi_2WO_6 heterojunctions demonstrated high photoactivity for reduction of CO_2 to CO reaching up to 17.39 $\mu mol \cdot g^{-1} \cdot h^{-1}$ production rate at 95.4% selectivity that is 18.0-fold and 4.2-fold higher compared to that of bulk-Bi_2WO_6 and Bi_2WO_6 nanosheets, respectively [133]. An epitaxial heterostructure 2D/2D BiOCl @ Bi_2O_3 photocatalyst was constructed using sonication-assisted etching of Cl in BiOCl nanosheets to form in-situ interfaced Bi_2O_3 nanosheets [134].

The heterojunction BiOCl @ Bi_2O_3 accelerated efficiently the migration of separated photoexcited carriers via the internal built-in electric field made by lateral interfaced chemical bonding of semiconductors for the enhanced CO_2 photoreduction activity to produce CO at 30 μmol g^{-1} h^{-1} rate. Bi_2WO_6 is a visible light photoactive bismuth-based semiconductors. However, its photocatalytic efficiency is hampered by high recombination rates of e^-/h^+, thus making it difficult to achieve its maximum photocatalytic performances in many applications. The interface of Bi_2WO_6 with semiconductors such as BiOI [135, 136], TiO_2 [137], Ti_3C_2 [138], C_2PbBr_3 [139] in the photocatalytic CO_2 photoreduction with enhanced photoactivity has been explored. In the interface of pre-synthesized Bi_2WO_6 2D nanosheets with 2D nanosheets of BiOI via chemical assembly, surface defects induced by the oxygen deficient Bi_2WO_6 were introduced in the 2D/2D heterointerface. The 2D/2D interface formed a *p-n* heterojunctions with internal electric field which promoted accelerated directional flow transfer of spatially separated photoexcited charge carriers, hence high activity performance for CO_2 photoreduction [135]. The surface induced oxygen vacancies sites acted as the trapping of electrons, which improved the active sites for the efficient CO_2 adsorption and its photoreduction to CH_4 over an extend near infra-red (NIR) region light absorption range [135]. The 2D nanosheets TiO_2 interface with 2D Bi_2WO_6 nanosheets and sandwiching 0D Pt NPs demonstrated a highly visible light photoactive heterojunction photocatalyst with efficient CO_2 photoreduction activity to CH_4 under H_2O conditions in a continuous fluidized-bed reactor gas phase reaction [137]. The face-to-face intimate contact of Bi_2WO_6 and TiO_2 provided an efficient spatial charge separation and migration facilitated by Pt NPs which acted as reservoir site for electron trapping, thus enhanced photoactivity for reduction of CO_2 to CH_4 [137]. The combination of the bismuth oxide (Bi_2MoO6) and its oxyhalide (BiOI) nanosheets afforded heterojunctions 2D/2D modified interfacial face-to-face contact with improved charge separation and transfer promoted by induced van der Waals contact effect of large interfacial area. The Bi_2MoO_6/BiOI photocatalysts heterojunction interface improved the CO_2 photoreduction efficiency to CH_4 and CO as main products due to the effect of the operating S-scheme carriers transfer mechanism with strong redox reactivity [136].

Recently, Ti_3C_2 is emerging as a promising 2D semiconductor material made up of metal carbide but with low photocatalytic efficiency on its own due to high charge recombination rates. The construction of 2D Ti_3C_2 nanosheets interfaced with 2D Bi_2WO_6 nanosheets provide short charge carrier migration distance due to strong and large interfacial contact area enhanced by the large surface-to-volume ration in the layer planar motifs which enhanced significantly the photoactivity for CO_2 reduction [138]. The layered interfaced 2D/2D Ti_3C_2/Bi_2WO_6 heterojunction achieved high photoactivity for CO_2 reduction to CH_4 and CH_3OH at evolution rates that are 4.6-fold higher to that of pure Bi_2WO_6 nanosheets [138]. Perovskite materials make are good sensitizing candidates to achieve photoactive light absorption in the visible to NIR range with improved photoactivity [140–142]. A heterojunction 2D/2D of $CsPbBr_3/Bi_2WO_6$ photocatalyst with direct Z-scheme carrier transfer mechanism was constructed via a simple electrostatic self-assembly method that allowed the creation of intimate contacted ultrathin nanosheets of several atomic layers [139].

This thin layers of heterointerfaced $CsPbBr_3/Bi_2WO_6$ composite possessed large interfacial area with reduced short charge carrier migration distance promoted by the Z-scheme mobility between Bi_2WO_6 and $CsPbBr_3$ with efficient separation. The heterostructure $CsPbBr_3/Bi_2WO_6$ demonstrated high photoactivity reduction performance for the conversion of CO_2. With the incorporation of 0D Pt NPs cocatalyst further improvement in the photoreduction activity of CO_2 conversion was obtained at stable yields of up to 1582.0 μmol g^{-1} and 8603.0 μmol g^{-1} for selective generation of CO (11.4%) and CH_4 (84.3%), respectively which was 12.2-fold greater than that of $CsPbBr_3$ nanosheets [139]. The combination of bismuth oxyhalides and metal sulfides makes a very compatible heterostructures with the interfacial junctions originating from heteroatoms of sulfur and chloride that are effective dopants to enhance charge carriers over an extend visible light accompanied by the efficient spatial separations and migrations for effective utilization in the redox photoreactions [143]. A 2D/2D heterointerfaced S-scheme heterojunctions boosting enhanced charge carries separation and transfer efficiency was constructed by coupling the semiconductors of different properties BiOBr/CdS aminated with diethylenetriamine and containing 40% mass ratio of BiOBr [144]. The diethylenetriamine aminated 2D/2D BiOBr/CdS heterojunction photocatalysts exhibited efficient CO_2 reduction photoactivity to produce CO evolution rate of up to 19.4 μmol h^{-1} g^{-1}, which was 3.68-fold greater than that of pure 2D CdS and 2D BiOBr. On the other hand, the non-aminated 2D/2D BiOBr/CdS showed low CO evolution rate of 11.2 μmol h^{-1} g^{-1}, which demonstrated the significant surface decoration of the 2D/2D BiOBr/CdS to overcome photocatalyst corrosion while inducing the efficient S-scheme charge carriers migration heterojunction [144].

A Schottky junction interface of 2D $Ti_3C_2T_x$ nanosheet with 2D NiAl-LDH was fabricated to with a core–shell structure via hydrothermal method [145]. The NiAl-LDH/$Ti_3C_2T_x$ 2D/2D heterojunction demonstrated superior photoactivity for reduction of CO_2 to CO at evolution rate of up to 2128.46 μmol h^{-1} g^{-1} and 90.2% selectivity, which was 8.6-fold higher than that of bare NiAl-LDH or $Ti_3C_2T_x$ [145]. The systematic construction of such Schottky heterojunction 2D/2D of nanosheets interface of NiAl-LDH and $Ti_3C_2T_x$ is schematically illustrated in Fig. 13.6. Another 2D/2D heterointerface of titania-sandwiched Ti_3C_2 MXenes multilayers with oxygen vacancy-enriched trimetallic CoAlLa-LDH was constructed for enhanced CO_2 photoreduction [146]. The presence of La^{3+} in the CoAlLa-LDH/$Ti_3C_2T_x$ induced the formation of coordinatively unsaturated metal centers with enhanced reductive sites. The intimate contact between $Ti_3C_2T_x$ and the trimetallic CoAlLa-LDH provide a S-scheme heterojunction pathways for the efficient charge carriers mobility with improved CO_2 reduction performance for the generation of CO and CH_4 at evolution rates of 46.32 and 31.02 μmol g^{-1} h^{-1}, respectively compared to low of the pristine catalysts [146].

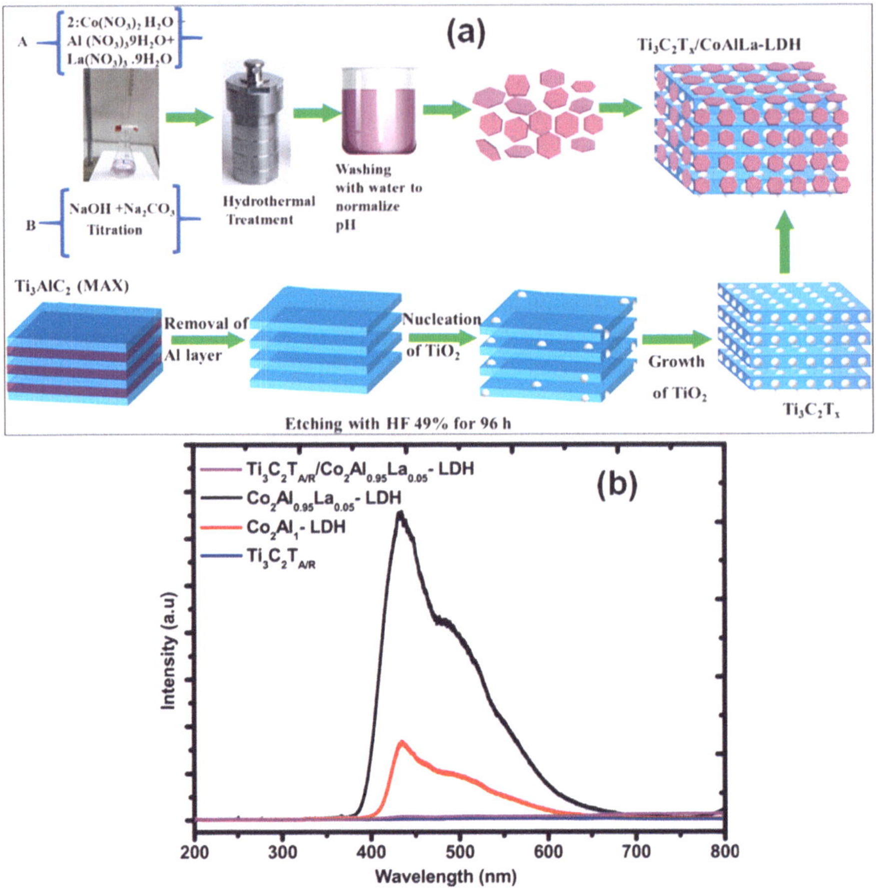

Fig. 13.6 **a** Synthesis of layered 2D/2D Ti_3C_2Tx/CoAlLa-LDH heterojunction photocatalyst. **b** PL charge recombination rates of Ti_3C_2Tx/CoAlLa-LDH based photocatalysts. Reproduced with permission from Ref. [145]. Copyright 2021, American Chemical Society

13.2.5 Heterojunction 2D-3D Metal Oxides Photocatalysts

Heterojunctions 2D/3D interfaced composites are very versatile in providing the design of high porous nanostructured materials with confinement effect that can improve the light absorption capacity, carriers separation efficiency and high migration rates to suppress the recombination of the e^-/h^+ for effective redox photocatalytic utilizations [147–153]. High photoexcited carrier separation efficiency is very critical in determining the rate of CO_2 photoreduction activity but achieving the directed flow transfer channels is still presenting a challenge. A heterostructure 2D

MoS_2 with few layers interfaced with 3D hierarchical flower-like Bi_2WO_6 microspheres was constructed via a simple two-step synthesis method [155]. The fabricated MoS_2/Bi_2WO_6 composite photocatalysts boasted excellent CO_2 adsorption capacity, extended visible-light absorption range, effective charge mobility facilitated by the intimate 2D/3D contact of MoS_2 and Bi_2WO_6 with improved separation of e^-/h^+ pairs, which lead to enhanced CO_2 reduction photoactivity [154]. Under optimized loading content of MoS_2 (0.4 wt%), the 2D/3D MoS_2/Bi_2WO_6 photocatalyst achieved the best photoactivity of CO_2 reduction compared to the neat Bi_2WO_6 with respective methanol and ethanol yields of 36.7 and 36.6 μmol $gcat^{-1}$ that were 1.94-fold greater than that Bi_2WO_6 [154]. In another study, the morphology-controlled synthesis of 3D/2D architectured heterojunction photocatalyst was fabricated from using the octahedrons Cu_2O (3D) interfaced with 2D WO_3 nanoflakes [155]. The 3D/2D Cu_2O/WO_3 composite photocatalyst consisted of cubic Cu_2O phase and hexagonal WO_3 phase forming a Type II band edges aligned interface leading to Z-scheme mechanism configuration of the photoinduced carriers (e^-/h^+) migrations. The morphology-dimensions interfaced constructed Cu_2O/WO_3 photocatalysts exhibited high photocatalytic activities in the reduction of CO_2 reaction with H_2O to produce CO and CH_4 at respective evolution yields rates of 7.04 and 8.42 μmol $gcat^{-1}$ [155]. The construction of full spectrum ranging from visible to NIR photoactive materials has emerged recently as promising effective strategies to alleviated limitations in realizing effective photocatalytic properties for efficient processes applications. Ji et al. [156] fabricated a defective 2D/3D heterojunction $Cu_{2-x}S$/Ni–Al-LDH photocatalyst using hydrothermal method. The multidimensional interfaced heterostructure composite composed the 3D LDHs with flower-ball structure that seems to have self-assembled from the randomly arranged 2D nanosheets to form porous nanostructures with excellent exposed surface-active sites.

The most photoactive composite $Cu_{2-x}S$/LDH photocatalyst with optimized 15%Cu produced CH_4 and CO with respective evolution rates of 14.2 μmol/g/h (2.83 μmol/g/h in the NIR region) and 5.3 μmol/g/h (2.99 μmol/g/h in the NIR region) (Fig. 13.7b). This high photoactivity for CO_2 photoreduction performance was proposed to be due to the formed p–n heterojunction with Z-scheme mechanism for carrier transport between the intimate contact of $Cu_{2-x}S$ and LDH and effective suppressed e^-/h^+ recombination (Fig. 13.7a, c and d) [156]. A Z-scheme configured 2D/2D architecture interface of $Bi_2O_2CO_3$/Bi/NiAl-LDH (BBOC/LDH) nanosheets was fabricated using hydrothermal method [166]. The heterojunction BBOC/LDH composites with 20%BBOC optimal content loading was highly active for photocatalytic reduction of CO_2 reaction to produce CH_4 at up to 56.64 μmol/g_{cat} rate that was 3.9-fold greater than neat LDH. The combination of larger surface area due to LDH, and the efficient separation of photoinduced e^-/h^+ from the influence of the presence of surface metal Bi SPR effect together with the efficient Z-scheme charge carrier migrations, enhanced the photocatalytic performance for CO_2 reduction [157].

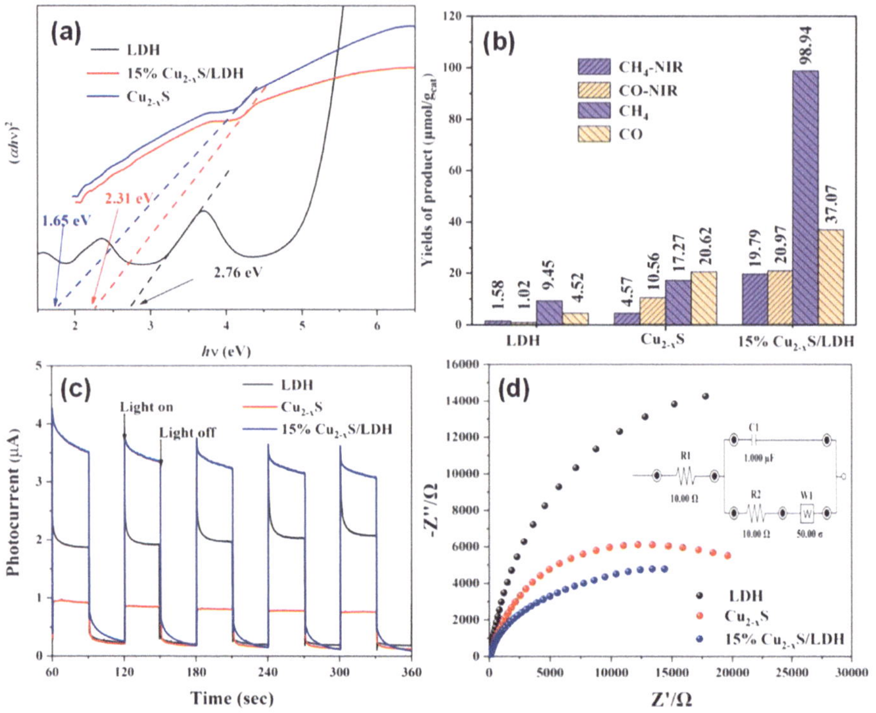

Fig. 13.7 **a** Electronic band energy value of LDH based photocatalysts, **b** Productions formation rates in CO_2 photoreduction, **c** Electrochemical impedance spectroscopy and **d** Photocurrent resistance. Reproduced with permission from Ref. [156]. Copyright 2022, American Chemical Society

Hollow structure materials possess enhanced physical textural properties such as high porosity and adsorption capacity due to their channels or cavity confinement effects [157, 158, 160]. These types of materials have been applied in photocatalysis widely with superior photocatalytic performances [168, 169]. Li et al. [161] constructed a hollow structured ZnO/Co_3O_4 catalyst with p-n heterojunction configuration following the pyrolysis of bimetallic ZnCo-ZIFs, then modification with polydopamine (PDA). By modulating the Zn/Co ratio composition and the surface content of PDA on ZnO/Co_3O_4, the high separation efficiency of the photoexcited e^-/h^+ pairs and maximized CO_2 adsorption capacity were achieved that lead to a superior photoactive $PDA_{15}/ZnO/Co_3O_4$ photocatalyst for CO_2 reduction to produce CO at rate of 537.5 μmol/g/h with 97.7% selectivity without using additional sacrificial agent and photosensitizer or cocatalyst. The hybrid interfaced PDA-ZnO/Co_3O_4 composite presented an effective strategy to construct a highly efficient photocatalyst with remarkable properties of light harvesting capacity, separating charge carriers, and CO_2 adsorption for efficient reduction [161]. The S-scheme heterojunction interfaces have shown to present enhanced photocatalytic properties comparable to those of Z-scheme and better than the common Type-II and Type-I mechanisms [162]. A S-scheme configured heterojunctions was synthesized from diethylenetriamine (DETA) ammoniated MnS (α-MnS) and Bi_2MoO_6 boosting the Bi surface plasmon resonance (SPR) [163]. This highly redox-active α-MnS/Bi_2MoO_6 heterostructure photocatalysts consisted of 5 wt% content of Bi (5%Bi-M/BMO). The presence of DETA which was protonated to form surface positively charged ammonium ions enabled enhanced adsorption of acid gas CO_2, thus reduced its activation energy to promote efficient CO_2 photoreduction to CO at 61.11 μmol g^{-1} h^{-1} generation rate, which was 2.42, 7.89 and 5.01-folds greater compared to that of 5%M/BMO, pure α-MnS and Bi_2MoO_6, respectively. The enhanced photoactivity was improved by the Bi SPR and S-scheme heterojunction effects that facilitated the separation of photoexcited e^-/h^+ pairs [163]. An active 0D BiO_{2-x} NPs on 3D Bi_2MoO_6 nanoflowers interfaced heterostructure photocatalyst was synthesized by *in-situ* H_2/Ar low temperature plasma method for improved CO_2 photoreduction [164]. The 0D/3D BiO_{2-x}/Bi_2MoO_6 heterostructure with Mo–O-Bi bonds between the Bi_2MoO_6 and BiO_{2-x} interfaces provided active sites acting as a charge transfer bridge, thus promoting the directional conveyance of electrons for the enhanced photoactivity of CO_2 photoreduction rates involving multiple electrons. A heterojunction Bi_2MoO_6/BiO_{2-x} showed good photostability and high photoactivity that was $\approx$3.0-fold higher than that of bare Bi_2MoO_6. The interatomic charge transfer bridges interfaces between Bi_2MoO_6 and BiO_{2-x} provided an effective capability to suppress the carrier recombination to realize high photoactive catalytic systems for efficient CO_2 reduction [164]. Both $ZnFe_2O_4$ and Bi_2MoO_6 are well-proven narrow band gap photoactive semiconductors in visible light range but are both suffering for high recombination rates individually [165, 166]. However, the heterojunction construction of interfaced $ZnFe_2O_4$/Bi_2MoO_6 photocatalyst consisting of 0D $ZnFe_2O_4$ NPs and 3D Bi_2MoO_6 nanospheres provides a versatile multidimensional heterostructure design approach to high photoactive materials which was achieved by a simple solvothermal route [167]. This interface composing various dimensions with narrow bandgap edges allows for improved light absorption capacity, create surface active sites environment at boundary interfaces to promoted carriers separation efficiency of the $ZnFe_2O_4$/Bi_2MoO_6 heterostructure with enhanced photocatalytic properties. With the 20% content loading of Bi_2MoO_6, the heterojunction $ZnFe_2O_4$/Bi_2MoO_6 achieved the highest photoactivity efficiency in the reduction of CO_2 to produce products yield of 47.1 μmol g^{-1} compared to pure $ZnFe_2O_4$ (14.79 μmol g^{-1}) and pristine Bi_2MoO_6 (19.01 μmol g^{-1}) [167]. Quantum dots (QDs) are known for the excellent visible light to NIR photoactivity light spectrum range due to the quantum size and LPRS effect phenomena [168]. Jiang et al. [169] designed 3D hollow nanostructures consisting self-assembled Bi_2WO_6 QDs via a one-pot solvothermal method for application in photoreduction of CO_2 to CH_3OH [169]. The hollow structure of 0D/3D Bi_2WO_6 showed higher adsorption of CO_2 which led to enhanced photoactivity for CO_2 reduction compared to the bulk-Bi_2WO_6 [169]. The modulation of semiconductors Fermi level shift in the valence band allows for band gap edges modification that can be suitable for the enhanced photocatalytic properties [170, 171].

To demonstrate this concept approach in designing highly photoactive CO_2 photocatalytic materials, Chen et al. [172] constructed a novel Bi_2WO_6/Si heterostructure photocatalyst. The shift of the Fermi level to the valence band can successfully reduce the difficulty of e^-/h^+ pair generation and enhances the ability to separate them, thus prolonging the electrons lifetime for effective redox utilization. The heterostructure Bi_2WO_6/Si with 1:1 ration (i.e. $Bi_2WO_6:Si = 1:1$) achieved high CO_2 photoreduction activity to produce selectivity ethanol at the rate of up to 402.76 μmol/L [172]. The recent emergence of Mxenes (Ti_3C_2) 2D materials with excellent optical and electronic conductivity properties has shown promising noble metal-free photoactive co-catalyst in the CO_2 reduction reaction. A 3D/2D architecture heterostructure photocatalyst consisting of 3D cube-like CeO_2 interfaced with Ti_3C_2-MXene nanosheet was synthesized by hydrothermal method to afford a superior photoactive catalyst for reduction of CO_2 [173]. The heterojunction CeO_2/Ti_3C_2-MXene with optimized ratio of Ti_3C_2-MXene showed enhanced photoactivity for CO_2 photoreduction at a higher rate of 1.5-fold than of pure CeO_2 due to the Schottky junction formed by the built-in electric field between CeO_2 and Ti_3C_2-MXene that promoted efficient carrier migrations of photoexcited electrons from CeO_2 to Ti_3C_2-MXene, thus expediting separation of the electrons and holes for efficient redox utilization [173]. 2D designed metal semiconductor can plays vital roles significant structural modifications in the electronic and charge distribution properties of the materials by oxygen deficient effect. This has been demonstrated by Dai et al. [174] in the designing oxygen vacancy-rich Bi_2MoO_6 which self-assembled its 2D ultrathin Bi_2MoO_6 nanoflakes into 3D nanospheres of Bi_2MoO_6 homojunction interface boosting isolated Mo atoms as binding active sites for CO_2 molecules trapping to lower its activation energy barrier to for enhanced CO_2 photoreduction to selectivity forming CH_3OH and CH_3CH_2OH with respective production yields of up to 106.5 and 10.3 $\mu mol\ g^{-1}$, which was higher than the bulk-Bi_2MoO_6 [174].

13.3 Conclusion

This chapter deliberated on the recent progress in the design and synthesis of two-dimensional (2D) all-metal oxide-based semiconductor materials and their applications in developing efficient solar CO_2 photoreduction reaction into fuel and chemicals. The chapter critically discussed significant structure properties of 2D metal oxides characterized by stacked nano-layers with high specific surface area, large interlayer spacing, tunable exfoliation into single or few layers to improve on mass transport and highly exposed surface-active sites. The explored literature results showed significant improvements in the hetero-interfaced 2D nanostructures with other multidimensional (e.g. 0D, 1D, and 3D) tunable by controlled physicochemical features including defects and vacancies, crystal facets or planes, redox, and opto-electronics. The control of these properties in the heterostructure interfaced 2D materials demonstrated high photocatalytic performance efficiencies in CO2 reduction reaction driven by extended visible to near-infra-red photo-light absorption response, improved photocarriers generation, separation, and mobility for effective redox utilization in photocatalytic CO_2 conversion. The overall data evaluation of the combined novel planar 2D nanostructures and their associate interfaces with modified physico-chemical and opto-electronic structures demonstrated promising strategies to develop high-performance photoactive materials with great potentials for possible future realization of efficient and sustainable commercial process for CO_2 solar conversion to fuel and chemicals.

Acknowledgements The author thanks CSIR for the support to publish this chapter.

References

1. K. Mak, J. Shan, Photonics and optoelectronics of 2D semiconductor transition metal dichalcogenides. Nat. Photon **10**, 216–226 (2016). https://doi.org/10.1038/nphoton.2015.282
2. T. Tan, X. Jiang, C. Wang, B. Yao, H. Zhang, 2D Material optoelectronics for information functional device applications: status and challenges. Adv. Sci. **7**, 2000058 (2020). https://doi.org/10.1002/advs.202000058
3. F. Wang, K. Pei, Y. Li, H. Li, T. Zhai, 2D Homojunctions for electronics and optoelectronics. Adv. Mater. **33**, 2005303 (2021). https://doi.org/10.1002/adma.202005303
4. X. Zhou, X. Hu, J. Yu, S. Liu, Z. Shu, Q. Zhang, H. Li, Y. Ma, H. Xu, T. Zhai, 2D Layered material-based van der Waals heterostructures for optoelectronics. Adv. Funct. Mater. **28**, 1706587 (2018). https://doi.org/10.1002/adfm.201706587
5. A. Manikandan, Y.Z. Chen, C.C. Shen, C.W. Sher, H.C. Kuo, Y.L. Chueh, A critical review on two-dimensional quantum dots (2D QDs): from synthesis toward applications in energy and optoelectronics. Prog. Quantum Electron. **68**, 100226 (2019). https://doi.org/10.1016/j.pquantelec.2019.100226
6. Z. Sun, N. Talreja, H. Tao, J. Texter, M. Muhler, J. Strunk, J. Chen, Catalysis of carbon dioxide photoreduction on nanosheets: fundamentals and challenges. Angew. Chem. Int. Ed. **57**, 7610–7627 (2018). https://doi.org/10.1002/anie.201710509
7. Y. Wang, B. Ren, J.Z. Ou, K. Xu, C. Yang, Y. Li, H. Zhang, Engineering two-dimensional metal oxides and chalcogenides for enhanced electro- and photocatalysis. Sci. Bull. **66**, 1228–1252 (2021). https://doi.org/10.1016/j.scib.2021.02.007
8. W.J. Ong, L.K. Putri, A.R. Mohamed, Rational design of carbon-based 2D nanostructures for enhanced photocatalytic CO_2 reduction: A dimensionality perspective. Chem. Eur. J. **26**, 9710–9748 (2020). https://doi.org/10.1002/chem.202000708
9. G.Q. Zhao, J. Hu, X. Long, J. Zou, J.G. Yu, F.P. Jiao, A critical review on black phosphorus-based photocatalytic CO_2 reduction application. Small **17**, 2102155 (2021). https://doi.org/10.1002/smll.202102155
10. M. Premila, J. Fatema, A. Alahmad, M.T. Salem, M. Tahir, Layered double hydroxide (LDH) nanomaterials with engineering aspects for photocatalytic CO_2 conversion to energy efficient fuels: fundamentals, recent advances, and challenges. J. Environ. Chem. Eng. **10**, 108151 (2022). https://doi.org/10.1016/j.jece.2022.108151
11. J. Shen, Z. Wu, C. Li, C. Zhang, A. Genest, G. Rupprechter, L. He L, Emerging applications of MXene materials in CO_2 photocatalysis. FlatChem **28**, 100252 (2021). https://doi.org/10.1016/j.flatc.2021.100252
12. A. Garg, S. Basu, N.P. Shetti, K.R. Reddy, 2D materials and its heterostructured photocatalysts: synthesis, properties, functionalization and applications in environmental remediation. J. Environ. Chem. Eng. **9**, 106408 (2021). https://doi.org/10.1016/j.jece.2021.106408

13. G. Guan, E. Ye, M. You, Z. Li, Hybridized 2D nanomaterials toward highly efficient photocatalysis for degrading pollutants: cCurrent status and future perspectives. Small **16**, 1907087 (2020). https://doi.org/10.1002/smll.201907087
14. V.S. Bhati, M. Kumar, R. Banerjee, Gas sensing performance of 2D nanomaterials/metal oxide nanocomposites: a review. J. Mater. Chem. C **9**, 8776–8808 (2021). https://doi.org/10.1039/D1TC01857D
15. X. Gan, D. Lei, K.Y. Wong, Two-dimensional layered nanomaterials for visible-light-driven photocatalytic water splitting. Mater. Energy Today **10**, 352–367 (2018). https://doi.org/10.1016/j.mtener.2018.10.015
16. K. Jiang, P. Xiong, J. Ji, J. Zhu, R. Ma, T. Sasaki, F. Geng, Two-dimensional molecular sheets of transition metal oxides toward wearable energy storage. Acc. Chem. Res. **53**(10), 2443–2455 (2020). https://doi.org/10.1021/acs.accounts.0c00483
17. N. Mahmood, I.A. De Castro, K. Pramoda, K. Khoshmanesh, S.K. Bhargava, K. Kalantar-Zadeh, Atomically thin two-dimensional metal oxide nanosheets and their heterostructures for energy storage. Energy Storage Mater. **16**, 455–480 (2019). https://doi.org/10.1016/j.ensm.2018.10.013
18. Y. Zhou, Z. Wang, L. Huang, S. Zaman, L. Lei, T. Yue, Z. Li, B. You, B.Y. Xia, Engineering 2D photocatalysts toward carbon dioxide reduction. Adv. Energy Mater. **11**, 2003159 (2021). https://doi.org/10.1002/aenm.202003159
19. G. Zhang, Y. Li, C. He, X. Ren, P. Zhang, H. Mi, Recent progress in 2D catalysts for photocatalytic and electrocatalytic artificial nitrogen reduction to ammonia. Adv. Energy Mater. **11**, 2003294 (2021). https://doi.org/10.1002/aenm.202003294
20. Y.K. Jo, J.M. Lee, S. Son, S.J. Hwang, 2D inorganic nanosheet-based hybrid photocatalysts: design, applications, and perspectives. J. Photochem. Photobiol. C: Photochem. Rev. **40**, 150–190 (2019). https://doi.org/10.1016/j.jphotochemrev.2018.03.002
21. M. Farbod, R. Taheri, A. Kosarian, Preparation, characterization and photocatalytic performance of phosphorene/MoS_2 as a 2D hybrid semiconductor. Mater. Sci. Semicond. Process. **123**, 105562 (2021). https://doi.org/10.1016/j.mssp.2020.105562
22. J. Li, B. Huang, Q. Guo, S. Guo, Z. Peng, J. Liu, Q. Tian, Y. Yang, Q. Xu, Z. Liu, B. Liu, Van der Waals heterojunction for selective visible-light-driven photocatalytic CO_2 reduction. Appl. Catal. B: Environ. **284**, 119733 (2021). https://doi.org/10.1016/j.apcatb.2020.119733
23. X. Zhang, R. Pan, T. Hou, S. Zhang, X. Wan, Y. Li, S. Liu, J. Liu, J. Zhang, Doping transition metal in $PdSeO_3$ atomic layers by aqueous cation exchange: a new doping protocol for a new 2D photocatalyst. Chinese Chem. Lett. **33**, 3739–3744 (2022). https://doi.org/10.1016/j.cclet.2021.10.088
24. L. Liu, S. Wang, H. Huang, Y. Zhang, T. Ma, Surface sites engineering on semiconductors to boost photocatalytic CO_2 reduction. Nano Energy **75**, 04959 (2020). https://doi.org/10.1016/j.nanoen.2020.104959
25. T. Bak, W. Li, J. Nowotny, A.J. Atanacio, J. Davis, Photocatalytic properties of TiO_2: evidence of the key role of surface active sites in water oxidation. J. Phys. Chem. A **119**(36), 9465–9473 (2015). https://doi.org/10.1021/acs.jpca.5b05031
26. J. Xu, J. Zhang, W. Zhang, C.S. Lee, Interlayer nanoarchitectonics of two-dimensional transition-metal dichalcogenides nanosheets for energy storage and conversion applications. Adv. Energy Mater. **7**, 1700571 (2017). https://doi.org/10.1002/aenm.201700571
27. F. Kishimoto, T. Wakihara, T. Okubo, Water-dispersible triplet–triplet annihilation photon upconversion particle: Molecules integrated in hydrophobized two–dimensional interlayer space of montmorillonite and their application for photocatalysis in the aqueous phase. ACS Appl. Mater. Interf. **12**(6), 7021–7029 (2020). https://doi.org/10.1021/acsami.9b15957
28. F.I. Alzakia, S.C. Tan, Liquid-exfoliated 2D materials for optoelectronic applications. Adv. Sci. **8**, 2003864 (2021). https://doi.org/10.1002/advs.202003864
29. M. Xiong, B. Chai, J. Yan, G. Fan, G. Song, Few-layer WS_2 decorating $ZnIn_2S_4$ with markedly promoted charge separation and photocatalytic H_2 evolution activity. Appl. Surf. Sci. **514**, 145965 (2020). https://doi.org/10.1016/j.apsusc.2020.145965

30. B. Tang, B. Che, M. Xu, Z.P. Ang, J. Di, H.-J. Gao, H. Yang, J. Zhou, Z. Liu, Recent advances in synthesis and study of 2D twisted transition metal dichalcogenide bilayers. Small Structures **2**, 2000153 (2021). https://doi.org/10.1002/sstr.202000153
31. X. Yu, Y. Su, W.W. Xu, J. Zhao, Efficient photoexcited charge separation at the interface of a novel 0D/2D heterojunction: a time-dependent ultrafast dynamic study. J. Phys. Chem. Lett. **12**(9), 2312–2319 (2021). https://doi.org/10.1021/acs.jpclett.1c00023
32. S.A. Abubshait, S. Iqbal, H.A. Abubshait, A.A. AlObaid, T.I. Al-Muhimeed, H.S.M. Abd-Rabboh, A. Bahadur, W.Z. Li, Effective heterointerface combination of 1D/2D Co-NiS/S-g-C_3N_4 heterojunction for boosting spatial charge separation with enhanced photocatalytic degradation of organic pollutants and disinfection of pathogens. Colloids Surf. A: Physicochem. Eng. Aspects **628**, 127390 (2021). https://doi.org/10.1016/j.colsurfa.2021.127390
33. X. Deng, D. Wang, H. Li, W. Jiang, T. Zhou, Y. Wen, B. Yu, G. Che, L. Wang, Boosting interfacial charge separation and photocatalytic activity of 2D/2D g-C_3N_4/$ZnIn_2S_4$ S-scheme heterojunction under visible light irradiation. J. Alloys Compds **894**, 162209 (2022). https://doi.org/10.1016/j.jallcom.2021.162209
34. N. Kaur, A. Ghosh, M. Ahmad, D. Sharma, R. Singh, B.R. Mehta, Increased visible light absorption and charge separation in ^{2}D–^{3}D In_2S_3-ZnO heterojunctions for enhanced photoelectrochemical water splitting. J. Alloys Compds **903**, 164007 (2022). https://doi.org/10.1016/j.jallcom.2022.164007
35. Y. Xiao, Z.Y. Peng, S. Zhang, Y. Jiang, X. Jing, X. Yang, J. Zhang, L. Ni, Z-scheme $CdIn_2S_4$/BiOCl nanosheet face-to-face heterostructure: in-situ synthesis and enhanced interfacial charge transfer for high-efficient photocatalytic performance. J. Mater. Sci. **54**, 9573–9590 (2019). https://doi.org/10.1007/s10853-019-03401-2
36. H. Hou, X. Zhang, Rational design of 1D/2D heterostructured photocatalyst for energy and environmental applications. Chem. Eng. J. **395**, 125030 (2020). https://doi.org/10.1016/j.cej.2020.125030
37. Y. Jin, D. Jiang, D. Li, P. Xiao, X. Ma, M. Chen, $SrTiO_3$ Nanoparticle/$SnNb_2O_6$ nanosheet 0D/2D heterojunctions with enhanced interfacial charge separation and photocatalytic hydrogen evolution activity. ACS Sustainable Chem. Eng. **5**(11), 9749–9757 (2017). https://doi.org/10.1021/acssuschemeng.7b01548
38. X.D. Xiong, L, Wang Q. Li Y, Jiang Q, Hu J., Synthesis and photocatalytic activity of BiOBr nanosheets with tunable exposed 010 facets. Appl. Catal. B: Environ. **188**, 283–291 (2016). https://doi.org/10.1016/j.apcatb.2016.02.018
39. Y. Mi, H. Li, Y. Zhang, N. Du, W. Hou, Synthesis and photocatalytic activity of BiOBr nanosheets with tunable crystal facets and sizes. Catal. Sci. Technol. **8**, 2588–2597 (2018). https://doi.org/10.1039/C8CY00143J
40. X. Dang, M. Xie, F. Dai, J. Guo, J. Liu, X. Lu, Ultrathin 2D/2D $ZnIn_2S_4$/g-C_3N_4 nanosheet heterojunction with atomic-level intimate interface for photocatalytic hydrogen evolution under visible light. Adv. Mater. Interf. **8**, 2100151 (2021). https://doi.org/10.1002/admi.202100151
41. S. Navalon, A. Dhakshinamoorthy, M. Alvaro, H. Garcia, Metal nanoparticles supported on two-dimensional graphenes as heterogeneous catalysts, coordination Chem. Rev. **312**, 99–148 (2016). https://doi.org/10.1016/j.ccr.2015.12.005
42. G. Eda, S.A. Maier, (2013) Two-dimensional crystals: managing light for optoelectronics. ACS Nano **7**(7), 5660–5665 (2013). https://doi.org/10.1021/nn403159y
43. H. Tao,Q. Fan, T. TaoMa, S. Liu, H. Gysling, J. Texter, F. Guo, Z. Sun Two-dimensional materials for energy conversion and storage. Progress in Mater. Sci. **111**, 100637 (2020). https://doi.org/10.1016/j.pmatsci.2020.100637
44. Q. Fu, X. Bao, Surface chemistry and catalysis confined under two-dimensional materials. Chem. Soc. Rev. **46**, 1842–1874 (2017). https://doi.org/10.1039/C6CS00424E
45. X. Wang, M. Sayed, O. Ruzimuradov, J. Zhang, Y. Fan, X. Li, X. Bai, ow J, A review of step-scheme photocatalysts. Appl. Mater. Today **29**, 101609 (2022). https://doi.org/10.1016/j.apmt.2022.101609

46. Y. Yuan, R.T. Guo, L.F. Hong, X.Y. Ji, Z.D. Lin, Z.S. Li, W.G. Pan, A review of metal oxide-based Z-scheme heterojunction photocatalysts: actualities and developments. Mater. Today Energy **21**, 100829 (2021). https://doi.org/10.1016/j.mtener.2021.100829
47. Y. Wang, Q. Wang, X. Zhan, F. Wang, M. Safdar, J. He, Visible light driven type II heterostructures and their enhanced photocatalysis properties: a review. Nanoscale **5**, 8326–8339 (2013). https://doi.org/10.1039/C3NR01577G
48. J. Fu, K. Jiang, X. Qiu, J. Yu, M. Liu, Product selectivity of photocatalytic CO_2 reduction reactions. Mater. Today **32**, 222–243 (2020). https://doi.org/10.1016/j.mattod.2019.06.009
49. J. Ran, M. Jaroniec, S.Z. Qiao, Cocatalysts in semiconductor-based photocatalytic CO_2 reduction: achievements, challenges, and opportunities. Adv. Mater. **30**, 1704649 (2018). https://doi.org/10.1002/adma.201704649
50. E. Gong, S. Ali, C.B. Hiragond, H.S. Kim, N.S. Powar, D. Kim, H. Kim, S.I. In, Solar fuels: research and development strategies to accelerate photocatalytic CO_2 conversion into hydrocarbon fuels. Energy Environ. Sci. **15**, 880–937 (2022). https://doi.org/10.1039/D1EE02714J
51. H. Abdullah, M.D.M.R. Khan, H.R. Ong, Z. Yaakob, Modified TiO_2 photocatalyst for CO_2 photocatalytic reduction: an overview. J. CO_2 Utilization **22**, 15–32 (2017). https://doi.org/10.1016/j.jcou.2017.08.004
52. Z. Li, S. Wang, J. Wu, W. Zhou, Recent progress in defective TiO_2 photocatalysts for energy and environmental applications. Renew. Sustain. Energy Rev. **156**, 111980 (2022). https://doi.org/10.1016/j.rser.2021.111980
53. J. Zhang, P. Zhou, J. Liu, J. Yu, New understanding of the difference of photocatalytic activity among anatase, rutile and brookite TiO_2. Phys. Chem. Chem. Phys. **16**, 20382–20386 (2014). https://doi.org/10.1039/C4CP02201G
54. R. Katal, S. Masudy-Panah, M. Tanhaei, M.H.D.A. Farahani, H. Jiangyong, A review on the synthesis of the various types of anatase TiO_2 facets and their applications for photocatalysis. Chem. Eng. J. **384**, 123384 (2020). https://doi.org/10.1016/j.cej.2019.123384
55. L. Ren, Y. Li, M. Mao, L. Lan, X. Lao, X. Zhao, Significant improvement in photocatalytic activity by forming homojunction between anatase TiO_2 nanosheets and anatase TiO_2 nanoparticles. Appl. Surf. Sci. **490**, 283–292 (2019). https://doi.org/10.1016/j.apsusc.2019.05.351
56. Z.W. Wang, Q. Wan, Y.Z. Shi, H. Wang, Y.Y. Kang, S.Y. Zhu, S. Lin, L. Wu, Selective photocatalytic reduction CO_2 to CH_4 on ultrathin TiO_2 nanosheet via coordination activation. Appl. Catal. B: Environ. **288**, 120000 (2021). https://doi.org/10.1016/j.apcatb.2021.120000
57. Z. He, J. Tang, J. Shen, J. Chen, S. Song, Enhancement of photocatalytic reduction of CO_2 to CH_4 over TiO_2 nanosheets by modifying with sulfuric acid. Appl. Surf. Sci. **364**, 416–427 (2016). https://doi.org/10.1016/j.apsusc.2015.12.163
58. Z. He, L. Wen, D. Wang, Y. Xue, Q. Lu, C. Wu, J. Chen, S. Song, Photocatalytic reduction of CO2 in aqueous solution on surface-fluorinated anatase TiO2 nanosheets with exposed 001 facets. Energy Fuels **28**(6), 3982–3993 (2014). https://doi.org/10.1021/ef500648k
59. W. Zhang, J. Xue, Q. Shen, S. Jia, J. Gao, X. Liu, H. Jia, Black single-crystal TiO2 nanosheet array films with oxygen vacancy on 001 facets for boosting photocatalytic CO2 reduction. J. Alloys Compds **870**, 159400 (2021). https://doi.org/10.1016/j.jallcom.2021.159400
60. L. Zhang, C. Ni, H. Jiu, H. Chen, G. Qi, Preparation of anatase/TiO_2(B) TiO_2 nanosheet for high performance of photocatalytic reduction of CO_2. J. Mater. Sci. Mater. Electron. **28**, 6601–6606 (2017). https://doi.org/10.1007/s10854-017-6350-7
61. H. Xu, S. Ouyang, P. Li, T. Kako, J. Ye, High-active anatase TiO2 nanosheets exposed with 95% 100 facets toward efficient H2 evolution and CO2 photoreduction. ACS Appl. Mater. Interf. **5**(4), 1348–1354 (2013). https://doi.org/10.1021/am302631b
62. O. Martin, A.J. Martín, C. Mondelli, S. Mitchell, T.F. Segawa, R. Hauert, C. Drouilly, D. Curulla-Ferré, J. Pérez-Ramírez, Indium oxide as a superior catalyst for methanol synthesis by CO_2 hydrogenation. Angew. Chem. Int. Ed. **55**, 6261–6265 (2016). https://doi.org/10.1002/anie.201600943

63. Y. Qi, L. Song, S. Ouyang, X. Liang, S. Ning, Q.Q. Zhang, J. Ye, Photoinduced defect engineering: enhanced photothermal catalytic performance of 2D black In_2O_{3-x} nanosheets with bifunctional oxygen vacancies. Adv. Mater. **32**, 1903915 (2020). https://doi.org/10.1002/adma.201903915
64. X. Chen, Y. Zhou, Q. Liu, Z. Li, J. Liu, Z. Zou, Ultrathin, single-crystal WO_3 nanosheets by two-dimensional oriented attachment toward enhanced photocatalytic reduction of CO_2 into hydrocarbon fuels under visible light. ACS Appl. Mater. Interf. **4**(7), 3372–3377 (2012). https://doi.org/10.1021/am300661s
65. Y. Zhou, Z. Tian, Z. Zhao, Q. Liu, J. Kou, X. Chen, J. Gao, S. Yan, Z. Zou, High-yield synthesis of ultrathin and uniform Bi_2WO_6 square nanoplates benefitting from photocatalytic reduction of CO_2 into renewable hydrocarbon fuel under visible light. ACS Appl. Mater. Interf. **3**(9), 3594–3601 (2011). https://doi.org/10.1021/am2008147
66. Y.Y. Li, J.S. Fan, R.Q. Tan, H.C. Yao, Y. Peng, Q.C. Liu, Z.J. Li, Selective photocatalytic reduction of CO_2 to CH_4 modulated by chloride modification on Bi_2WO_6 nanosheets. ACS Appl. Mater. Interf. **12**(49), 54507–54516 (2020). https://doi.org/10.1021/acsami.0c11551
67. Y.W. Teh, C.C. Er, X.Y. Kong, B.J. Ng, S.T. Yong, S.P. Chai, Charge modulation at atomic-level through substitutional sulfur doping into atomically thin Bi_2WO_6 toward promoting photocatalytic CO_2 reduction. Chemsuschem **15**, e202200471 (2022). https://doi.org/10.1002/cssc.202200471
68. R. Verma, R. Belgamwar, V. Polshettiwar, Plasmonic photocatalysis for CO_2 conversion to chemicals and fuels. ACS Mater. Lett. **3**(5), 574–598 (2021). https://doi.org/10.1021/acsmaterialslett.1c00081
69. N.-N. Vu, S. Kaliaguine, T.-O. Do, Plasmonic photocatalysts for sunlight-driven reduction of CO_2: details, developments, and perspectives. Chemsuschem **13**, 3967–3991 (2020). https://doi.org/10.1002/cssc.202000905
70. C. Lu, X. Li, Q. Wu, J. Li, L. Wen, Y. Dai, B. Huang, B. Li, Z. Lou, Constructing surface plasmon resonance on Bi_2WO_6 to boost high-selective CO_2 reduction for methane. ACS Nano **15**(2), 3529–3539 (2021). https://doi.org/10.1021/acsnano.1c00452
71. S. Li, L. Bai, N. Ji, S. Yu, S. Lin, N. Tian, H. Huang, Ferroelectric polarization and thin-layered structure synergistically promoting CO_2 photoreduction of Bi_2MoO_6. J. Mater. Chem. A **8**, 9268–9277 (2020). https://doi.org/10.1039/D0TA02102D
72. J. Di, X. Zhao, C. Lian, M. Ji, J. Xia, J. Xiong, W. Zhou, X. Cao, Y. She, H. Liu, K.P. Loh, S.J. Pennycook, H. Li, Z. Liu, Atomically-thin Bi_2MoO_6 nanosheets with vacancy pairs for improved photocatalytic CO_2 reduction. Nano Energy **61**, 54–59 (2019). https://doi.org/10.1016/j.nanoen.2019.04.029
73. Y. Zhang, Z. Xu, Q. Wang, W. Hao, X. Zhai, X. Fei, X. Huang, Y. Bi, Unveiling the activity origin of ultrathin BiOCl nanosheets for photocatalytic CO_2 reduction. Appl. Catal. B: Environ. **299**, 120679 (2021). https://doi.org/10.1016/j.apcatb.2021.120679
74. L. Ye, X. Jin, X. Ji, C. Liu, Y. Su, H. Xie, C. Liu, Facet-dependent photocatalytic reduction of CO_2 on BiOI nanosheets. Chem. Eng. J. **291**, 39–46 (2016). https://doi.org/10.1016/j.cej.2016.01.032
75. W. Jiao, Y. Xie, F. He, K. Wang, Y. Ling, Y. Hu, J. Wang, H. Ye, J. Wu, Y. Hou, A visible light-response flower-like La-doped BiOBr nanosheets with enhanced performance for photoreducing CO_2 to CH_3OH. Chem. Eng. J. **418**, 129286 (2021). https://doi.org/10.1016/j.cej.2021.129286
76. H. Yu, H. Huang, K. Xu, W. Hao, Y. Guo, S. Wang, X. Shen, S. Pan, Y. Zhang, Liquid-phase exfoliation into monolayered BiOBr nanosheets for photocatalytic oxidation and reduction. ACS Sustainable Chem. Eng. **5**(11), 10499–10508 (2017). https://doi.org/10.1021/acssuschemeng.7b02508
77. J. Di, C. Chen, C. Zhu, P. Song, J. Xiong, M. Ji, J. Zhou, Q. Fu, M. Xu, W. Hao, J. Xia, S. Li, H. Li, Z. Liu, Bismuth vacancy-tuned bismuth oxybromide ultrathin nanosheets toward photocatalytic CO_2 reduction. ACS Appl. Mater. Interf. **11**(34), 30786–30792 (2019). https://doi.org/10.1021/acsami.9b08109

78. L. Tan, Z. Wang, Y. Zhao, Y.-F. Song, Recent progress on nanostructured layered double hydroxides for visible-light-induced photoreduction of CO_2. Chem. Asian J. **15**, 3380–3389 (2020). https://doi.org/10.1002/asia.202000963
79. M.J. Wu, J.Z. Wu, J. Zhang, H. Chen, J.Z. Zhou, G.R. Qian, Z.P. Xu, Z. Du, Q.L. Rao, A review on fabricating heterostructures from layered double hydroxides for enhanced photocatalytic activities. Catal. Sci. Technol. **8**, 1207–1228 (2018). https://doi.org/10.1039/C7CY02314F
80. G. Zhang, X. Zhang, Y. Meng, G. Pan, Z. Ni, S. Xia, Layered double hydroxides-based photocatalysts and visible-light driven photodegradation of organic pollutants: a review. Chem. Eng. J. **392**, 123684 (2020). https://doi.org/10.1016/j.cej.2019.123684
81. S. Bai, T. Li, H.W. Ling, T.Y. Zhao, Y.F. Song, Scale-up synthesis of monolayer layered double hydroxide nanosheets via separate nucleation and aging steps method for efficient CO_2 photoreduction. Chem. Eng. J. **419**, 129390 (2021). https://doi.org/10.1016/j.cej.2021.129390
82. X. Xiong, Y. Zhao, R. Shi, W. Yin, Z.Y. Waterhouse, GIN, Zhang T, Selective photocatalytic CO_2 reduction over Zn-based layered double hydroxides containing tri or tetravalent metals. Sci. Bulletin **65**, 987–994 (2020). https://doi.org/10.1016/j.scib.2020.03.032
83. R. Wang, Z. Qiu, S. Wan, Y. Wang, Q. Liu, J. Ding, Q. Zhong, Insight into mechanism of divalent metal cations with different d-bands classification in layered double hydroxides for light-driven CO_2 reduction. Chem. Eng. J. **427**, 130863 (2022). https://doi.org/10.1016/j.cej.2021.130863
84. S. Gong, X. Teng, Y. Niu, X. Liu, M. Xu, C. Xu, L. Ji, Z. Chen, Construction of S-scheme 0D/2D heterostructures for enhanced visible-light-driven CO_2 reduction. Appl. Catal. B: Environ. **298**, 120521 (2021). https://doi.org/10.1016/j.apcatb.2021.120521
85. H. Shi, S. Long, S. Hu, J. Hou, W. Ni, C. Song, K. Li, G.G. Gurzadyan, X. Guo, Interfacial charge transfer in 0D/2D defect-rich heterostructures for efficient solar-driven CO_2 reduction. Appl. Catal. B: Environ. **245**, 760–769 (2019). https://doi.org/10.1016/j.apcatb.2019.01.036
86. H. Shi, J. Du, J. Hou, W. Ni, C. Song, K. Li, G.G. Gurzadyan, X. Guo Solar-driven CO_2 conversion over Co^{2+} doped 0D/2D TiO_2/g-C_3N_4 heterostructure: Insights into the role of Co^{2+} and cocatalyst. J. CO_2 Utilizat. **38**, 16–23 (2020). https://doi.org/10.1016/j.jcou.2020.01.005
87. S. Sinha, H. Kim, A.W. Robertson, Preparation and application of 0D–2D nanomaterial hybrid heterostructures for energy applications. Mater. Today Adv. **12**, 100169 (2021). https://doi.org/10.1016/j.mtadv.2021.100169
88. D. Tan, J. Zhang, J. Shi, S. Li, B. Zhang, X. Tan, F. Zhang, L. Liu, D. Shao, B. Han, Photocatalytic CO_2 transformation to CH_4 by Ag/Pd bimetals supported on N-doped TiO_2 nanosheet. ACS Appl. Mater. Interf. **10**(29), 24516–24522 (2018). https://doi.org/10.1021/acsami.8b06320
89. R. Wang, J. Shen, K. Sun, H. Tang, Q. Liu, Enhancement in photocatalytic activity of CO_2 reduction to CH_4 by 0D/2D Au/TiO_2 plasmon heterojunction. Appl. Surf. Sci. **493**, 1142–1149 (2019). https://doi.org/10.1016/j.apsusc.2019.07.121
90. S. Zhu, S. Liang, Y. Tong, X. An, J. Long, X. Fu, X. Wang, Photocatalytic reduction of CO2 with H2O to CH4 on Cu(i) supported TiO2 nanosheets with defective 001 facets. Phys. Chem. Chem. Phys. **17**, 9761–9770 (2015). https://doi.org/10.1039/C5CP00647C
91. D. Chen, J. Ye, Hierarchical WO_3 hollow shells: dendrite, sphere, dumbbell, and their photocatalytic properties. Adv. Funct. Mater. **18**, 1922–1928 (2008). https://doi.org/10.1002/adfm.200701468
92. V. Dutta, S. Sharma, P. Raizada, V.K. Thakur, A.A.P. Khan, V. Saini, A.M. Asiri, P. Singh, An overview on WO_3 based photocatalyst for environmental remediation. J. Environ. Chem. Eng. **9**, 105018 (2021). https://doi.org/10.1016/j.jece.2020.105018
93. S.G. Kumar, K.S.R.K. Rao, Comparison of modification strategies towards enhanced charge carrier separation and photocatalytic degradation activity of metal oxide semiconductors (TiO_2, WO_3 and ZnO). Appl. Surf. Sci. **391**, 124–148 (2017). https://doi.org/10.1016/j.apsusc.2016.07.081
94. K. Aravindraj, S.M. Roopan, WO_3-based materials as heterogeneous catalysts for diverse organic transformations: a mini-review. Synthet. Commun.1457–1476 (2022). https://doi.org/10.1080/00397911.2022.2089588

95. P.R. Makgwane, S.S. Ray, Efficient room temperature oxidation of cyclohexane over highly active hetero-mixed WO_3/V_2O_5 oxide catalyst. Catal. Commun. **54**, 118–123 (2014). https://doi.org/10.1016/j.catcom.2014.05.031
96. Z. Xie, Y. Xu, D. Li, L. Chen, S. Meng, D. Jiang, M. Chen, Construction of CuO quantum Dots/WO_3 nanosheets 0D/2D Z-scheme heterojunction with enhanced photocatalytic CO_2 reduction activity under visible-light. J. Alloys Compnds **858**, 157668 (2021). https://doi.org/10.1016/j.jallcom.2020.157668
97. W. Shi, X. Guo, C. Cui, K. Jiang, Z. Li, L. Qu, J.C. Wang, Controllable synthesis of Cu_2O decorated WO_3 nanosheets with dominant (001) facets for photocatalytic CO_2 reduction under visible-light irradiation. Appl. Catal. B: Environ. **243**, 236–242 (2019). https://doi.org/10.1016/j.apcatb.2018.09.076
98. D.P.H. Tran, M.T. Pham, X.T. Bui, Y.F. Wang, S.J. You, CeO_2 as a photocatalytic material for CO_2 conversion: a review. Sol. Energy **240**, 443–466 (2022). https://doi.org/10.1016/j.solener.2022.04.051
99. A.A. Fauzi, A.A. Jalil, N.S. Hassan, F.F.A. Aziz, M.S. Azami, I. Hussain, R. Saravanan, D.V.N. Vo, A critical review on relationship of CeO_2-based photocatalyst towards mechanistic degradation of organic pollutant. Chemosphere **286**, 131651 (2022). https://doi.org/10.1016/j.chemosphere.2021.131651
100. M.A. Mavuso, P.R. Makgwane, S.S. Ray, Construction of heterojunctions CeO_2 interfaced Nb, Sn, Ti, Mo and Zn metal oxide catalysts for photocatalytic oxidation of α-pinene inert C-H. Inorg. Chem. Commun. **137**, 109199 (2022). https://doi.org/10.1016/j.inoche.2022.109199
101. M.A. Mavuso, P.R. Makgwane, S.S. Ray, Morphology modulated photocatalytic activity of CeO_2 nanostructures for selective oxidation of biobased alpha-pinene to oxygenates. Chem. Select **5**, 12940–12951 (2020). https://doi.org/10.1002/slct.202002285
102. M.A. Mavuso, P.R. Makgwane, S.S. Ray, Heterostructured CeO_2–M (M = Co, Cu, Mn, Fe, Ni) oxide nanocatalysts for the visible-light photooxidation of pinene to aroma oxygenates. ACS Omega **5**(17), 9775–9788 (2020). https://doi.org/10.1021/acsomega.9b04396
103. Y. Sun, Z. Wang, JZu.XWu.Y. Zhu, S. Shang, P. Ling, P. Qiao, C. Liu, J. Hu, Y. Pan, J. Zhu, Y. Xie, Selective CO_2 photoreduction to CH_4 via Pd^{δ}+-assisted hydrodeoxygenation over CeO_2 nanosheets. Angew. Chem. Int. Ed. **61**, e202203249 (2022). https://doi.org/10.1002/anie.202203249
104. G. Ren, S. Liu, Z. Li, H. Bai, X. Hu, X. Meng, Highly selective photocatalytic reduction of CO_2 to CO Over Ru-modified Bi_2MoO_6. Solar RRL 2200154 (2022). https://doi.org/10.1002/solr.202200154
105. Z. Xu, C. Zhang, Y. Zhang, Y. Gu, Y. An, BiOCl-based photocatalysts: Synthesis methods, structure, property, application, and perspective. Inorg. Chem. Commun. **138**, 109277 (2022). https://doi.org/10.1016/j.inoche.2022.109277
106. A. Malathi, J. Madhavan, M. Ashokkumar, P. Arunachalam, A review on $BiVO_4$photocatalyst: Activity enhancement methods for solar photocatalytic applications. Appl. Catal. A: Gen. 555:47–74 (2018). https://doi.org/10.1016/j.apcata.2018.02.010
107. X. Liu, S. Gu, Y. Zhao, G. Zhou, W. Li, $BiVO_4$, Bi_2WO_6 and Bi_2MoO_6 photocatalysis: a brief review. J. Mater. Sci. Technol. **56**, 45–68 (2020). https://doi.org/10.1016/j.jmst.2020.04.023
108. M. Arumugam, M.Y. Choi, Recent progress on bismuth oxyiodide (BiOI) photocatalyst for environmental remediation. J. Ind. Eng. Chem. **81**, 237–268 (2020). https://doi.org/10.1016/j.jiec.2019.09.013
109. T. Chen, L. Liu, C. Hu, H. Huang, Recent advances on Bi_2WO_6-based photocatalysts for environmental and energy applications, Chinese. J. Catal. **42**, 1413–1438 (2021). https://doi.org/10.1016/S1872-2067(20)63769-X
110. S. Chen, D. Huang, M. Cheng, L. Lei, Y. Chen, C. Zhou, R. Deng, B. Li, Surface and interface engineering of two-dimensional bismuth-based photocatalysts for ambient molecule activation. J. Mater. Chem. A **9**, 196–233 (2021). https://doi.org/10.1039/D0TA08165E
111. A. Sreedhar, Q. Thanh, H. Ta, J.S. Noh, Role of p-n junction initiated mixed-dimensional 0D/2D, 1D/2D, and 2D/2D BiOX (X = Cl, Br, and I)/TiO_2 nanocomposite interfaces for environmental remediation applications: a review. Chemosphere **305**, 135478 (2022). https://doi.org/10.1016/j.chemosphere.2022.135478

112. Q. Wang, K. Wang, L. Zhang, H. Wang, W. Wang, Photocatalytic reduction of CO_2 to methane over PtOx-loaded ultrathin Bi_2WO_6 nanosheets. Appl. Surf. Sci. **470**, 832–839 (2019). https://doi.org/10.1016/j.apsusc.2018.11.197
113. J. Wang, J. Wang, N. Li, X. Du, J. Ma, C. He, Z. Li, Direct Z-scheme 0D/2D heterojunction of $CsPbBr_3$ quantum dots/B_i2W_O6 nanosheets for efficient photocatalytic CO_2 reduction. ACS Appl. Mater. Interf. **12**(28), 31477–31485 (2020). https://doi.org/10.1021/acsami.0c08152
114. S. Kumar, M.A. Isaacs, R. Trofimovaite, L. Durndell, C.M.A. Parlett, R.E. Douthwaite, B. Coulson, M.C.R. Cockett, K. Wilson, A.F. Lee, P25@CoAl layered double hydroxide heterojunction nanocomposites for CO_2 photocatalytic reduction. Appl. Catal. B: Environ. **209**, 394–404 (2017). https://doi.org/10.1016/j.apcatb.2017.03.006
115. H. Jiang, K.I. Katsumata, J. Hong, A. Yamaguchi, K. Nakata, C. Terashima, N. Matsushita, M. Miyauchi, A. Fujishima, Photocatalytic reduction of CO_2 on Cu_2O loaded Zn-Cr layered double hydroxides. Appl. Catal. B: Environ. **224**, 783–790 (2018). https://doi.org/10.1016/j.apcatb.2017.11.011
116. J. Xu, X. Liu, Z. Zhou, L. Deng, L. Liu, M. Xu, Platinum nanoparticles with low content and high dispersion over exfoliated layered double hydroxide for photocatalytic CO_2 reduction. Energy Fuels **35**(13), 10820–10831 (2021). https://doi.org/10.1021/acs.energyfuels.1c00820
117. Y. Zhong, C. Peng, Z. He, D. Chen, H. Jia, J. Zhang, H. Ding, X. Wu, Interface engineering of heterojunction photocatalysts based on 1D nanomaterials. Catal. Sci. Technol. **11**, 27–42 (2021). https://doi.org/10.1039/D0CY01847C
118. Z. He, J. Zhang, X. Li, S. Guan, M. Dai, S. Wang, 1D/2D Heterostructured photocatalysts: from design and unique properties to their environmental applications. Small **16**, 2005051 (2020). https://doi.org/10.1002/smll.202005051
119. J.K. Qin, C. Wang, L. Zhen, L.J. Li, C.Y. Xu, Y. Chai, Van der Waals heterostructures with one-dimensional atomic crystals. Prog. Mater. Sci. **122**, 100856 (2021). https://doi.org/10.1016/j.pmatsci.2021.100856
120. Q. Guo, Q. Zhang, H. Wang, Z. Liu, Z. Zhao, Core-shell structured ZnO@Cu-Zn–Al layered double hydroxides with enhanced photocatalytic efficiency for CO_2 reduction. Catal. Commun. **77**, 118–122 (2016). https://doi.org/10.1016/j.catcom.2016.01.019
121. B. Yu, Y. Wu, F. Meng, Q. Wang, X. Jia, M.W. Khan, C. Huang, S. Zhang, L. Yang, H. Wu, Formation of hierarchical Bi_2MoO_6/In_2S_3 S-scheme heterojunction with rich oxygen vacancies for boosting photocatalytic CO_2 reduction. Chem. Eng. J. **429**, 132456 (2022). https://doi.org/10.1016/j.cej.2021.132456
122. M. Zhang, K. Zhao, J. Xiong, Y. Wei, C. Han, W. Li, G. Cheng, A 1D/2D WO_3 nanostructure coupled with a nanoparticulate CuO cocatalyst for enhancing solar-driven CO_2 photoreduction: the impact of the crystal facet. Sustain. Energy Fuels **4**, 2593–2603 (2020). https://doi.org/10.1039/D0SE00034E
123. C.P. Sajan, S. Wageh, A.A. Al-Ghamdi, J. Yu, S. Cao, TiO2 nanosheets with exposed 001 facets for photocatalytic applications. Nano Res. **9**, 3–27 (2016). https://doi.org/10.1007/s12274-015-0919-3
124. Z. Xiong, Z. Lei, Y. Li, L. Dong, Y. Zhao, J. Zhang, A review on modification of facet-engineered TiO_2 for photocatalytic CO_2 reduction. J. Photochem. Photobiol. C: Photochem. Rev. **36**, 24–47 (2018). https://doi.org/10.1016/j.jphotochemrev.2018.07.002
125. W.J. Ong, L.L. Tan, S.P. Chai, S.T. Yong, A.R. Mohamed, Facet-dependent photocatalytic properties of TiO_2-based composites for energy conversion and environmental remediation. Chemsuschem **7**, 690–719 (2014). https://doi.org/10.1002/cssc.201300924
126. M. Ge, C. Cao, J. Huang, S. Li, Z. Chen, K.Q. Zhang, S.S. Al-Deyab, Y. Lai, A review of one-dimensional TiO_2 nanostructured materials for environmental and energy applications. J. Mater. Chem. A **4**, 6772–6801 (2016). https://doi.org/10.1039/C5TA09323F
127. J. Tian, Z. Zhao, A. Kumar, R.I. Boughton, H. Liu, Recent progress in design, synthesis, and applications of one-dimensional TiO_2 nanostructured surface heterostructures: a review. Chem. Soc. Rev. **43**, 6920–6937 (2014). https://doi.org/10.1039/C4CS00180J
128. J. Zhang, G. Xiao, F.X. Xiao, B. Liu, Revisiting one-dimensional TiO_2 based hybrid heterostructures for heterogeneous photocatalysis: a critical review. Mater. Chem. Front. **1**, 231–250 (2017). https://doi.org/10.1039/C6QM00141F

129. E. Kim, K.H. Do, J. Wang, Y. Hong, A.P. Rangappa, D.A. Reddy, D.P. Kumar, T.K. Kim, Construction of 1D TiO_2 nanotubes integrated ultrathin 2D $ZnIn_2S_4$ nanosheets heterostructure for highly efficient and selective photocatalytic CO_2 reduction. Appl. Surf. Sci. **587**, 152895 (2022). https://doi.org/10.1016/j.apsusc.2022.152895
130. F. Xu, B. Zhu, B. Cheng, J. Yu, J. Xu, 1D/2D TiO_2/MoS_2 Hybrid nanostructures for enhanced photocatalytic CO_2 reduction. Adv. Optical Mater. **6**, 1800911 (2018). https://doi.org/10.1002/adom.201800911
131. L. Biswal, S. Nayak, K. Parida, Recent progress on strategies for the preparation of 2D/2D MXene/g-C_3N_4 nanocomposites for photocatalytic energy and environmental applications. Catal. Sci. Technol. **11**, 1222–1248 (2021). https://doi.org/10.1039/D0CY02156C
132. I. Ahmad, S. Shukrullah, M.Y. Naz, M. Ahmad, E. Ahmed, Y. Liu, A. Hussain, S. Iqbal, S. Ullah, Recent advances and challenges in 2D/2D heterojunction photocatalysts for solar fuels applications. Adv. Colloid Interf. Sci. **304**, 102661 (2022). https://doi.org/10.1016/j.cis.2022.102661
133. H. Hou, X. Zeng, X. Zhang, 2D/2D heterostructured photocatalyst: rational design for energy and environmental applications. Sci. China Mater. **63**, 2119–2152 (2020). https://doi.org/10.1007/s40843-019-1256-0
134. B. Zhu, B. Cheng, J. Fan, W. Ho, J. Yu, g-C_3N_4-based 2D/2D composite heterojunction photocatalyst. Small Struct. **2**, 2100086 (2021). https://doi.org/10.1002/sstr.202100086
135. M.Z. Qin, W.X. Fu, H. Guo, C.G. Niu, D.W. Huang, C. Liang, Y.Y. Yang, H.Y. Liu, N. Tang, Q.Q. Fan, 2D/2D Heterojunction systems for the removal of organic pollutants: a review. Adv. Colloid Interf. Sci. **297**, 102540 (2021). https://doi.org/10.1016/j.cis.2021.102540
136. X. Liu, Q. Zhang, D. Ma, Advances in 2D/2D Z-scheme heterojunctions for photocatalytic applications. Solar RRL **5**, 20003970 (2021). https://doi.org/10.1002/solr.202000397
137. Z. Xie, Y. Xu, D. Li, S. Meng, M. Chen, D. Jiang, Covalently bonded Bi_2O_3 nanosheet/Bi_2WO_6 network heterostructures for efficient photocatalytic CO_2 reduction. ACS Appl. Energy Mater. **3**(12), 12194–12203 (2020). https://doi.org/10.1021/acsaem.0c02252
138. L. Wang, X. Zhao, D. Lv, C. Liu, W. Lai, C. Sun, Z. Su, X. Xu, W. Hao, S.X. Dou, Y. Du, Promoted photocharge separation in 2D lateral epitaxial heterostructure for visible-light driven CO_2 photoreduction. Adv. Mater. **32**, 2004311 (2020). https://doi.org/10.1002/adma.202004311
139. X.Y. Kong, W.Q. Lee, A.R. Mohamed, S.P. Chai, Effective steering of charge flow through synergistic inducing oxygen vacancy defects and p-n heterojunctions in 2D/2D surface-engineered Bi_2WO_6/BiOI cascade: towards superior photocatalytic CO_2 reduction activity. Chem. Eng. J. **372**, 1183–1193 (2019). https://doi.org/10.1016/j.cej.2019.05.001
140. Z. Wang, B. Cheng, L. Zhang, J. Yu, Y. Li, S. Wageh, A.A. Al-Ghamdi, S-Scheme 2D/2D Bi_2MoO_6/BiOI van der Waals heterojunction for CO_2 photoreduction. Chinese J. Catal. **43**, 1657–1666 (2022). https://doi.org/10.1016/S1872-2067(21)64010-X
141. S. Murcia-López, V. Vaiano, M.C. Hidalgo, J.A. Navío, D. Sannino, Photocatalytic reduction of CO_2 over platinised Bi_2WO_6-based materials. Photochem. Photobiol. Sci. **14**, 678–685 (2015). https://doi.org/10.1039/c4pp00407h
142. S. Cao, B. Shen, T. Tong, J. Fu, J. Yu, 2D/2D heterojunction of ultrathin MXene/Bi_2WO_6 nanosheets for improved photocatalytic CO_2 reduction. Adv. Funct. Mater. **28**, 1800136 (2018). https://doi.org/10.1002/adfm.201800136
143. Y. Jiang, H.Y. Chen, J.Y. Li, J.F. Liao, H.H. Zhang, X.D. Wang, D.B. Kuang, Z-Scheme 2D/2D heterojunction of $CsPbBr_3$ /$B_{i2}W_{O6}$ for improved photocatalytic C_{O2} reduction. Adv. Funct. Mater. **30**, 2004293 (2020). https://doi.org/10.1002/adfm.202004293
144. K.S. Schanze, P.V. Kamat, P. Yang, J. Bisquert, Progress in perovskite photocatalysis. ACS Energy Lett. **5**(8), 2602–2604 (2020). https://doi.org/10.1021/acsenergylett.0c01480
145. M. Zhu, X. Cai, M. Fujitsuka, J. Zhang, T. Majima, A Au/$La_2Ti_2O_7$ Nanostructures sensitized with black phosphorus for plasmon-enhanced photocatalytic hydrogen production in visible and near-infrared light. Ang. Chem. Int. Ed. **56**, 2064–2068 (2017). https://doi.org/10.1002/anie.201612315

146. E. Grabowska, Selected perovskite oxides: characterization, preparation and photocatalytic properties - a review. Appl. Catal. B: Environ. **186**, 97–126 (2016). https://doi.org/10.1016/j.apcatb.2015.12.035
147. Y. Huang, J. Zhang, K. Dai, C. Liang, G. Dawson, Efficient solar-driven CO_2 reduction on aminated 2D/2D BiOBr/CdS-diethylenetriamine S-scheme heterojunction. Ceram. Int. **48**, 8423–8432 (2022). https://doi.org/10.1016/j.ceramint.2021.12.050
148. S. Zhao, D. Pan, Q. Liang, M. Zhou, C. Yao, S. Xu, Z. Li, Ultrathin NiAl-layered double hydroxides grown on 2D $Ti_3C_2T_x$ MXene to construct core–shell heterostructures for enhanced photocatalytic CO_2 reduction. J. Phys. Chem. C **125**(19), 10207–10218 (2021). https://doi.org/10.1021/acs.jpcc.1c00017
149. A.A. Khan, M. Tahir, Construction of an S-Scheme heterojunction with oxygen-vacancy rich trimetallic CoAlLa-LDH anchored on titania-sandwiched Ti_3C_2 multilayers for boosting photocatalytic CO_2 reduction under visible light. Ind. Eng. Chem. Res. **60**(45), 16201–16223 (2021). https://doi.org/10.1021/acs.iecr.1c03242
150. M.S. Sajna, S.M. Simon, N.V. Unnikrishnan, K.K. Sadasivuni, An overview of graphene-based 2D/3D nanostructures for photocatalytic applications. Topics Catal (2022). https://doi.org/10.1007/s11244-021-01539-5
151. Y. Xia, B. Cheng, J. Fan, J. Yu, G. Liu, Near-infrared absorbing 2D/3D $ZnIn_2S_4$/N-doped graphene photocatalyst for highly efficient CO_2 capture and photocatalytic reduction. Sci. China Mater. **63**, 552–565 (2020). https://doi.org/10.1007/s40843-019-1234-x
152. W. Xing, Y. Zhang, J. Zou, T. Zhang, C. Liu, G. Wu, G. Chen, Sulfur-doped 2D/3D carbon nitride-based van der Waals homojunction with superior photocatalytic hydrogen evolution and wastewater purification. Int. J. Hydrogen Energy **47**, 12559–12568 (2022). https://doi.org/10.1016/j.ijhydene.2022.02.006
153. F. Mu, B. Dai, Y. Wu, G. Yang, S. Li, L. Zhang, J. Xu, Y. Liu, W. Zhao, 2D/3D S-scheme heterojunction of carbon nitride/iodine-deficient bismuth oxyiodide for photocatalytic hydrogen production and bisphenol a degradation. J. Colloid Interf. Sci. **612**, 722–736 (2022). https://doi.org/10.1016/j.jcis.2021.12.196
154. R. Cao, H. Yang, S. Zhang, X. Xu, Engineering of Z-scheme 2D/3D architectures with $Ni(OH)_2$ on 3D porous g-C_3N_4 for efficiently photocatalytic H_2 evolution. Appl. Catal. B: Environ. **258**, 117997 (2019). https://doi.org/10.1016/j.apcatb.2019.117997
155. N. Li, Y. Tian, J. Zhao, J. Zhang, W. Zuo, L. Kong, H. Cui, Z-scheme 2D/3D g-C_3N_4@ZnO with enhanced photocatalytic activity for cephalexin oxidation under solar light. Chem. Eng. J. **352**, 412–422 (2018). https://doi.org/10.1016/j.cej.2018.07.038
156. X. Zhao, J. Guan, J. Li, X. Li, H. Wang, P. Huo, Y. Yan, CeO_2/3D g-C_3N_4heterojunction deposited with Pt cocatalyst for enhanced photocatalytic CO_2 reduction. Appl. Surf. Sci. **537**, 147891 (2021). https://doi.org/10.1016/j.apsusc.2020.147891
157. M. Sayed, B. Zhu, P. Kuang, X. Liu, B. Cheng, A.A. Al Ghamdi, S. Wageh, L. Zhang, J. Yu, EPR investigation on electron transfer of 2D/3D g-C_3N_4/ZnO S-scheme heterojunction for enhanced CO_2 photoreduction. Adv. Sustain. Syst. **6**, 2100264 (2022). https://doi.org/10.1002/adsu.202100264
158. W. Dai, J. Yu, Y. Deng, X. Hu, T. Wang, X. Luo, Facile synthesis of MoS_2/Bi_2WO_6nanocomposites for enhanced CO_2 photoreduction activity under visible light irradiation. Appl. Surf. Sci. **403**, 230–239 (2017). https://doi.org/10.1016/j.apsusc.2017.01.171
159. W. Shi, X. Guo, J.C. Wang, Y. Li, L. Liu, Y. Hou, Y. Li, H. Lou, Enhanced photocatalytic 3D/2D architecture for CO_2 reduction over cuprous oxide octahedrons supported on hexagonal phase tungsten oxide nanoflakes. J. Alloys Compnds **830**, 154683 (2020). https://doi.org/10.1016/j.jallcom.2020.154683
160. X.Y. Ji, R.T. Guo, J.Y. Tang, Y.F. Miao, Z.D. Lin, L.F. Hong, Y. Yuan, Z.S. Li, W.G. Pan, Construction of full solar-spectrum-driven $Cu_{2-x}S$/Ni-Al-LDH heterostructures for efficient photocatalytic CO_2 reduction. ACS Appl. Energy Mater. **5**(3), 2862–2872 (2022). https://doi.org/10.1021/acsaem.1c03538

161. Y.F. Miao, R.T. Guo, J.W. Gu, Y.Z. Liu, G.L. Wu, C.P. Duan, W.G. Pan, Z-Scheme $Bi/Bi_2O_2CO_3$/layered double-hydroxide nanosheet heterojunctions for photocatalytic CO_2 reduction under visible light. ACS Appl. Nano Mater. **4**(5), 4902–4911 (2021). https://doi.org/10.1021/acsanm.1c00458
162. J. Hu, M. Chen, X. Fang, L. Wu, Fabrication and application of inorganic hollow spheres. Chem. Soc. Rev. **40**, 5472–5491 (2011). https://doi.org/10.1039/C1CS15103G
163. Y. Kuang, H. Li, Targeted engineering of metal@hollow carbon spheres as nanoreactors for biomass hydrodeoxygenation. Renew. Sustain. Energy Rev. **151**, 111582 (2021). https://doi.org/10.1016/j.rser.2021.111582
164. Z. Wang, S.A. Monny, L. Wang, Hollow structure for photocatalytic CO_2 reduction. ChemNanoMat **6**, 881–888 (2020). https://doi.org/10.1002/cnma.202000157
165. M. Xiao, Z. Wang, M. Lyu, B. Luo, S. Wang, G. Liu, H.M. Cheng, L. Wang, Hollow nanostructures for photocatalysis: advantages and challenges. Adv. Mater. **31**, 1801369 (2019). https://doi.org/10.1002/adma.201801369
166. X. Liu, M. Sayed, C. Bie, B. Cheng, B. Hu, J. Yu, L. Zhang, Hollow CdS-based photocatalysts. J. Materiom. **7**, 419–439 (2021). https://doi.org/10.1016/j.jmat.2020.10.010
167. M. Li, S. Zhang, L. Li, J. Han, X. Zhu, Q. Ge, H. Wang, Construction of highly active and selective polydopamine modified hollow ZnO/Co_3O_4 p-n heterojunction catalyst for photocatalytic CO_2 reduction. ACS Sustain. Chem. Eng. **8**(20), 11465–11476 (2020). https://doi.org/10.1021/acssuschemeng.0c04829
168. L. Zhang, J. Zhang, H. Yu, J. Yu, Emerging S-Scheme photocatalyst. Adv. Mater. **34**, 2107668 (2021). https://doi.org/10.1002/adma.202107668
169. L. Liu, K. Dai, J. Zhang, L. Li, Plasmonic Bi-enhanced ammoniated α-MnS/Bi_2MoO_6 S-scheme heterostructure for visible-light-driven CO_2 reduction. J. Colloid Interf. Sci. **604**, 844–855 (2021). https://doi.org/10.1016/j.jcis.2021.07.064
170. X. Zhu, Z. Wang, K. Zhong, Q. Li, P. Ding, Z. Feng, J. Yang, Y. Du, Y. Song, Y. Hua, J. Yuan, Y. She, H. Li, H. Xu, Mo-O-Bi bonds as interfacial electron transport bridges to fuel CO_2 photoreduction via in-situ reconstruction of black Bi_2MoO_6/BiO_{2-x} heterojunction. Chem. Eng. J. **429**, 132204 (2022). https://doi.org/10.1016/j.cej.2021.132204
171. S.S. Sonu, V. Dutta, P. Raizada, A. Hosseini-Bandegharaei, V. Thakur, V.H. Nguyen, Q. Van Le, P. Singh, An overview of heterojunctioned $ZnFe_2O_4$ photocatalyst for enhanced oxidative water purification. J. Environ. Chem. Eng. **9**, 105812 (2021). https://doi.org/10.1016/j.jece.2021.105812
172. X. Yang, X. Xu, J. Wang, T. Chen, S. Wang, X. Ding, H. Chen, Insights into the surface/interface modifications of Bi_2MoO_6: Feasible strategies and photocatalytic applications. Solar RRL **5**, 2000442 (2021). https://doi.org/10.1002/solr.202000442
173. Y. Zhou, W. Jiao, Y. Xie, F. He, Y. Ling, Q. Yang, J. Zhao, H. Ye, Y. Hou, Enhanced photocatalytic CO_2-reduction activity to form CO and CH_4 on S-scheme heterostructured $ZnFe_2O_4/Bi_2MoO_6$ photocatalyst. J. Colloid Interf. Sci. **608**, 2213–2223 (2022). https://doi.org/10.1016/j.jcis.2021.10.053
174. M. Jouyandeh, S.S.M. Khadem, S. Habibzadeh, A. Esmaeili, O. Abida, V. Vatanpour, N. Rabiee, M. Bagherzadeh, S. Iravani, M.R. Saeb, R.S. Varma, Quantum dots for photocatalysis: synthesis and environmental applications. Green Chem. **23**, 4931–4954 (2021). https://doi.org/10.1039/D1GC00639H
175. Z. Jiang, X. Liang, H. Zheng, Y. Liu, Z. Wang, P. Wang, X. Zhang, X. Qin, Y. Dai, M.H. Whangbo, B. Huang, Photocatalytic reduction of CO_2 to methanol by three-dimensional hollow structures of Bi_2WO_6 quantum dots. Appl. Catal. B: Environ. **219**, 209–215 (2017). https://doi.org/10.1016/j.apcatb.2017.07.023
176. M. Jakob, H. Levanon, P.V. Kamat, Charge distribution between UV-Irradiated TiO_2 and gold nanoparticles: determination of shift in the Fermi level. Nano Lett. **3**(3), 353–358 (2003). https://doi.org/10.1021/nl0340071
177. C. Liu, S. Mao, M. Shi, X. Hong, D. Wang, F. Wang, M. Xia, Q. Chen, Enhanced photocatalytic degradation performance of $BiVO_4/BiOBr$ through combining Fermi level alteration and oxygen defect engineering. Chem. Eng. J. **449**, 137757 (2022). https://doi.org/10.1016/j.cej.2022.137757

178. P. Chen, T. Du, H. Jia, L. Zhou, Q. Yue, H. Wang, Y. Wang, A novel Bi_2WO_6/Si heterostructure photocatalyst with Fermi level shift in valence band realizes efficient reduction of CO_2 under visible light. Appl. Surf. Sci. **585**, 152665 (2022). https://doi.org/10.1016/j.apsusc.2022.152665
179. J. Shen, J. Shen, W. Zhang, X. Yu, H. Tang, M. Zhang, L.Q. Zulfiqar, Built-in electric field induced CeO_2/Ti_3C_2-MXene Schottky-junction for coupled photocatalytic tetracycline degradation and CO_2 reduction. Ceram. Int. **45**, 24146–24153 (2019). https://doi.org/10.1016/j.ceramint.2019.08.123
180. W. Dai, J. Long, L. Yang, S. Zhang, Y. Xu, X. Luo, J. Zou, S. Luo, Oxygen migration triggering molybdenum exposure in oxygen vacancy-rich ultra-thin Bi_2MoO_6 nanoflakes: dual binding sites governing selective CO_2 reduction into liquid hydrocarbons. J. Energy Chem. **61**, 281–289 (2021). https://doi.org/10.1016/j.jechem.2021.01.009

Chapter 14
Nano-engineered 2D Materials for CO_2 Capture

Neeraj Kumar, Rashi Gusain, and Suprakas Sinha Ray

Abstract Recently, nano-engineered two-dimensional (2D) materials have gained immense interest in various applications, including CO_2 capture. The precise atomic structure of 2D nanomaterials introduced various significant characteristics required for specific applications. Increasing levels of CO_2 in the environment is a concerning topic for surviving a sustainable life on Earth. Therefore, CO_2 capture and conversion into useful products have been recognized as the best approach to reduce the CO_2 level in the atmosphere. To capture CO_2, several materials have been studied and emphasised about their advantages and disadvantages. The recent progress in 2D materials, especially graphene-based materials, has shown their potential in CO_2 capture. Graphene-based materials, transition metal dichalcogenides (TMDCs), 2D transition metal oxides (TMOs), MXenes, boron nitrides, carbon nitrides, 2D metal–organic frameworks (MOFs) etc., are the various examples of 2D materials, which have been investigated for CO_2 capture. This chapter aims to provide a brief overview of the recent advantages in the nano-engineering of the various 2D materials for CO_2 capture. In particular, the recent development of emerging strategies such as doping, defects engineering, hetero-structural designing, and architectural functionalization of 2D nanomaterials for enhanced CO_2 capture are discussed thoroughly. The challenges and future outcomes have also been highlighted, which will open the directions for future research.

N. Kumar (✉) · R. Gusain · S. S. Ray
Department of Chemical Sciences, University of Johannesburg, Doornfontein 2028, South Africa
e-mail: ynk.neeraj@gmail.com

S. S. Ray
e-mail: rsuprakas@csir.co.za

Centre for Nanostructures and Advanced Materials, DSI-CSIR Nanotechnology Innovation Centre, Council for Scientific and Industrial Research, Pretoria 0001, South Africa

N. Kumar et al. (eds.), *Two-Dimensional Materials for Environmental Applications*, Springer Series in Materials Science 332, https://doi.org/10.1007/978-3-031-28756-5_14

14.1 Introduction

Industrial modernisation and increased living standards have deteriorated environmental health. The ever-increasing demand and consumption of energy leading to the sharp rise of CO_2 concentration in the environment globally, which is directly or indirectly responsible for the deteriorating climate situation and global warming. CO_2 is one of the prime greenhouse gases (GHGs), and its high levels are not beneficial for the health of planet Earth. Several agencies noted that the CO_2 level in the atmosphere has surpassed the 415 ppm, which is the maximum record yet and alarming the current situation [1]. IPCC (International Panel on Climate Change) has also predicted that by 2100, the level of CO_2 will reach 590 ppm, which will raise the global temperature by 1.90 °C [2]. This exponential rise of CO_2 in the atmosphere can cause severe adverse effects on the planet, such as glaciers melting, and ice melting on Earth's poles, which further raise the sea level [3]. The constant rise in global temperature can significantly affect the living survival on Earth. Therefore, there is a need to control the emission of CO_2 in the atmosphere [4] and to improve the research and development techniques for CO_2 capture and storage (CCS) [5] and convert the captured CO_2 into value-added products via different conversion routes [6].

Industrial CO_2 emission is the major source of CO_2 in the environment via pre-combustion or post-combustion treatments. To control the environmental CO_2 concentration, there is a foremost need to develop technologies to capture it from the atmosphere. Adsorption, absorption and membrane separation are the most applicable methods for CO_2 capture. However, except for adsorption, all other methods feature drawbacks such as corrosion, long processing period, high cost, and the generation of other harmful side products [7, 8]. Therefore, CO_2 capture via adsorption has attracted the attention of researchers globally. Several materials such as ionic liquids [9], polymers [10], carbon-based materials [11], zeolites [12], metal oxides [13], MXenes [14] and MOFs [15] have been proposed as favourable sorbent materials to capture the CO_2 from the atmosphere. Among all the class of 2D nanomaterials, for example, graphene oxide (GO), metal dichalcogenides, carbon nitrides, metal–organic frameworks, boron nitrides, MXenes etc., has gained immense interest in various applications, including CO_2 capture due to its outstanding characteristics such as high active surface area, high aspect ratio, excellent optical and mechanical properties and ease of functionalization [16]. Figure 14.1 represents the various examples of 2D nano-engineered materials used in CO_2 adsorption with their unique characteristics. In the last few decades, there is an exponential growth in the advancement of the nano-engineering of 2D materials using new strategies. 2D materials were fabricated with tunable characteristics, controlled orientations, and doped or combined with several other nanomaterials to synthesize 2D based nanocomposite materials with improved CO_2 capturing capacity.

Several review articles have been published on CO_2 capture and conversion with a prime focus on the various technologies or using specific materials [5, 17–20]. For instance, the progress in using various 2D nano-engineered materials for CO_2 capture has been addressed. Recently, the significance of 2D nanomaterials such as

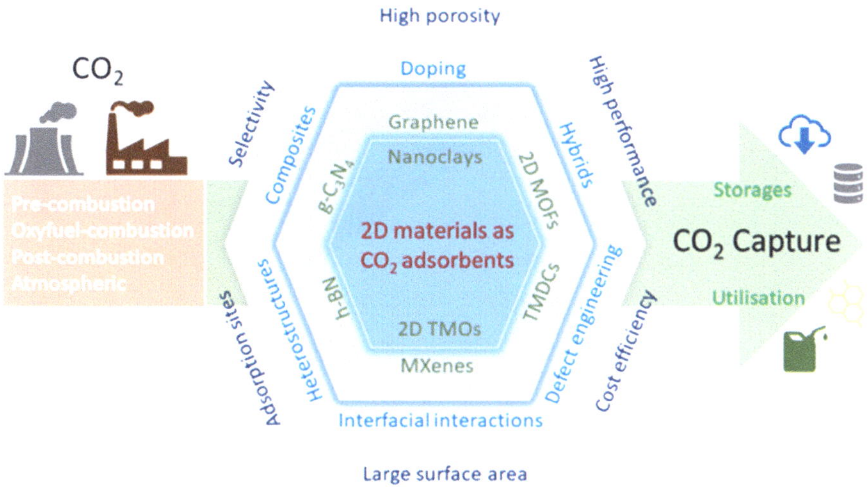

Fig. 14.1 Examples of 2D materials, various ways to functionalize 2D materials and their specific characteristics to capture the CO_2

graphene, boron nitrides, carbon nitrides, MXenes and metal chalcogenides have been proven to be successful candidates for the CO_2 capture. Further, in several reports, the analysis of the engineering aspects of 2D materials for improved CO_2 capture has also been reported [16]. Therefore, there is a need to put the application of various 2D nano-engineered materials in one frame for better understanding for the researchers. This chapter aims to methodically review the critical developments in applying different nano-engineered 2D materials in CO_2 capture. The functionalization of 2D nanomaterials or preparation of 2D nanomaterial-based nanocomposites can potentially improve the CO_2 capture aptitude of parent 2D nanomaterials. Additionally, the current scenario, opportunities, and challenges of applying nano-engineered 2D materials have also been discussed.

14.2 CO_2: Sources and Hazardous Effect

CO_2 gas is one of the major gases of GHGs emissions and causes global warming and climate change. CO_2 is one of the significant environmental emissions that can be through natural sources or anthropogenic activities. Figure 14.2 shows the possible sources of CO_2 emission in the environment [3]. Major natural sources of CO_2 emission include the CO_2 release from decomposition of biomass and vegetation, respiration, venting volcanos, ocean release and natural wildfires. Other than this, most of the CO_2 is released into the environment through human activities such as burning fossil fuels, deforestation, burning solid waste and biological matters, transportation, power generation, agricultural practices and industrial release. In one

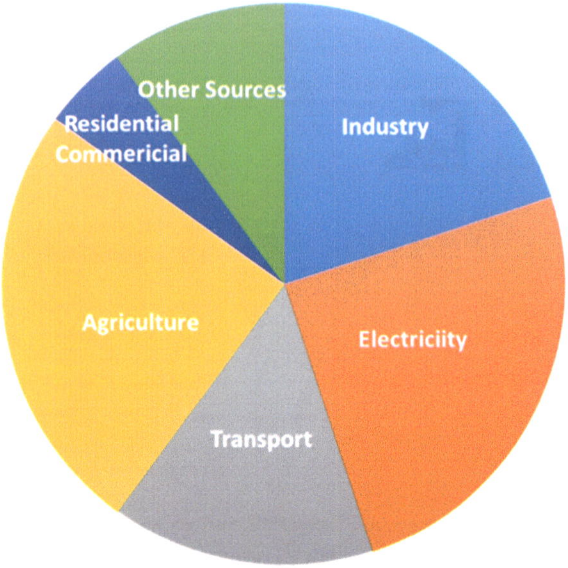

Fig. 14.2 Examples of several sources of CO_2 emission in the environment. Reproduced with permission from ref. [3]. Copyright 2022 CRC PRESS, Taylor, and Francis

survey (2014), total global CO_2 emission was noticed to be 32.4 Gt CO_2, out of which the energy sector contributed majorly (82%), and the burning of fossil fuels was the prime cause. Amongst fossil fuels, oil consumption was the leading cause of CO_2 emission (34%), followed by coal (46%), gas fuels (1%), and the remaining was from the geothermal, nuclear, solar, hydro, biofuels, wind and waste [21]. Therefore, the ever-increasing demand for energy with the increase in population is one of the major causes of CO_2 release. Increasing CO_2 levels can be life-threatening to the planet, and there is a need to curb the CO_2 emissions or GHGs emissions to save life on Earth.

The regular CO_2 emission via natural and human activities is causing global warming and increasing the global temperature. IIPCC has stated that to avoid any devastating cost of CO_2 emission, the increase in global temperature should be restricted to 2 °C and it is most likely achieved 1.5 °C already [22]. Global warming and drastic climate changes can negatively impact the terrestrial and ocean ecosystem, water supply, food chain, health, weather condition and economic growth (Fig. 14.3). Acid rain is one of the disastrous consequences of increased CO_2 levels and can damage trees and the environment [23–25]. Global warming aids in melting the glaciers, particularly in Greenland and Antarctica, which adds water to oceans and seas. This exercise increases the sea level, which can cause floods, agricultural soil contamination with salt, destructive erosion and destruct habitat for living creatures such as fish, plants and birds. It is also apparent that the gaseous exchange between air and oceans will increase the ocean CO_2 concentration with the increase in atmospheric CO_2. This will increase the dissolved inorganic carbon and consequently decrease the ocean surface pH, resulting in ocean acidification [26]. Ocean acidification is threatening the habitat of marine creatures and will make a negative

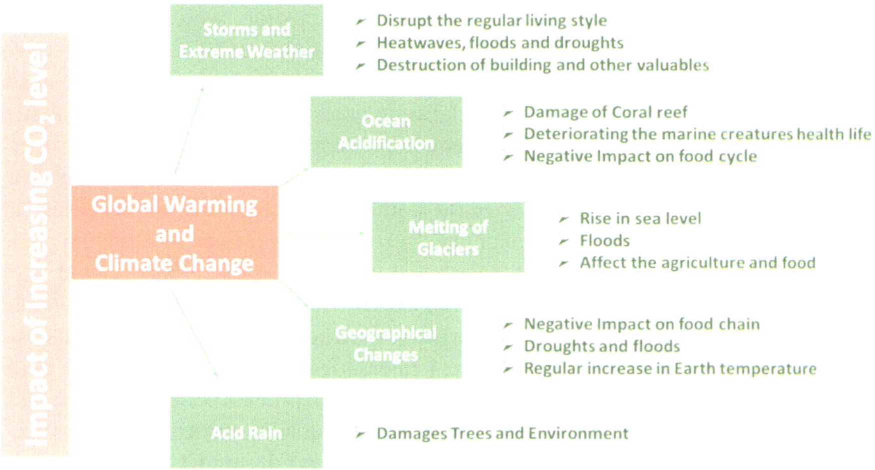

Fig. 14.3 An illustration of the impact of increasing CO_2 concentration in the environment

impact on the food chain. In one study, it has been noticed that the CO_2 concentration has risen from 378 to 410 ppm in the span of 14 years (i.e. 2005–2020), which is a clear indication that more drastic weather conditions are about to come [27, 28].

Therefore, there is an urgent need to curb CO_2 or GHGs emissions into the environment. Nowadays, several strategies are adopted to reduce atmospheric CO_2 levels, such as (a) absorption/adsorption of atmospheric CO_2, (b) progressive adaptation of renewable energy sources, (c) CCS and (c) CO_2 capture and utilization (CCU). Various porous materials, such as zeolites, can be considered as possible adsorbents for CO_2 capture, but the poor selectivity limits their industrial applications. In this regard, 2D nanomaterials have been fabricated to improve CO_2 adsorption/absorption and selectivity. The following sections will describe the fundamentals and mechanism of CO_2 capture using various 2D nanomaterials.

14.3 Mechanism of Carbon Dioxide Capture

Carbon dioxide capture is an important process that has been developed to reduce the amount of carbon dioxide emissions released into the atmosphere. Carbon dioxide capture involves trapping and storing carbon in various ways, including capturing it from industrial sources before it is emitted into the atmosphere or by directly removing it from ambient air. This technology can help mitigate climate change and reduce our dependence on fossil fuels for energy production. The most widely used form of carbon dioxide capture is direct air capture (DAC). DAC systems use filters to trap CO_2 molecules as they pass through them, which are then stored in tanks or other containers until they can be processed further for reuse or disposal. Other methods include pre-combustion techniques and oxyfuel combustion, which uses chemical absorption processes like amine scrubbing. These technologies have proven effective at reducing CO_2 emissions but require significant investment in infrastructure costs due to their complexity and costliness relative to other emission reduction strategies like renewable energy sources or efficiency improvements within existing facilities.

Overall, while there are many different mechanisms available for capturing carbon dioxide emissions before they enter Earth's atmosphere, each one requires careful consideration when evaluating its potential effectiveness against environmental goals while also taking into account any associated economic implications related with implementation costs & benefits over time. Currently, adsorption is the most viable technology for CO_2 capture due to its straightforward operation and low cost. Adsorption occurs in two ways- strong chemical interactions (chemisorption) and weak physical interactions (physisorption) [29, 30]. Solid adsorbents benefit from the lack of substantial amounts of water/solvent compared to aqueous absorbents (use amine-based solvents), which results in less energy being used during CO_2 desorption. In addition, solid adsorbents have lower heat capacities than aqueous ones, reducing the energy needed to heat the adsorbent to desorb the CO_2. The solid adsorbents also offer fast kinetics of CO_2 adsorption/desorption and effortless operation [29]. Because of these advantages, development of novel adsorbents with high efficiency are utmost importance for remediation of CO_2 from the atmosphere. Figure 14.4 illustrates the various kinds of interactions taking place during CO_2 adsorption.

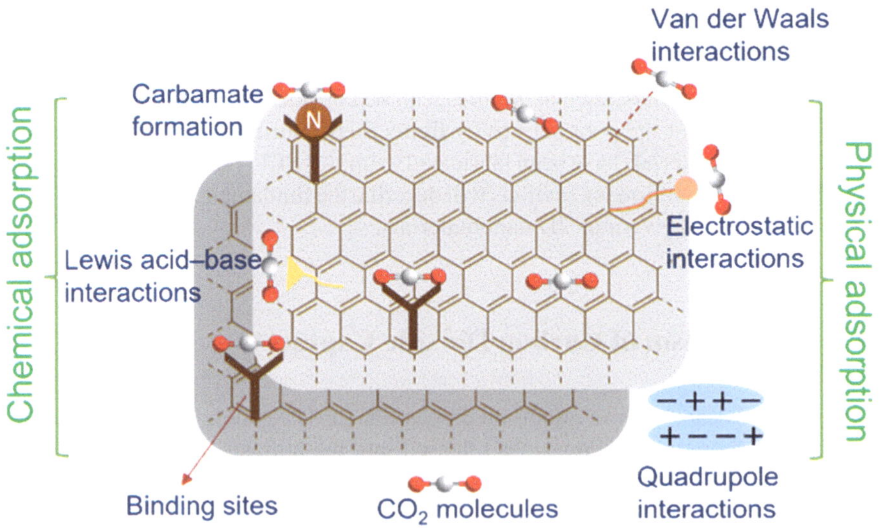

Fig. 14.4 Adsorption of CO_2 on the 2D materials-based adsorbents via physical and chemical adsorption

Solid adsorbents are further classified into three ways: low-temperature adsorbents (less than 200 °C), mid-temperature adsorbents (200–400 °C), and high-temperature adsorbents (greater than 400 °C) [31]. The low temperature adsorbents follow physisorption mechanism through electrostatic or van der Waals forces. Because of low interaction energy, desorption and active site regeneration can be done at low temperature conditions, resulting in lower energy and cost requirements. However, the major drawback of low temperature adsorbent is poor selectivity towards CO_2 [30]. Moreover, flue gas stream contaminants (such as NO_X, SO_X, particulate matter and water vapour) can deactivate the sorption site for CO_2 or out-compete CO_2. Furthermore, water vapours can selectively adsorb over CO_2 on low-temperature adsorbents, resulting in poor adsorbent stability due to deactivation [30]. Contrarily, high-temperature adsorbents offer better CO_2 selectivity and thus, they have been used in DAC conditions [32]. However, to replenish these adsorbents, a cyclic carbonation/decarbonation process is needed at high temperatures. In contrast, mid-temperature adsorbents provide reasonable selectivity as well as high performance for CO_2 capture. Additionally, these adsorbents are suitable for effective CO_2 capture and sequential catalytic conversion as they operate in the same temperature window [33].

14.4 Capture of CO_2 Using Nano-engineered 2D Materials

Regular emission of toxic and greenhouse gases into the environment has become a threat to the healthy planet. CO_2 is one of the major greenhouse gases released into the atmosphere via several natural and anthropogenic sources. The healthy management of these gases in the atmosphere is one of the essential requirements to balance survival. Several routes have been adopted to lower the CO_2 level, such as (a) control of the CO_2 emission, (b) CCS and (c) CCU conversion and utilization. Various 2D nano-engineered materials have been proven as suitable candidates for CO_2 capture and conversion. Lately, 2D materials such as graphene and graphene-based nanomaterials, 2D TMDCs, h-BN, 2D TMOs, MXenes, 2D MOFs, 2D covalent organic frameworks (COFs), g-C_3N_4, borophene, phosphorenes, nanoclays, etc. have been employed for CO_2 capture, (Fig. 14.1) [34]. The subsequent sections will discuss role of these 2D materials for CO_2 capture in details.

14.4.1 Graphene and Graphene-Based Nanomaterials

Since the discovery of graphene, it has been considered as an active and exciting material in wide range of applications such as lubrication, catalysis, supercapacitors, tissue engineering, 3D printing, sensors, desalination, water and air purification, drug delivery, solar cells and molecular imaging [35–38]. Graphene based materials such as graphene oxide (GO) and reduced graphene oxide (rGO) have also been widely used in the CO_2 capture as adsorbents and absorbents due to its favourable properties [39]. Graphene exhibit a 2D hexagonal morphology, which resembles more like a honeycomb network and provides extraordinary physicochemical, mechanical, thermal and electrical properties for its high efficiency in a wide range of applications [40, 41]. Theoretical graphene nanosheets demonstrate a very high active surface area (~2630 $m^2.g^{-1}$), which is one of the key characteristics for adsorption/absorption applications. Additionally, the high porosity of graphene-based materials, high thermal and chemical stability, and ease of functionalization of graphene nanosheets open the infinite possibilities to construct an adsorbent/absorbent with high CO_2 capture efficiency and selectivity [42]. The oxygen functionalities on the GO make its surface basic in nature, which attracts the CO_2, which is acidic in nature and helps in CO_2 adsorption. Moreover, the oxygen functionalities on GO surface open the prominent platform for easy modification of the adsorbent surface to enhance CO_2 adsorption.

Graphite and GO have been modified via several routes with different functional groups to enhance their capability for gas storage. Exfoliated rGO is one of the popular graphene derivatives in CO_2 capture and exists in the stacked few layers of defected graphene sheets with fewer oxygen functionalities. The route of GO reduction to prepare rGO also significantly affects the CO_2 adsorption capacity. As chemical/thermal reduction reduces the CO_2 adsorption capacity of rGO by 1.5 $mmol.g^{-1}$ than GO at ambient conditions [43], whereas on reducing GO hydrothermally, the CO_2 adsorption capacity enhances by 2.4 $mmol.g^{-1}$ at ambient pressure and 0 °C [44]. Additionally, on exfoliating the graphene or graphene derivatives sheets, the CO_2 adsorption efficiency increases due to exposure to a more active surface. Molecular dynamics simulation studies have also proven that under the assistance of H_2O molecules and the presence of active functional groups on the GO, surface enhances the CO_2 adsorption from the environment [45]. Further, the adsorption of CO_2 can be promoted by the functionalization and doping of the graphene sheets. CO_2 adsorption efficiency of rGO was compared with boron-doped rGO (B/rGO) under identical conditions. Doped boron significantly improves the adsorption characteristics of rGO from 1.3 $mmol.g^{-1}$ to 1.8 $mmol.g^{-1}$ [43]. Theoretical simulation studies reveal that the CO_2 molecule adsorbed on the graphene sheets with a parallel orientation and the distance between the graphene sheets and CO_2 was found to be 0.345–0.36 nm, which is attributed to the van der Waals interactions.

Doping or functionalizing graphene and its derivatives with nitrogen or nitrogen-based functional groups is another way to improve CO_2 adsorption efficiency. Polyaniline/hydrogen exfoliated graphene (PANI/HEG) based nanocomposite materials have excellent applications as CO_2 adsorbents [46]. PANI/HEG could adsorb 75 $mmol.g^{-1}$ CO_2 at room temperate and 11 bar pressure, which was almost 3.5 times higher than the CO_2 adsorbed by HEG (21.6 $mmol.g^{-1}$) only [47]. FTIR studies of CO_2-adsorbed PANI/HEG revealed that the interactions between the CO_2 and nitrogen atom present on adsorbent are majorly chemical interactions. Additionally, PANI/HEG also showed excellent recyclability, with 2–3% less adsorption efficiency. Amine functionalized graphene is one of the most popular graphene based adsorbent material, which has been investigated to capture the CO_2. Nucleophilic nature of amines strongly interacts with the electrophilic CO_2 and improve the CO_2 adsorption efficiency of GO. Shin et al. used polyethyleneimine (PEI) as an amine source to functionalize the GO and used it as an adsorbent material for CO_2 capture [48]. The amine-functionalized GO has shown an excellent adsorption capacity (84 $mg.g^{-1}$) for CO_2 adsorption, which was attributed to the appropriate PEI concentration and improved the electron donor and acceptor interaction between the adsorbate and adsorbent. Additionally, the loading of PEI on GO surface also enhances the selectivity of CO_2 adsorption over N_2 (Fig. 14.5). However, the excess loading of PEI further reduces the CO_2 capacity, which might be due to the unavailability of the active sites of the GO surface.

In another report, TEPA (Tetraethylenepentamine) was employed as an amine precursor to functionalize the GO to improve the CO_2 uptake [49]. Ultrasonic waves were used to activate the TEPA functionalized GO for CO_2 adsorption studies. Ultrasonic-activated TEPA-GO has shown higher CO_2 adsorption capacity (1.2 $mmol.g^{-1}$) than bare GO (0.3 $mmol.g^{-1}$) under identical conditions. This might be due to ultrasonic waves assisted in the exfoliation of graphene layers in the adsorbent material and provided a large number of active sites with functional groups for CO_2 adsorption. Consequently, the nucleophilic amine in the adsorbent material (TEPA-GO) interacts with electrophilic CO_2 efficiently and performs excellently.

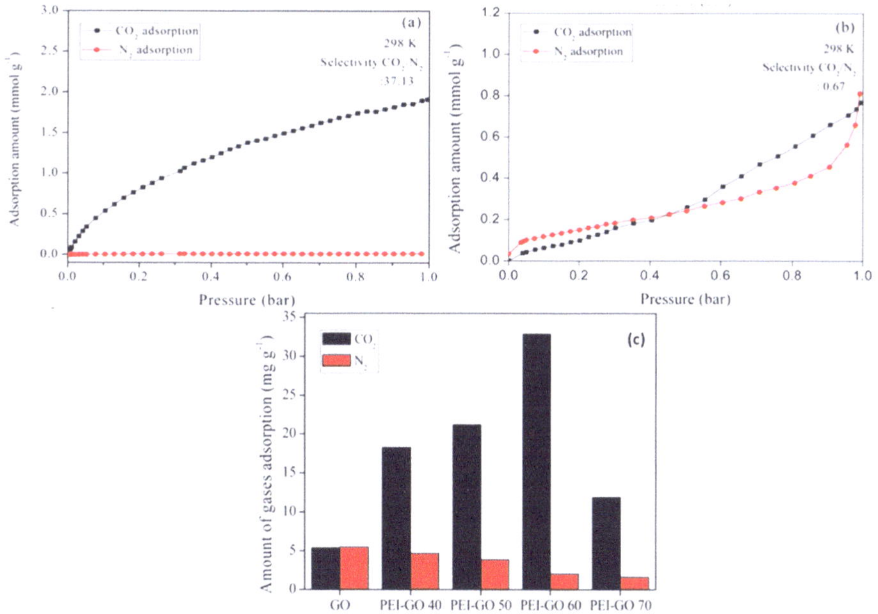

Fig. 14.5 Adsorption curves of **a** GO and **b** PEI-GO for CO_2 and N_2 gases and **c** amount of CO_2 and N_2 gases adsorbed on GO and various PEI functionalized GO. Reproduced with permission from ref. [48]. Copyright 2016, Elsevier Publications

TEPA-GO also display excellent recycling stability, which is suitable for practical applications.

Various graphene-based nanocomposite materials have also been investigated for CO_2 capture. In one study, Pokhrel et al. designed novel nanocomposite compromises of GO, Zeolite imidazolate framework (ZIF-8), and amine functional groups, which was present due to functionalization with 3-aminopropyl triethylsilane (APTES), PEI and ethylene diamine (ED) [50]. The amine groups inserted in the GO layers increased the interlayer spacing, which helped in improving CO_2 adsorption. Among all amine, functional groups precursors, APTES-modified GO-based nanocomposite showed the highest adsorption efficiency with a 36% increase in CO_2 adsorption capacity compared to unmodified GO. Additionally, the adsorption efficiency of APTES-modified GO nanocomposite increased in pre-adsorbed water conditions (10% RH), and successfully adsorbed 33% extra CO_2 than in dry adsorption conditions. This might be due to the generation of bicarbonates in the presence of moisture which provides more active sites of CO_2 adsorption. On comparing the ZIF-8/GO amine functionalized adsorbent with amine-modified ZIF-8,

Other than functionalized and doped graphene, 3D graphene, such as graphene-based hydrogels and aerogels, has also been studied for CO_2 adsorption. The 3D structure of the graphene-based adsorbent provides a more porous architecture for the adsorption studies. Hsan et al. prepared the chitosan (CS) embedded GO aerogels (CSGO) for the adsorption of CO_2 [51]. To construct crosslinking between the CS and GO, (1-ethyl-3-(3-dimethylaminopropyl) carbodiimide hydrochloride and N-hydroxy succinimide crosslinkers were used and followed by the freeze-drying process to obtain aerogel structure. CSGO was found to be excellent adsorbent material with 0.25 $mmol.g^{-1}$ adsorption capacity for CO_2 at 298 K and 1 bar pressure in comparison to raw CS (0.05 $mmol.g^{-1}$). The presence of nitrogen contents on the aerogel skeleton improves the selective CO_2 adsorption capacity. Therefore, introducing the nitrogen functionalities on the exfoliated graphene and its derivatives improved the CO_2 capture efficiency.

14.4.2 2D Transition Metal Oxide-Based Nanomaterials

2D transition metal oxide (TMOs)-based nanomaterials have been explored for various applications including electrochemistry, photochemistry, energy storage, tribology, catalysis, and environmental remediation [52–56]. Due to their remarkable properties such as high surface area, a large number of active sites, cost-effectiveness, ease of modification, abundance availability and microstructural features, metal oxide-based nanomaterials have been considered potent competitors in adsorption applications [53]. Additionally, oxide-based adsorbent material have been found highly selective towards CO_2 adsorption at high temperature conditions [57]. In dry atmospheric conditions, TMOs surface is usually terminated with O^{2-} anions, which are typically bigger than the cationic M^+. This mismatching of size creates a lower coordination of cations on the surface and results in plenty of adsorption sites

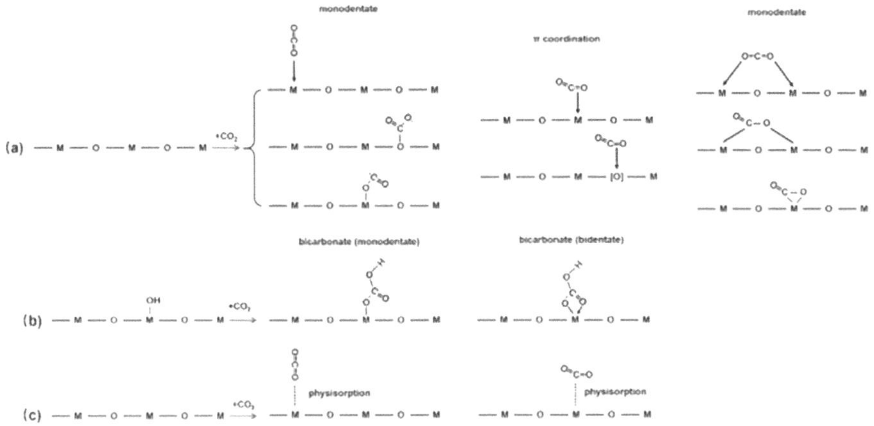

Fig. 14.6 CO_2 adsorption on metal oxides via **a** chelation, **b** chemisorption, and **c** physisorption. Reproduced with permission from ref [33]. Copyright 2022, Elsevier Publications

[58]. Therefore, several metal oxide-based nanomaterials are used in gas adsorption. TMOs can adsorb the CO_2 via physisorption or chemisorption or both (Fig. 14.6) [33]. Metal oxides usually exhibit a stronger affinity towards CO_2 and form a strong chemical interaction with CO_2 in the presence of other co-existing gases, which are adsorbed by weak physio-sorption only. This improves the selective adsorption application, which is indispensable for practical approaches.

Magnesium oxide (MgO) and Calcium oxide (CaO) are the two most successful metal oxides for CO_2 capture. Other metal oxides such as CuO, MnO_2, and CeO_2 were also used in CO_2 capture but were comparatively less explored. For the adsorption of CO_2, MgO exhibits appropriate surface morphology with oxygen, which improves the adsorption capacity and lowers the regeneration energy consumption [59, 60]. Recently Hu et al. published a detailed review focused on the strategy development of MgO-based adsorbents for CO_2 capture [61]. CO_2 adsorption on MgO is majorly an acid–base type reaction followed by physical interactions. The acidic CO_2 reacts with the basic sites of MgO to form a complex (MgO–O-C = O) to get chemically adsorbed. Othman et al. have incorporated several metal oxide nanoparticles (MgO, MnO_2, ZnO, and CaO) on activated carbon nanofibers (ACNFs) via electrospinning and pyrolysis process and examined for the CO_2 capture [62]. MgO incorporated ACNFs were found to be with highest surface area (413 $m^2.g^{-1}$) and micropore volume (0.1777 $cm^3.g^{-1}$) than other metal oxide ACNFs and pristine ACNFs. CO_2 adsorption capacity of MgO-ACNFs was also the highest (60 $cm^3.g^{-1}$) at 298 K. Also, the incorporation of MgO nanoparticles on the activated carbon-based bamboo (BAC) enhanced the BAC CO_2 adsorption capacity by 112% [63]. The major driven force for the adsorption of CO_2 on MgO-BAC was physical adsorption forces than chemical interactions because the chemical interactions between MgO and CO_2 to form MgO–O-C = O ($MgCO_3$) was slow than physical attractions.

Recently, alkali metal nitrates have been mixed with MgO-based nanoadsorbents to improve their efficiency in CO_2 capture. Alkali metals in MgO melt during the CO_2 capture process and avoid the $MgCO_3$ formation of MgO surface [64]. Figure 14.7 demonstrates the application of molten alkali nitrates in MgO to enhance CO_2 capture via carbonation [64]. $NaNO_3$ and KNO_3 are the popular nitrate promoters used in MgO as they do not react with CO_2 directly [65]. However, alkali metal carbonates can also be used in place of alkali metal nitrates to improve the CO_2 capture efficiency of MgO-based adsorbents. Carbonate promotors in MgO enable easier melting during the CO_2 capture process, allowing the faster transport of CO_2 through the carbonate layer. Kwak et al. mixed the triple eutectic alkali carbonate (TEC) (Li_2CO_3, Na_2CO_3 and K_2CO_3) with MgO for enhanced CO_2 capture [66]. The TEC amount in MgO significantly affects the kinetics of CO_2 capture. The CO_2 capture process was classified into two key steps: (a) fast and large and (b) slow and small. Higher the amount of TEC promotes the step 'a' but does not affect step 'b'.

Other than MgO, CaO is also extensively examined metal oxide material for CO_2 capture due to its fast carbonation and easy regeneration. CaO is generally used for CO_2 storage and reacts with CO_2 reversibly, as shown in Eq. 14.1

$$CaO + CO_2 \leftrightarrow CaCO_3 \tag{14.1}$$

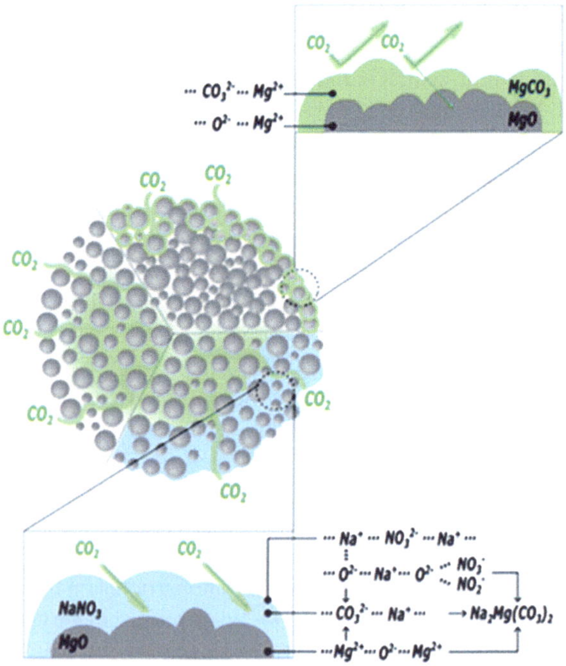

Fig. 14.7 Diagrammatic illustration of MgO sorbent role with nitrate salts in enhancing the CO_2 capture capacity. Reproduced with permission from ref. [64]. Copyright 2022, Royal Society of Chemistry

The adsorption of CO_2 takes place at 500 °C, and the exothermic reaction knows as carbonation. However the desorption of CO_2 takes place at higher temperature (800–950 °C) and the endothermic reaction is known as calcination or decarbonation [67]. The reversible adsorption and desorption of CO_2 using CaO are known as chemical looping. CaO-based adsorbents are commercially used to capture the CO_2 from cement plants. However, mechanical failure and excessive sintering are the significant drawbacks of CaO in CO_2 capture. Once CO_2 uses the initial layer of CaO to form the $CaCO_3$, the adsorption process slows down. Several works have been proposed to improve the adsorption efficiency of CaO. On using nanosized CaO, the number of active surface sites is increased for the adsorption studies. Florin et al. derived the CaO from nanosized $CaCO_3$ and used it for CO_2 capture [68]. The prepared CaO was exposed for 5 carbonation cycles (each cycle was run for 24 h) with no morphological impediment. Also, the material was used for 100 CO_2 carbonation and decarbonisation cycles (20 min for each cycle), representing the potential of nano CaO for CO_2 capture [68]. In summary, MgO and CaO are the most explored transition metal oxide-based nanomaterials for CO_2 capture, whereas there is a need to examine other metal oxides also.

14.4.3 MXenes

Beyond graphene, transition metal carbon/nitride (MXene) is another class of 2D material that has drawn a lot of interest from research communities. By carefully etching the A layer from MAX phases, where M is a transition metal, A is a IIIA or IVA element, and X is C or N, MXenes have been synthesized [69]. Due to the high corrosivity of hydrofluoric acid, it has been employed to etch the covalently bonded IIIA or IVA element [70]. To avoid the toxicity of hydrofluoric acid, alternative environmentally friendly processes have also been implemented, including alkali treatment, Lewis acid etching, electrochemical etching and so on [71]. MXenes have received enormous attention for their functionality and characteristics in electrochemical charge storage, environmental remediation, electromagnetic interference shielding, catalysts and other applications [72]. Owing to high chemical and mechanical stability with outstanding thermal and electrical conductivities, it offers requisite features for CO_2 activation/conversion, capture and storage. Various termination functionalities (−O, OH, halogens, chalcogens or their mixtures) of MXenes make them attractive materials for adsorption processes. Additionally, these materials can be produced with high surface areas in range of 250–1000 m^2/g that provide high reactive sites and porous functionalities for CO_2 capture and conversion [70]. Importantly, it was noticed that MXenes have higher adsorption energies compare to other 2D materials [73].

CO_2 adsorption on MXenes depends on many factors such as the thickness of sheets, defects, specific surface area, lone pair electrons and presence of CO_2-philic groups on the surface [74]. For example, Viñes et al. used 2D MXene, which has the general formula $M_{n+1}X_nT_x$ ($M = Ti, Zr, Hf, V, Nb, Ta, Cr, Mo, W$, $n = 1, 2, 3, 4$), for effective

CO_2 removal [75]. The maximum CO_2 abatement rate could be 8.25 mol CO_2 per kg MXene. Studied MXenes followed the given order for CO_2 capture: $Ti_2CT_x > V_2CT_x > Zr_2CT_x > Nb_2CT_x > Mo_2CT_x > Hr_2CT_x > Ta_2CT_x > W_2CT_x$. They exhibited high adsorption capacity despite of low CO_2 partial pressure and high temperature. The higher adsorption energy, large surface area, MXene/CO_2 charge transfer, and abundant adsorption sites could be the causes of the greater ability of MXenes to capture CO_2. The study summarized that the adsorption strength and conformation had the greatest influence on CO_2 desorption. In another study, dimethylsulfoxide intercalated $Ti_3C_2T_x$ was prepared via wet chemistry using sodium fluoride and hydrochloric acid [76]. It demonstrated a high surface area of 66 $m^2.g^{-1}$ with a high volume capacity of 502 $V.v^{-1}$ and exhibited a high CO_2 adsorption uptake of 5.79 $mmol.g^{-1}$. However, this adsorption capacity was lower than the theoretical capacity (44.2 $mmol.g^{-1}$) of $Ti_3C_2T_x$ with a specific surface area of 496 $m^2.g^{-1}$ [76]. Pristine $Ti_3C_2T_x$ showed a lower surface area (21 $m^2.g^{-1}$) and lower CO_2 adsorption capacity. It is noteworthy that tailoring the surface area of MXenes is a potential approach for enhancing the CO_2 uptake. In another study, individual sheets of $Ti_3C_2T_x$ carbides were employed to achieve efficient CO_2 adsorption (12 mol/kg) [77].

Numerous theoretical studies have been conducted to understand the insights of the CO_2 adsorption process on MXenes. Using density functional theory, Morales-Garcia et al. proposed that there was insignificant effect of a number of MXene sheets on CO_2 adsorption [78]. Using first-principles calculations, Sun et al. proposed the surface lone pair of MXenes account for high CO_2 absorption [79]. They selected nine MXene for the study such as Ti_2CT_x, Zr_2CT_x, Hf_2CT_x, V_2CT_x, Nb_2CT_x, Ta_2CT_x, Cr_2CT_x, Mo_2CT_x, and W_2CT_x. The CO_2 adsorption by MXenes followed the IVB > VB > VIB trend, which was linked to the lone pair electrons of MXene that shift to CO_2 during the absorption process. Thus, the occurrence of surface lone pairs on metals of MXenes could play a significant role in CO_2 adsorption. Another study used DFT calculations to examine the CO_2 adsorption and desorption on 2D M_2N materials (M = Ti, Zr, Hf, V, Nb, Ta, Cr, Mo, W) [80]. M_2N is better than Mo_2C for effective CO_2 uptake due to the presence of high electronegativity N atom layers and substantial adsorption energy (−3.13 eV).

Furthermore, MXene-based composites have been prepared to enhance the adsorption capacity for capturing CO_2 gas at the different sources. For example, MXene-based solid nanocomposite (Ti_3C_2/PEI/BO) adsorbent was fabricated using $Ti_3C_2(OH)_x$, polyethyleneimine and 1,2-butylene Oxide [81]. Ti_3C_2/PEI/BO adsorbent demonstrated high CO_2 adsorption capacity (2.60 mmol/g) during CO_2 (100%) as a purge gas at a high temperature (120 °C). Furthermore, adsorbent also showed long-term cyclic stability and superior attrition resistance. In another study, thin-film matrix membranes were prepared by incorporating $Ti_3C_2T_x$ nanosheets in Pebax-1657 (Pebax is the trade name of polyether block amide) for effective mixed gases (CO_2/N_2 and CO_2/H_2) separation [82]. The separation performance of the composite membranes exceeds Robeson's upper bounds, while working within the target range for economical CO_2 remediation. The uniform galleries of MXene produced via H-bonding interaction between MXene and rubbery Pebax might be responsible for the high CO_2 capacity. Fast and specialized CO_2 transport was made possible by

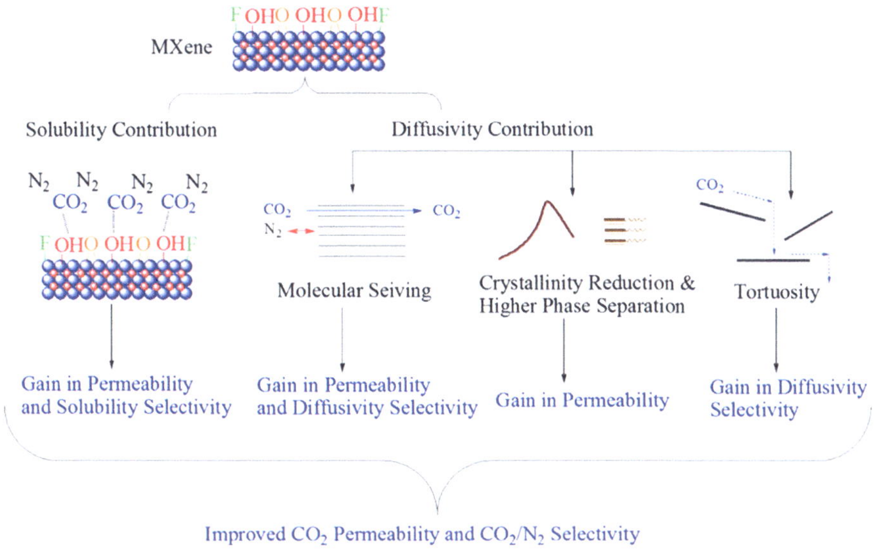

Fig. 14.8 Proposed mechanism for transporting and separating CO_2 and N_2 by $Ti_3C_2T_x$/Pebax membranes. Reproduced with permission from ref. [82]. Copyright 2020, American Chemical Society

interfacial interactions and selective $Ti_3C_2T_x$ composite nanochannels. The performance mechanism of MXene/Pebax membrane for selective capturing of CO_2 and N_2 can be depicted in Fig. 14.8. Because CO_2 has a higher quadrupole moment than N_2, it more easily affiliates with the MXene surface. The surface hydroxyl groups of MXene improved CO_2 capture efficiency and also resulted in highly selective adsorption of CO_2. MXene galleries were ~0.35 nm in diameter, bigger than CO_2 but smaller than N_2 [82]. It demonstrated that N_2 was constrained while CO_2 was qualified to infiltrate and diffuse across the membrane. Additionally, the adsorption capacity of various 2D materials-based adsorbents at different conditions is discussed in Table 14.1.

14.4.4 Hexagonal Boron Nitride (h-BN)

Hexagonal boron nitride (h-BN) nanosheet is made of sp^2-conjugated boron and nitrogen atoms that create a 2D structure. It has high thermal stability (up to 800 °C), excellent thermal conductivity (300–2000 $Wm.K^{-1}$), high Young's modulus (0.8 TPa) and superior fracture strength (165 GPa) [99]. h-BN nanosheet has been prepared using both top-down and bottom-up approaches. Tunning and functionalization of pores and structures of BN via selecting appropriate precursors during synthesis is a crucial strategy to achieve excellent CO_2 adsorbent. For instance, a

Table 14.1 Comparison of CO_2 adsorption capacities of various 2D materials

2D materials adsorbent	Surface area ($m^2.g^{-1}$)	Adsorption capacity ($mmol.g^{-1}$)	Pressure (bar)	Temperature (°C)	References
g-C_3N_4	23.04	8.54	15	25	[8])
g-C_3N_4-ionic liquid	182.93	42.93	15	25	[83]
C_3N_4-porous rGO aerogel	450	0.43	0.1	27	[84]
C_3N_4-ZnO hybrid	36.3	6.11	1	25	[85]
Graphene	42.87	21.6	11	25	[47]
PPy/rGO	1650	6.8	1	0	[86]
Fe_3O_4/graphene	98.2	60	11	25	[87]
rGO/N-porous carbon composite	865.1	5.77	5	25	[88]
$Ti_3C_2T_x$	66	5.79	4	25	[76]
MXene, M_2N	–	7.96	1	727	[75]
Mo_2C-Li	8.9	3.66	10	25	[89]
Porous h-BN	1900	1.68	1–20	25	[90]
C doped h-BN	–	5.5	1	0	[91]
Porous BN	1585	1.22	1	25	[92]
Micropore-rich BN fibers	1042	2.85	1	0	[93]
Cu-Loaded Porous BN Nanofiber	653	2.77	1	0	[94]
FBNNSs/ZnO superstructures	71.2	2.79	0–1	0	[95]
rGO-MoS_2/PPy-600	8.43	0.57	1	25	[96]
2D Cu MOF	169.5	1.78	1	0	[97]
MMT/rGO	50.77	0.5	1	25	[98]

few layered h-BN (1–4 layers) with a high surface area (927 $m^2.g^{-1}$) were produced chemically using boric acid and urea and found to be advantageous for CO_2 adsorption [100]. Yang et al. synthesized three-dimensional (3D) flower-like BN nanosheets (FBNNSs) by template-free cylinder compressing route using boron oxide and guanidine hydrochloride as starting precursors [101]. FBNNSs demonstrated efficient CO_2 adsorption up to 74.4 $mg.g^{-1}$ at 1 bar, which was ascribed to the large surface area (1114 $m^2.g^{-1}$), pore volume (0.7 $cm^3.g^{-1}$), hierarchical pore availability, and plentiful edge functionalities (−OH/−NH_2). The CO_2 adsorption mechanism on nanosheets was governed by hydrogen bonding, and weak van der Waals interaction. FBNNSs showed high stability and performance even after 10 cycles of recyclability. Additionally, in another study, a new type of micropore-rich BN fibres was produced by introducing a hexamethylenetetramine surfactant in the conventional porous BN precursor (melamine-diborate) [93]. Because of excessive micropores, micropore-rich BN fibres showed better CO_2 capture capacity (2.85 $mmol.g^{-1}$) than original porous BN fibres (3.19). These thermally stable BN fibres can be reused in several cycles without loss their adsorption property. Moreover, large surface area (1900 $m^2.g^{-1}$) BN can be produced using multiple N precursors (mixing biuret or melamine with urea), exhibiting high adsorption capacity of 1.6 $mmol.g^{-1}$ at 1 bar, 25 °C and 8.3 $mmol.g^{-1}$ at 20 bar, 25 °C. More notably, CO_2 adsorption capacity decreased to 1.1 $mmol.g^{-1}$ (1 bar and 25 °C) after palletization [90]. C-doped BN fibre was also prepared using melamine and boric acid with final treatment at 1100 °C [102]. Excellent sorption property (3.71 $mmol.g^{-1}$ at 273 K) of doped fibre was attributed to micropores (formed by doping of C) and more negative charges on BN (induced by structural defects), resulting in high CO_2 affinity via chemisorption interaction. Thus, it is noteworthy that the adsorption performance of BN sheets/fibres can be improved by slight modification in their structure and composition.

To improve CO_2 capture performance, BN can be engineered with metallic nanoparticles. For instance, in-situ Cu-incorporated BN nanofibers (BNNF) were prepared using BN precursors (melamine and boric acid) and Cu source (Copper dichloride dihydrate) for efficient uptake of CO_2 (Fig. 14.9a) [94]. It was observed that Cu^{2+} ions speed up the nucleation process during the reaction between melamine and boric acid, which is discovered to be a crucial component for producing Cu@BNNF with uniform and tiny diameters (Fig. 14.9b–d). Interestingly, the presence of Cu on BNNF improves the CO_2 affinity, the adsorption capacity of the BN nanofibers enhanced from 1.34 to 2.77 $mmol.g^{-1}$ (1 bar and 273 K) (Fig. 14.9e). The Cu@BNNF also exhibited high recycling stability, unlike several other BN-based absorbents (Fig. 14.9f). Moreover, CO_2 adsorption mechanism on Cu-loaded BNNS was investigated with density-functional theory calculations. It was discovered that adding Cu on BNNF can lead to enhanced electron interaction between BNNF and the CO_2 molecule, resulting in better CO_2 affinity and adsorption capacity. In another study, Yang et al. synthesized 3D hierarchically cubic/spherical morphologies of functionalized BN nanosheets (FBNNSs)/ZnO superstructures using the evaporation-mediated solvothermal approach [95]. These materials offer several benefits, including scalability, inexpensive technology and high throughput. As-produced FBNNSs/ZnO superstructures showed a high CO_2 adsorption performance of 124.5 $mg.g^{-1}$ from 0 to 1 bar at 273 K and presented reasonable reusability of 10 cycles with an average adsorption capacity up to 115.7 $mg.g^{-1}$. The superior adsorption performance of superstructure for CO_2 removal was described by van der Waals interaction, chemisorption and hydrogen bonds from the ZnO and amino groups of FBNNs.

14.4.5 Transition Metal Dichalcogenides

Transition metal dichalcogenides (TMDCs) have recently gained attention for their potential application in carbon dioxide and other gases capture [103]. TMDCs possess

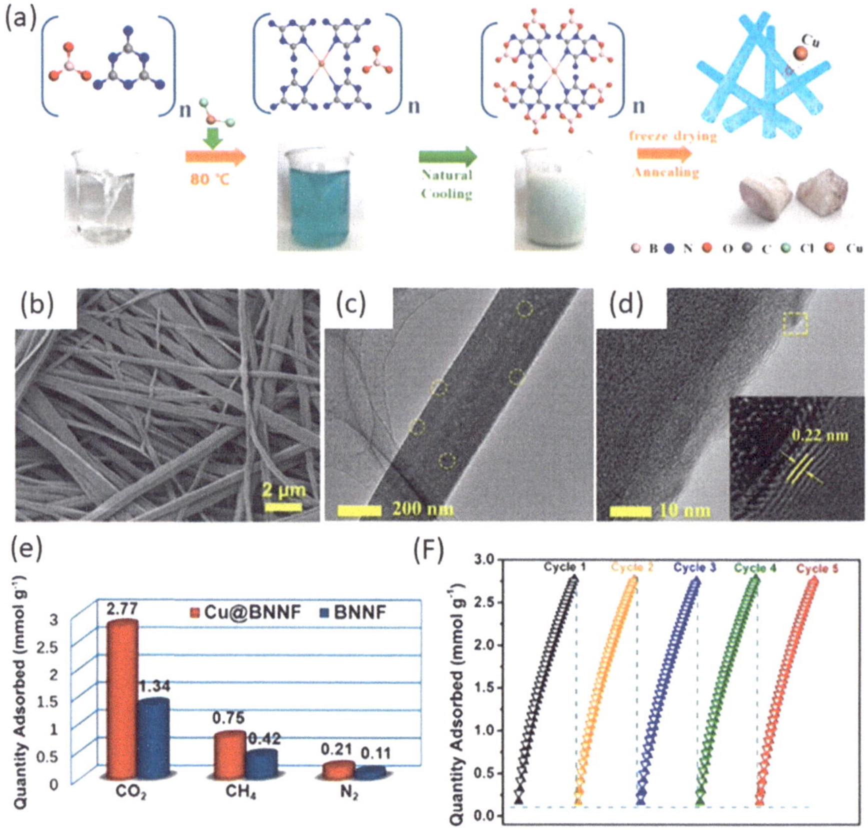

Fig. 14.9 **a** Synthesis of Cu@BNNF; SEM **b** and TEM images **c–d** exhibit uniform and smaller diameter of Cu@BNNF; **e** adsorption quantity of Cu@BNNF and BNNF; **f** CO_2 adsorption and desorption cycling for the Cu@BNNF. Reproduced with permission from ref. [94]. Copyright 2020, American Chemical Society

unique physical and chemical properties, such as high surface area and strong adsorption abilities, which make them ideal candidates for capturing CO_2 from air or exhaust streams. In addition to their excellent adsorptive properties, they also exhibit a low cost of production compared to other materials used in carbon capture technologies [104, 105].

The most common type of TMDC used for CO_2 capture is molybdenum disulfide (MoS_2). This material has been found to be highly effective at trapping CO_2 molecules due to its large surface area and strong affinity towards the gas molecule [106]. MoS_2 can be produced synthetically via chemical vapour deposition methods and direct chemical synthesis using S and Mo precursors or extracted from natural sources like shale deposits [107–110]. The synthesis approaches are relatively inexpensive, reducing the cost of CO_2 adsorption compared with traditional methods used in CO_2 removal systems such as cryogenic distillation or amine scrubbing processes [111]. Additionally, this material does not require any post-treatment steps after use, making it an attractive option for industrial applications where cost savings is essential. Recently, various studies have been reported for CO_2 capture using pristine or modified MoS_2. According to Sun et al., the strength of the applied electric field affects how strongly CO_2 adsorb on the MoS_2 [112]. When an electric field is applied, CO_2 is adsorbed on the material and desorbed when the electric force is relaxed. It becomes a possible carbon capture material because of this distinctive quality. The Cu-modified MoS_2 nanosheets were prepared for the CO_2 reduction process by Shi et al. [113]. The Cu/MoS_2 hybrid (0.44 cm^3/g) showed better adsorption capacity compared to pristine MoS_2 (0.22 $cm^3.g^{-1}$). Thus, Cu nanoparticles have improved the adsorption capacity since CO_2 can now be adsorbed on both Cu and host MoS_2 nanosheets.

In addition, the surface of MoS_2 can be tuned to obtain high reactive edge sites via doping or creating defects for the adsorption of a specific gas. Wu et al. studied the role of nitrogen doping on MoS_2 (defect-rich and defect-free structure) reactivity towards CO_2 adsorption [114]. N-doped MoS_2 with Mo defects demonstrated the highest CO_2 adsorption capacity because of more electrostatic and covalent interactions with the CO_2. On the contrary, the defect-free MoS_2 followed weak van der Waals interactions, which resulted in weak CO_2 adsorption. Also, the defective and N-doped MoS_2 showed high selectivity of CO_2 over N_2 [114]. In another study, MoS_2 nanocomposites were prepared using rGO and polypyrrole (PPy) to enhance the reactive sites for CO_2 adsorption [96]. Nanocomposite with the highest PPy concentration exhibited the highest CO_2 adsorption capacity (13 $cm^3.g^{-1}$) compared to MoS_2 (6 $cm^3.g^{-1}$) due to the presence of excessive nitrogen binding sites (Fig. 14.10). Therefore, the adsorption capability of MoS_2 nanosheets can be improved by structural modifications using doping/defect or making composite with other active species.

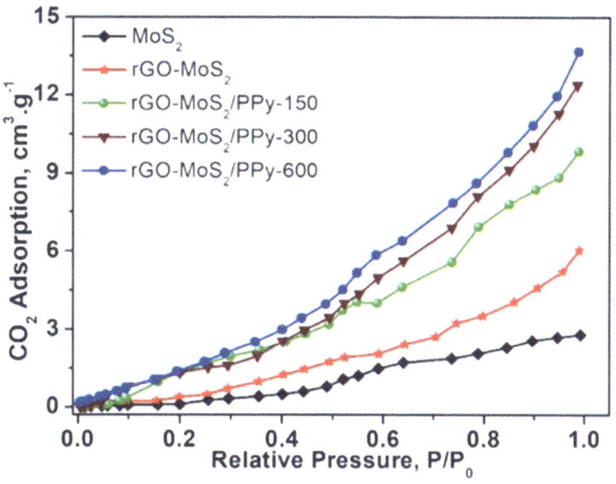

Fig. 14.10 CO_2 adsorption of the MoS_2 nanosheets, rGO–MoS_2, rGO–MoS_2/PPy-600, rGO–MoS_2/PPy-300, and rGO–MoS_2/PPy-150 nanocomposites. Reproduced with permission from ref. [96]. Copyright 2015, American Chemical Society

14.4.6 Carbon Nitride

Like graphene, graphitic carbon nitride (g-C_3N_4) is made of hexagonally ordered heptazine (tri-s-triazine) units bonded across tertiary amines. The carbon nitride term is basically used for CN material with elemental compositions C/N (x) of less than 1.0, in which the main constituting elements are carbon and nitrogen atoms are either substituted for carbon atoms or bonded as nitrogen functionalities. In the literature, the materials with CNx with x > 1.0 are frequently referred to as N-rich carbon nitrides [115]. Because of its excellent physicochemical characteristics, g-C_3N_4 has gained significant attention for energy and environmental applications [116, 117]. Unlike other 2D materials, g-C_3N_4 can be commonly synthesized by direct thermal decomposition of N-rich precursors such as urea or melamine. Conventional top-down exfoliation of g-C_3N_4 is not popular because of densely packed heptazine units. The affinity of pristine g-C_3N_4 and CO_2 is low due to the poor alkalinity of N species in g-C_3N_4, which result in limited utilization in CO_2 removal. Nevertheless, amine functionalization of carbon nitride is the way to enhance CO_2 uptake through improving adsorbent-adsorbate interactions. For instance, polyethylenimine (PEI)-functionalized g-C_3N_4 was synthesized using the physical impregnation method to enhance the adsorption capacity and selectivity of g-C_3N_4 towards CO_2 [118]. Due to the presence of excessive amine groups and stacked pores, the highest CO_2 adsorption capacity of PEI-g-C_3N_4 was found to be 3.77 mmol.g^{-1} at 100 °C and ambient pressure, which was better than pristine g-C_3N_4. In another study, N-rich ionic liquid functionalization of g-C_3N_4 was performed with ionic liquid 1-Butyl-3-Methylimidazolium bis(trifluoromethyl sulfonyl)imide ([BMIM][TFSI]) for enhanced uptake of CO_2 [83]. The NS-g-C_3N_4/[BMIM][TFSI] demonstrated excellent adoption capacity to 42.93 mmol.g^{-1} under nearly ideal conditions, better than bulk g-C_3N_4 (8.54 mmol.g^{-1}). The high sorption uptake was ascribed to high N content (56.6 atomic%) that acts as reactive sites for CO_2 interactions. At the same time, ionic liquid molecules on the g-C_3N_4 surface offer extra physicochemical adsorption mechanism thus increasing the CO_2 uptake.

Additionally, more active sites on g-C_3N_4 can be generated by making a composite of it with other nanomaterials. Carbon nitride-ZnO hybrid adsorbent was prepared for CO_2 adsorption. The hybrid material showed a remarkable adsorption capacity of 6.11 mmol.g^{-1} (1 atm and 25 °C), which was three times better than g-C_3N_4 [85]. It was summarized that the high CO_2 affinity of hybrid material is governed by basicity, amount of amine precursors, framework stability, surface area and surface porosity. However, Kim et al. used RGO aerogel as a template for the evolution of carbon nitride with a large surface area for CO_2 uptake [84]. The schematic of the preparation of carbon nitride aerogels for CO_2 capture is described in Fig. 14.11. The as-prepared structure exhibited high adsorption properties (0.43 mmol.g^{-1}) at ambient conditions with high selectivity against the N_2 and excellent regeneration capacity. The increased CO_2 capture capacity of carbon nitride aerogels was attributed to the microporous

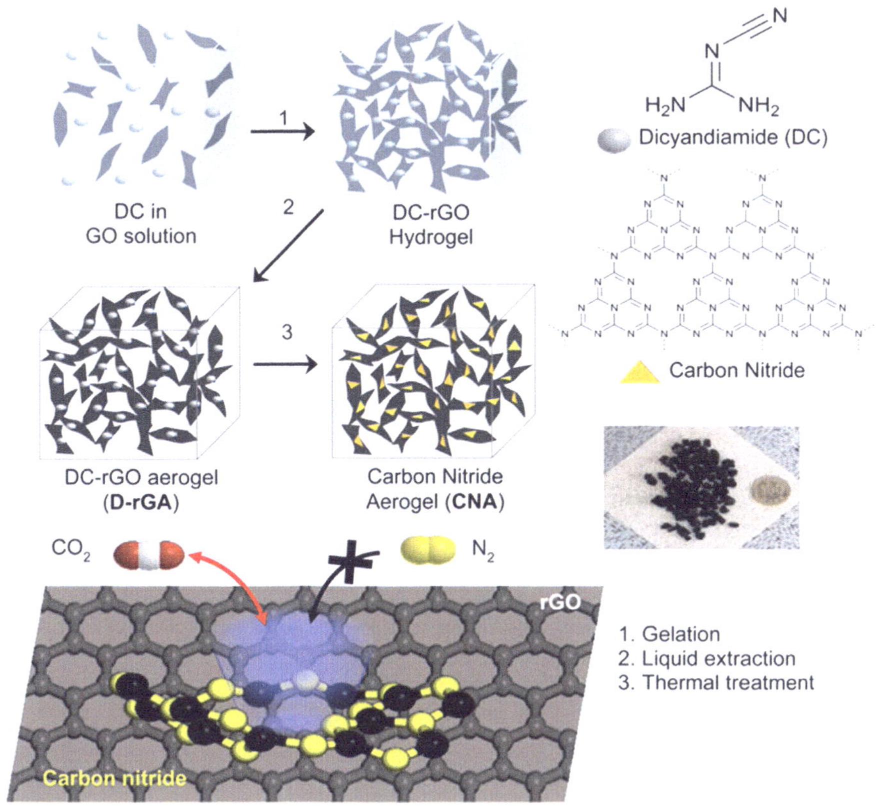

Fig. 14.11 Synthesis of carbon nitride aerogels for the selective adsorption of CO_2. Reproduced with permission from ref. [84]. Copyright 2015, American Chemical Society

edges of carbon nitride that provide the optimal CO_2 affinity via dipole-induced dipole interactions.

14.4.7 2D Metal–Organic Frameworks

Metal–organic frameworks (MOFs) are periodic network-structured porous coordination polymers, created by carefully coordinating the congregation of metal ions or clusters with organic ligands. Due to the diversities of metal ions/clusters and organic ligands, MOFs have shown a large surface area, high porosity, unique framework structure and tailorable pore, and open metal sites, which make them suitable for married applications, including CO_2 capture [119]. When the MOFs dimension is reduced to two-dimension, ultrathin 2D MOFs layers have more outward remarkable characteristics, offering various scientific applications [120]. Like graphene,

2D MOFs can be prepared using either top-down or bottom-up methods [41, 120]. For example, Zhang et al. applied a bottom-up room temperature approach for preparing 2D MOFs using Cu(II) propeller and different ligands in a dimethylformamide/water solvent system [97]. The size and surface functionalization can be adjusted by changing the solvent ratio and ligands, respectively (Fig. 14.12). The findings of the gas sorption experiments indicated that nitro and amine-functionalized 2D MOFs exhibited better CO_2 sorption selectivity over CH_4 and N_2, indicating that these materials may also be applied for carbon capture from flue gas (CO_2/N_2). The highest CO_2 adsorption capacity was found to be 1.78 mmol.g^{-1} (1 atm, 273 K) for $[Cu_2(nbdc)_2(dabco)]_n$. Sun et al. synthesized Co based 2D Co-MOF-UPC-32 with persistent porosity for high CO_2 and H_2 adsorption [121].

Particularly in hot and humid environments, structural instability of MOFs is a major issue because of brittle coordination bonds. This restricts their capability to capture CO_2, especially in pre- or post-combustion processes. On the other hand, covalent organic frameworks (COFs) are joined by covalent bonds and are structurally stable to be used in harsh CO_2 capture settings. In fabricating membranes for the capture/separation of CO_2 gas, 2D COFs are primarily used as active constituents [122–125]. Lately, 3D MOFs and 2D COFs have been applied to fabricate dual-layered membranes for CO_2/H_2 separation [122]. 2D COFs were accommodated on vertical binding sites of MOFs, producing composite membranes with exceptional H_2/CO_2 selectivity (32.9) and superior permeability.

14.4.8 Other 2D Materials

Nanoclays consist of layered mineral silicate with excellent mechanical properties. Due to its high abundance and low cost, it has been used as a solid adsorbent for environmental remediation. Clay-based materials have also been studied for their potential to adsorb CO_2, but they have a major drawback [98, 126, 127]. The efficiency of CO_2 uptake under wet conditions is low due to the presence of diffused H_2O, which avert the trap of gas molecules. In dry conditions, however, high uptake can be expected because of nano-channels present in clay that allow easy intercalation of CO_2. The moisture issue has led researchers to investigate ways in which this could be overcome, such as by modification or impregnation with other compounds like polymers and surfactants or by making composites with other nanomaterials such as carbon nanotubes and graphene oxide sheets. For example, montmorillonite (MMT) clay was first modified with polyphosphoric acid and used to make a composite with rGO for CO_2 capture [98]. The MMT/rGO hybrid exhibited a reasonable CO_2 adsorption capacity of 0.5 mmol.g^{-1} at 1 bar and 25 °C. In another method, porous MMT was modified with octadecylamine to adsorb CO_2, which showed excellent adsorption of CO_2 (7.16 mmol.g^{-1}) compared to pristine MMT (only 3.47 mmol.g^{-1}) at room temperature and 50 bar [127].

Other 2D materials, such as borophene [128, 129], phosphorene [130] and graphyne (comprised of both sp and sp^2 carbons) [131], have mainly been studied

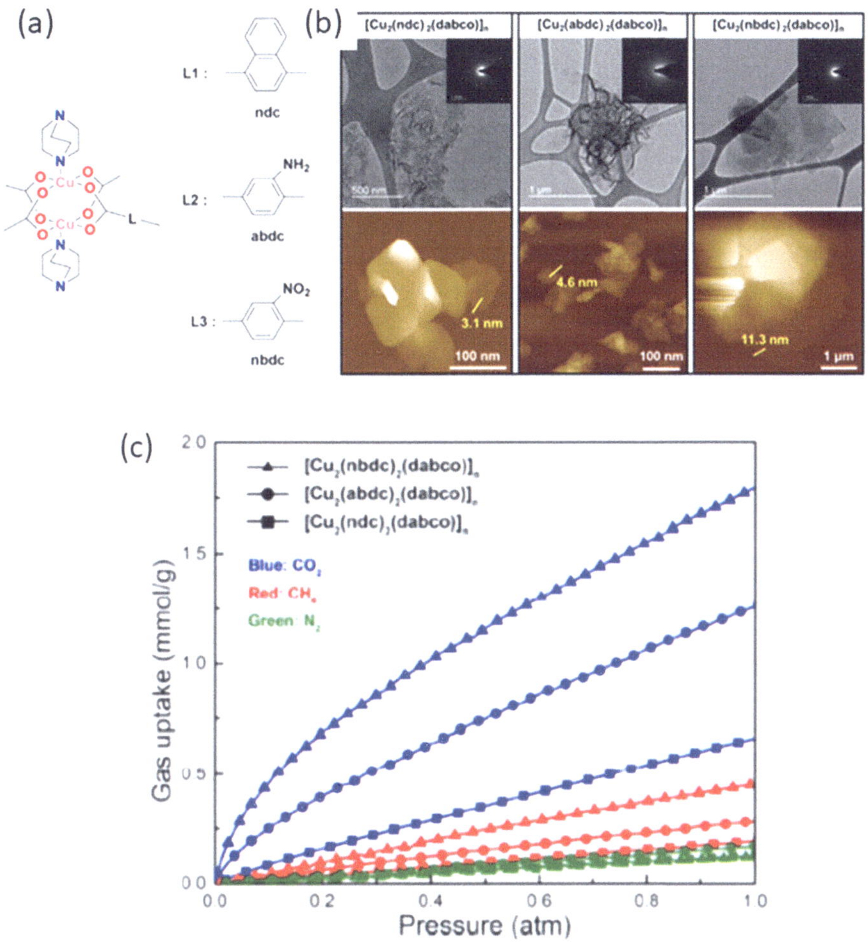

Fig. 14.12 **a** Schematic shows the structure of 2D MOFs; **b** TEM images (with electron diffraction patterns in insets) and AFM topographic images of synthesized 2D MOFs; **c** CO_2 sorption capacity for CO_2, N_2 and CH_4. Reproduced with permission from ref. [97]. Copyright 2018, American Chemical Society

theoretically for CO_2 capture and separation. The adsorption of gas molecules (CO, CO_2, NH_3, NOx, and CH_4) on these 2D materials was investigated by DFT and grand canonical Monte Carlo (GCMC) calculations. The findings suggest they are potential candidates for excellent adsorption of CO_2 at ambient conditions.

14.5 Conclusion and Future Directions

The distinct and extraordinary properties of 2D materials (e.g., high surface area and porosity, tailorable structure, surface functionalities, chemical and thermal stability) have been fundamental driving forces for researchers to explore their potential in CO_2 capture and utilisation. Lately, 2D materials such as graphene and graphene-based nanomaterials, TMDCs (e.g., MoS_2), h-BN, 2D TMOs, MXenes, 2D MOFs, 2D COFs, g-C_3N_4, borophene, phosphorene, and nanoclays have been employed for CO_2 capture. This chapter summarised an overview of various sources of CO_2 and commonly employed mechanisms of CO_2 removal technologies, as well as recent developments related to the use of 2D materials for capturing CO_2. Usually, the CO_2 adsorption efficiency is controlled by external factors, including pressure, temperature, and contaminations. Various strategies, such as doping, defect engineering, and the formation of composites and hybrids, have been employed to improve the CO_2 adsorption capability of 2D materials. The importance of controlling and functionalising pores of adsorbent for CO_2 uptake is discussed thoroughly. Adsorption performance could also be improved by functionalizing 2D materials and their composites with CO_2-philic components. Moreover, new 2D materials such as phosphenes, borophene, 2D MOFs/COFs, and graphyne are mostly theoretically studied for CO_2 capture. Therefore, more research is required to establish their CO_2 capture and conversion potential.

There are still many limitations of commercializing 2D materials-based adsorbents technology for CO_2 capture. For CO_2-containing mixed gases, more focus should be placed on the high selectivity and high-performance CO_2 adsorption at moderate temperatures and low CO_2 concentration. The design and development of multi-functional 2D materials are critical to future technologies, where materials can concurrently capture CO_2 from gas streams and convert it into value-added chemicals using in-situ heterogeneous catalysis. The selectivity over other gases and low performance are the major troublesome for industrial adoption of 2D materials for CO_2 capture and conversion. Much research efforts are needed to match the industrial scale efficiencies. Still, there is a high preparation cost and non-uniform quality of the materials at large-scale production. Therefore, dedicated research is needed for the sustainable, affordable synthesis and processing of 2D materials-based technologies. Additionally, machine learning tools can be employed to fast-track the discovery of advanced 2D materials, experimental designs and to achieve the best possible performance.

Acknowledgements The authors are thankful to the Council for Scientific and Industrial Research (C6ACH20), and the University of Johannesburg (086310) for their financial support.

References

1. T. Zhang, W. Zhang, R. Yang, Y. Liu, M. Jafari, CO_2 capture and storage monitoring based on remote sensing techniques: a review. J. Clean. Prod. **281**, 124409 (2021)
2. B. Netz, O.R. Davidson, P.R. Bosch, R. Dave, L.A. Meyer, Climate change 2007: Mitigation. Contribution of Working Group III to the Fourth Assessment Report of the Intergovernmental Panel on Climate Change. Summary for Policymakers. Climate change 2007: Mitigation Contribution of Working Group III to the Fourth Assessment Report of the Intergovernmental Panel on Climate Change Summary for Policymakers (2007)
3. R. Gusain, N. Kumar, S.S. Ray, Metal oxide-based nanocomposites for photocatalytic reduction of CO_2 Adv. Mater. Sustain. Environ., pp. 293–315 (2022). CRC Press
4. S. Dey, G.C. Dhal, Materials progress in the control of CO and CO_2 emission at ambient conditions: an overview. Mater. Sci. Energy Technol. **2**(3), 607–623 (2019)
5. N.S. Sifat, Y. Haseli, A critical review of CO_2 capture technologies and prospects for clean power generation. Energies **12**(21), 4143 (2019)
6. A. Saravanan, D.-V.N. Vo, S. Jeevanantham, V. Bhuvaneswari, V.A. Narayanan, P. Yaashikaa, et al. A comprehensive review on different approaches for CO_2 utilization and conversion pathways. Chem. Eng. Sci., **236**, 116515 (2021).
7. A.A. Olajire, CO_2 capture and separation technologies for end-of-pipe applications–a review. Energy **35**(6), 2610–2628 (2010)
8. J. Pires, F. Martins, M. Alvim-Ferraz, M. Simões, Recent developments on carbon capture and storage: an overview. Chem. Eng. Res. Des. **89**(9), 1446–1460 (2011)
9. S. Lian, C. Song, Q. Liu, E. Duan, H. Ren, Y. Kitamura, Recent advances in ionic liquids-based hybrid processes for CO_2 capture and utilization. J. Environ. Sci. **99**, 281–295 (2021)
10. A. Sattari, A. Ramazani, H. Aghahosseini, M.K. Aroua, The application of polymer containing materials in CO_2 capturing via absorption and adsorption methods. J. CO_2 Utilization; **48**, 101526 (2021)
11. U. Kamran, S.-J. Park, Chemically modified carbonaceous adsorbents for enhanced CO_2 capture: a review. J. Clean. Prod. **290**, 125776 (2021)
12. D. Bonenfant, M. Kharoune, P. Niquette, M. Mimeault, R. Hausler, Advances in principal factors influencing carbon dioxide adsorption on zeolites. Sci. Technol. Adv. Mater. **9**(1), 013007 (2008)
13. N. Mat, S.N. Timmiati, L.P. The, Recent development in metal oxide-based core–shell material for CO_2 capture and utilisation. Appl. Nanosci., 1–21 (2022)
14. Y. Chen, C. Liu, S. Guo, T. Mu, L. Wei, Y. Lu, CO_2 capture and conversion to value-added products promoted by MXene-based materials. Green Energy Environ.
15. C.A. Trickett, A. Helal, B.A. Al-Maythalony, Z.H. Yamani, K.E. Cordova, O.M. Yaghi, The chemistry of metal–organic frameworks for CO_2 capture, regeneration and conversion. Nat. Rev. Mater. **2**(8), 1–16 (2017)
16. E. Jelmy, N. Thomas, D.T. Mathew, J. Louis, N.T. Padmanabhan, V. Kumaravel et al., Impact of structure, doping and defect-engineering in 2D materials on CO_2 capture and conversion. Reaction Chem. Eng.g. **6**(10), 1701–1738 (2021)
17. E.I. Koytsoumpa, C. Bergins, E. Kakaras, The CO_2 economy: review of CO_2 capture and reuse technologies. J. Supercritical Fluids. **132**, 3–16 (2018)
18. T.N. Borhani, M. Wang, Role of solvents in CO_2 capture processes:tThe review of selection and design methods. Renew. Sustain. Energy Rev. **114**, 109299 (2019)
19. B. Shao, Y. Zhang, Z. Sun, J. Li, Z. Gao, Z. Xie et al., CO_2 capture and in-situ conversion: recent progresses and perspectives. Green Chem. Eng. **3**(3), 189–198 (2022)
20. M. Usman, N. Iqbal, T. Noor, N. Zaman, A. Asghar, M.M. Abdelnaby et al., Advanced strategies in metal-organic frameworks for CO_2 capture and separation. Chem. Rec. **22**(7), e202100230 (2022)
21. M. Aghaie, N. Rezaei, S. Zendehboudi, A systematic review on CO_2 capture with ionic liquids: Current status and future prospects. Renew. Sustain. Energy Rev. **96**, 502–525 (2018)

22. M. Haaf, R. Anantharaman, S. Roussanaly, J. Ströhle, B. Epple, CO_2 capture from waste-to-energy plants: Techno-economic assessment of novel integration concepts of calcium looping technology. Resour. Conserv. Recycl. **162**, 104973 (2020)
23. E. Paoletti, F. Manes, Effects of elevated carbon dioxide and acidic rain on the growth of holm oak. Developments in Environmental Science. 3: Elsevier; p. 375–89 (2003)
24. M. Cellura, F. Guarino, S. Longo, G. Tumminia, Climate change and the building sector: modelling and energy implications to an office building in southern Europe. Energy Sustain. Dev. **45**, 46–65 (2018)
25. J. Kahl, Effect of acid rain on building material of the El Tajín archaeological zone in Veracruz. Mexico. Environ. Pollut. **144**(2), 655–660 (2006)
26. H. Kurihara, Y. Shirayama, Effects of increased atmospheric CO_2 on sea urchin early development. Mar. Ecol. Prog. Ser. **274**, 161–169 (2004)
27. J.M. Kolle, M. Fayaz, A. Sayari, Understanding the effect of water on CO_2 adsorption. Chem. Rev. **121**(13), 7280–7345 (2021)
28. W. Gao, S. Liang, R. Wang, Q. Jiang, Y. Zhang, Q. Zheng et al., Industrial carbon dioxide capture and utilization: state of the art and future challenges. Chem. Soc. Rev. **49**(23), 8584–8686 (2020)
29. B.P. Spigarelli, S.K. Kawatra, Opportunities and challenges in carbon dioxide capture. J. CO_2 Utilization **1**, 69–87 (2013)
30. M. Younas, M. Sohail, L.K. Leong, M.J. Bashir, S. Sumathi, Feasibility of CO_2 adsorption by solid adsorbents: a review on low-temperature systems. Int. J. Environ. Sci. Technol. **13**(7), 1839–1860 (2016)
31. L.-P. Merkouri, T.R. Reina, M.S. Duyar, Closing the carbon cycle with dual function materials. Energy Fuels **35**(24), 19859–19880 (2021)
32. H. Sun, C. Wu, B. Shen, X. Zhang, Y. Zhang, J. Huang, Progress in the development and application of CaO-based adsorbents for CO_2 capture—a review. Mater. Today Sustain. **1–2**, 1–27 (2018)
33. L. Yang, J. Heinlein, C. Hua, R. Gao, S. Hu, L. Pfefferle et al., Emerging dual-functional 2D transition metal oxides for carbon capture and utilization: a review. Fuel **324**, 124706 (2022)
34. E.J. Jelmy, N. Thomas, D.T. Mathew, J. Louis, N.T. Padmanabhan, V. Kumaravel et al., Impact of structure, doping and defect-engineering in 2D materials on CO_2 capture and conversion. Reaction Chem. Eng. **6**(10), 1701–1738 (2021)
35. A. Razaq, F. Bibi, X. Zheng, R. Papadakis, S.H.M. Jafri, H. Li, Review on graphene-, graphene oxide-, reduced graphene oxide-based flexible composites: from fabrication to applications. Materials **15**(3), 1012 (2022)
36. R. Gusain, P. Kumar, O.P. Sharma, S.L. Jain, O.P. Khatri, Reduced graphene oxide–CuO nanocomposites for photocatalytic conversion of CO_2 into methanol under visible light irradiation. Appl. Catal. B **181**, 352–362 (2016)
37. R. Gusain, H.P. Mungse, N. Kumar, T.R. Ravindran, R. Pandian, H. Sugimura et al., Covalently attached graphene–ionic liquid hybrid nanomaterials: synthesis, characterization and tribological application. J. Mater. Chem. A **4**(3), 926–937 (2016)
38. N. Mukwevho, R. Gusain, E. Fosso-Kankeu, N. Kumar, F. Waanders, S.S. Ray, Removal of naphthalene from simulated wastewater through adsorption-photodegradation by ZnO/Ag/GO nanocomposite. J. Ind. Eng. Chem. **81**, 393–404 (2020)
39. A. Ali, R. Pothu, S.H. Siyal, S. Phulpoto, M. Sajjad, K.H. Thebo, Graphene-based membranes for CO_2 separation. Mater. Sci. Energy Technol. **2**(1), 83–88 (2019)
40. P. Li, H.C. Zeng, Hierarchical nanocomposite by the integration of reduced graphene oxide and amorphous carbon with ultrafine MgO nanocrystallites for enhanced CO_2 capture. Environ. Sci. Technol. **51**(21), 12998–13007 (2017)
41. N. Kumar, R. Salehiyan, V. Chauke, O. Joseph Botlhoko, K. Setshedi, M. Scriba et al., Top-down synthesis of graphene: a comprehensive review. FlatChem. **27**, 100224 (2021)
42. L. Jiang, Z. Fan, Design of advanced porous graphene materials: from graphene nanomesh to 3D architectures. Nanoscale **6**(4), 1922–1945 (2014)

43. J. Oh, Y.-H. Mo, V.-D. Le, S. Lee, J. Han, G. Park et al., Borane-modified graphene-based materials as CO_2 adsorbents. Carbon **79**, 450–456 (2014)
44. Z.-Y. Sui, B.-H. Han, Effect of surface chemistry and textural properties on carbon dioxide uptake in hydrothermally reduced graphene oxide. Carbon **82**, 590–598 (2015)
45. D. Kim, D.W. Kim, H.-K. Lim, J. Jeon, H. Kim, H.-T. Jung et al., Intercalation of gas molecules in graphene oxide interlayer: The role of water. Jo. Phys. Chem. C. **118**(20), 11142–11148 (2014)
46. A.K. Mishra, S. Ramaprabhu, Nanostructured polyaniline decorated graphene sheets for reversible CO_2 capture. J. Mater. Chem. **22**(9), 3708–3712 (2012)
47. A.K. Mishra, S. Ramaprabhu, Carbon dioxide adsorption in graphene sheets. AIP Adv. **1**(3), 032152 (2011)
48. G.-J. Shin, K. Rhee, S.-J. Park, Improvement of CO_2 capture by graphite oxide in presence of polyethylenimine. Int. J. Hydrogen Energy **41**(32), 14351–14359 (2016)
49. Y. Liu, B. Sajjadi, W.-Y. Chen, R. Chatterjee, Ultrasound-assisted amine functionalized graphene oxide for enhanced CO_2 adsorption. Fuel **247**, 10–18 (2019)
50. J. Pokhrel, N. Bhoria, S. Anastasiou, T. Tsoufis, D. Gournis, G. Romanos et al., CO_2 adsorption behavior of amine-functionalized ZIF-8, graphene oxide, and ZIF-8/graphene oxide composites under dry and wet conditions. Microporous Mesoporous Mater. **267**, 53–67 (2018)
51. N. Hsan, P. Dutta, S. Kumar, R. Bera, N. Das, Chitosan grafted graphene oxide aerogel: Synthesis, characterization and carbon dioxide capture study. Int. J. Biol. Macromol. **125**, 300–306 (2019)
52. R. Gusain, K. Gupta, P. Joshi, O.P. Khatri, Adsorptive removal and photocatalytic degradation of organic pollutants using metal oxides and their composites: a comprehensive review. Adv. Coll. Interface. Sci. **272**, 102009 (2019)
53. K. Gupta, P. Joshi, R. Gusain, O.P. Khatri, Recent advances in adsorptive removal of heavy metal and metalloid ions by metal oxide-based nanomaterials. Coord. Chem. Rev. **445**, 214100 (2021)
54. R. Gusain, O.P. Khatri, Ultrasound assisted shape regulation of CuO nanorods in ionic liquids and their use as energy efficient lubricant additives. J. Mater. Chem. A **1**(18), 5612–5619 (2013)
55. Y. Ren, Z. Ma, P.G. Bruce, Ordered mesoporous metal oxides: synthesis and applications. Chem. Soc. Rev. **41**(14), 4909–4927 (2012)
56. M.S.S. Danish, A. Bhattacharya, D. Stepanova, A. Mikhaylov, M.L. Grilli, M. Khosravy et al., A systematic review of metal oxide applications for energy and environmental sustainability. Metals **10**(12), 1604 (2020)
57. A. Azmi, A. Ruhaimi, M. Aziz, Efficient 3-aminopropyltrimethoxysilane functionalised mesoporous ceria nanoparticles for CO_2 capture. Mater. Today Chem. **16**, 100273 (2020)
58. J.C. Védrine, Metal oxides in heterogeneous oxidation catalysis: State of the art and challenges for a more sustainable world. Chemsuschem **12**(3), 577–588 (2019)
59. Y. Guo, C. Tan, P. Wang, J. Sun, W. Li, C. Zhao et al., Magnesium-based basic mixtures derived from earth-abundant natural minerals for CO_2 capture in simulated flue gas. Fuel **243**, 298–305 (2019)
60. P. Li, R. Chen, Y. Lin, W. Li, General approach to facile synthesis of MgO-based porous ultrathin nanosheets enabling high-efficiency CO_2 capture. Chem. Eng. J. **404**, 126459 (2021)
61. Y. Hu, Y. Guo, J. Sun, H. Li, W. Liu, Progress in MgO sorbents for cyclic CO_2 capture: a comprehensive review. J. Mater. Chem. A **7**(35), 20103–20120 (2019)
62. F.E.C. Othman, N. Yusof, S. Samitsu, N. Abdullah, M.F. Hamid, K. Nagai, et al. Activated carbon nanofibers incorporated metal oxides for CO_2 adsorption: Effects of different type of metal oxides. J. CO_2 Utilization **45**, 101434 (2021)
63. W.N.R. Wan Isahak, Z.A.C. Ramli, M.W. Mohamed Hisham, M.A. Yarmo (eds.), Magnesium oxide nanoparticles on green activated carbon as efficient CO_2 adsorbent. AIP Conference Proceedings; 2013: American Institute of Physics
64. R. Chang, X. Wu, O. Cheung, W. Liu, Synthetic solid oxide sorbents for CO_2 capture: state-of-the art and future perspectives. J. Mater. Chem. A (2022)

65. B.W. Hwang, J.H. Lim, H.J. Chae, H.-J. Ryu, D. Lee, J.B. Lee et al., CO_2 capture and regeneration properties of MgO-based sorbents promoted with alkali metal nitrates at high pressure for the sorption enhanced water gas shift process. Process Saf. Environ. Prot. **116**, 219–227 (2018)
66. J.-S. Kwak, K.-R. Oh, K.-Y. Kim, J.-M. Lee, Y.-U. Kwon, CO 2 absorption and desorption characteristics of MgO-based absorbent promoted by triple eutectic alkali carbonate. Phys. Chem. Chem. Phys. **21**(37), 20805–20813 (2019)
67. L.K.G. Bhatta, U.M. Bhatta, K. Venkatesh, Metal oxides for carbon dioxide capture, in Inamuddin, Asiri A.M., Lichtfouse, E. (eds.), *Sustainable Agriculture Reviews 38: Carbon Sequestration Vol 2 Materials and Chemical Methods*. Cham: Springer International Publishing, p. 63–83 (2019)
68. N.H. Florin, A.T. Harris, Reactivity of CaO derived from nano-sized CaCO3 particles through multiple CO_2 capture-and-release cycles. Chem. Eng. Sci. **64**(2), 187–191 (2009)
69. O. Folorunso, N. Kumar, Y. Hamam, R. Sadiku, S.S. Ray, Recent progress on 2D metal carbide/nitride (MXene) nanocomposites for lithium-based batteries. FlatChem. **29**, 100281 (2021)
70. M. Naguib, O. Mashtalir, J. Carle, V. Presser, J. Lu, L. Hultman et al., Two-dimensional transition metal carbides. ACS Nano **6**(2), 1322–1331 (2012)
71. M.W. Barsoum, Y. Gogotsi, Removing roadblocks and opening new opportunities for MXenes. Ceramics Int. (2022)
72. L. Wang, M. Han, C.E. Shuck, X. Wang, Y. Gogotsi, Adjustable electrochemical properties of solid-solution MXenes. Nano Energy **88**, 106308 (2021)
73. V. Parey, B.M. Abraham, S.H. Mir, J.K. Singh, High-throughput screening of atomic defects in Mxenes for CO_2 capture, activation, and dissociation. ACS Appl. Mater. Interfaces. **13**(30), 35585–35594 (2021)
74. Y. Chen, C. Liu, S. Guo, T. Mu, L. Wei, Y. Lu, CO_2 capture and conversion to value-added products promoted by MXene-based materials. Green Energy Environ **7**(3), 394–410 (2022)
75. Á. Morales-García, A. Fernández-Fernández, F. Viñes, F. Illas, CO_2 abatement using two-dimensional MXene carbides. J. Mater. Chem. A **6**(8), 3381–3385 (2018)
76. B. Wang, A. Zhou, F. Liu, J. Cao, L. Wang, Q. Hu, Carbon dioxide adsorption of two-dimensional carbide MXenes. J. Adv. Ceramics **7**(3), 237–245 (2018)
77. I. Persson, J. Halim, H. Lind, T.W. Hansen, J.B. Wagner, L.-Å. Näslund et al., 2D transition metal carbides (MXenes) for carbon capture. Adv. Mater. **31**(2), 1805472 (2019)
78. Á. Morales-García, M. Mayans-Llorach, F. Viñes, F. Illas, Thickness biased capture of CO_2 on carbide MXenes. Phys. Chem. Chem. Phys. **21**(41), 23136–23142 (2019)
79. Z. Guo, Y. Li, B. Sa, Y. Fang, J. Lin, Y. Huang et al., M2C-type MXenes: Promising catalysts for CO_2 capture and reduction. Appl. Surf. Sci. **521**, 146436 (2020)
80. R. Morales-Salvador, Á. Morales-García, F. Viñes, F. Illas, Two-dimensional nitrides as highly efficient potential candidates for CO_2 capture and activation. Phys. Chem. Chem. Phys. **20**(25), 17117–17124 (2018)
81. F.-Q. Liu, X. Liu, L. Sun, R. Li, C.-X. Yin, B. Wu, MXene-supported stable adsorbents for superior CO_2 capture. J. Materials Chem. A **9**(21), 12763–12771 (2021)
82. A.A. Shamsabadi, A.P. Isfahani, S.K. Salestan, A. Rahimpour, B. Ghalei, E. Sivaniah et al., Pushing rubbery polymer membranes to be economic for CO_2 separation: embedment with Ti3C2Tx MXene nanosheets. ACS Appl. Mater. Interfaces **12**(3), 3984–3992 (2020)
83. S. Ghosh, S. Ramaprabhu, High-pressure investigation of ionic functionalized graphitic carbon nitride nanostructures for CO_2 capture. J. CO_2 Utilization **21**, 89–99 (2017)
84. Y. Oh, V.-D. Le, U.N. Maiti, J.O. Hwang, W.J. Park, J. Lim et al., Selective and regenerative carbon dioxide capture by highly polarizing porous carbon nitride. ACS Nano **9**(9), 9148–9157 (2015)
85. S.A. Anuar, K.N. Ahmad, A. Al-Amiery, M.S. Masdar, W.N.R. Wan Isahak, Facile preparation of carbon nitride-ZnO hybrid adsorbent for CO_2 capture: the significant role of amine source to metal oxide ratio. Catalysts **11**(10), 1253 (2021)

86. B. Szczęśniak, S. Borysiuk, J. Choma, M. Jaroniec, Mechanochemical synthesis of highly porous materials. Mater. Horiz. **7**(6), 1457–1473 (2020)
87. A.K. Mishra, S. Ramaprabhu, Enhanced CO_2 capture in Fe3O4-graphene nanocomposite by physicochemical adsorption. J. Appl. Phys. **116**(6), 064306 (2014)
88. J. Xiao, Y. Wang, T.C. Zhang, S. Yuan, rGO/N-porous carbon composites for enhanced CO_2 capture and energy storage performances. J. Alloy. Compd. **857**, 157534 (2021)
89. S. Jin, Y. Guo, J. Wang, L. Wang, Q. Hu, A. Zhou, Carbon dioxide adsorption of two-dimensional Mo2C MXene. Diam. Relat. Mater. **128**, 109277 (2022)
90. S. Marchesini, C.M. McGilvery, J. Bailey, C. Petit, Template-free synthesis of highly porous boron nitride: insights into pore network design and impact on gas sorption. ACS Nano **11**(10), 10003–10011 (2017)
91. S. Chen, P. Li, S. Xu, X. Pan, Q. Fu, X. Bao, Carbon doping of hexagonal boron nitride porous materials toward CO_2 capture. J. Mater. Chem. A **6**(4), 1832–1839 (2018)
92. R. Shankar, M. Sachs, L. Francàs, D. Lubert-Perquel, G. Kerherve, A. Regoutz et al., Porous boron nitride for combined CO_2 capture and photoreduction. J. Mater. Chem. A **7**(41), 23931–23940 (2019)
93. D. Wang, Y. Xue, C. Wang, J. Ji, Z. Zhou, C. Tang, Improved capture of carbon dioxide and methane via adding micropores within porous boron nitride fibers. J. Mater. Sci. **54**(14), 10168–10178 (2019)
94. J. Liang, Q. Song, J. Lin, G. Li, Y. Fang, Z. Guo et al., In Situ Cu-loaded porous boron nitride nanofiber as an efficient adsorbent for CO_2 capture. ACS Sustain. Chem. Eng. **8**(19), 7454–7462 (2020)
95. C. Yang, D. Liu, Y. Chen, C. Chen, J. Wang, Y. Fan et al., Three-dimensional functionalized boron nitride nanosheets/ZnO superstructures for CO_2 capture. ACS Appl. Mater. Interfaces. **11**(10), 10276–10282 (2019)
96. N. Kumar, S. Kumar, R. Gusain, N. Manyala, S. Eslava, S.S. Ray, Polypyrrole-Promoted rGO–MoS2 nanocomposites for enhanced photocatalytic conversion of CO_2 and H_2O to CO, CH_4, and H_2 products. ACS Appl. Energy Mater. **3**(10), 9897–9909 (2020)
97. J. Zha, X. Zhang, Room-Temperature synthesis of two-dimensional metal-organic frameworks with controllable size and functionality for enhanced CO_2 sorption. Cryst. Growth Des. **18**(5), 3209–3214 (2018)
98. S. Stanly, E.J. Jelmy, C.P.R. Nair, H. John, Carbon dioxide adsorption studies on modified montmorillonite clay/reduced graphene oxide hybrids at low pressure. J. Environ. Chem. Eng. **7**(5), 103344 (2019)
99. M.G. Rasul, A. Kiziltas, B. Arfaei, R. Shahbazian-Yassar, 2D boron nitride nanosheets for polymer composite materials. npj 2D Mater. Appl. **5**(1), 56 (2021)
100. A. Nag, K. Raidongia, K.P.S.S. Hembram, R. Datta, U.V. Waghmare, C.N.R. Rao, Graphene analogues of BN: novel synthesis and properties. ACS Nano **4**(3), 1539–1544 (2010)
101. C. Yang, J. Wang, Y. Chen, D. Liu, S. Huang, W. Lei, One-step template-free synthesis of 3D functionalized flower-like boron nitride nanosheets for NH3 and CO_2 adsorption. Nanoscale **10**(23), 10979–10985 (2018)
102. Y. Li, L. Liu, H. Yu, Y. Zhao, J. Dai, Y. Zhong et al., Synergy of developed micropores and electronic structure defects in carbon-doped boron nitride for CO_2 capture. Sci. Total Environ. **811**, 151384 (2022)
103. X. Zhang, S.Y. Teng, A.C.M. Loy, B.S. How, W.D. Leong, X. Tao, Transition metal dichalcogenides for the application of pollution reduction: a review. Nanomaterials **10**(6), 1012 (2020)
104. N. Kumar, E. Fosso-Kankeu, S.S. Ray, Achieving Controllable MoS2 nanostructures with increased interlayer spacing for efficient removal of Pb(II) from aquatic systems. ACS Appl. Mater. Interfaces. **11**(21), 19141–19155 (2019)
105. S. Pandey, E. Fosso-Kankeu, M.J. Spiro, F. Waanders, N. Kumar, S.S. Ray et al., Equilibrium, kinetic, and thermodynamic studies of lead ion adsorption from mine wastewater onto MoS2-clinoptilolite composite. Mater. Today Chem. **18**, 100376 (2020)

106. N. Aguilar, S. Aparicio, Theoretical insights into CO_2 adsorption by MoS2 nanomaterials. J. Phys. Chem. C **123**(43), 26338–26350 (2019)
107. N. Kumar, B.P.A. George, H. Abrahamse, V. Parashar, J.C. Ngila, Sustainable one-step synthesis of hierarchical microspheres of PEGylated MoS2 nanosheets and MoO3 nanorods: their cytotoxicity towards lung and breast cancer cells. Appl. Surf. Sci. **396**, 8–18 (2017)
108. L. Seravalli, M. Bosi, A review on chemical vapour deposition of two-dimensional MoS2 flakes. Materials **14**(24), 7590 (2021)
109. R. Gusain, N. Kumar, F. Opoku, P.P. Govender, S.S. Ray, MoS2 Nanosheet/ZnS composites for the visible-light-assisted photocatalytic degradation of oxytetracycline. ACS Appl. Nano Mater. **4**(5), 4721–4734 (2021)
110. A. Parviainen, K. Loukola-Ruskeeniemi, Environmental impact of mineralised black shales. Earth Sci. Rev. **192**, 65–90 (2019)
111. T.E. Rufford, S. Smart, G.C.Y. Watson, B.F. Graham, J. Boxall, J.C. Diniz da Costa et al., The removal of CO_2 and N2 from natural gas: A review of conventional and emerging process technologies. J. Petroleum Sci. Eng., 94–95,123–54 (2012)
112. Q. Sun, G. Qin, Y. Ma, W. Wang, P. Li, A. Du et al., Electric field controlled CO_2 capture and CO_2/N_2 separation on MoS2 monolayers. Nanoscale **9**(1), 19–24 (2017)
113. G. Shi, L. Yu, X. Ba, X. Zhang, J. Zhou, Y. Yu, Copper nanoparticle interspersed MoS2 nanoflowers with enhanced efficiency for CO_2 electrochemical reduction to fuel. Dalton Trans. **46**(32), 10569–10577 (2017)
114. F.M. Enujekwu, Y. Zhang, C.I. Ezeh, H. Zhao, M. Xu, E. Besley et al., N-doping enabled defect-engineering of MoS2 for enhanced and selective adsorption of CO_2: A DFT approach. Appl. Surf. Sci. **542**, 148556 (2021)
115. M. Inagaki, M. Toyoda, Y. Soneda, T. Morishita, Nitrogen-doped carbon materials. Carbon **132**, 104–140 (2018)
116. S.S. Ray, R. Gusain, N. Kumar, Chapter ten—Two-dimensional carbon nanomaterials-based adsorbents, in Ray S.S., Gusain, R., Kumar, N. (eds.) *Carbon Nanomaterial-Based Adsorbents for Water Purification*: Elsevier; pp. 225–73 (2020)
117. N. Mukwevho, N. Kumar, E. Fosso-Kankeu, F. Waanders, J.R. Bunt, R.S. Sinha, Visible light-excitable ZnO/2D graphitic-C3N4 heterostructure for the photodegradation of naphthalene. Desalin. Water Treat. **163**, 286–296 (2019)
118. H.-L. Peng, F.-Y. Zhong, J.-B. Zhang, J.-Y. Zhang, P.-K. Wu, K. Huang et al., Graphitic carbon nitride functionalized with polyethylenimine for highly effective capture of carbon dioxide. Ind. Eng. Chem. Res. **57**(32), 11031–11038 (2018)
119. Z. Li, P. Liu, C. Ou, X. Dong, Porous metal-organic frameworks for carbon dioxide adsorption and separation at low pressure. ACS Sustain. Chem. Eng. **8**(41), 15378–15404 (2020)
120. W. Wang, Y. Yu, Y. Jin, X. Liu, M. Shang, X. Zheng et al., Two-dimensional metal-organic frameworks: from synthesis to bioapplications. J. Nanobiotechnol. **20**(1), 207 (2022)
121. W. Fan, Y. Wang, Z. Xiao, Z. Huang, F. Dai, R. Wang et al., Two-dimensional cobalt metal-organic frameworks for efficient C3H6/CH4 and C3H8/CH4 hydrocarbon separation. Chin. Chem. Lett. **29**(6), 865–868 (2018)
122. S. Das, T. Ben, S. Qiu, V. Valtchev, Two-Dimensional COF–three-dimensional MOF dual-layer membranes with unprecedentedly high H2/CO_2 selectivity and ultrahigh gas permeabilities. ACS Appl. Mater. Interfaces. **12**(47), 52899–52907 (2020)
123. Z. Kang, Y. Peng, Y. Qian, D. Yuan, M.A. Addicoat, T. Heine et al., Mixed matrix membranes (MMMs) comprising exfoliated 2D covalent organic frameworks (COFs) for efficient CO_2 separation. Chem. Mater. **28**(5), 1277–1285 (2016)
124. Y.B. Apriliyanto, N. Darmawan, N. Faginas-Lago, A. Lombardi, Two-dimensional diamine-linked covalent organic frameworks for CO_2/N_2 capture and separation: theoretical modeling and simulations. Phys. Chem. Chem. Phys. **22**(44), 25918–25929 (2020)
125. Y. Cheng, X. Wang, C. Jia, Y. Wang, L. Zhai, Q. Wang et al., Ultrathin mixed matrix membranes containing two-dimensional metal-organic framework nanosheets for efficient CO_2/CH_4 separation. J. Membr. Sci. **539**, 213–223 (2017)

126. N. Chouikhi, J.A. Cecilia, E. Vilarrasa-García, S. Besghaier, M. Chlendi, F.I. Franco Duro et al., CO_2 adsorption of materials synthesized from clay minerals: a review. Minerals [Internet] **9**(9) (2019)
127. M. Atilhan, S. Atilhan, R. Ullah, B. Anaya, T. Cagin, C.T. Yavuz et al., High-pressure methane, carbon dioxide, and nitrogen adsorption on amine-impregnated porous montmorillonite nanoclays. J. Chem. Eng. Data **61**(8), 2749–2760 (2016)
128. X. Tan, H.A. Tahini, S.C. Smith, Borophene as a promising material for charge-modulated switchable CO_2 capture. ACS Appl. Mater. Interfaces. **9**(23), 19825–19830 (2017)
129. T. Liu, Y. Chen, M. Zhang, L. Yuan, C. Zhang, J. Wang et al., A first-principles study of gas molecule adsorption on borophene. AIP Adv. **7**(12), 125007 (2017)
130. T. Kaewmaraya, L. Ngamwongwan, P. Moontragoon, W. Jarernboon, D. Singh, R. Ahuja et al., Novel green phosphorene as a superior chemical gas sensing material. J. Hazard. Mater. **401**, 123340 (2021)
131. S. Zhou, M. Wang, S. Wei, S. Cao, Z. Wang, S. Liu et al., First-row transition-metal-doped graphyne for ultrahigh-performance CO_2 capture and separation over N2/CH4/H2. Materials Today Physics. **16**, 100301 (2021)

The manufacturer's authorised representative in the EU is Springer Nature Customer Service Centre GmbH, Europaplatz 3, 69115 Heidelberg, Germany. If you have any concerns regarding our products, please contact ProductSafety@springernature.com

Printed and bound by CPI Group (UK) Ltd, Croydon, CR0 4YY

01/10/2026

02225852-0018